Aruba Certified Switching Associate
OFFICIAL CERTIFICATION STUDY GUIDE (EXAM HPE6-A41)

First Edition

Miriam Allred

HPE Press
660 4th Street, #802
San Francisco, CA 94107

Aruba Certified Switching Associate
Official Certification Study Guide
(Exam HPE6-A41)
Miriam Allred

Published by:

Hewlett Packard Enterprise Press
660 4th Street, #802
San Francisco, CA 94107

ISBN: 978-1-942741-52-7

Printed in Mexico

WARNING AND DISCLAIMER

This book provides information about the topics covered in the Aruba Certified Switching Associate certification exam (HPE6-A41). Every effort has been made to make this book as complete and as accurate as possible, but no warranty or fitness is implied.

The information is provided on an "as is" basis. The author and Hewlett Packard Enterprise Press shall have neither liability nor responsibility to any person or entity with respect to any loss or damages arising from the information contained in this book or from the use of the discs or programs that may accompany it.

The opinions expressed in this book belong to the author and are not necessarily those of Hewlett Packard Enterprise Press.

"Per current HPE trademark policy, the trademark acknowledgment sentence should be deleted unless there is an agreement requiring HPE to specifically acknowledge the trademark owner(s)."

GOVERNMENT AND EDUCATION SALES

This publisher offers discounts on this book when ordered in quantity for bulk purchases, which may include electronic versions. For more information, please contact U.S. Government and Education Sales 1-855-447-2665 or email sales@hpepressbooks.com.

Feedback Information

At HPE Press, our goal is to create in-depth reference books of the best quality and value. Each book is crafted with care and precision, undergoing rigorous development that involves the expertise of members from the professional technical community.

Readers' feedback is a continuation of the process. If you have any comments regarding how we could improve the quality of this book, or otherwise alter it to better suit your needs, you can contact us through email at hpepress@epac.com. Please make sure to include the book title and ISBN in your message.

We appreciate your feedback.

Publisher: Hewlett Packard Enterprise Press

HPE Contributors: Don McCracken and Gerhard Roets

HPE Press Program Manager: Michael Bishop

About the Author

Miriam Allred has spent the last 11 years configuring, testing, and troubleshooting HPE wired and wireless networks. Miriam combines this wide range of technical expertise with pedagogy and instructional design training, allowing her to create technical training courses for both advanced and entry-level networking professionals. Miriam Allred has a Masters degree from Cleveland State University and a Bachelors degree from Brigham Young University.

Introduction

This book helps you study for the Applying Aruba Switching Fundamentals for Mobility exam (HPE6-A41) to achieve the Aruba Certified Switching Associate (ACSA) certification. If you have an existing HPE ATP certification, you may be able to take the delta exam, HPE2-Z40.

The ACSA certification validates that you have the networking skills and expertise to design, implement and manage the modern network, based on Aruba wired and wireless solutions for small to mid-sized businesses.

Areas of study include the ability to:

- Perform basic configuration, access security and setup on Aruba switches
- Configure Layer 2 technologies such as Spanning Tree Protocol, Link Aggregation and VLANs
- Configure two Aruba 5400R zl Series Switches to form a Virtual Switching Fabric (VSF)
- Configure basic IP Routing with static and dynamic routing technologies (OSPF)
- Perform basic configuration tasks on Aruba Instant Access Points (IAPs) and create a cluster
- Manage and monitor networks with Aruba AirWave software

Certification and Learning

Hewlett Packard Enterprise Partner Ready Certification and Learning provides end-to-end continuous learning programs and professional certifications that can help you open doors and succeed in the idea economy. We provide: continuous learning activities and job-role based learning plans to help you keep pace with the demands of the dynamic, fast paced IT industry; professional sales and technical training and certifications to give you the critical skills needed to design, manage and implement the most sought-after IT disciplines; and training to help you navigate and seize opportunities within the top IT transformation areas that enable business advantage today.

As a Partner Ready Certification and Learning certified member, your skills, knowledge, and real-world experience are recognized and valued in the marketplace. To continue your professional and career growth, you have access to our large HPE community of world-class IT professionals, trend-makers and decision-makers. Share ideas, best practices, business insights, and challenges as you gain professional connections globally.

To learn more about HPE Partner Ready Certification and Learning certifications and continuous learning programs, please visit http://certification-learning.hpe.com

Audience

This book is designed for presales solution architects, consultants, installation technicians and other IT professionals involved in supporting Aruba networking solutions and products.

Assumed Knowledge

This is a foundational guide to Aruba wired and wireless solutions for switching and mobility. Before reading this book, however, you should be familiar with IP addressing, broadcast domains, and the Open Systems Interconnection (OSI) model.

Minimum Qualifications

To pass the Applying Aruba Switching Fundamentals for Mobility (HPE6-A41) exam, you should have at least one year of experience in designing small and medium-sized networks with intermediate switching, basic routing and wireless technologies. Exams are based on an assumed level of industry standard knowledge that may be gained from training, hands-on experience, or other prerequisites.

Relevant Certifications

After you pass the Applying Aruba Switching Fundamentals for Mobility (HPE6-A41) exam (or the HPE2-Z40 delta exam), your achievement may be applicable toward more than one certification. To determine which certifications can be credited with this achievement, log in to The Learning Center and view the certifications listed on the exam's More Details tab. You might be on your way to achieving additional certifications.

Preparing for the HPE6-A41 Exam

This self-study guide does not guarantee that you will have all the knowledge you need to pass the exam. It is expected that you will also draw on real-world experience and would benefit from completing the hands-on lab activities provided in the instructor-led training.

Recommended Training

Recommended training to prepare for each exam is accessible from the exam's page in The Learning Center. See the exam attachment, "Supporting courses," to view and register for the courses.

Obtain Hands-on Experience

You are not required to take the recommended, supported courses, and completion of training does not guarantee that you will pass the exam. Hewlett Packard Enterprise strongly recommends a combination of training, thorough review of courseware and additional study references, and sufficient on-the-job experience prior to taking an exam.

Exam Registration

To register for an exam, go to http://certification-learning.hpe.com/tr/certification/learn_more_about_exams.html

CONTENTS

1 Introduction to Aruba Solutions

EXAM OBJECTIVES

✓ Describe market trends that are leading customers to implement a digital workplace

✓ Describe how Aruba, a Hewlett Packard Enterprise company, delivers the digital workplace

ASSUMED KNOWLEDGE

Before reading this chapter, you should have a basic understanding of:

- IP addressing
- Open Systems Interconnection (OSI) model

INTRODUCTION

In this chapter, you will consider the trends that are driving companies to move to a digital workplace.

Balancing diverse demands in the idea economy

Ideas have always fueled business success. Ideas have built companies, markets, and industries. However, there is a difference today.

Businesses operate in the idea economy, which is also sometimes called the digital, application, or mobile economy. Doing business in the idea economy means turning an idea into a new product, capability, business, or industry. This has never been easier or more accessible—for you and for your competitors.

Today, an entrepreneur with a good idea has access to the infrastructure and resources that a traditional Fortune 1000 company would have. That entrepreneur can rent compute capacity on demand, implement a software-as-a-service enterprise resource planning system, use PayPal or Square for transactions, market products and services using Facebook or Google, and have FedEx or UPS run the supply chain.

Companies such as Vimeo, One Kings Lane, Dock to Dish, Uber, Pandora, Salesforce, and Airbnb used their ideas to change the world with very little start-up capital. Uber had a dramatic impact after launching its application connecting riders and drivers in 2009. Three years after its founding, the company expanded internationally. Without owning a single car, Uber serves more than 473 cities in 76 countries (as of July 28, 2016, according to the web site http://uberestimator.com/cities).

In a technology-driven world, it takes more than just ideas to be successful, however. Success is defined by how quickly ideas can be turned into value.

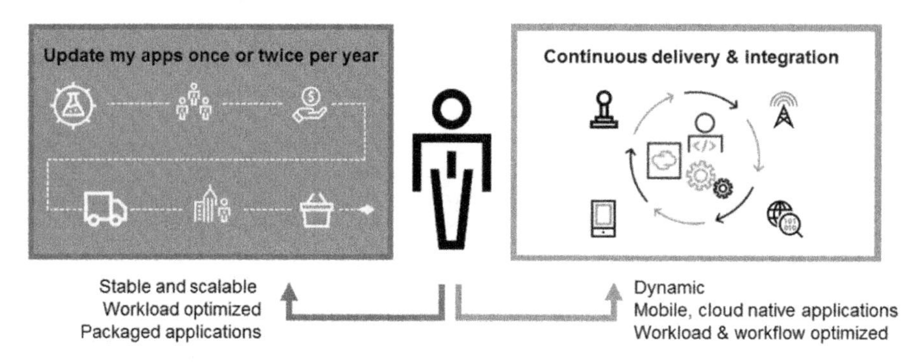

Figure 1-1: Balancing diverse demands in the idea economy

The idea economy presents an opportunity and a challenge for most enterprises. On one hand, cloud, mobile, big data, and analytics give businesses the tools to accelerate time to value. This increased speed allows organizations to combine applications and data to create dramatically new experiences, even new markets.

On the other hand, most organizations were built with rigid IT infrastructures that are costly to maintain. This rigidity makes it difficult, if not impossible, to implement new ideas quickly.

Creating and delivering new business models, solutions, and experiences requires harnessing new types of applications, data, and risks. It also requires implementing new ways to build, operate, and consume technology. This new way of doing business no longer just supports the company—it becomes the core of the company.

To respond to the disruptions created by the idea economy, IT must transform from a cost center to a value creator.

HPE Transformation Areas

To succeed in the idea economy companies must transform. After talking to customers and researching their needs and requirements, HPE has found that customers consider four areas of transformation most important.

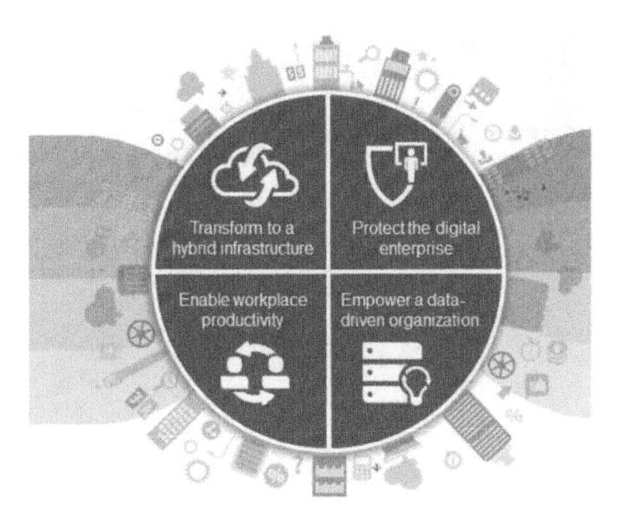

Figure 1-2: HPE transformation areas

- **Transforming to a hybrid infrastructure**—A hybrid infrastructure enables customers to get better value from the existing infrastructure and delivers new value quickly and continuously from all applications. This infrastructure should be agile, workload optimized, simple, and intuitive.
- **Protecting the digital enterprise**—Customers consider it a matter of when, not if, their digital walls will be attacked. The threat landscape is wider and more diverse than ever before. A complete risk management strategy involves security threats, backup and recovery, high availability, and disaster recovery.
- **Empowering the data-driven organization**—Customers are overwhelmed with data; the solution is to obtain value from information that exists. Data-driven organizations generate real-time, actionable insights.
- **Enabling workplace productivity**—Many customers are increasingly focused on enabling workplace productivity. Delivering a great digital workplace experience to employees and customers is a critical step.

The transformation areas are described in more depth on the following pages.

Transform to a hybrid infrastructure

An organization might see cloud services as a key component to access the IT services they need, at the right time and the right cost. A hybrid infrastructure is based on open standards, is built on a common architecture with unified management and security, and enables service portability across deployment models.

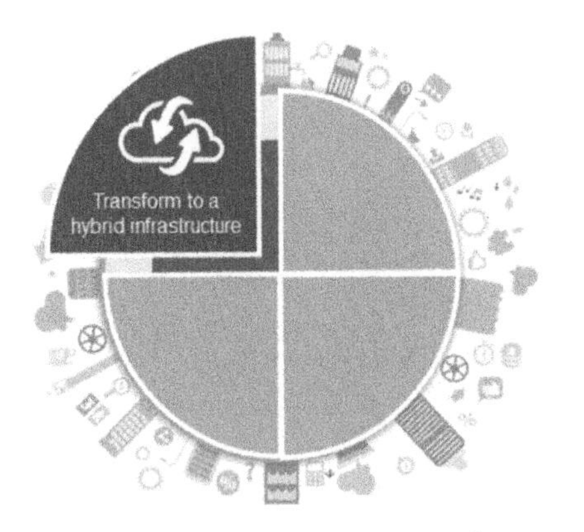

Figure 1-3: Transform to a hybrid infrastructure

Getting the most out of hybrid infrastructure opportunities requires planning performance, security, control, and availability strategies. For this reason, organizations must understand where and how a hybrid infrastructure strategy can most effectively be applied to their portfolio of services.

Protect the digital enterprise

Protecting a digital enterprise requires alignment with key IT and business decision makers for a business-aligned, integrated, and proactive strategy to protect the hybrid IT infrastructure and data-driven operations, as well as enable workplace productivity. By focusing on security as a business enabler, HPE brings new perspectives on how an organization can transform from traditional static security practices to intelligent, adaptive security models to keep pace with business dynamics.

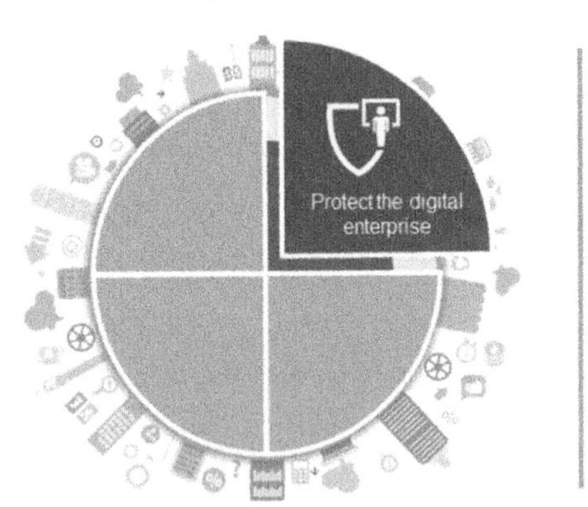

Figure 1-4: Protect the digital enterprise

HPE solutions help customers protect their data in a variety of ways. HPE StoreOnce delivers simple and secure data backup and recovery for the entire enterprise. HPE ProLiant Gen9 servers support options such as UEFI Secure Boot to prevent untrusted, potential malware from booting.

Empower the data-driven organization

A data-driven organization leverages valuable feedback that is available consistently from both internal and external sources.

Harness 100% of data, regardless of source or scale, and generate actionable insights at the speed of business to drive superior outcomes

Figure 1-5: Empower the data-driven organization

By harnessing insights from data in the form of information, organizations can determine the best strategies to pave the way for seamless integration of agile capabilities into an existing environment. Because both technical and organizational needs must be considered, HPE helps organizations define the right ways to help ensure that processes, security, tools, and overall collaboration are addressed properly for successful outcomes.

Enable workplace productivity

Organizations seeking to improve their efficiency and speed place a premium on creating a desirable work environment for their employees, including offering technology employees want and need. They believe they must enable employees to work how, where, and whenever they want.

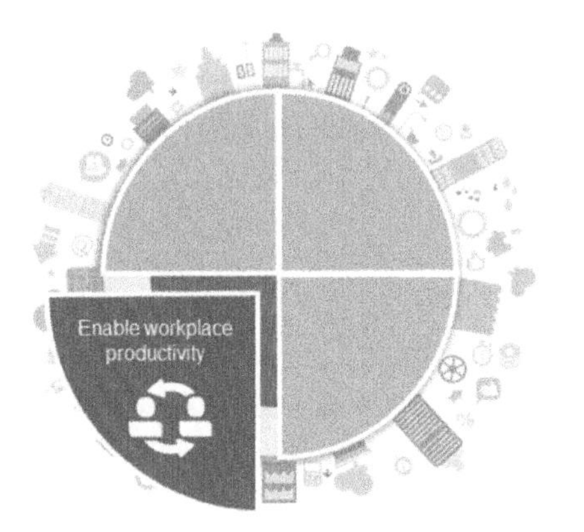

Figure 1-6: Enable workplace productivity

HPE solutions for the workplace provide secure, easy, mobile collaboration and anywhere, anytime access to data and applications for better productivity and responsiveness. To achieve the expected potential, spending growth for mobile resources is expected to be twice the level of IT spending growth in general, according to 2014 IDC survey results.

Digital workplace

As you know only too well, the nature of work is changing. Part of this change is occurring because the workforce itself is changing. According to an article in *Time* magazine, millennials will make up three-fourths of the global workforce by 2025. ("Millennials vs. Baby Boomers: Who Would You Rather Hire?" Time.com, Mar. 2012.) These GenMobile workers are tech savvy and like to do their work using a variety of devices. They thrive when they can collaborate with new tools in an open environment. To be productive and experience job satisfaction, these new workers need a network that provides access for any device, anywhere they want to work.

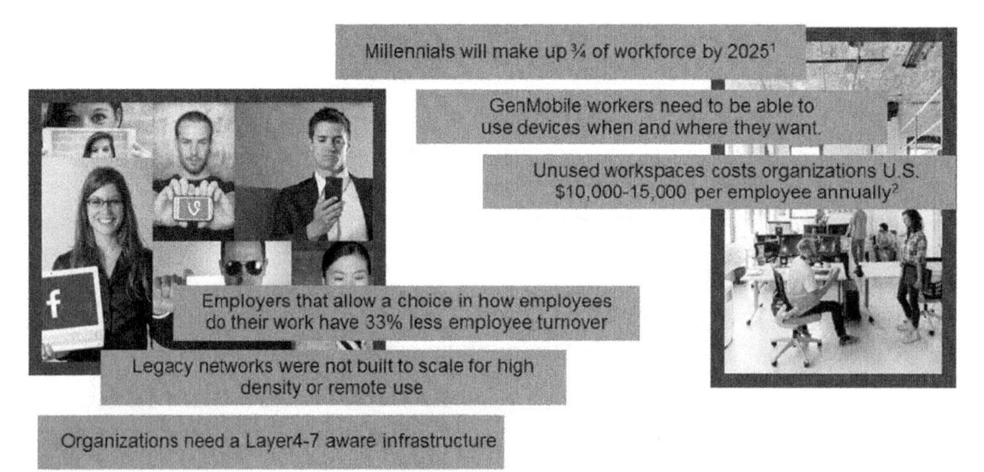

Figure 1-7: Digital workplace

Employees also frequently telecommute—either from home or on the road. The government estimates that unused workspaces cost organizations between $10,000 and $15,000 per employee each year. ("Workspace Utilization and Allocation Benchmark," GSA Office of Government-wide Policy, Jul. 2012.)

Employers that want to attract and retain the best of this new workforce must pay attention to employees' need to work where and how they want. Companies are already seeing the clear advantage of providing GenMobile workers with the work environment they need. A study by Cornell University reports 33% less attrition rate within companies that allow their workforce to work wherever they roam.

Unfortunately, however, legacy networks were not built to support high density wireless deployments or remote access. They are either not equipped or configured to recognize delay-sensitive traffic and prioritize it properly. And although organizations want the high speeds 802.11ac Access Points (APs) promise, their wired network may not be able to handle the higher traffic load from these APs.

Aruba Mobile First Network

The Aruba Mobile First Network helps adapt to the challenges that attend mobile and Internet of Things (IoT), Bring Your Own Device (BYOD), and Unified Communications (UC.) At the heart of the Mobile First network is the Aruba Mobile First platform, the intelligent software layer between the Aruba technology ecosystem and the Aruba Mobile First infrastructure.

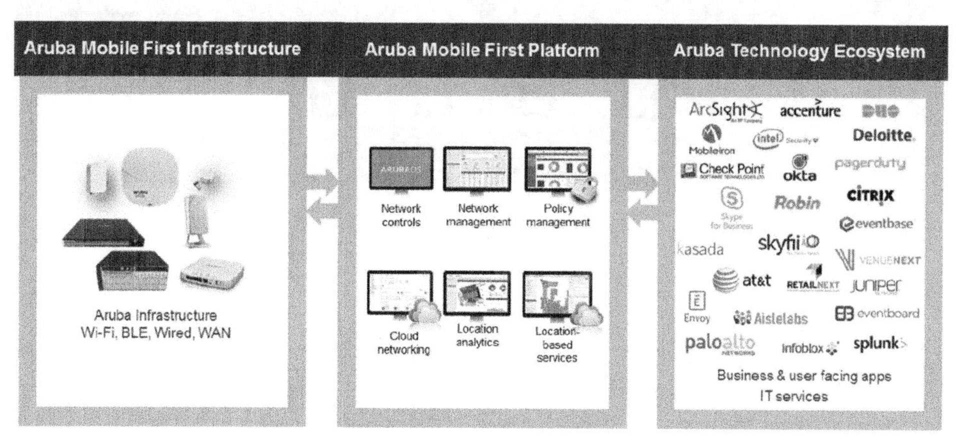

Figure 1-8: Aruba Mobile First Network

The Mobile First platform makes the underlying Aruba Mobile First infrastructure smarter, enabling it to adapt to idea-economy innovations in real time. And because all products within the Mobile First infrastructure share a single architecture, your customers can deploy a single set of hardware components across the entire company and manage them all from anywhere with the Mobile First platform.

Designed to enrich the mobile experience and secure IoT for operational technologies, the Mobile First platform acts as a rich source of insights for business and IT. The platform's focus on mobile and IoT clearly differentiates Aruba and helps companies accelerate innovation within the Aruba technology ecosystem, which speaks to Aruba's commitment to open standards.

Consisting of third party IT services as well as business and end-user facing applications, the Aruba technology ecosystem allows developers to freely innovate and customize networking functions dynamically in real time by using the rich contextual data that mobile and IoT devices collect.

Aruba WLAN solutions

Aruba offers a wide range of wireless LAN (WLAN) solutions. In this guide, you will learn about some of these solutions, but you should visit the Aruba web site (www.arubanetworks.com) to see the full range of solutions and to obtain detailed information about each one.

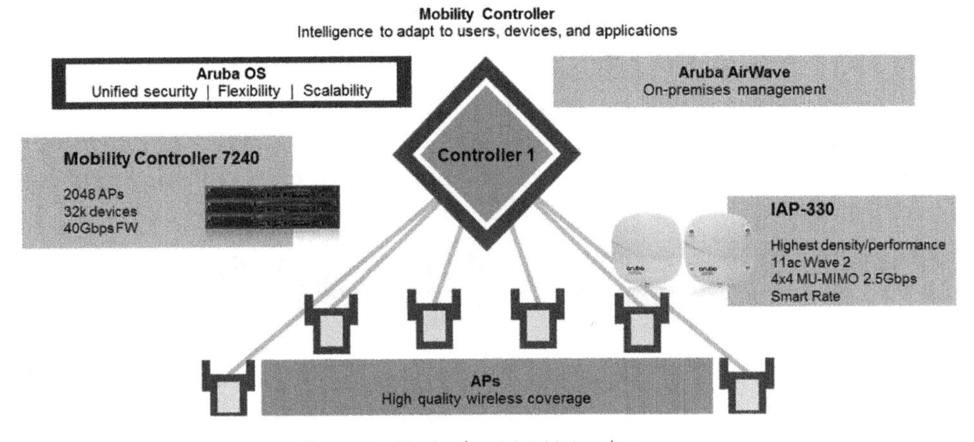

Figure 1-9: Aruba WLAN solutions

Aruba APs

To provide the kind of wireless performance organizations need, Aruba provides APs that support the fastest 802.11 standard available—802.11ac Wave 2. You will learn more about 802.11 standards and some of the features that enable 802.11ac Wave 2 to deliver these faster transmission speeds in Chapter 10.

For example, the Aruba 330 Series APs are dual radio APs, with one radio supporting up to 1,733 Mbps in the 5 GHz band and one radio supporting up to 800 Mbps in the 2.4 GHz band. The APs also support Multi-User Multi-Input Multi-Output (MU-MIMO), which you will learn more about in Chapter 10.

These APs also provide ClientMatch technology, enabling them to detect 802.11ac Wave 2 devices and automatically collect these devices under an 802.11ac Wave 2 radio. This enables the devices to achieve the highest transmission rates possible. The 330 Series APs also support Smart Rate uplinks to Ethernet networks, thereby supporting up to 5 Gbps.

Additional capabilities include:

- Quality of Service (QoS) features , such as priority handing, policy enforcement, and deep packet inspection to classify, block, or prioritize traffic
- Adaptive Radio Management (ARM), which automatically assigns channel and power settings, ensures "airtime fairness," and mitigates RF interference
- Spectrum analysis, which allows APs to monitor radio bands for interference
- Advanced Cellular Coexistence (ACC), which enables peak efficiency by minimizing interference from 3G or 4G LTE networks
- Security features such as integrated wireless intrusion protection; the ability to identify and classify malicious files, URLs, and IPs; and integrated trusted platform (TPM), which securely stores credentials and digital keys

For a complete list of capabilities, visit the Aruba web site (www.arubanetworks.com)

In this book, we will reference the Aruba 320 Series APs.

The Aruba 320 Series APs are also dual-radio APs: one radio supports 802.11ac, delivering wireless data rates of up to 1.733 Gbps in the 5-GHz radio band and one radio supports 802.11n clients, delivering wireless data rates of up to 800 Mbps in the 2.4-GHz radio band. Other capabilities include:

- ClientMatch technology
- ACC
- QoS
- RF management
- Spectrum analysis
- Security features such as integrated wireless intrusion protection; the ability to identify and classify malicious files, URLs, and IPs; and integrated trusted platform (TPM), which securely stores credentials and digital keys

Aruba APs also supports Multi-Input Multi-Output (MIMO) technology, which allows a wireless device to send and receive different data streams over different antennas. The 320 Series APs support 4x4 MIOM on both radios (four transmission streams and four receiving streams).

Aruba APs are available as controlled APs, which are managed by a mobility controller, or as Instant APs, which run in standalone mode. This guide focuses on Instant APs.

Aruba Mobility Controllers

Aruba also offers a range of mobility controllers, which provide simple WLAN management and secure access for users, independent of their location, device, or application. For example, the Aruba 7200 Series Mobility Controllers are optimized for mobile applications, providing the best experience for mobile users. The Aruba AppRF technology makes these applications highly visible to IT administrators, showing the applications each user is accessing.

Designed for enterprise networks, the Aruba 7200 Series Mobility Controllers includes four models that scale to different capacities. The 7205 provides firewall throughput (or FW) of up to 12 Gigabits per second and can support up to 8,192 user devices on 256 APs. The 7210 supports twice as many user devices and APs, while the 7220 supports up to 24,576 users and 1,024 APs. The 7240 provides the highest capacity, providing FW throughput of 40 Gigabits per second and supporting up to 32,768 user devices on 2048 APs.

The 7200 Series also manages:

- VPN connections
- IPv4 and IPv6 routing
- Policy enforcement firewall
- Adaptive radio management
- Spectrum analysis
- Wireless intrusion protection

The Aruba 7200 Series Mobility Controllers are also architected for high availability, supporting hot-swappable redundant power supplies, hot-swappable fan trays, and redundant network connections.

Aruba switch solutions

Aruba offers a range of switches as well. This guide focuses on two switch series: Aruba 5400R zl2 Switch Series and Aruba 3810 Switch Series. For more information about these and other switches, visit the Aruba web site (www.arubanetworks.com).

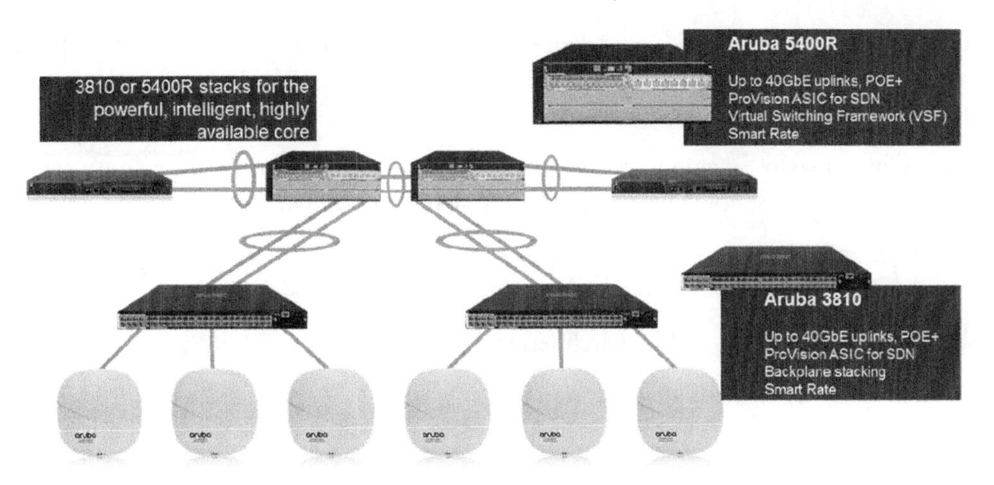

Figure 1-10: Aruba switch solutions

The **Aruba 5400R zl2 Switch Series**, which can be deployed at the access or core of a campus network, consists of advanced intelligent switches available in either a 6-slot or 12-slot chassis. (If you are familiar with the HPE switch portfolio, this switch was formerly called the HPE 5400R zl2 Switch Series.) The foundation for the switch series is a purpose-built, programmable ProVision ASIC that allows the most demanding networking features, such as QoS and security, to be implemented in a scalable yet granular fashion.

Aruba 5400R zl2 Switch Series supports redundant management modules as well as redundant power supplies, providing high availability for environments that cannot tolerate down time. The switch series supports any combination of 10/100, Gigabit Ethernet, 10 Gigabit Ethernet, and 40 Gigabit Ethernet interfaces and full Power over Ethernet Plus (PoE+) on all ports simultaneously. The switch series also supports Smart Rate ports, which deliver both high speed and power for 802.11ac Wave 2 devices, using existing CAT5e and CAT6 twisted pair wiring.

In addition, the 5400R zl2 Switch Series provides Layer 3 features such as OSPF, BGP, VRRP, and PIM, and support for emerging technologies such as Software Defined Networking (SDN).

Finally, the switch series supports Virtual Switching Framework (VSF), which enables you to combine two switches into one virtual switch called a fabric. You will learn more about VSF in Chapter 9.

The **Aruba 3810 Switch Series** is a family of fully managed Gigabit Ethernet switches available in 24-port and 48-port models, with or without PoE+ and with either SFP+, 10GbE, or 40GbE uplinks. One model in the series also supports Smart Rate ports.

Like the 5400R zl2 Switch Series, the 3810 Switch Series is built on the latest ProVision ASIC technology and advances in hardware engineering to deliver one of the most resilient and energy-efficient switches in the industry. Backplane stacking technology is implemented in the 3810 Switch Series to deliver chassis-like resiliency in a flexible, stackable form factor, providing

- Virtualized switching or stacking—when stacked, switches appear as a single chassis, providing simplified management (can eliminate the need for spanning tree in a switched network and reduce the number of routers in a routed network)
- High-performance for stacking—provides up to 336 Gbps of stacking throughput; each 4-port stacking module can support up to 42 Gbps in each direction per stacking port
- Ring, chain and mesh topologies for stacking—supports up to a 10-member ring or chain and 5-member fully meshed stacks; meshed topologies offer increased resiliency vs. a standard ring
- Fully managed Layer 3 stackable switch series
- Low-latency, highly resilient architecture

Skype for Business integration

Unified communications (UC) are no longer for the early adopters. UC has become a necessity due to the significant savings realized with using apps on smart phones and other mobile devices. Customers that utilize UC capabilities need to make sure Quality of Service (QoS) is reliable across their wired and wireless network. Aruba gives customers peace-of-mind that the network is optimized for their solution. Using HPE Network Optimizer, a software-defined networking (SDN) application, organizations can receive the QoS required for voice capabilities. This SDN application works with an SDN controller and OpenFlow enabled APs and switches to automatically and dynamically create prioritized traffic flows for Skype for Business traffic.

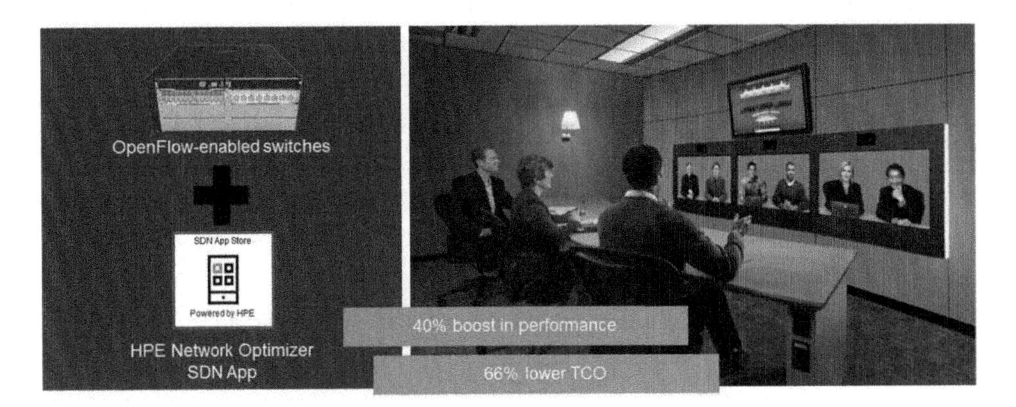

Figure 1-11: Skype for Business integration

Aruba AirWave ensures that the network is managed appropriately with options such as the UCC dashboard. Application prioritization and UCC optimization allows for fluid communication, which increases employee productivity, reduces costs, and improves collaboration.

HPE runs the largest Skype for Business (formerly Microsoft Lync) deployment in the world with more than 310,000 seats on Aruba and HPE networking equipment. Microsoft recognizes Aruba as a preferred partner for Skype for Business solutions. With UC solutions from Aruba, customers have reported a 40% boost in performance and a 66% lower TCO.

Unified wired and wireless management

Aruba AirWave is a multivendor wired and wireless network management solution that gives organizations deep visibility into the access network. Available as software or a combined hardware and software appliance, AirWave reduces cost and complexity, improves service quality, and enables IT administrators to make intelligent, well-informed decisions about network design and usage.

With a single interface, IT administrators can view the entire network with real-time performance monitoring of every device, every user, and every application running on the wired and wireless network, as well as other network services impacting user experience. Granularity, end-to-end visibility and predictive insight make AirWave unique in the industry and perfect for mobile-first access networks.

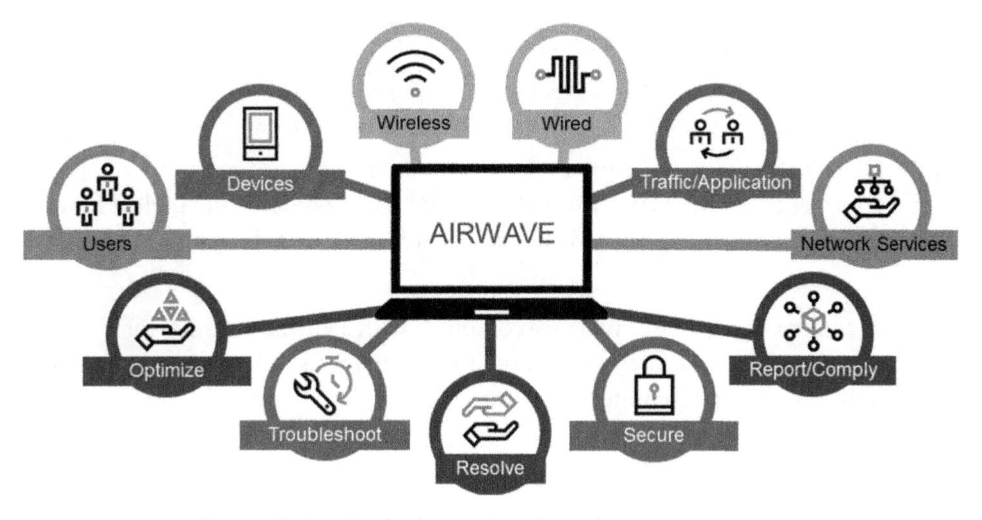

Figure 1-12: Unified wired and wireless management

Aruba AirWave is ideal for mid to large enterprise networks. The management solution supports all Aruba controller-managed and controller-based APs as well as Mobility and Cloud Controllers. AirWave even supports Aruba campus switches. More importantly, Airwave supports third-party wired and wireless networks from vendors such as Cisco, Brocade, Alcatel-Lucent, Meru, Motorola, and more.

Unified, policy-based access management: Aruba ClearPass

Aruba ClearPass is an enterprise-ready, scalable policy management solution that secures network access for a mobile world. The comprehensive Policy Manager includes ClearPass Onboard to manage device and employee access, ClearPass OnGuard to manage device health, and ClearPass Guest to manage guest access with a customizable portal.

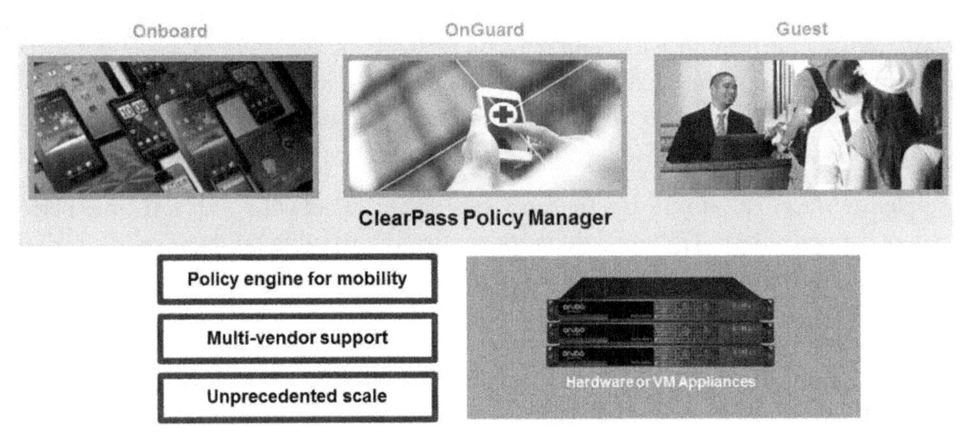

Figure 1-13: Unified, policy-based access management: Aruba ClearPass

In the world of virtual workplaces, mobile devices, and wireless connectivity, ClearPass provides secure, high-performance connectivity. This policy engine for mobility is the perfect foundation for network security in any organization with unprecedented scale and integration for multi-vendor architectures.

Aruba ClearPass comes as a hardware appliance combined with a virtual machine (or VM) or as a VM that can be deployed on the customer's hardware. Three different-sized appliances are available and each can be purchased for use in remote locations.

Summary

In this chapter you reviewed the demands that employees are placing on IT as they look for a digital workplace, which allows them to use any device anywhere to access their data. You also learned how the Aruba Mobile First Network delivers this digital workplace, providing the WLAN, switch, and management solutions that companies need.

- Digital workplace
- Aruba Mobile First Network
- Aruba WLAN solutions
- Aruba switch solutions
- Aruba Airwave and ClearPass

Figure 1-14: Summary

Learning check

Answer these questions to assess what you have learned in this chapter.

1. Which Aruba switch supports Virtual Switching Fabric?

2. Which Aruba solution provides policy-based access management?

Answers to Learning check

1. Which Aruba switch supports Virtual Switching Fabric?

   Aruba 5400R zl2 Series

2. Which Aruba solution provides policy-based access management?

   Aruba ClearPass

2 Basic Switch Setup

EXAM OBJECTIVES

✓ Describe out-of-band management

✓ Complete the initial setup on ArubaOS switches

✓ Verify your configuration settings

ASSUMED KNOWLEDGE

Before reading this chapter, you should have a basic understanding of:

- IP addressing
- Ethernet
- OSI model

INTRODUCTION

This chapter describes out-of-band management and how to navigate the ArubaOS switch command line interface (CLI), and begin configuring the switches.

Out-of-band management

When you initially configure a switch, you will typically use out-of-band management on a console port. For this type of management, you must have physical access to the switch. You connect your management station to the switch's console port using the serial cable that ships with the switch. This connection is dedicated to the management session. Out-of-band management does not require the switch to have network connectivity.

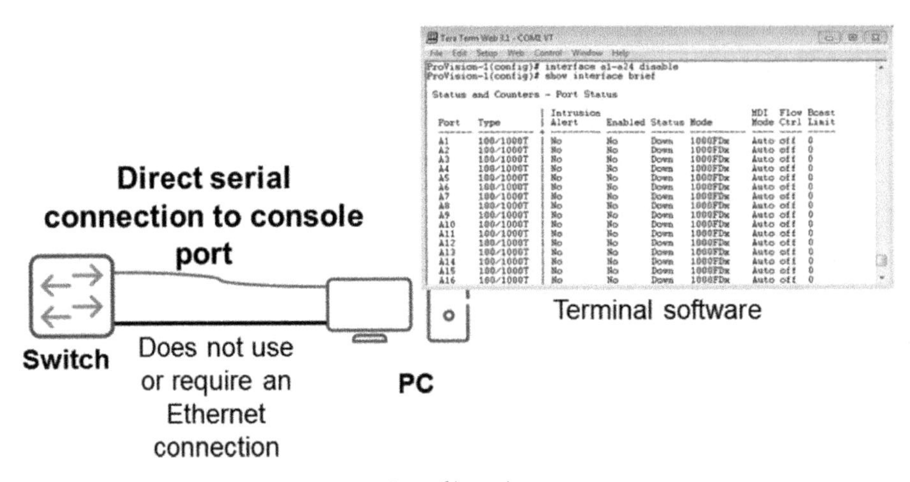

Figure 2-1: Out-of-band management

To open a management session with the switch and access the command line interface (CLI), you use terminal software such as Tera Term or PuTTY, which are available as freeware, or Microsoft HyperTerminal, which is available free from Microsoft.

The switches use default settings for the terminal emulation software:

- 9600 bps
- 8 data bits
- No parity
- 1 stop bit
- No flow control

ArubaOS switch CLI contexts

The ArubaOS switch CLI is organized into different contexts, or levels. You can determine your current context from the switch prompt, as shown in the figure below.

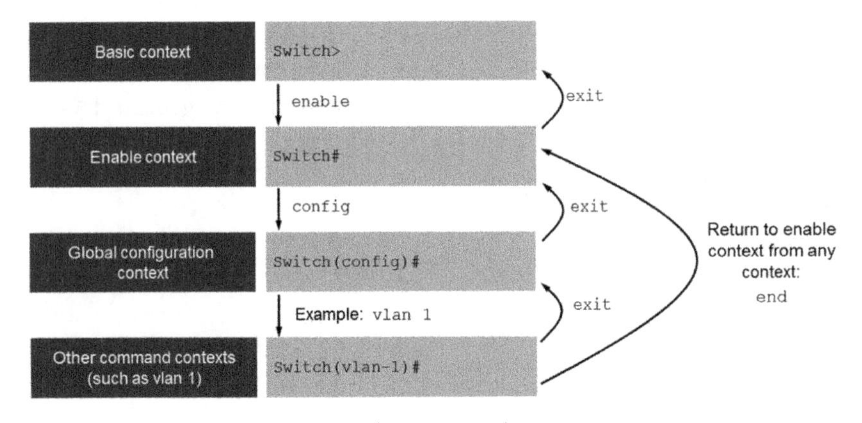

Figure 2-2: ArubaOS switch CLI contexts

```
Switch>
```

The > symbol in the switch prompt indicates you are at the basic level. At this level you can view statistics and configuration information.

To move to the enable context, enter `enable`.

```
Switch#
```

The # symbol in the switch prompt appears at the enable context. From this context, you can view additional information about the switch and begin configuring the switch. For example, you can update the switch software.

To move to the global configuration context, enter `config`.

```
Switch(config)#
```

The word "(config)" in the switch prompt indicates you are at the global configuration context. At this context, you can make configuration changes to the system's software features.

```
Switch(<context>)#
```

From the global configuration context, you can access other command contexts by entering the appropriate commands. In the switch prompt, <context> will be replaced with a context such as VLAN, a routing protocol such as Open Shortest Path First (OSPF), or port contexts:

```
Switch(vlan-1)#
```

```
Switch(ospf)#
```

You can then make configuration changes to that specific context.

Table 2-1 summarizes these contexts.

Table 2-1: Contexts in the ArubaOS CLI

Context	CLI Prompt	Description
Basic	`Switch>`	View a limited number of statistics and configuration settings.
Enable	`Switch#`	Begin switch configuration (such as updating system software).
Global configuration	`Switch(config)#`	Make configuration changes to the system's software features.

Other command contexts	`Switch(<content>)#`
	Examples:
	`Switch(vlan-1)#`
	`Switch(rip)#`
		Make configuration changes within a specific context, such as to a VLAN, one or more ports, or routing protocols.

Note that at factory default settings, you have unrestricted access to the switch; no passwords are configured. To protect your network environment, you should control access, as you will learn in Chapter 3.

Help keys

The ArubaOS switch CLI provides a number of help keys, which help you to remember and complete commands. The table below provides a reference for help keys. Note that, if you enter a string and ? without a space, you see commands that start with those letters. If you enter a command and ? after a space, you see options for that command.

<Tab> auto completes a command or command option, which can be useful for verifying that you are entering the correct command. However, you do not need to enter the full command; the CLI will accept the command as long as you enter enough characters to identify that command uniquely.

– ? to get lists of commands and options

– <Tab> to auto complete commands

Figure 2-3: Help keys

Table 2-2: Context-sensitive help

CLI	Description
`? or help`	See a brief description for all available commands at your context or view.
`<string>?`	See commands that start with certain letters.
`<command> ?`	See options for the command and a brief description of each option.
`<string>`**`<Tab>`**	Auto complete a command or command option: Type as many characters as necessary to identify the command uniquely and press **<Tab>.**

Accessing ArubaOS switch interfaces

On a fixed port switch, the interfaces are numbered from 1 to the maximum number of ports supported. On a modular switch, each module installed in the switch is assigned a letter, based on the slot in which it is installed. The ports in the module each have a unique number. You reference each port with the letter of the slot and the number of the port. For example, if the module is installed in the first slot and you want to configure settings on port 10, you would enter **a10**.

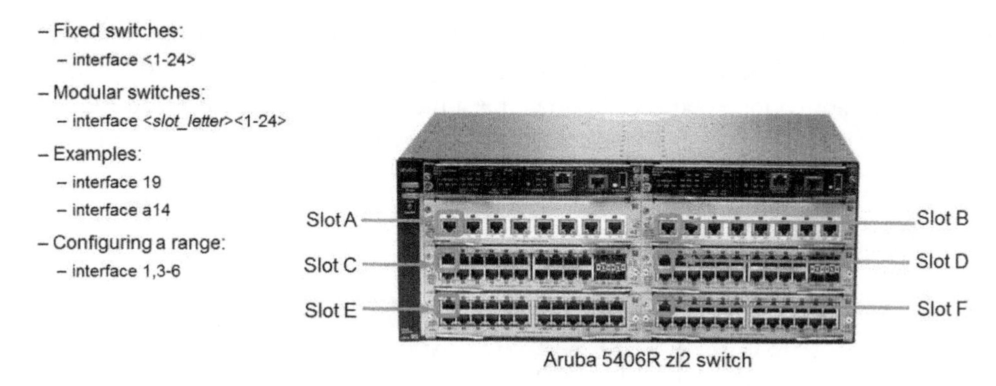

Figure 2-4: Accessing ArubaOS switch interfaces

You can also configure settings on a range of interfaces at the same time. Simply specify multiple IDs rather than the individual IDs. Use , (comma) to separate port IDs and - (dash) to indicate a range.

For example, enter:

```
Switch(config)# interface 1,3-6
```

You then access a prompt that indicates the range of interfaces (such as **Switch(eth-1,3-6)#**), and you can configure the settings.

Completing basic configuration tasks on a test network

Now that you have a basic understanding of the ArubaOS switch CLI, you will learn more about how to use the CLI to complete the following tasks:

- Access the ArubaOS switch CLI
- Return all the switches to factory default settings
- Configure a hostname on the ArubaOS switches
- Disable and enable interfaces
- Practice using CLI help commands
- Save the configuration settings

Throughout this study guide, you will see step-by-step instructions for practicing how to implement technologies and products on a test network. If you have ArubaOS switches and Aruba Instant Access Points (IAPs), you can set up this small test network and complete the step-by-step instructions. If you do not have access to this equipment, you can read the instructions to get an overview of how the technologies are implemented on ArubaOS switches and IAPs. To pass the related exam, however, HPE recommends that you have hands-on experience.

Figure 2-5 shows the basic topology for this test network, which includes two Aruba 5400R zl2 Series switches, two Aruba 3800 Series switches, a Windows server, and a Windows PC. On your local test network, you will not connect the switches together until you are instructed to do so. To make it easier to complete the tasks, you should use the ports shown in the figure to connect the switches.

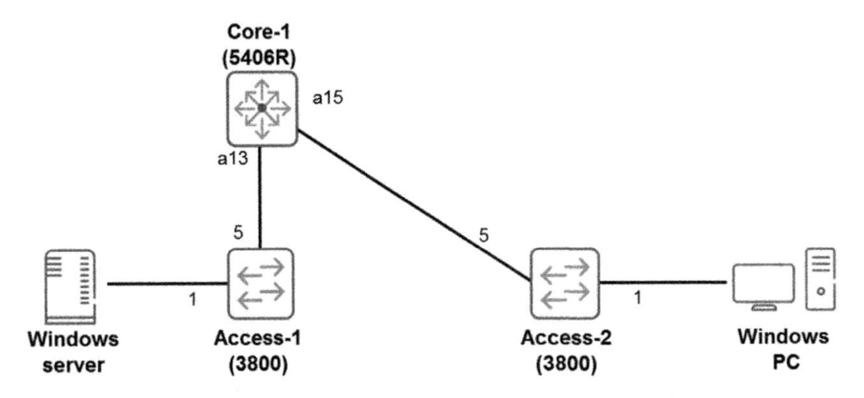

Figure 2-5: Basic topology for the test network

Table 2-3: Switch ports used in the test network

Local switch	Local port	Remote device	Remote port
Core-1	a13	Access-1	5
Core-1	a15	Access-2	5
Access-1	1	Windows server	n/a
Access-1	5	Core-1	a13
Access-2	1	Windows PC	n/a
Access-2	5	Core-1	a15

Command reference

Table 2-4 lists the commands you will use for the basic configuration tasks.

Table 2-4: Command reference

Description	ArubaOS switch commands
Access the enable context	`enable`
Access the global configuration	`configure terminal`
Return to previous context	`exit`
Define hostname on switch	`hostname`
Access an interface	`interface <int-id>`
Disable an interface	`interface <int-id> disable`

Table 2-4: Continued

Enable an interface	`interface <int-id> enable`
Access Menu interface	`menu`
Reboot the switch	`reload`
View the configuration	`show running-config`
View CPU utilization	`show cpu`
Verify an interface's operation	`show interface <int-id>`
View a summary of the interfaces	`show interface brief`
View a list of previously executed commands	`show history`
View version information	`show version`
Save the configuration	`write memory`

Task 1: Initialize switches

In the first task, you will reset the switches to factory default settings and disable all of their interfaces. Disabling the interfaces prevents loops and broadcast storms in environments that have hard -cabled connections between switches.

Make sure you complete this task on all **four** switches.

Core-1

1. Establish a console connection to the Core-1 switch, which is the first 5400R in the test network.

Important

You will be establishing sessions with each switch in your topology. Move the Core-1 session to a specific location on your desktop, preferably corresponding to the switch's location in the topology. Carefully note the Telnet port indicated in the session. These steps will help you keep track of which switch you are configuring.

2. Press **<Enter>** until you see a prompt.

3. Erase the startup-config. (Note that the hostname might be different from that indicated below.) When prompted, press **y**.

```
Core-1# erase startup-config
```

4. Return to the session with Core-1.

5. Press **[Enter]** twice until the switch hostname is displayed.

6. What prompt is displayed on the switch?

7. Move to global configuration mode.

```
HP-Switch-5406Rzl2# configure
```

Notice how the switch prompt changes.

8. Disable all of the interfaces.

```
HP-Switch-5406Rzl2(config)# interface a1-a24 disable
```

9. Assign the switch the hostname Core-1.

```
HP-Switch-5406Rzl2(config)# hostname Core-1
```

Notice the switch name changes:

```
Core-1(config)#
```

Core-2

10. Access Core-2, which is the second 5400R switch (the switch shown on the right of the topology).

11. Erase the startup-config. (Note that the hostname might be different from that indicated below.) When prompted, press **y**.

```
Core-2# erase startup-config
```

12. As soon as the switch finishes rebooting, move to global configuration mode and disable all of the interfaces.

```
HP-Switch-5406Rzl2# config

HP-Switch-5406Rzl2(config)# interface a1-a24 disable
```

Access-1

13. Access Access-1, which is the first 3800 switch (the switch on the left side of the topology).

14. Erase the startup-config. (Note that the hostname might be different from that indicated below.) When prompted, press **y**.

```
Access-1# erase startup-config
```

15. As soon as the switch finishes rebooting, move to global configuration mode, disable all of the interfaces, and set the hostname to Access-1.

```
HP-Switch-3800-24G-2SFP+# config

HP-Switch-3800-24G-2SFP+(config)# interface 1-26 disable

HP-Switch-3800-24G-2SFP+(config)# hostname Access-1
```

As soon as the switch finishes rebooting, move to global configuration mode and disable all of the interfaces.

```
HP-Switch-3800-24G-2SFP+# config
HP-Switch-3800-24G-2SFP+(config)# interface 1-26 disable
```

Important

If you are using an Aruba 3500 instead of an Aruba 3800, you must instead enter **interface 1-24 disable**.

Access-2

16. Access Access-2, which is the second 3800 switch (the switch on the right).
17. Erase the startup-config. When prompted, press **y**.

```
Access-2# erase startup-config
```

18. As soon as the switch finishes rebooting, move to global configuration mode, disable all of the interfaces, and set the hostname to Access-2.

```
HP-Switch-3800-24G-2SFP+# config
HP-Switch-3800-24G-2SFP+(config)# interface 1-26 disable
HP-Switch-3800-24G-2SFP+(config)# hostname Access-2
```

Important

If you are using an Aruba 3500 instead of an Aruba 3800, you must instead enter **interface 1-24 disable**.

Task 2: Configure initial settings

You will set a few initial settings on the switches.

Core-1

1. Access your terminal session with Core-1.
2. Examine the status of the Core-1 interfaces. You should see that all ports are disabled.

```
Core-1(config)# show interfaces brief
```

```
Status and Counters - Port Status

                 | Intrusion                      MDI  Flow Bcast
 Port Type       | Alert Enabled Status Mode      Mode Ctrl Limit
 ---- --------- + ----- ------- ------ ------- ---- ---- -----
 A1   100/1000T | No    No      Down   1000FDx NA   off  0
 A2   100/1000T | No    No      Down   1000FDx Auto off  0
 A3   100/1000T | No    No      Down   1000FDx Auto off  0
 A4   100/1000T | No    No      Down   1000FDx Auto off  0
 A5   100/1000T | No    No      Down   1000FDx Auto off  0
 <-output omitted->
```

3. Enable just the ports that this switch is using for the test network (see Figure 2-5 or Table 2-3). If your test network uses the ports shown in Figure 2-5, enter the following command (otherwise, replace the interface IDs with the proper ones for your test network).

```
Core-1(config)# interface a13,a15 enable
```

4. Add descriptions to the interfaces.

```
Core-1(config)# interface a13 name Access-1
Core-1(config)# interface a15 name Access-2
```

5. Exit to manager level, and view the commands that you have executed so far.

```
Core-1(config)# exit
Core-1# show history
8       configure
7       interface a1-a24 disable
6       hostname Core-1
5       show interfaces brief
4       interface a13,a15 enable
3       interface a13 name Access-1
2       interface a15 name Access-2
1       exit
```

6. Select the output and copy using **<Alt+C>**.
7. Open Notepad on the machine that you are using to reach the switches. (You should be able to click Start and see the Notepad application.) Paste the commands into the Notepad document. You will use them later.
8. Return to the terminal session with Core-1. Verify that the two correct interfaces on the Core-1 switch are enabled.

```
Core-1(config)# show interfaces brief
```

For example, if your test network uses the ports shown in Figure 2-5, the output would look like this (check the Enabled column but not the Status for now):

```
Core-1# show interfaces brief

Status and Counters - Port Status

                  | Intrusion                           MDI  Flow Bcast
  Port  Type      | Alert Enabled Status Mode    Mode Ctrl Limit
 ------ --------- + ----- ------- ------ ------- ---- ---- -----
  A1    100/1000T | No    No      Down   1000FDx NA   off  0
  <-output omitted->
  A13   100/1000T | No    Yes     Down   1000FDx Auto off  0
  A14   100/1000T | No    No      Down   1000FDx Auto off  0
  A15   100/1000T | No    Yes     Down   1000FDx MDI  off  0
  <-output omitted->
```

Tip

If your output wraps around to the next line, you can raise the terminal width. For example, enter **terminal width 150**. You might also need to drag the corner of the terminal window to increase its size.

Access-1

9. Return to your Notepad document. Adjust the text to:
   - Remove the numbers.
   - Remove the **interface a1-a24 disable** and **show interfaces brief** commands.
   - Customize for Access-2 which has:

- Different hostname
- Different interfaces enabled and different interface descriptions

The text in the Notepad file should look like this when you are done:

```
configure
hostname Access-1
interface 1,5 enable
interface 1 name Server
interface 5 name Core-1
exit
```

10. Select all of the commands (**<Ctrl+A>**) and copy them (**<Ctrl+C>**).
11. Establish a console connection to the Access-1 switch (first 3800 switch) and access the terminal.
12. Paste the commands **(<Alt+V>** or **<Right click>**) into the terminal.

Tip

Copying and pasting commands can be a powerful tool.

13. Verify that only the appropriate interfaces on the Access-1 switch are enabled. Only interfaces 1 and 5 should be enabled and up.

```
Access-1# show interfaces brief
 Status and Counters - Port Status

                  | Intrusion                         MDI  Flow Bcast
  Port Type       | Alert Enabled Status Mode         Mode Ctrl Limit
  ---- --------- + ----- ------- ------ ------- ---- ---- -----
  1    100/1000T  | No    Yes     Up     1000FDx MDIX off  0
  2    100/1000T  | No    No      Down   1000FDx Auto off  0
  3    100/1000T  | No    No      Down   1000FDx Auto off  0
  4    100/1000T  | No    No      Down   1000FDx Auto off  0
  5    100/1000T  | No    Yes     Up     1000FDx MDI  off  0
  <-output omitted->
```

Access-2

14. Repeat the steps in the previous section to configure Access-2. The Notepad text should look like this:

```
configure

hostname Access-2

interface 1,5 enable

interface 1 name Client

interface 5 name Core-1

exit
```

15. After you have pasted in the commands, check the interface status. Only interfaces 1 and 5 should be enabled and up.

```
Access-2(config)# show interfaces brief

 Status and Counters - Port Status

                 | Intrusion                           MDI  Flow Bcast

  Port  Type      | Alert Enabled Status Mode           Mode Ctrl Limit

  ----- --------- + ----- ------- ------ -------  ---- ---- -----

  1     100/1000T | No    Yes     Up     1000FDx MDIX off  0

  2     100/1000T | No    No      Down   1000FDx Auto off  0

  3     100/1000T | No    No      Down   1000FDx Auto off  0

  4     100/1000T | No    No      Down   1000FDx Auto off  0

  5     100/1000T | No    Yes     Up     1000FDx MDI  off  0

  <-output omitted->
```

Task 3: Explore the ArubaOS switch CLI

In this task, you will become more familiar with the ArubaOS switch CLI. Do not be afraid to try out different commands on the CLI: you will learn by experimenting.

1. Access the console port of one of the switches. The commands will show Access-1 for the host-name, but it doesn't matter which of the switches you use.

2. You should be in enable mode, but if you are in the global configuration mode, enter the exit command to return to the enable mode.

3. Enter the ? command:

```
Access-1# ?
```

Page through the commands available at this level. Some important commands available at this level include.

- **show**, which enables you to examine current configuration parameters
- **copy**, which enables you to back up the switch configuration
- **ping** and **traceroute**, which are connectivity test tools

4. List the parameters available for the **show** command.

```
Access-1# show ?
```

5. Scroll through or abort the **show** command output (**<Ctrl+C>**).

6. What options are available for the **show cpu** command?

```
Access-1# show cpu ?
process              Display the process usage statistics for active
                     management module or specified interface modules.
 slot                Display module CPU statistics.
 <1-300>             The time in seconds over which to average CPU
                     utilization.
 <cr>
```

Notice the <cr> at the end—this means that you can execute the command without supplying any further parameters.

7. View the CPU utilization on the switch.

```
Access-1# show cpu
1 percent busy, from 300 sec ago
1 sec ave: 1 percent busy
5 sec ave: 1 percent busy
1 min ave: 1 percent busy
```

Important

If you have any Layer-2 loops (because STP is not enabled on ArubaOS switches, by default), the CPU utilization would be very high.

8. Type **exit** and press **<Enter>**.

   How has the prompt changed?

   This indicates you have entered the Operator level.

9. Press ?.

```
Access-1> ?
```

   What are the main differences between commands available at the enable (manager) and basic (operator) levels?

10. Return to the manager level (or enable mode).

```
Access-1> enable
```

11. Use the tab help key to list commands that start with "co":

```
Access-1# co<tab>
```

   What does the CLI display?

12. Enter:

```
Access-1# con<tab>
```

   What does the CLI display?

   If only one available command begins with the string entered into the CLI, you can complete the command by hitting **<Tab>**.

Tip

You can execute any command as soon as you have entered an unambiguous character string. For instance, **conf [Enter]** will have the same effect as **configure [Enter]** at the Manager level.

Next, you will enter a command that is invalid, and then fix issues with it by using the command-recall feature.

13. Enter this command exactly as shown: `show hitory`.

```
Access-1# show hitory
Invalid input: hitory
```

14. Recall the command by pressing the <Up> arrow key.

15. Go to the beginning of the command with the <CTRL>a shortcut.

16. Go to the end of the command line with the <CTRL>e shortcut.

17. With the <Left> and <Right> arrow keys, move you cursor to the correct position in "hitory" and place the letter "s."
18. Press the <Enter> key at any time (no matter where your cursor is) to execute the command.

Tip

Repeating commands can be a useful way to enter similar commands more quickly, as well as to correct mistakes in commands.

19. Access the menu interface.

```
Access-1# menu
```

20. You will be prompted to save the configuration. Press **<y>**.
21. You will now see the menu.

```
Access-1                                              4-Jun-2016   4:08:36

======================- CONSOLE - MANAGER MODE -======================

                              Main Menu

   1. Status and Counters…

   2. Switch Configuration…

   3. Console Passwords…

   4. Event Log

   5. Command Line (CLI)

   6. Reboot Switch

   7. Download OS

   8. Run Setup

   0. Logout

Provides the menu to display configuration, status, and counters.

To select menu item, press item number, or highlight item and press <Enter>.
```

Tip

When using the menu, you can use the **<Up>** and **<Down>** arrow keys to make a selection or just type in the number of the selection.

22. Select **1. Status and Counters**, and then **1. General System Information**.
23. Exit the Menu. Enter the option for Back until you reach the main screen. Then enter the option for returning to the Main Menu (0) and logging out (0).
24. When prompted whether you want to logout, press **<y>** and then **<Enter>**.

Task 4: Save configurations

You must save the configuration on each of the switches.

```
Core-1# write memory

Core-2# write memory

Access-1# write memory

Access-2# write memory
```

Learning check

You will now review what you have learned about the ArubaOS switch CLI. As you prepare to take the HPE6-A41 exam, you should use this and other learning checks in this book to ensure that you have a thorough understanding of the concepts being tested.

1. What is the Menu interface?

2. How do you think the Menu interface could be used?

3. How can you determine the current context in the ArubaOS switch CLI?

4. Which help commands can you use in the ArubaOS switch CLI?

5. What shortcuts did you learn to use in the ArubaOS switch CLI?

Answers to Learning check

1. What is the Menu interface?

   The Menu interface lets you navigate through menus and choose options from those menus. You can use the interface to obtain information about the switch and configure some settings.

2. How do you think the Menu interface could be used?

   The Menu interface is designed to help you set up initial settings very quickly.

3. How can you determine the current context in the ArubaOS switch CLI?

   You can look at the prompt. **Hostname(config)#** indicates global configuration mode, **Hostname(vlan-X)#** indicates a VLAN context, and so on.

4. Which help commands can you use in the ArubaOS switch CLI?

   You can use ? and <Tab> for help.

5. What shortcuts did you learn to use in the ArubaOS switch CLI?

   You learned a number of shortcuts such as **<Ctrl+a>** and **<Ctrl+e>** for moving in the line. You learned how to use the **<Up>** arrow to recall commands. You might mention other shortcuts as well.

Link Layer Discovery Protocol (LLDP)

In a heterogeneous network, devices from different vendors need to be able to discover one another and exchange configuration information. To enable this exchange of information, the Internet Engineering Task Force (IETF) defined LLDP in IEEE 802.1AB. The protocol operates at the Data Link layer, enabling directly connected devices to exchange information.

- Devices use this Layer 2 protocol to send periodic announcements about:
  - Major device functions (such as bridging and routing)
  - Management IP address
  - Device identifier (such as sysname, MAC address, and chassis ID)
  - Port identifier (port ID and description)
- How do you think you could use LLDP in a lab environment?
- What are other uses for LLDP?

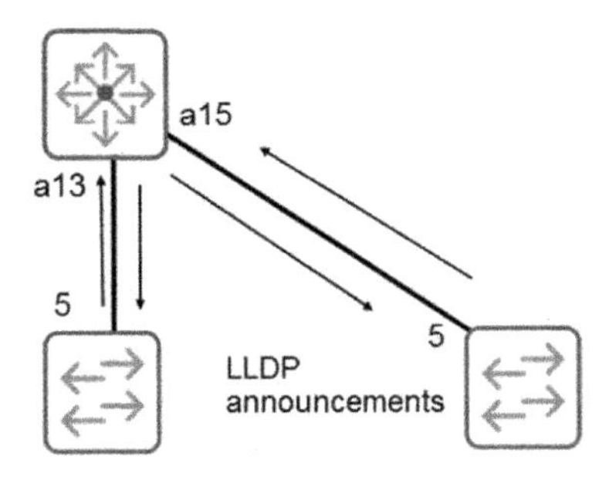

Figure 2-6: Link Layer Discovery Protocol (LLDP)

With LLDP, devices exchange local device information such as its major functions, management IP address, device ID, and port ID. Each device sends this information as type, length, and value (TLV) in LLDP data units (LLDPDUs) to directly connected devices. At the same time, the device receives LLDPDUs from neighbors that support LLDP. The local device saves the information it receives in a standard management information base (MIB). Simple Network Management Protocol (SNMP) solutions can use the LLDP information stored in MIBs to quickly detect Layer 2 network topology changes and identify each change.

All current ArubaOS switches support LLDP. Aruba APs support LLDP as well.

Learning check

1. How do you think you could use LLDP in a test environment?

2. What are other uses for LLDP?

Answers to Learning check

1. How do you think you could use LLDP in a test environment?

   You can use LLDP information to determine how devices are connected in the test topology, which is particularly important when you do not have physical access to the switches. You can verify connectivity and check the port connections and other basic configuration settings (such as IP address).

2. What are other uses for LLDP?

   LLDP can be used for troubleshooting and to locate devices on the network and gather information about those devices. SNMP programs and other applications can use LLDP to discover devices on the network and create a Layer 2 topology of the environment.

Assigning IP addresses to switches

On ArubaOS switches, you typically configure IP addresses on a virtual LAN (VLAN) basis. The switches have a default VLAN, VLAN 1, which is often used to manage the switches. You will learn more about using VLANs later in this guide.

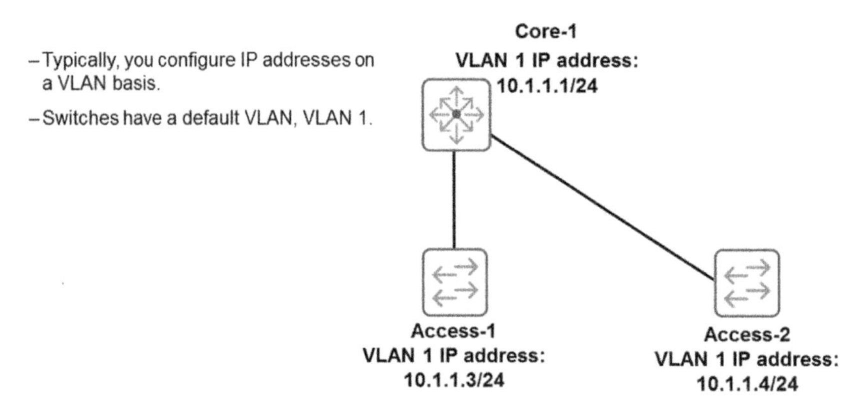

Figure 2-7: Assigning IP addresses to switches

Using DHCP to assign IP address on ArubaOS switches

By default, ArubaOS switches are configured to accept an IP address from a Dynamic Host Configuration Protocol (DHCP) server for VLAN 1. When the switch is booted, it immediately begins sending DHCP request packets on the network. The figure shows an example in which Access-2 is using DHCP on VLAN 1.

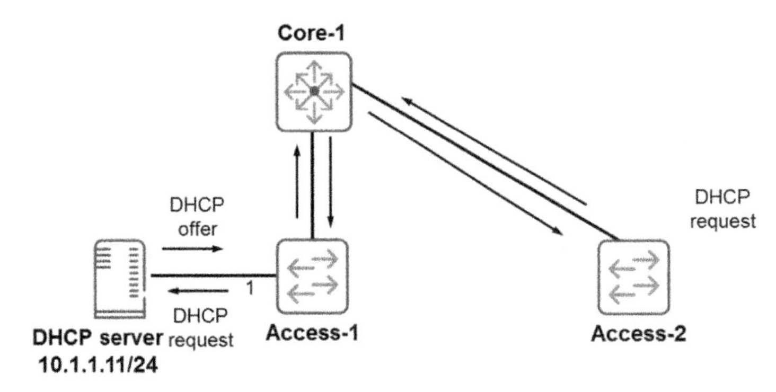

Figure 2-8: Using DHCP to assign IP address on ArubaOS switches

You can configure your DHCP server to send the switches an IP address (using the DHCP offer packet). You should configure the DHCP server to send the ArubaOS switches a particular IP address so that you know what that IP address is and the switch always receives the same IP address.

Because the ArubaOS switches support DHCP assignment of IP addresses on VLAN 1, this gives you another option for initially accessing and configuring the switch. You can use in-band management, which you will learn more about in the next chapter.

If the switch does not receive a reply to its DHCP requests, it continues to periodically send request packets, but with decreasing frequency.

Task 1: Configure IP addresses

In this task, you will configure IP addresses on all of the switches and then verify connectivity by entering:

- **ping** commands
- **LLDP** commands
- **show** commands

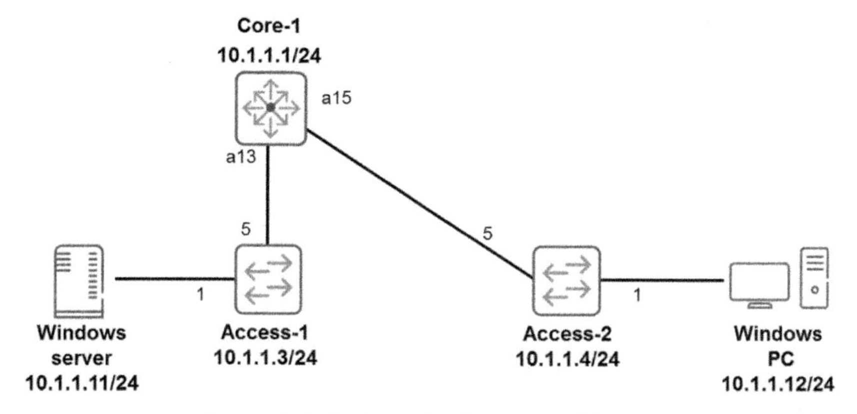

Figure 2-9: Task 1: Configure IP addresses

If you did not complete the tasks in the "Completing basic configuration tasks on a test network" section or if you lost the configuration on one or more of your test network switches, you can complete the last task, which provides instructions for loading the appropriate configurations.

In this task, you will configure IP addresses on the three switches in your test network. You will also change the Windows 2008 server static IP address so that it is in the same subnet as the switches.

Table 2-5 lists the IP addresses for these devices in VLAN 1. A 24-bit mask (255.255.255.0) will be used. Because the pods in your test network are *not* connected to each other, each pod will use the same IP addresses.

Table 2-5: IP addressing in VLAN 1

Device	IP address
Core-1	10.1.1.1
Access-1	10.1.1.3
Access-2	10.1.1.4
Windows server	10.1.1.11
Windows PC	DHCP Often, 10.1.1.12

Core-1

1. Access the terminal session with Core-1.
2. Assign VLAN 1 the IP address 10.1.1.1 with a 24-bit mask.

```
Core-1# configure
Core-1(config)# vlan 1 ip address 10.1.1.1/24
Core-1(config)# exit
```

3. Verify the configuration.

```
Core-1# show ip
Internet (IP) Service
 IP Routing : Disabled

 Default Gateway :
 Default TTL     : 64
 Arp Age         : 20
 Domain Suffix   :
 DNS server      :

                 |                                               Proxy ARP
    VLAN         | IP Config  IP Address Subnet Mask     Std  Local
    ------------ + --------- ---------- -------------   ---- ----
    DEFAULT_VLAN | Manual     10.1.1.1   255.255.255.0   No   No
```

Access-1

4. Repeat this process for Access-1, assigning VLAN 1 the IP address 10.1.1.3/24.

```
Access-1# configure
Access-1(config)# vlan 1 ip address 10.1.1.3/24
```

Access-2

5. Repeat this process for Access-2, assigning VLAN 1 the IP address 10.1.1.4/24.

```
Access-2# configure
Access-2(config)# vlan 1 ip address 10.1.1.4/24
```

Windows server

6. Add the necessary Ethernet connections from the Windows 2008 server to the Access-1 switch.

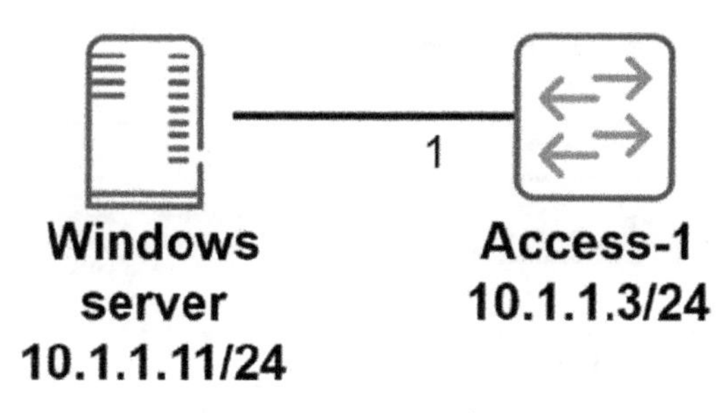

Figure 2-10: Test network topology

7. Access the Windows server desktop.
8. Change the IP address on the Windows 2008 server to 10.1.1.11 with a subnet mask of 255.255.255.0 and default gateway of 10.1.1.1. If you need help completing this task, use the steps below.

   a. Open the Network and Sharing Center from the Control Panel.

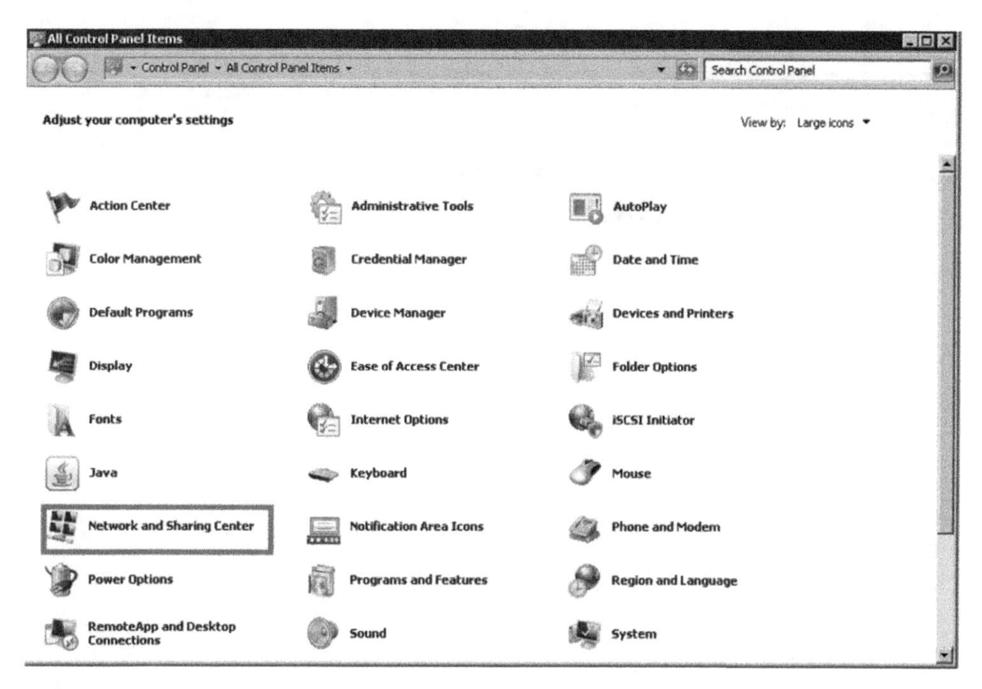

Figure 2-11: Open Network and Sharing Center

b. Click Change adapter settings.

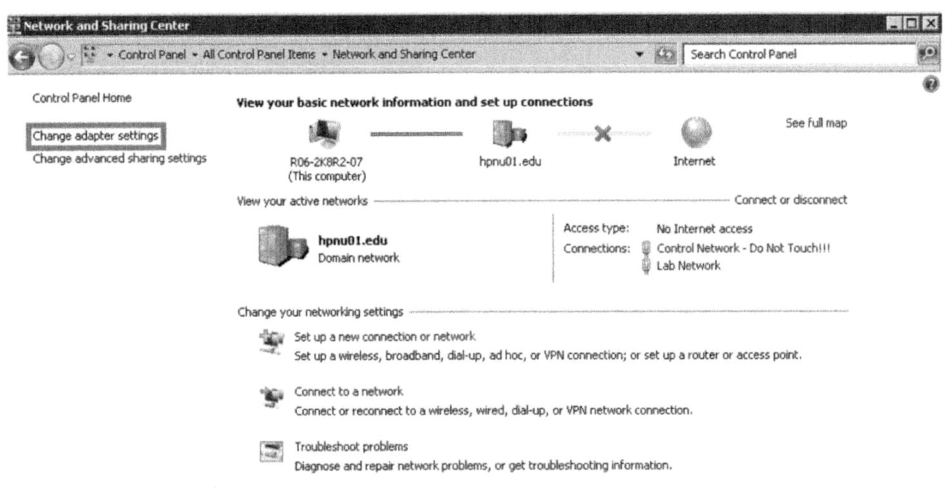

Figure 2-12: Change adapter settings

c. Right-click the Windows Server NIC that connects to the test network and select Create shortcut.

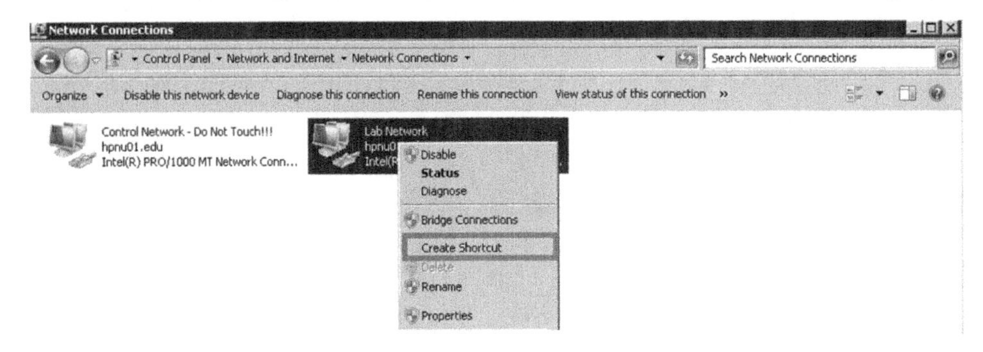

Figure 2-13: Create shortcut

d. When prompted, indicate that you want to create the shortcut on the desktop.

e. Right click the NIC that connects to the test network and select Properties.

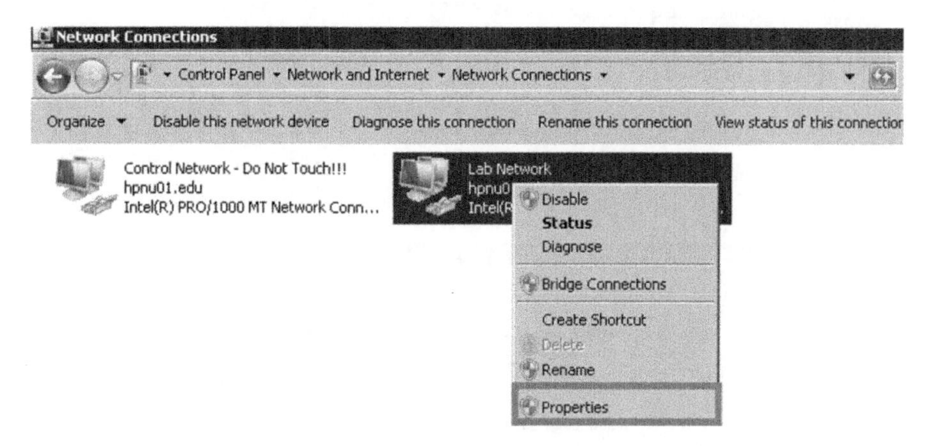

Figure 2-14: Select Properties

f. Select Internet Protocol version 4 (TCP/IPv4) and click **Properties**.

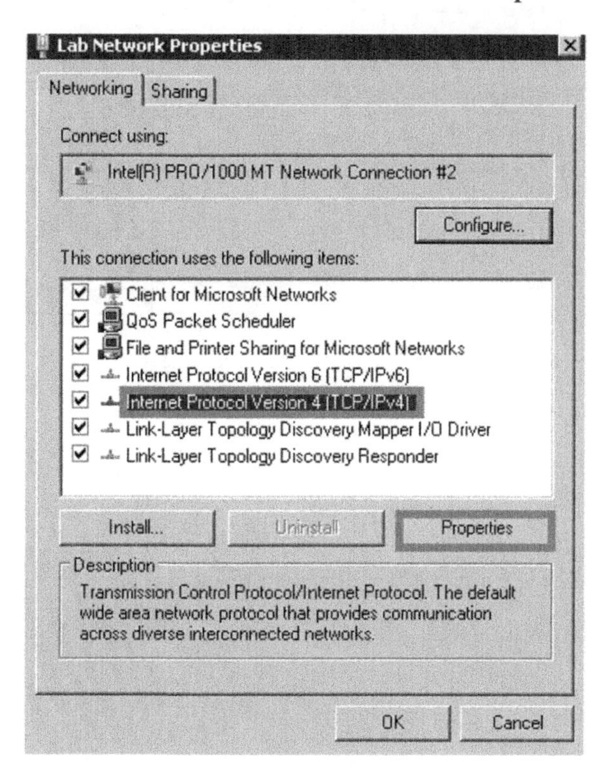

Figure 2-15: Select Internet Protocol version 4 (TCP/IPv4)

g. Enter these settings:

  - IP address = 10.1.1.11
  - Mask = 255.255.255.0
  - Default gateway = 10.1.1.1

h. Under Use the following DNS server addresses, make sure that 127.0.0.1 is listed.

i. Click **OK**.

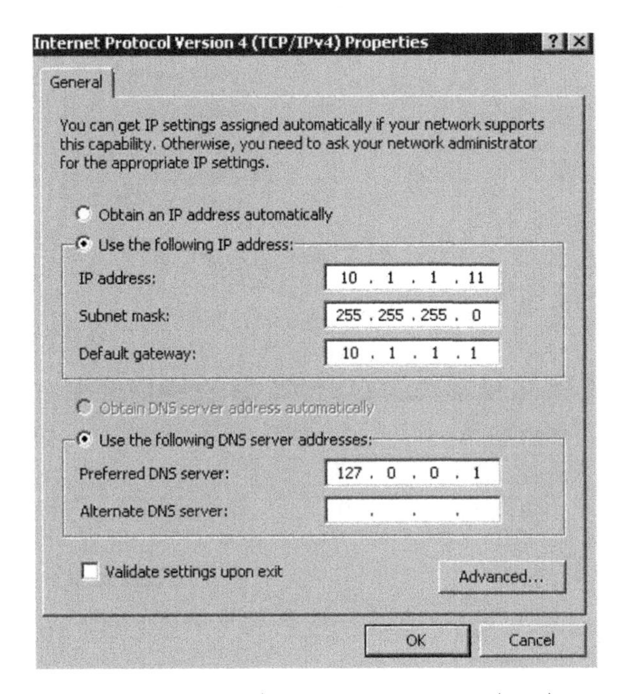

Figure 2-16: Make sure 127.0.0.1 is listed

j. Click **Close**.

9. Verify the IP addressing on the server by opening a Windows command (cmd) prompt and entering the ipconfig /all command. Check the IP and MAC addresses for the correct Ethernet adapter. Here's an example from the Windows 2008 server:

```
C:\Users\Admin> ipconfig /all

Windows IP Configuration
```

```
   Host Name . . . . . . . . . . . . : T02-Win7-Cli-1
   Primary Dns Suffix  . . . . . . . :
   Node Type . . . . . . . . . . . . : Hybrid
   IP Routing Enabled. . . . . . . . : No
   WINS Proxy Enabled. . . . . . . . : No

Ethernet adapter Test Network:

   Connection-specific DNS Suffix  . :
   Description . . . . . . . . . . . : Intel(R) PRO/1000 MT Network
Connection # 2
   Physical Address. . . . . . . . . : 00-50-56-97-7C-9D
   DHCP Enabled. . . . . . . . . . . : No
   Autoconfiguration Enabled . . . . : Yes
   Link-local IPv6 Address . . . . . : fe80::e876:9487:cb15:cdeb%12
(Preferred)
   IPv4 Address. . . . . . . . . . . : 10.1.1.11(Preferred)
   Subnet Mask . . . . . . . . . . . : 255.255.255.0
Default Gateway . . . . . . . . .    : 10.1.1.1
```

Windows 7 PC

10. Establish a connection between the Windows 7 PC and Access-2.

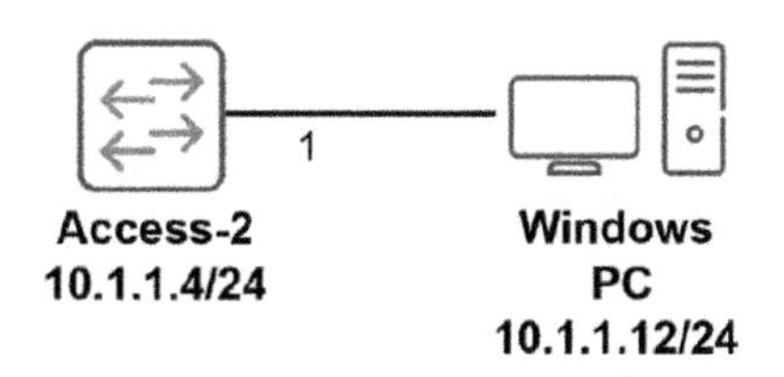

Figure 2-17: Test network topology

11. Access a remote desktop session with the Windows 7 PC (the Windows client).
12. If you have not already done so, make sure that the PC is set up to use a DHCP address. If you need help completing this task, use the steps below.

a. Open the Network and Sharing Center from the Control Panel.

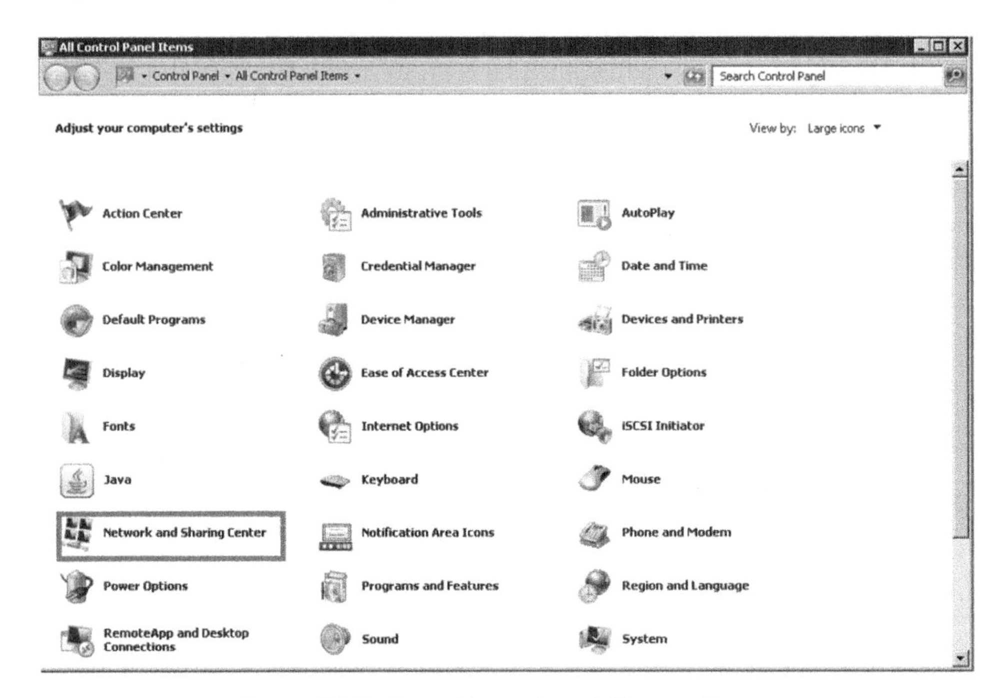

Figure 2-18: Open Network and Sharing Center

b. Click Change adapter settings.

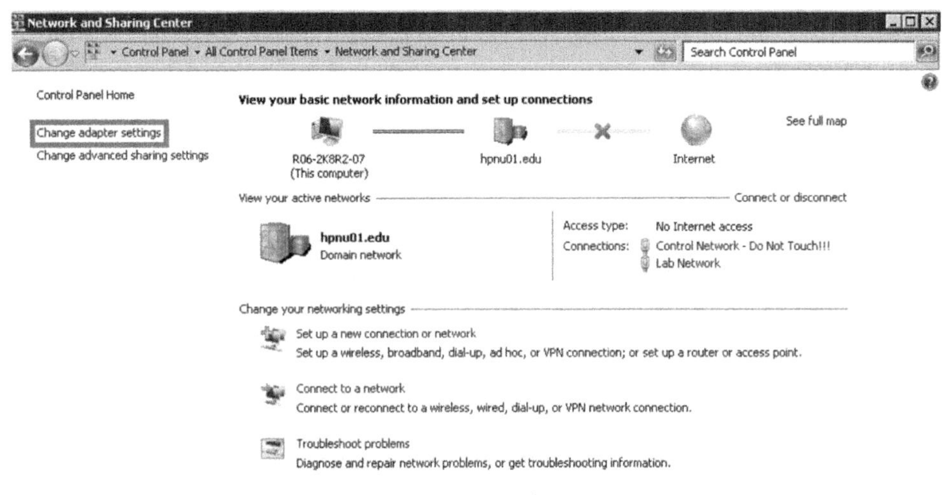

Figure 2-19: Change adapter settings

c. Right-click the NIC that connects to the test network and select Create shortcut.

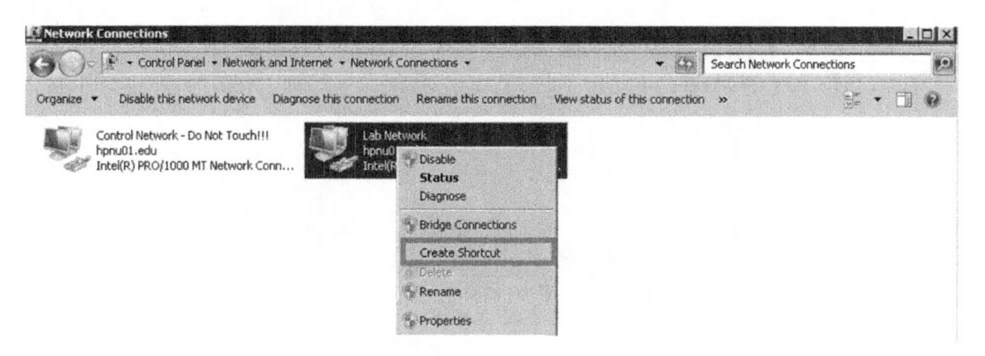

Figure 2-20: Create shortcut

d. When prompted, indicate that you want to create the shortcut on the desktop. You can use this shortcut in later tasks to reach the test network NIC properties window more quickly.

e. Right click the NIC that connects to the test network and select **Properties**.

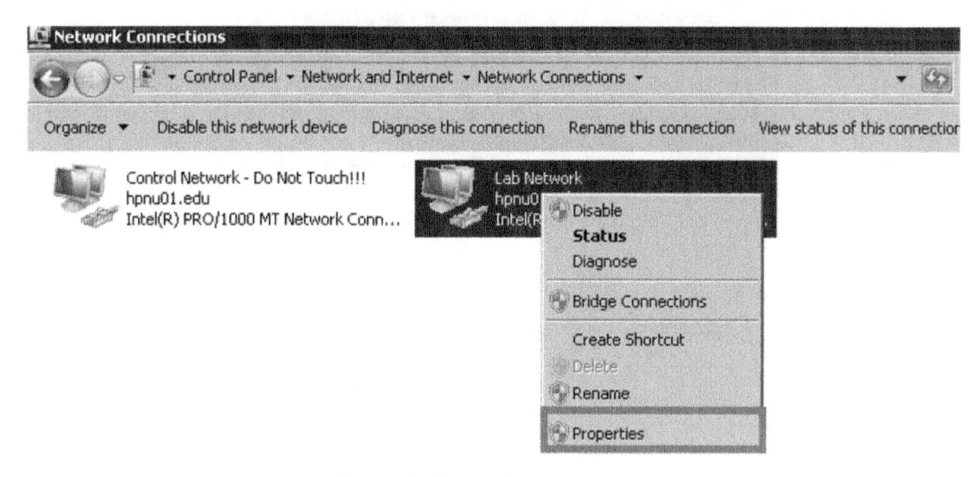

Figure 2-21: Select Properties

f. Select Internet Protocol version 4 (TCP/IPv4) and click **Properties**.

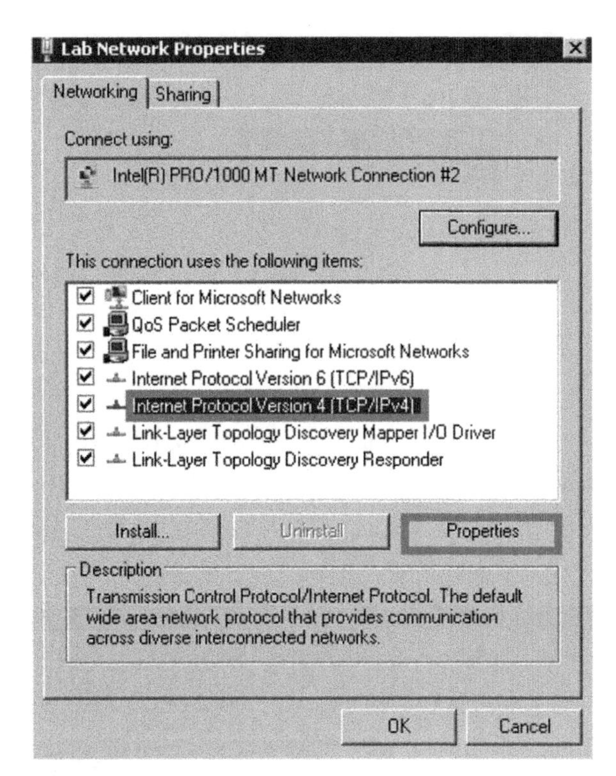

Figure 2-22: Select Internet Protocol version 4 (TCP/IPv4)

g. Select Obtain an IP address automatically and Obtain DNS server address automatically.

h. Click **OK**.

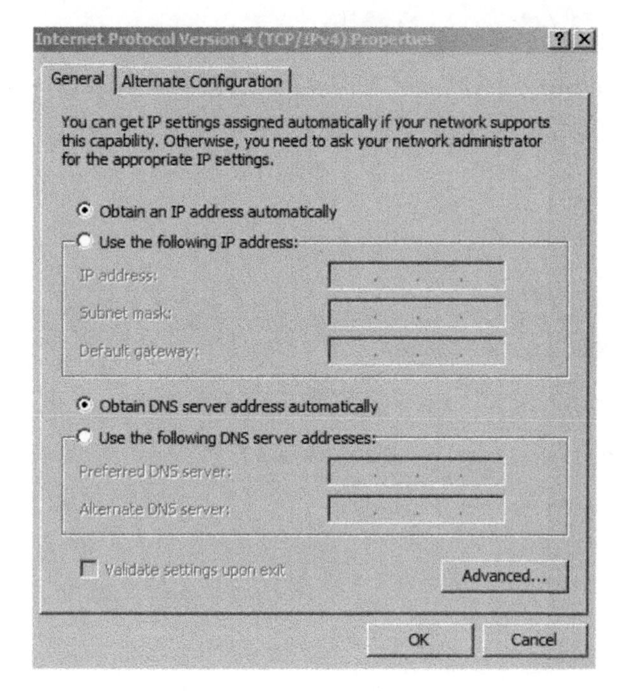

Figure 2-23: Select Obtain an IP address automatically and Obtain DNS server address automatically

i. Click **Close**.

13. Open the Windows command (cmd) prompt and enter the **ipconfig /all** command to verify that the client has received a DHCP address. Typically, it should receive 10.1.1.12. However, if it has received a different IP address in the 10.1.1.0/24 subnet, note that IP address now, so that you can use it in future commands.

Note

If the client already was using DHCP, enter **ipconfig/release** and **ipconfig/renew**. If the client does not receive an IP address, assign it a static IP address of 10.1.1.32 255.255.255.0. Then ping the server and switches to test connectivity. If the client still cannot ping, you must check that interfaces are enabled.

Task 2: Verify connectivity

In this task, you will verify connectivity among the switches.

Core-1

1. Verify Layer-2 connectivity on Core-1 by examining its Link Layer Discovery Protocol (LLDP) table. First, list the switch's LLDP neighbors:

```
Core-1# show lldp info remote-device
```

In the output, you should see Access-1 and Access-2 as LLDP neighbors.

```
 LLDP Remote Devices Information

  LocalPort | ChassisId                 PortId PortDescr  SysName
  --------- + ------------------------- ------ ---------- --------
  A13       | 6c 3b e5 62 08 c0         5      5          Access-1
  A15       | 84 34 97 02 23 c0         5      5          Access-2
```

2. View detailed information about the switch connected to A13 on Core-1:

```
Core-1# show lldp info remote-device a13

LLDP Remote Device Information Detail

  Local Port    : A13
  ChassisType   : mac-address
  ChassisId     : 6c 3b e5 62 08 c0
  PortType      : local
  PortId        : 5
  SysName       : Access-1
  System Descr  : HP J9575A 3800-24G-2SFP+ Switch, revision KA.16.01.0006,
                  ...
  PortDescr     : 5
  Pvid          : 1
  System Capabilities Supported  : bridge, router
  System Capabilities Enabled    : bridge

  Remote Management Address
     Type    : ipv4
     Address : 10.1.1.3

  <-output omitted->
```

3. Verify that Core-1 can ping Access-1 (10.1.1.3).

```
Core-1# ping 10.1.1.3
10.1.1.3 is alive, time = 1397 ms
```

4. Use the ping command to verify that Core-1 can ping all other devices:
   - Access-2 (10.1.1.4)
   - Windows 2008 server (10.1.1.11)
   - Windows 7 PC (10.1.1.12)
5. Examine the ARP table on the Core-1 switch.

```
Core-1# show arp
```

6. Use the output to fill in the following table:

Table 2-6: ARP table

Device	IP Address	MAC Address	Port
Access-1	10.1.1.3		
Access-2	10.1.1.4		
Windows server	10.1.1.11		
Windows PC	10.1.1.12		

7. Examine the MAC address table on Core-1.

```
Core-1# show mac-address
```

8. As you see, this command shows just the MAC address and the associated forwarding port. In this small test network, every MAC address learned in the MAC forwarding table is also in the ARP table. However, this is not always the case. (Switches also learn MAC addresses in VLANs in which they do not have IP addresses and do not listen for ARP replies.)

Task 3: Save your configurations

Save the configuration on all three switches using the **write memory** command.

Learning check

1. Did you have any issues verifying connectivity? How did you troubleshoot the problem?
2. What is an ARP table?
3. How can you use the ARP table to troubleshoot?

Answers to Learning check

1. Did you have any issues verifying connectivity?

   The answer to this question depends on your experience. You can troubleshoot with show commands such as:

   – **show interface brief**

   – **show lldp info remote-device [port]**

2. What is an ARP table?

   An ARP table shows the MAC address that is associated with an IP address. A switch learns the MAC addressees associated with IP addresses that it needs to process. For example, if you ping a device in the same subnet, the switch will use ARP to resolve the device's IP address. The ARP table might also show the forwarding interface for reaching the MAC address, which the switch learns from its MAC forwarding table.

3. How can you use the ARP table to troubleshoot?

   You can see whether the switch is able to resolve IP addresses. You can see whether the resolved MAC address is the device that you think owns that IP address. You can see which forwarding interface the switch has learned for a particular IP address and MAC address.

Summary

Take a minute to think about the configuration settings you learned about in this chapter. You do not need to remember the exact command syntax because you now know how to use help commands to discover the syntax of the commands you want to enter. Instead, focus on the tasks you completed. For example, you might remember that you disabled interfaces.

– What configuration settings did you learn?

Figure 2-24: Summary

Learning check

The following questions will help you measure your understanding of the material presented in this chapter.

1. On an ArubaOS switch, what tasks can you complete at the basic context? At the enable context?

2. What character do you type to use the context-sensitive help?

3. Of the following information, which cannot be seen in the LLDP table of a switch for neighboring devices?

   a. Major device functions

   b. Current users logged in

   c. Management IP address

   d. Device identifier

   e. Local port identifier

Answers to Learning check

1. On an ArubaOS switch, what tasks can you complete at the basic context? At the enable context?

   At the basic context, you can view statistics and configure information. At the enable context, you can begin configuring the switch. You can also access the global configuration context, where you configure more settings and access other contexts.

2. What character do you type to use the context-sensitive help?

   You type a question mark (?).

3. Of the following information, which cannot be seen in the LLDP table of a switch for neighboring devices?

   a. Major device functions

   b. Current users logged in

   c. Management IP address

   d. Device identifier

   e. Local port identifier

3 Protecting Management Access

EXAM OBJECTIVES

✓ Protect in-band and out-of-band management access to ArubaOS switches using local authentication

✓ Control the privilege level allowed to managers

ASSUMED KNOWLEDGE

Before reading this chapter, you should be able to:

- Access the ArubaOS switch command line interface (CLI) out of band
- Navigate the ArubaOS switch CLI
- Complete basic management tasks on ArubaOS switches, including returning switches to factory default settings, disabling or enabling switch interfaces, and configuring an IP address on VLAN 1
- Verify connectivity between switches

INTRODUCTION

This chapter explains how to control basic management access to Aruba switches, focusing on local authentication. You will learn how to implement simple password protection for various forms of CLI access.

In-band management

In the previous chapter, you learned about accessing a switch's command line interface (CLI) out-of-band using a serial connection to the switch's console port. You will now learn about an alternative way to reach the CLI: in-band.

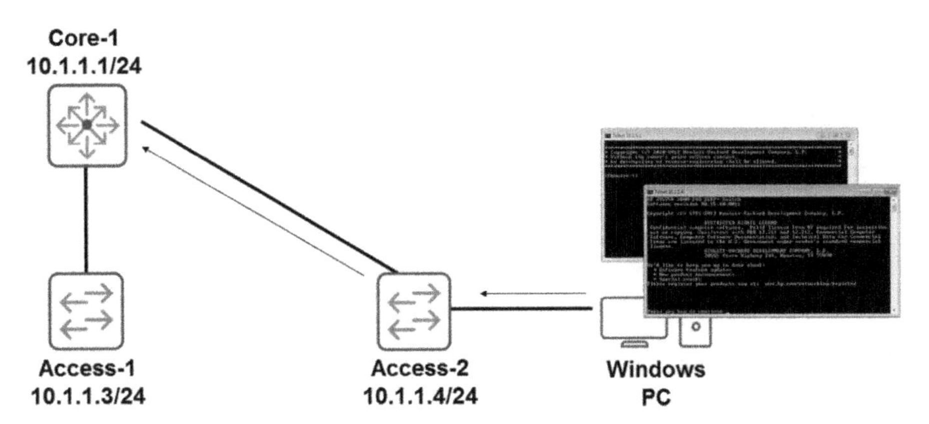

Figure 3-1: In-band management

With in-band management, your management communications run over network connections. You require IP connectivity to the networking device through a direct or indirect Ethernet connection. To open an in-band management session to access the network device's CLI, you must use terminal emulation software, such as Putty or Tera Term. This software communicates with the switch using a network protocol:

- Telnet—Carries the terminal session in plaintext
- Secure Shell (SSH)—Establishes a secure tunnel for the session using authentication and encryption

This chapter focuses on Telnet, but you should prefer using SSH in the real world because SSH provides security.

To manage a switch through in-band (networked) access, you must configure the switch with an IP address and subnet mask that is reachable on your network. You might also need to configure various settings, which are described in the following pages.

Learning check

Compare in-band management with out-of-band console management.

1. What are advantages of in-band management?

2. What are disadvantages of in-band management?

Answers to Learning check

1. In-band management is convenient. You can access any network device from any location that has IP connectivity to that device. You can open multiple sessions and manage multiple devices at the same time. Sometimes you need to access devices remotely so that you can respond to problems and troubleshoot in a more timely fashion, shrinking network downtime and providing better availability.

2. On the other hand, in-band management is less secure. Console out-of-band management is the most secure method because a manager needs physical access to the device. With in-band management, any user who can reach a network infrastructure device's IP address can attempt to log into the device. A hacker who obtains access to infrastructure devices can collect a great deal of valuable information, cause denial of service (DoS) attacks, redirect traffic, eavesdrop, and so on. Therefore, you must be careful to secure in-band management.

Out-of-band management with an Ethernet OOBM port

Many ArubaOS switches offer another option for out-of-band management: an out-of-band management (OOBM) Ethernet port. Like a console port, this port connects directly to the switch's management plane, which hosts functions such as the CLI, Web management interface, SNMP agents, and Syslog. At the same time, the OOBM port provides the convenience that was traditionally associated with in-band access. You can establish an Ethernet management network to connect managers to the switch remotely.

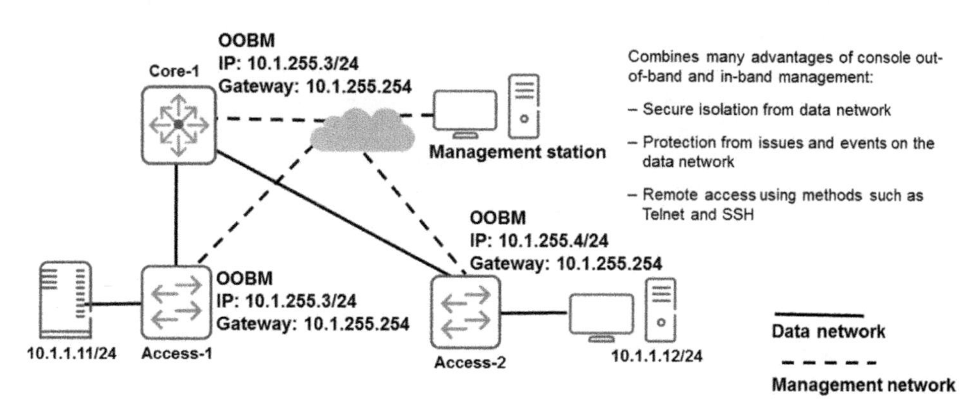

Figure 3-2: Out-of-band management with an Ethernet OOBM port

The Ethernet management and data networks are completely isolated from each other. By default, the OOBM port is set up to receive a DHCP address. Alternatively, you can assign a static IP address directly to the OOBM interface rather than to a VLAN. You can also assign the OOBM interface a gateway address, to which it sends any traffic that it needs routed to a remote management station. In either case, devices on the data network cannot reach the IP address assigned to the OOBM port.

Isolating the data and management networks helps to protect management access even if a problem occurs that affects connectivity in the data network. It also helps to secure the switch from unauthorized access. When a switch has an OOBM connection and it acts only at Layer 2, it does not always require an IP address on the data network, in which case unauthorized users in this network cannot access the switch through Telnet or another protocol. (Sometimes, though, the switch does require an IP address on the data network, in which case, you should make sure to secure in-band access.)

As you will learn in later chapters, devices acting at Layer 3 require IP addresses on their VLANs. But you can still prevent users on the data network from reaching the switch's management functions if you want. For every management function such as Telnet, you can choose whether the switch listens for and responds to incoming requests on the OOBM port, the data plane, or both. By default, it listens on both, but you can restrict listening to the OOBM port to lock out access from the data network.

OOBM is not compatible with a management VLAN or management stacking. If you are using the OOBM port, you do not need to use the management VLAN. A management VLAN restricts the switch to accepting incoming management requests from that VLAN only, creating a somewhat isolated management network. However, the OOBM port already establishes a management network that is completely separate from the data plane and can also use routing. If you want to use management stacking, you should disable OOBM. You can enable and disable OOBM globally on the switch. Note that this is different from enabling and disabling the OOBM port, which you can also do.

Operator and manager roles

Whatever form of access you are using, it is best practice to force users to submit credentials before they can log in to the CLI.

Role	Access
Operator	Basic mode (limited read-only commands)
Manager	Enable mode (full read-write access)

Figure 3-3: Operator and manager roles

ArubaOS switches support two default roles for users:

- **Operator**: An operator has access to operator mode, which provides limited show commands. The operator can only view statistics and limited configuration information (for example, the operator cannot view the configuration files).
- **Manager**: A manager can access enable mode and from enable mode, global configuration mode. So the manager has complete read-write access to the switch. As a manager you can make configuration changes as well as view information.

When you have not set up any protections on the switch, all users log in as managers.

 **Warning**

At factory default settings, there are no passwords, and any user that contacts the switch at its IP address can log in as a manager.

Applying local passwords

You can protect management access to the switch's CLI by setting a manager username and password, an operator username and password, or both.

- Set manager and operator passwords to restrict access to that role:
  - One username and password per-role
  - Applies to in-band and out-of-band management
  - If no username set, default username is used (manager or operator)
- Passwords are encrypted and not included in the config:
  - Omitted from configs (local and backed up); retained when the config is reset
  - Exception: include-credentials enabled

Figure 3-4: Applying local passwords

When you set either type of credential, users who establish a console or Telnet session (or SSH or HTTP session, as well) are prompted to enter a username and password. If they enter the correct credentials for the operator role, they enter the operator or basic context, from which they can enter limited **show** commands. If they enter the correct credentials for the manager role, they enter enable mode and receive complete read-write access.

If you only set a manager password, though, users can press **<Enter>** to skip past the authentication prompt. They then receive operator access without logging in.

A user can move from operator mode to enable mode by entering **enable**. If no manager password is set, the user accesses enable mode without entering any credentials. But if a manager password is set, the user must enter the valid manager username and password.

When you use these passwords, you can set just *one* username and password for the manager role and *one* username and password for the operator role. By default, the passwords apply to all forms of accessing the switch CLI, including in-band and out-of-band.

When you enter the commands for setting the password, you are not required to specify a username. If you do not, the switch keeps the default user names, which are **manager** for manager and **operator** for operator. But you can enter your own usernames.

By default, the credentials are not included in the configuration for security purposes. The same holds true for other credentials listed in Table 3-1. Because the credentials are not stored in the configuration, they are not included if you save the configuration to a file and download that file from the switch. Also if you reset the configuration to factory defaults, the passwords are still applied. You must manually remove the passwords by using the **no** form of the **password** command.

Table 3-1: Credentials not included in the flash by default

Credentials
Local manager and operator usernames and passwords
SNMP security credentials
Usernames and passwords used for 802.1X credentials
TACACS+ encryption keys
RADIUS shared secret
SSH public keys for authorizing managers

If you want to store the credentials in the internal flash and view them in the config, you should enter the **include-credentials** command. You can then:

- View credentials in the running-config and saved configurations.
- Save credentials in configurations and load them to other devices.
- Erase credentials when you erase the startup-config.

If you want to have the include-credentials behavior but protect the credentials from unauthorized viewers, you can use the **encrypt-credentials** command.

- The **include-credentials** has two options that you can add to it:
- The **tacacs-radius** only option configures the switch to store TACACS+ encryption keys and RADIUS shared secrets in the internal flash and show them in the config. However, all other credentials are saved separately as at the factory default settings.
- Entering the **include-credentials store-in-config** command has exactly the same effect as entering the **include-credentials** command. However, the **no include-credentials** and **no include-credentials store-in-config** commands have different effects:
  - The **no include-credentials** command removes the credentials from the running-config. However, the switch continues to use any credentials that you have configured, as well as any SSH keys.
  - The **no include-credentials store-in-config** command also returns the switch to the factory default behavior for not storing credentials in the normal configuration files. But it also entirely deletes the manager and operator username and password and the SSH keys for authorizing managers (SSH keys are described in more detail in the next section.)

SSH

In the real world, administrators should always use SSH to manage switches. Telnet transmits authentication credentials, as well as the terminal session itself, in plaintext, leaving data exposed to unauthorized users. Telnet also fails to provide any authentication of the switch (Telnet server) to the management station (Telnet client), which could expose management users to man-in-the-middle attacks, in which they connect to rogue devices.

SSH, on the other hand, uses asymmetric keys to establish a secure tunnel between the management station (SSH client) and the switch (SSH server). This tunnel provides data integrity and privacy for both the authentication process and the terminal session itself. That is, an unauthorized party cannot tamper with data, and the data is encrypted to prevent eavesdroppers from viewing it. SSH keys also provide for authentication of the server to the client, preventing man-in-the-middle attacks.

- Create SSH key pair on the switch.
- Load switch public key on management station SSH client (optional)
- Choose the type of authentication:
  - Password (default)
    - Same setup as for Telnet
    - Focus for this course
  - Key
    - Management user public keys loaded on switch
- Make sure SSH is enabled on the switch.
- Disable Telnet for higher security.

Secure tunnel
Authentication
Session
Switch
Public/private
Key pair 1
Public key 2
Management station
Public key 1
Public/private
Key pair 2

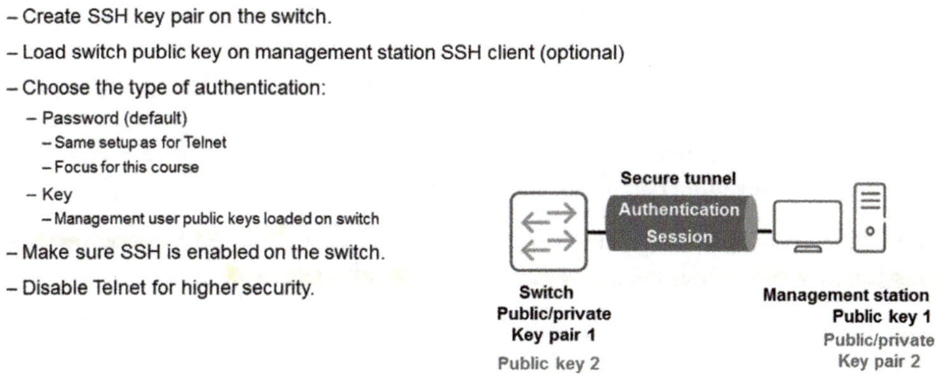

Figure 3-5: SSH

You do not need to understand the details of asymmetric cryptography. However, you should understand that an SSH key is actually a key pair with a private key and a public key. The private and public key have a special relationship. Each decrypts data that the other encrypts; however, you cannot derive the private key from the public key. The private key is retained as a secret key on the device that owns in while the public key can be distributed freely. The SSH server "signs" data by encrypting it with its private key. Any device that decrypts the data with the corresponding public key knows that only the server—the only device that knows the correct private key—could have sent it. (Note that the SSH private key authenticates the server but doesn't encrypt the actual data sent over the tunnel. Otherwise, any device with the public key could read it. Instead, the server and client use a secure key exchange process to generate a key for the session.)

ArubaOS switches enable SSH by default. The switch requires an SSH host key for authenticating the switch to SSH clients and creating the secure tunnel. The switch has an SSH host key automatically generated on it at the factory default settings. Therefore, you could access a switch through SSH without completing any set up. However, it is recommended that you generate a new SSH keypair on your ArubaOS switch. After you create the key pair, it remains permanently on the switch and persists across configuration resets. The **crypto key <generate | zeroize>** ssh commands create and delete the switch's SSH host key.

If you choose, you can view the public key portion of SSH host key, copy it to file, and load the public key on management stations' SSH clients. The clients will then trust the ArubaOS switch as a valid SSH server. This step is optional for many SSH clients. Instead, you can choose to connect to the SSH client to the switch in a controlled environment for the first time and automatically trust the key. (Refer to the switch documentation for details on loading keys on SSH clients if you want to use that option.)

The next step is to choose the form of authentication for SSH clients. At default settings, ArubaOS switches enforce password authentication for SSH, controlled in the same manner as password authentication for Telnet. By default, this authentication is local, so SSH users will log in with the local manager and operator passwords that you just learned how to create. If you want to retain this behavior, you do not need to enter any commands.

Alternatively, you can set up key authentication for the SSH clients. This option requires you to generate or install SSH key pairs on the clients and then load the public keys for these pairs as authorized manager keys on the switch. The keys used to authenticate clients are installed with **ip ssh public-key** commands. These keys are stored separately from the config and persist across resets, by default. However, the **include-credentials** command changes this behavior as explained earlier. Refer to the switch documentation for more details on this second option.

SSH becomes disabled when the switch does not have an SSH key. You should always make sure that SSH is enabled after you create the key by entering the **ip ssh** command.

As a final step, you should typically disable Telnet to force managers to use the secure SSH option.

Set manager and operator passwords

You will now set manager and operator passwords to secure management access to one of your ArubaOS switches.

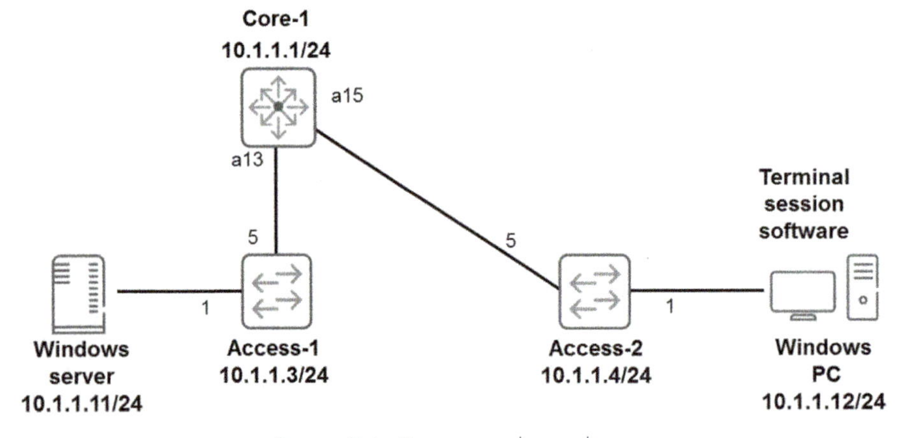

Figure 3-6: Test network topology

Task 1: Restrict Operator and Manager Access (default usernames)

In this task you will set up and test password restrictions to the manager and operator accounts.

1. As an optional step, before you begin configuring passwords, you can test that Telnet and SSH are enabled by default and do not require a password. From the PC, launch Tera Term. Establish a Telnet session to 10.1.1.4. See that you can log in without a password. Then use Tera Term to open an SSH session to 10.1.1.4. Accept the key and see that you can log in without a password.
2. Access a terminal session with the Access-2 switch.
3. Assign an operator password. For the sake of this activity, use **hpe** as the password. Please remember that passwords are case sensitive.

```
Access-2# configure

Access-2(config)# password operator

New password for operator: hpe

Re-enter the new password for operator: hpe

DHCP-based image file download from a TFTP server is disabled when an operator or manager password is set.

DHCP-based config file download from a TFTP server is disabled when an operator or manager password is set.

The TFTP Server is disabled when an operator or manager password is set.

Autorun is disabled when an operator or manager password is set.
```

4. The Telnet server is enabled on ArubaOS switches by default, so you are ready to test Telnet access.
5. Access the PC desktop.
6. Launch a Telnet client such as Tera Term. (You should be able to find this client in the Start menu or in the Kits\Software folder.) Set the host to 10.1.1.4 (or the IP address of the switch that you are using), choose Telnet, and open the connection.

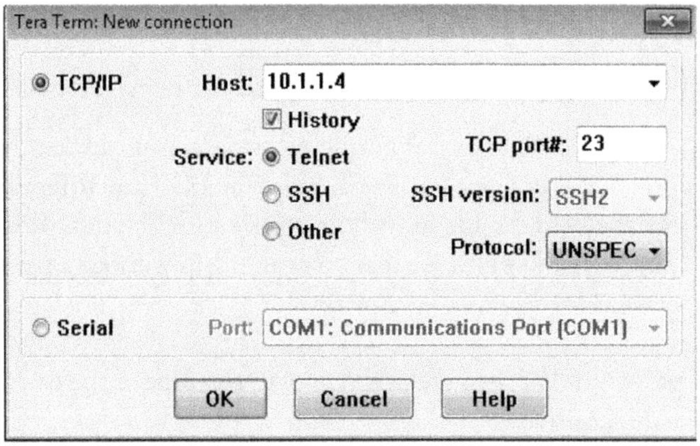

Figure 3-7: Launch Telnet client

7. You should see output such as this in the window:

```
HP J9575A 3800-24G-2SFP+ Switch

Software revision KA.16.01.0006

(C) Copyright 2016 Hewlett Packard Enterprise Development LP

                   RESTRICTED RIGHTS LEGEND

 Confidential computer software.  Valid license from  Hewlett Packard

<-output omitted->
```

8. You will also be prompted for a username. If you did not set a username, the default is used. Enter operator.

```
Username: operator
```

9. You will then be prompted for a password. Enter **hpe**. (You will not see the letters that you type.)

```
Password: hpe
```

10. Note that you are in the operator prompt. Explore the commands that are available to you by entering a question mark, **?**.

11. Try to move to enable mode.

```
Access-2> enable
```

12. As you see, you are allowed to move to enable mode because you have not set a manager password.

13. Set a manager password now.

```
Access-2# config

Access-2(config)# password manager
```

```
New password for manager: hpe
Re-enter the
new password for manager: hpe
```

14. Log out of your session, making sure to save your changes.

```
Access-2(config)# logout
Do you want to log out [y/n]? y
Do you want to save current configuration [y/n/^C]? y
```

15. Use Tera Term to open the Telnet session again.

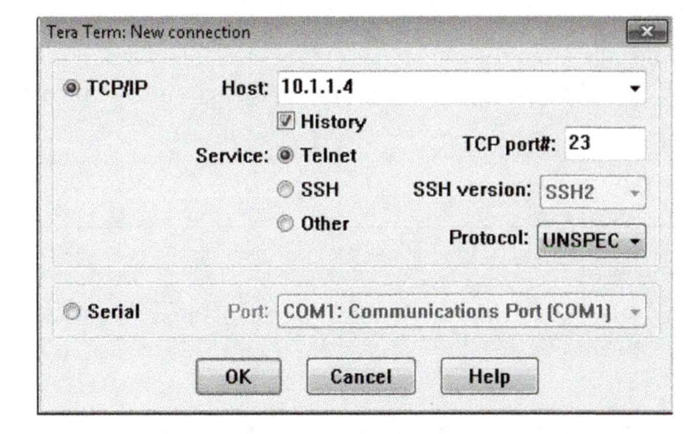

Figure 3-8: Reopen Telnet session

16. Log in as the operator.

```
  <-output omitted->
Username: operator
Password: hpe
```

17. Try to move to enable mode.

```
Access-2> enable
```

18. As you see, you are prompted for credentials because manager access is restricted by a password. Enter manager for the username and password.

```
Username: manager
Password: hpe
```

19. From enable mode, you now have complete access to the switch CLI.

Task 2: Enable SSH

You will now enable SSH access to the switch.

1. Return to the terminal session with Access-2 (you can use either your Telnet session or the session established through the console port). Move to global configuration mode.

2. Create a key for SSH.

```
Access-2(config)# crypto key generate ssh rsa bits 2048
Installing a new key pair.  If the key/entropy cache is
depleted, this could take up to a minute.
The installation of a new key pair is successfully completed.
```

3. SSH is enabled by default when the switch has an SSH key. However, if your switch did not have a key, SSH would have been disabled on it. Enter this command to make sure that SSH is enabled:

```
Access-2(config)# ip ssh
```

Important

In the real world, you should disable Telnet so that users must log in with SSH only.

Task 3: Restrict operator and manager access to ArubaOS switches (non-default usernames)

You will now practice changing credentials and using a non-default username.

Table 3-2: Switch credentials

Role	Username	Password
manager	admin	hpe

1. You should be in global configuration mode in the Access-2 CLI.

2. Change the manager credentials to the ones shown in the table. Remember that passwords are case-sensitive! (You will see ** rather than the password letters.)

```
Access-2(config)# password manager user-name admin
New password for manager: hpe
Please retype new password for manager: hpe
```

3. You could also set a non-default username for the operator account.

Task 4: Test SSH access and the new username

You will now log in with SSH and test the new username.

1. If you are currently logged into the Telnet session on the PC, log out.

```
Access-2(config)# logout
Do you want to log out [y/n]? y
Do you want to save current configuration [y/n/^C]? y
Connection to host lost.
```

2. In the terminal client such as TeraTerm, choose SSH and enter the IP address of your switch.

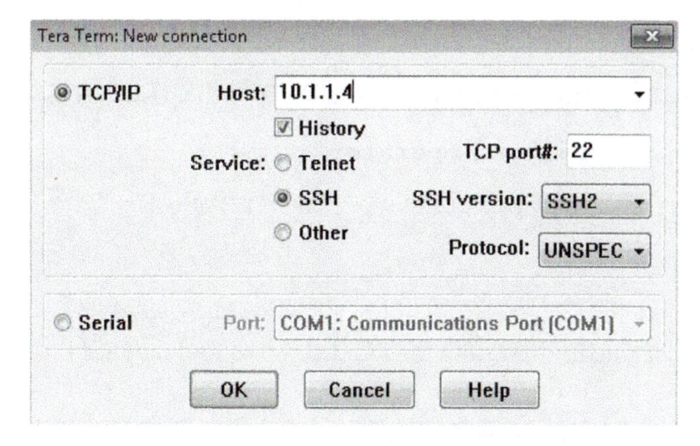

Figure 3-9: Choose SSH and enter switch IP address

3. You will be prompted to accept a key. Do so.

   This is the key that you created on the switch. It authenticates the switch to the client. You can install the key on the client in advance (see the switch documentation for details on doing so) or accept the key on the first connection as you are doing now. If the key changes later, then you know that you might be connecting to a rogue device.

4. Log in as the manager (admin).

```
Username: admin
Password: hpe
```

5. As you see, when you log in with manager credentials, you are placed in enable mode immediately.

6. Log out. (If you are prompted to save, press **y**.)

```
Access-2# logout
Do you want to log out [y/n]? y
```

7. Open a Telnet session again. Try to log in with the **manager** username and password.

```
<-output omitted->
Username: manager
Password: hpe
```

8. The login fails because the new credentials have replaced the old ones, and the old ones no longer work. The switch supports just one username and password for each type of account configured in this way.

9. Close the SSH session.

10. Return to your console port session with Access-2. Remove the password for the operator account.

```
Access-2(config)# no password operator
Password protection for operator will be deleted.
Continue [y/n]?  y
```

11. When the switch has a manager password but no operator password, can users obtain operator privileges without logging in? Return to the PC desktop to find out. SSH (or Telnet) to the switch.

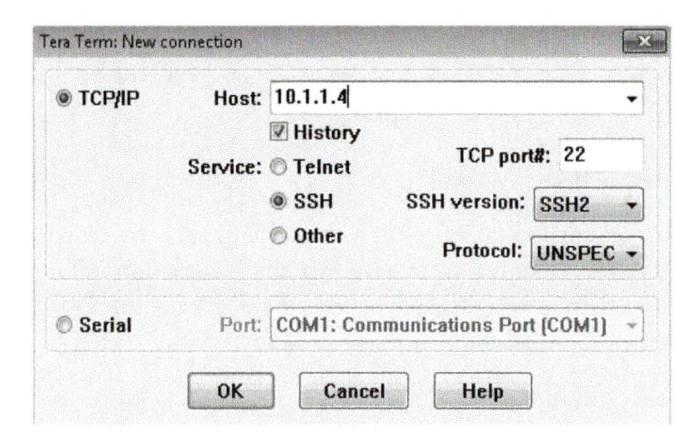

Figure 3-10: SSH (or Telnet) to the switch

12. As you see, you are still prompted to log in.

```
<-output omitted->
Username:
```

13. Try pressing **[Enter]**.
14. But you can skip past the authentication and obtain access to Basic mode.
15. Try to move to enable mode.

```
Access-2> enable
```

16. You are again prompted to log in. Try pressing **[Enter]** and observe that you now *cannot* skip past the authentication. If you enter manager credentials (admin, hpe), you can move to enable mode.
17. Return to the console session with Access-2. Remove the manager password.

```
Access-2(config)# no password manager

Password protection for manager will be deleted, continue [y/n]?  y
```

18. You can verify that the password is removed by returning to your Telnet session and entering **enable**. You should move to enable mode without being prompted to log in.

Learning check

1. Assume that you have set only a manager password on an ArubaOS switch. Can users receive operator access without logging in? How do users log in and receive privileges in this scenario?

2. Assume that you have set only an operator password on an ArubaOS switch. Can users receive manager level access? How do users log in and receive privileges in this scenario?

3. What happened when you entered a new manager username and password, admin and hpe? How many manager usernames and passwords does an ArubaOS switch support? How many operator usernames and passwords does an ArubaOS switch support?

4. ArubaOS switches do not include the manager and operator passwords in the config—nor other credentials, passwords, and keys—unless you enter the **include-credentials** command. What is a use case for using the **include-credentials** command to add these credentials to the configuration?

Answers to Learning check

1. Assume that you have set only a manager password on an ArubaOS switch. Can users receive operator access without logging in? Explain how users log in and receive privileges in this scenario.

As soon as you set one password, users are prompted to log in. However, when you have only set a manager password, users can press <Enter> to skip past authentication and receive basic mode access. When a user successfully logs in, though, the user is placed in enable mode and has manager privileges.

2. Assume that you have set only an operator password on an ArubaOS switch. Can users receive manager level access? Explain how users log in and receive privileges in this scenario.

   Yes, users can receive manager access. In this scenario, users must log in with the operator credentials. They are then placed in operator mode. But they can receive manager level access by entering **enable**. Because no manager password is set, the user moves to enable mode and receives full privileges without entering credentials. In other words, in this scenario, users are again either denied access or can easily obtain manager-level access.

3. What happened when you entered a new manager username and password, admin2 and hpe? How many manager usernames and passwords does an ArubaOS switch support? How many operator usernames and passwords does an ArubaOS switch support?

   The new credentials wrote over the old credentials. An ArubaOS switch supports a single manager account (username and password) and a single operator account (username and password).

   Note that you can use local-user accounts and authorization groups to add more accounts, which is covered later in this chapter.

4. ArubaOS switches do not include the manager and operator passwords in the config—nor other credentials, passwords, and keys—unless you enter the **include-credentials** command. What is a use case for using the include-credentials command to add these credentials to the configuration?

   You might want to use the **include-credentials** command when you are first setting up a base configuration for your switches:

   - You can more easily see the credentials, verify them, and test them.
   - You can save the credentials in the switch's startup configuration and then upload the configuration to a TFTP server. You can then use that configuration as a baseline configuration for other switches. Because the credentials are included, you do not have to reconfigure the credentials on all of your switches separately.
   - You can store different credentials in different config files. (ArubaOS switches running the latest software support three config files.) You can test new credentials by loading a new config and revert to the old credentials by loading the previous config.

   After you test the configuration and deploy it to all your switches, you might want to disable the include-credentials command again for security.

Adding Role-based Access Control (RBAC)

Some environments call for more granular control over management users than that provided by the operator and manager accounts. Perhaps the company has administrators who should be able to configure the switch, but should not have control over other administrators' credentials. Perhaps an administrator can control VLANs and routing, but shouldn't control port-based authentication. Or maybe different VLANs are used by different departments, and some administrators should be able to configure some VLANs but not others.

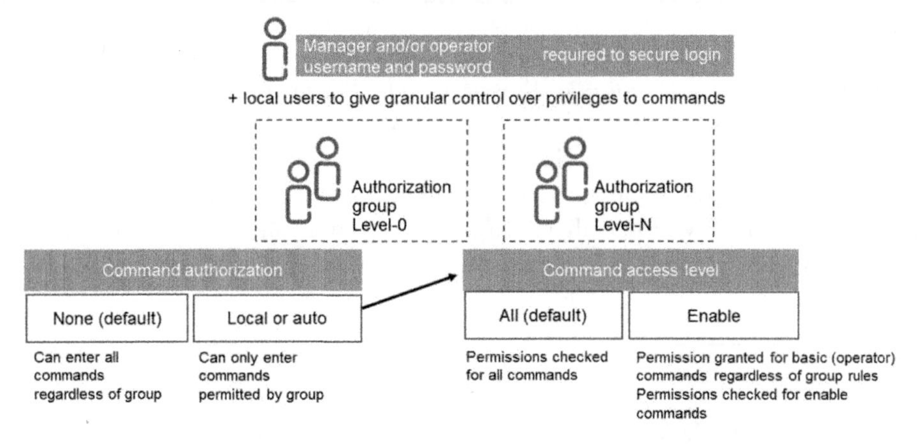

Figure 3-11: Adding Role-based Access Control (RBAC)

You can add local user accounts to the manager and operator accounts to fulfill use cases such as these. When you add a local user, you must assign that user to an authorization group, and you typically set a password as well. A user can then use those credentials to log into the switch using the console port or Telnet (or SSH or the Web user interface [UI], if you've set up those forms of access). At the factory default settings, command authorization is set to none, which means that the user's authorization group does not actually take effect. The user is granted full manager-level access. You could leave this behavior if you simply wanted to create different manager credentials for different managers—which could help in revoking the credentials if an administrator leaves the organization.

Usually, though, you want to use the authorization groups to control the users' privileges. In that case, you must set the **aaa authorization command** option to **local**. Then, when a local user logs in, that user receives access to commands based on rules set in the user's authorization group. By default, the switch checks the privileges for all commands. However, you can use the **aaa authorization access-level** command to set the switch to check the privileges for just enable commands. This option permits users to enter basic mode commands even if their group doesn't explicitly permit them.

Note

You can also configure the switch to authenticate console, Telnet, SSH, or Web UI users using an external authentication server (RADIUS or TACACS+). In that case, you can also configure the switch to receive information about the user's privilege level from the external authentication server. You would use the **radius** or **tacacs** options for the **aaa authorization command**. Or you could use the **auto** option, which configures the switch to use the same option for authorizing the user for commands that it uses to authenticate the user.

It is important for you to understand that RBAC and local users are designed for use *with* the manager and operator accounts, *not instead* of those accounts. If you configure only local users and no manager or operator password, the switch won't be protected from unauthorized access. Users will be prompted to log in, but they can press **[Enter]** and receive full manager access without logging in. Always set a manager password to protect management access to the switch. The manager and operator accounts are associated with default manager and operator authorization groups in the background. You cannot view or alter these groups, but they give the manager and operator the privileges discussed earlier.

Note

This feature requires K/KA/KB16.01 software or above. Check your switch documentation to verify that your switch supports the feature.

Authorization groups and rules

An authorization group contains up to 1000 rules, which define the commands that a user in this group can use. Each rule specifies:

- The authorization group to which the rule is being added
- The sequence for the rule
- Commands are matched to rules in ascending order, and only the first match applies.
- The rule expression
- Permit or deny, which indicates whether the user is permitted to enter the command or not

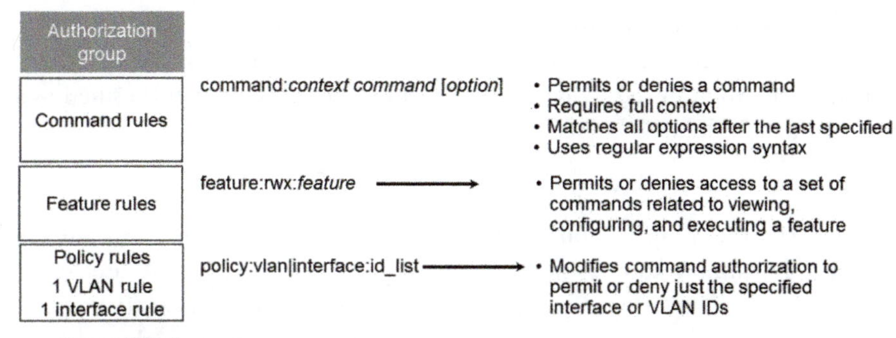

Figure 3-12: Authorization groups and rules

You can configure three types of rules.

Command rules

A command rule defines a command that a user is either permitted to enter or denied entering.

The rule must include the complete context for the command. For example, do not specify "ip routing enable." Instead specify "configure ip routing enable."

You can include regular expressions to create flexibility in matches. A useful regular expression is .*.

- . = any character
- * = any number of repetitions of the previous character

So .* matches any word.

For example, you can specify **"vlan .*" permit** to permit a user to enter the VLAN command with any VLAN ID.

A rule matches commands up to the last option specified. A **"configure ip routing enable" permit** rule allows a user to enter **ip routing enable**. However, on its own this rule does *not* permit the user to enter **configure** and access global configuration mode. Whenever you set up a group that permits access to configuration commands, you should also enter this rule: **"configure" permit**.

Similarly, **"configure vlan .* untagged" permit** allows users to enter the **vlan <ID> untagged** command from global configuration mode. However, it does *not* allow them to enter **vlan <ID>** and then the **untagged** command.

Rules permit any options *after* the last one specified. So, as a better strategy, you might use a rule such as: **"configure vlan .*" permit**, which allows users to enter the VLAN context and enter any command as well as to enter the **vlan** commands from global configuration mode. (For an even simpler approach, you could use the feature rules described below.) If you want to grant users access to some **vlan** commands, but not others, you can specify deny rules with the specific commands using lower sequence IDs and then command rules with higher IDs.

Keep a few more guidelines in mind:

- When you specify the complete command in the rule, users can still use truncated commands. For example, the authorization group has a **configure permit** rule. The user can enter **conf** because this is simply a shortened version of the command, which the switch still processes as **configure**. Do *not* use the shortened version in the command rule, though, unless you add the correct regular expression syntax for matching the complete command.
- The examples above list just the command string and permission. The complete syntax for a command to create a rule is:

```
aaa authorization group <group_name> <sequence_number> match-command
<command_string> <permit | deny> [log]
```

For example:

```
aaa authorization group Level-5 10 match-command "configure vlan .*"
permit
```

- You can also include **"command:** before the command string:

```
aaa authorization group Level-5 10 match-command "command: configure vlan
.*" permit
```

- The CLI does *not* validate command strings. You should test your rules carefully.

Feature rules

The ArubaOS switch has many pre-defined features. A feature lists all the rules related to a feature such as OSPF or VLANs. Each rule under the feature is classified with one of three keywords:

- r = a read rule, which allows users to enter commands related to viewing the feature
- w = a write rule, which allows users to enter commands related to configuring the feature
- x = an execution rule, which allows users to enter commands related to testing or using the feature (such as a ping or traceroute)

Rather than define many command rules, you can specify a feature rule as a simple and failsafe way to give an authorization group privileges to that feature. When you specify the feature, you also specify which type of rules in that feature users can enter. For example, you can authorize the group for only read rules. The syntax for entering a feature rule is:

```
aaa authorization group <group_name> <sequence_number> match-command
"feature:[r|w|x|rw|rx|wx|rwx]:feature" <permit | deny> [log]
```

For example, enter:

```
aaa authorization group Level-5 100 match-command "feature:rwx:vlan"
permit
```

Or enter:

```
aaa authorization group Level-2 100 match-command "feature:rx:vlan"
permit
```

To see the rules associated with a feature, enter this command:

```
show authorization feature <feature> detailed
```

Every rule associated with the specific feature and classification (such as r:vlan) counts as one rule toward the 1000 rule limit. In other words, when you enter this **rule aaa authorization group Level-5 100 match-command "feature:rwx:vlan" permit**, you have actually added many rules to the group.

Policy rules

Each authorization group can include one interface policy rule and one VLAN policy rule. These rules limit the interface IDs or VLAN IDs to which the group is granted privileges. These rules help you to easily grant administrators access to only the switch interfaces or VLANs that are relevant to their group. For example, a switch in a multi-tenant environment might assign VLAN ranges to different tenants. You could create an authorization group for each tenant and use a VLAN policy to ensure that tenant administrators configure only their VLANs.

The command syntax for a policy rule is:

```
aaa authorization group <group_name> <sequence_number> match-command
"policy:<vlan|interface>:<vlan or interface_ID_list> <permit | deny>
[log]
```

You must still enter command or feature rules to allow the users to enter commands related to configuring interfaces and privileges. The command and feature rules give the users privileges to enter the commands, and then the policy rules define which interfaces or VLANs those commands can specify. (If the authorization group has no policy rule, users can enter authorized commands for any interface or VLAN IDs.) For example, you might enter these rules:

```
aaa authorization group Level-5 100 match-command "feature:rwx:vlan"
permit
```

```
aaa authorization group Level-5 200 match-command "policy:vlan:10-19"
permit
```

An authorization group can have *only* one VLAN policy and one interface policy. The last policy rule entered overwrites any previous policy rule. If you want to add IDs to the policy, you must create a new policy rule that specifies the complete range of IDs. For example, authorization group Level-5 has the VLAN policy above. You know want to add VLANs 110-119 to the policy. Enter this command:

```
aaa authorization group Level-5 200 match-command "policy:vlan:10-19,110-119"
permit
```

Pre-defined and default groups

ArubaOS switches that support RBAC have 16 pre-defined groups that help you begin authorizing users for various privilege levels:

- Level-0 has pre-defined rules to give access to just a few diagnostic commands such as ping and traceroute. You can alter the rules for this group.
- Level-1 has pre-defined rules to give access to diagnostic commands and show command. You can alter the rules for this group.
- Level-9 permits access to all commands except ones related to configuring credentials (such as aaa commands). You cannot alter this group in any way.
- Level-15 permits access to all commands. You cannot alter this group in any way.
- Level-2 to Level-8 and Level-10 to Level-14 are pre-defined groups for you to customize as you desire. These groups by default have a single rule with sequence ID 999 that denies all commands. You can add and delete rules for these groups as you desire.

Note that ArubaOS switches also have three default groups:

- manager, which allows the manager user defined with the password manager command access to all commands. You cannot delete, view, or alter this group in anyway. You also cannot add users to it.
- operator, which allows the operator user defined with the password operator command access to basic mode commands. You cannot delete, view, or alter this group in anyway. You also cannot add users to it.
- default-security-group, which helps the switches meet the US government requirements (DoD UCR [UC-APL/JITC]). This regulation requires Department of Defense approved devices to provide a separate security role for reading security logs. You cannot delete the default rule within this group (which permits use of the **security-logging** rule), but you can add rules to the group. You can also assign users to it.

You cannot delete the default and pre-defined groups, but you can create another 45 groups of your own, up to 64 total.

Summary

In this chapter, you have learned about using Telnet to establish in-band management access, as well as access on OOBM ports. You learned how to secure management access on ArubaOS switches locally using operator and manager passwords, as well as RBAC and local users.

- In-band access with Telnet
- Networked OOBM access with Telnet
- Securing management access with local passwords and user accounts
- Using RBAC

Figure 3-13: Summary

Learning check

Review what you have learned in this chapter by answering these questions.

1. What type of management access permits remote access to an IP address but isolates management traffic from data traffic?
   a. In-band Telnet
   b. In-band SSH
   c. Out-of-band Telnet on an Ethernet OOBM port
   d. Out-of-band on a serial console port

2. How does an ArubaOS switch handle Telnet at factory default settings?
   a. It disables Telnet access.
   b. It enables Telnet access and enforces password authentication. But no password is set, so all access is denied.
   c. It enables Telnet access and enforces password authentication with a default password.
   d. It enables Telnet access with no password required.

Answers to Learning check

1. What type of management access permits remote access to an IP address but isolates management traffic from data traffic?
    a. In-band Telnet
    b. In-band SSH
    c. Out-of-band Telnet on an Ethernet OOBM port
    d. Out-of-band on a serial console port

2. How does an ArubaOS switch handle Telnet at factory default settings?
    a. It disables Telnet access.
    b. It enables Telnet access and enforces password authentication. But no password is set, so all access is denied.
    c. It enables Telnet access and enforces password authentication with a default password.
    d. It enables Telnet access with no password required.

4 Managing Software and Configurations

EXAM OBJECTIVES

✓ Manage software and upgrade the operating system on ArubaOS switches

✓ Manage configuration files on ArubaOS switches

ASSUMED KNOWLEDGE

Before reading this chapter, you should be able to:

- Access and navigate the ArubaOS switch CLI
- Control management access to the ArubaOS switch

INTRODUCTION

This chapter explains how to manage software and configuration files on ArubaOS switches.

Software file management

ArubaOS switches feature two flash memory locations for storing switch software image files:

- **Primary flash**—Default storage for a switch software image.
- **Secondary flash**—Additional storage for either a redundant or an alternate switch software image.

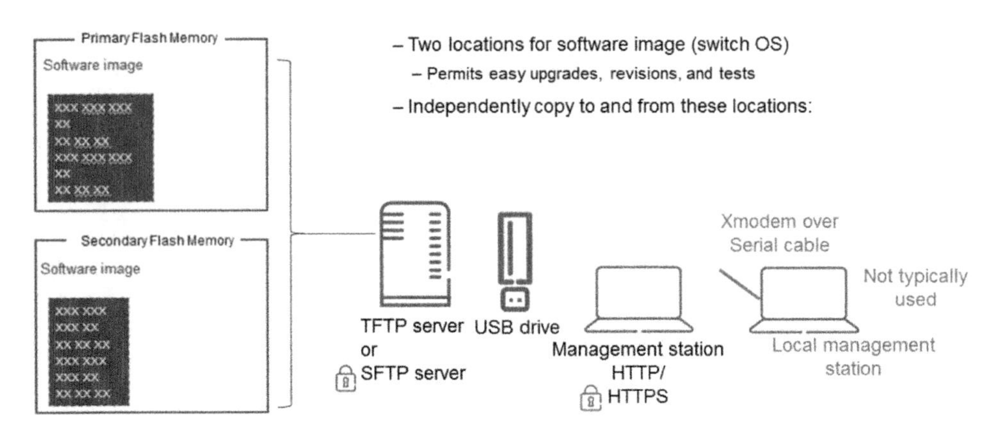

Figure 4-1: Software file management

The switch boots from and uses only one image at a time. You can choose the location for the software from which the switch boots. This location is not shown in the switch configuration but is rather stored in a separate file in the flash. After you change the location in the configuration, the switch must reboot in order for the switch to boot the software in the new location. An administrator with console access to the switch can also choose the software to boot during the boot process, overriding the default location selected in the configuration.

Having two flash memory locations for the software image helps you to manage the switch more easily. You can test a new image in your system without having to replace the previous image; if necessary, you can revert to the previous image easily. You can also use the two image locations for troubleshooting. For example, you can copy a problem image into secondary flash for later analysis and place another, proven image in primary flash to run your system.

 **Note**

Before you load new software, you should make sure that a switch has backup software in flash in case the new software has problems. You can then easily boot from the old software without having to reinstall it.

You can copy a new software image to either the primary or secondary flash location in several different ways:

- From a TFTP server or an SFTP server

  As indicated by the lock icon in the figure above, SFTP is a more secure option that provides encryption. (The SFTP copy options work with both SFTP and SCP servers.) You must set up SSH and use the **ip ssh filetransfer** command to allow SFTP to work. When you enter **ip ssh filetransfer**, the switch's TFTP client is automatically disabled, preventing insecure copying of plaintext files to and from a TFTP server. However, if you want, you can re-enable the TFTP client (**tftp client** command) after enabling the secure SFTP transfers.

- From a storage device attached to the USB port, if the switch has such a port
- From a management station, which is logged into the switch's Web browser interface

  This guide focuses on the CLI, but you should understand that you can use a Web browser interface to manage ArubaOS switches as well. HTTP access to the interface is enabled by default. Before you can enable the more secure HTTPS, you must either create a self-signed certificate on the switch or install a CA-signed certificate. The minimum number of commands for setting up HTTPS access are provided below. For more detailed steps and options, see the switch management and configuration guide:

```
Switch(config)# crypto pki enroll-self-signed-certificate-name <cert-
   name> subject common-name <switch FQDN or IP address>

Switch(config)# web-management ssl
```

  By default, the web interface is secured by local authentication (the same usernames and passwords that you learned how to set up for Telnet and SSH access in the previous chapter).

- From a management station, using Xmodem over a console connection

  You should not typically use Xmodem over the console connection because the update process is very slow. The fastest clock rate of the console connection is 115K, so an administrator would have to wait a very long time for a large software image to copy. (Xmodem over Telnet would have a higher speed.)

  Only use Xmodem when absolutely necessary: when the switch cannot reach a TFTP or SFTP server and you do not have physical access to the USB slot. (Usually, if you have console access to a switch, you also have physical access. But it is possible to receive a remote connection to the console port through a Telnet-to-serial server.)

It is best practice to copy new software to a different location from the software that is currently in use. The switch runs a check when it copies the software, verifying that the software image is valid for this switch.

You can also copy software images in the primary or secondary flash to any of the locations listed above. In this way, you can backup and archive images.

The job aids at the end of this chapter give you steps for upgrading ArubaOS software.

Manage software images on ArubaOS switches

Next, you will learn how to manage the software image files used by the ArubaOS switches.

1. Access a terminal session with the Access-2 switch CLI.
2. Examine the version of the OS the switch is running. Notice which boot image it used when it booted last.

```
Access-2# show version

Image stamp:

/ws/swbuildm/rel_richmond_qaoff/code/build/tam(swbuildm_rel_richmond_
qaoff_rel_ richmond)

                Mar 23 2016 11:47:03

                KA.16.01.0006

                432

Boot Image:     Primary

Boot ROM Version:   KA.15.09

Active Boot ROM:    Primary
```

3. Examine the primary and secondary image files in the switch's flash. Note which image is set as the primary image.

   Also note the boot ROM version. Sometimes you can upgrade a switch's software without upgrading the boot ROM, but sometimes you must upgrade the boot ROM. The release notes for the new software version will indicate the required boot ROM version, so you should keep track of which version your switch is using.

```
Access-2# show flash

Image                Size (bytes) Date     Version

-----------------    ------------ -------- --------------

Primary Image     :      15540244 03/23/16 KA.16.01.0006

Secondary Image   :      15890772 08/24/15 KA.15.15.0014

Boot ROM Version

----------------

Primary Boot ROM Version     : KA.15.09

Secondary Boot ROM Version : KA.15.09

Default Boot Image      : Primary

Default Boot ROM        : Primary
```

4. Reconfigure the switch to boot from the other software location, which should be secondary.

   The command reboots the switch immediately. Confirm the reboot and save if prompted.

```
Access-2# boot system flash secondary

This will reboot the system from the secondary image.

Continue (y/n)? y

Do you want to save current configuration [y/n/^C]?  y
```

5. Notice that as the switch reboots, its default boot profile is "secondary."

```
Boot Profiles:

0. Monitor ROM Console

1. Primary Software Image   [KA.16.01.0006]

2. Secondary Software Image [KA.15.15.0014]

Select profile (secondary):
```

6. After the switch reboots, examine the version information to see which flash image was used on the last reboot.

```
Access-2# show version

Image stamp:

/ws/swbuildm/rel_nashville_qaoff/code/build/tam(swbuildm_rel_nash-
ville_qaoff_re l_nashville)

               Aug 24 2015 12:19:50

               KA.15.15.0014

               2038

Boot Image:    Secondary
```

7. Did the switch use the secondary image for just this reboot or will it use it for subsequent reboots?

8. Verify your answer the **show flash** command, which shows the software selected for subsequent reboots.

```
Access-2# show flash

Image                Size (bytes) Date     Version

----------------- ------------ -------- --------------

Primary Image      :     15540244 03/23/16 KA.16.01.0006

Secondary Image    :     15890772 08/24/15 KA.15.15.0014
```

```
Boot ROM Version : KA.15.09
Default Boot     : Secondary
```

9. As you see, the default boot image is now secondary. Change the default flash location to primary. Confirm the change with **y**.

```
Access-2# boot set-default flash primary
This command changes the location of the default boot. This command will
change the default flash image to boot from primary image.
Hereafter, 'reload' and 'boot' commands will boot from primary image.
Do you want to continue [y/n]?  y
```

10. Verify the change.

```
Access-2# show flash
Image                 Size (bytes) Date     Version
----------------- ------------ -------- --------------
Primary Image     :     15540244 03/23/16 KA.16.01.0006
Secondary Image   :     15890772 08/24/15 KA.15.15.0014

Boot ROM Version : KA.15.09
Default Boot     : Primary
```

11. Has the current software version changed? View the version information.

```
Access-2# show version
Image stamp:
/ws/swbuildm/rel_nashville_qaoff/code/build/tam(swbuildm_rel_nashville_
qaoff_re l_nashville)
                    Aug 24 2015 12:19:50
                    KA.15.15.0014
                    2038
Boot Image:         Secondary
```

12. As you see the switch will use the primary image for the next reboot, but it does not reboot immediately. It is still using the secondary image, from which it booted before.

13. Initiate a reboot now.

```
Access-2# reload
```

```
System will be rebooted from primary image. Do you want to continue
[y/n]?  y
```

14. After the switch reboots, verify that it has booted from the primary image with the **show version** command.

Learning check

You will now answer several questions, based on what you have learned about managing software images on ArubaOS switches.

1. What are the similarities and differences between these ArubaOS commands:
   - **show version**
   - **show flash**

2. What are the similarities and differences between these ArubaOS commands:
   - **boot system flash <primary | secondary>**
   - **boot set-default flash <primary | secondary>**

Answers to Learning check

1. What are the similarities and differences between these ArubaOS switch commands:
   - **show version**
   - **show flash**

   Both commands give information about the software images on the switch. The **show version** command gives details about the software from which the switch last booted. The **show flash** command displays information about *both* software images on the switch. It indicates which software image location will be used for *subsequent* boots. This location might be different from the location in the **show version** command if you have changed the default and not yet rebooted the switch.

2. What are the similarities and differences between these ArubaOS commands:
   - **boot system flash <primary | secondary>**
   - **boot set-default flash <primary | secondary>**

Both commands configure which software image location the switch uses for subsequent boots.

The first command initiates an immediate, full reboot.

The second command does not initiate a reboot. To make the new software take effect, you must enter **boot** (full reboot) or **reload** (slightly faster reboot with fewer subsystem tests). If the switch has redundant management modules, this command could also cause the switch to failover from the current management module to the standby one.

Configuration file management

Network infrastructure devices have:

- **Running-config**—The current configurations that are running on the device and are stored in RAM (volatile memory)
- **Startup-config**—A configuration file that is stored in flash (non-volatile) memory and loaded when the switch boots

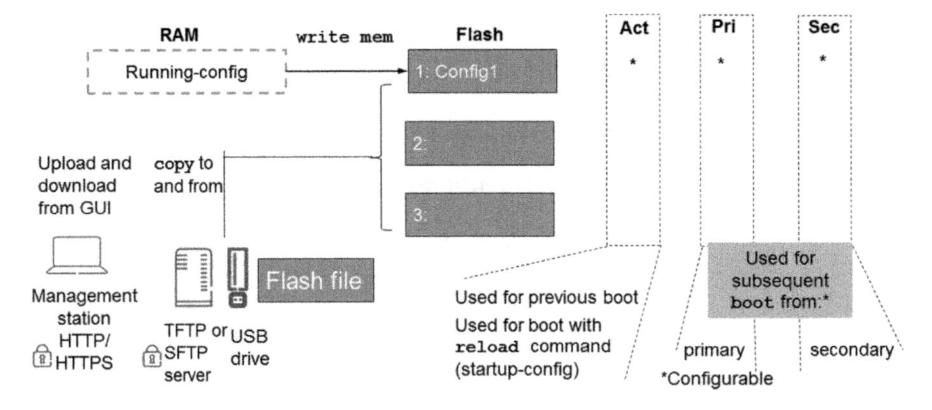

Figure 4-2: Configuration file management

The running-config file exists in volatile memory and controls switch operation. If no configuration changes have been made in the CLI since the switch was last booted, the running-config file is identical to the startup-config file. When you make configuration changes, those changes apply to the running-config. If you then reboot the switch without saving those changes, your changes are lost. The **write memory** command saves the changes to the startup-config, preserving the configuration across reboots. When you boot a switch, it applies the startup-config file as the new running-config file.

ArubaOS switches have three config file slots in their flash memory. When you save the running-config with the **write memory** command, the switch, by default, saves the config to a file in slot 1, which is named config1. This file is the active file, and it is used as the startup-config file by both the primary and secondary software.

You can copy config files to any slot in a number of ways using the CLI **copy** command:

- From another config file
- From a TFTP server or SFTP server
- From a USB device

You can alternatively use the switch's Web browser interface to copy files to and from your management station.

You can use Xmodem as well, but this method is not commonly used.

You can configure the switch to use a new config file as the startup-config file for subsequent system boots. You can choose the startup-config file separately for the primary and for the secondary software. For example, you might have a config file with new features that is associated with new software; a config file without those features is associated with older software to which you might need to revert.

When you view the config files with the show config files command, you will see three columns:

- **Act**—An asterisk (*) in this column indicates that the switch booted with this config file most recently. A **write memory** command saves the running-config to this file.
- **Pri**— An asterisk (*) in this column indicates that the switch is configured to boot with this config file when it boots the primary software.
- **Sec**— An asterisk (*) in this column indicates that the switch is configured to boot with this config file when it boots the secondary software.

If you are using the primary software, and the active file is different from the primary file, the **boot** command boots the primary file, and the **reload** command boots the active file. The same holds true if you are using the secondary software.

The job aids at the end of this chapter provide commands for managing config files.

Manage configuration files on ArubaOS switches

In this section, you will learn how to manage the configuration files used by the ArubaOS switches.

1. Access a terminal session with the Access-2.
2. You practiced changing the software in the previous task. Make sure that the switch is now using the K16.01 software, which is typically the primary image.

```
Access-2# show version

Image stamp:

/ws/swbuildm/rel_richmond_qaoff/code/build/tam(swbuildm_rel_richmond_qaoff_rel_richmond)
```

```
            Mar 23 2016 11:47:03
            KA.16.01.0006
            432
Boot Image:    Primary

Boot ROM Version:  KA.15.09
Active Boot ROM:   Primary
```

3. View the config files in the switch's flash.

```
Access-2# show config files
Configuration files:
 id | act pri sec | name
 ---+-------------+-----------------------------------------------------
  1 |  *   *   *  | config1
  2 |             |
  3 |             |
```

Important

Your output might not match this output. In that case, make the changes indicated below.

- If the file with the * in the **act** column is not named config1 and no other file is named config1, enter this command to change the name: **rename config <name> config1**.
- If the file with the * in the **act** column is not named config1 and another file *is* named config1, enter these commands: **delete config1** and **rename config <name> config1**.
- If config1 has * in the **act** column but not in both the **pri** and **sec** columns, enter this command: **startup-default config config1**.
- If, at this point, your switch has three config files, delete the extra configs with this command: **erase config <filename>** (or if the software version in KA15.10 or later, use the **delete <filename>** command).

4. Move to global configuration view.
5. Change the switch's hostname to "Test" and save the configuration.

```
Access-2(config)# hostname Test
Test(config)# write memory
```

6. Copy the config1 file in the first slot to a file called config2 and verify the copy.

```
Test# copy config config1 config config2

Test# show config files


Configuration files:

 id | act pri sec | name

 ---+-------------+------------------------------------------

   1 |  *   *   *  | config1

   2 |             | config2

   3 |             |
```

7. Change the current configuration's hostname back to "Access-2" and save.

```
Test(config)# hostname Access-2

Access-2(config)# write memory
```

8. At this point, the config2 file has a hostname of "Test" and the config1 file, the default configuration boot file, has a hostname of "Access-2."

   If you want, you can confirm this with the **show config config1** and **show config config2** commands.

9. ArubaOS switches give you two ways to boot from a config file:

   – Set the config file as the default file used with the primary or secondary software (**startup-default** command)

   – Boot from primary or secondary software and use the config file just for this first reboot (**boot system** command)

10. You will use the second option. Boot the switch from its current software, primary, and from config file, config2. If prompted to save the configuration, do not.

```
Access-2(config)# boot system flash primary config config2

This will reboot the system from the primary image.

Continue (y/n)? y
```

11. After the switch reboots, notice its prompt.

```
Test#
```

12. View the configuration files. Notice that:

   – config2 is the active file

- But config2 is not a primary or secondary configuration file. The boot system command specified a one-time boot of this file.

```
Test# show config files

Configuration files:

 id | act pri sec | name

 ---+-------------+--------------------------------------------------

  1 |       *   * | config1

  2 |  *          | config2

  3 |             |
```

13. If you make a change to the running-config and save it, the change saves to config2. Test now by making and saving a simple change and then viewing config2.

```
Test# config

Test(config)# hostname Access-Y

Access-Y(config)# write memory

Access-Y(config)# show config config2

Startup configuration:

; J9575A Configuration Editor; Created on release #KA.16.01.0006

; Ver #0c:10.34.59.34.6b.fb.ff.fd.ff.ff.3f.ef:07

hostname "Access-Y"

<output omitted>
```

14. When you fully boot the switch from the primary software, the primary config file, config1, will boot. Test now.

Important

Make sure to use the **boot** command. The **reboot** command will boot the active config (config2).

```
Access-Y(config)# boot

This will reboot the system from the primary image.

Continue (y/n)? y
```

15. After the switch has rebooted, verify that the prompt has returned to the config1 prompt.

```
Access-2#
```

Tip

Keep what you've learned in mind when you use the **boot system** command to load a config once. When you make changes to that config, save them, and boot, the changes seem to disappear because the default config boots instead. But the changes are saved in the config file that you used for the one-time boot.

16. Verify that config1 is listed as the active configuration file.

```
Access-2# show config files

Configuration files:

 id | act pri sec | name

 ---+-------------+------------------------------------------

  1 |  *   *   *  | config1

  2 |             | config2

  3 |             |
```

Learning check

1. To which config file are changes saved when you enter **write memory** on an ArubaOS switch?

2. How do you configure an ArubaOS switch to boot a config named **newConfig** once but keep the existing config file for other reboots? (Boot from the primary software.)

3. How do you configure an ArubaOS switch to always boot a config named **newConfig** when using the primary software?

4. You want to test out a new configuration on a switch. You first need to back up the existing configuration locally. You will then make the changes, which you want to be able to save. After you try out the changes, you want to revert to the backup configuration. Describe how you would complete this task on an ArubaOS switch.

Answers to Learning check

1. To which config file are changes saved when you enter **write memory** on an ArubaOS switch?

   The config file that is marked as active (the last configuration with which the switch booted).

2. How do you configure an ArubaOS switch to boot a config named **newConfig** once but keep the existing config file for other reboots? (Boot from the primary software.)

   You use this command: **boot system flash primary config newConfig**.

3. How do you configure an ArubaOS switch to always boot a config named **newConfig** when using the primary software?

   You use this command: **startup-default primary config newConfig**.

4. You want to test out a new configuration on a switch. You first need to back up the existing configuration locally. You will then make the changes, which you want to be able to save. After you try out the changes, you want to revert to the backup configuration. Describe how you would complete this task on an ArubaOS switch.

   You should follow a procedure like this:

   - Save the current configuration (**write memory**).
   - View the config files and find the name of the current startup-config (**show config files**), often config1. You should also check that a slot is free for the backup config. If no slot is free, you must delete a file. You might want to back up that file using TFTP or a USB first.
   - Copy the currently active config to a file named backup (**copy config config1 config backup**). You could also back up this file to a TFTP server or USB device.
   - Make the configuration changes, save them (**write memory**), and test them.
   - You have a few choices for reverting to the backup config.
     - You could configure the switch to use backup as its default primary, secondary, or both software config (**startup-default [primary | secondary] config backup**). You would then reboot the switch so that the backup config takes effect (**boot**). (Note that the **reload** command would boot the current config, not the backup one.)
     - You could archive the test configuration using another config slot or TFTP or USB (**copy config config1 <destination>**). You could then execute a one-time boot from the backup config (**boot system flash [primary | secondary] config backup**). You could then copy the backup config back over config1, which is still being used for primary and secondary boots (**copy config backup config config1**). Next time you reboot the switch, it will use the config1 file but have the configurations from the backup.

Planning a software upgrade

What should you plan and what steps should you take before you complete a software upgrade in the real world? You can use Job Aid 4-1 for ideas.

Summary

In this chapter, you have learned how to manage software and configuration files on ArubaOS switches. You learned about the file system and focused on practicing simple management tasks such as copying files, selecting the software image used for booting the switch, and selecting the startup-config file.

– Software file management
– Configuration file management

Figure 4-3: Summary

Learning check

The following questions will help you measure your understanding of the material presented in this chapter.

1. How many operating systems can you install in flash on an ArubaOS switch?
   a. One
   b. Two
   c. Three
   d. Unlimited

2. Which command specifies the primary flash location as the default boot file?

3. Which command erases the current startup configuration file and reboots the switch at factory default settings?

Answers to Learning check

1. How many operating systems can you install in flash on an ArubaOS switch?
   a. One
   b. Two
   c. Three
   d. Unlimited

2. Which command specifies the primary flash location as the default boot file?

   boot set-default flash primary

3. Which command erases the startup configuration file and reboots the switch at factory default settings?

   erase startup-config

Software and configuration job aids

Job Aid 4-1: Prework before a software update

	Step	ArubaOS switch command
☐	Schedule down time.	
☐	Check release notes to determine whether a boot ROM update is required.	
☐	Check release notes for known issues or gotchas like documented changes in system defaults.	
☐	Have a recovery plan in case the update doesn't go according to plan.	
☐	Consider testing the upgrade in a test network or less critical segment.	
☐	Back up the current configuration on all switches, making sure to save the running-config.	write memory copy start tftp <IP address> <filename>
☐	Make sure that you have a copy or backup of the current software image.	copy flash primary tftp <IP address> <filename>

Job Aid 4-2: Upgrade ArubaOS switch software (USB)

	Step	Command	Notes
1	Complete the pre-work checklist.		See Job Aid 4-1
2	Copy the new software image to the USB device.		
3	Insert the USB device in the switch's USB port.		
4	Log into the switch CLI.		Make sure that you are in enable mode.
5	Check the location of the current software image (primary or secondary).	show version	Look for Primary or Secondary next to **Boot Image**:
6	View files on the USB device and note the image name.	dir	Depending on your terminal client, you can copy the image name.
7	Backup the current configuration and associate this file with the current software.	write memory copy config <startup-config filename> config <backup filename> startup-default <primary | secondary> config <backup filename>	Configurations with the new software might not be backward compatible with the current software, so this step will make it easier for you to revert to the current software if necessary. If you saw Primary, in step 5, enter primary. If you saw Secondary, in step 5, enter secondary.

Job Aid 4-2: Continued

<table>
<tr><th></th><th>Step</th><th>Command</th><th>Notes</th></tr>
<tr><td>8</td><td>Copy the new image to the software location that is not currently in use.</td><td>copy usb flash <image name> <primary | secondary></td><td>Depending on your terminal client, you can paste the image name.
If you saw Primary, in step 5, enter secondary.
If you Secondary, in step 5, enter primary.</td></tr>
<tr><td>9</td><td>You will be prompted:
The Primary | Secondary OS Image will be deleted, continue [y/n]</td><td>y</td><td></td></tr>
<tr><td>10</td><td>Wait as the software is validated and copied.</td><td></td><td>The process might take a few minutes. When the process is complete, you will see this message:</td></tr>
<tr><td>11</td><td>Choose whether you want to:
• Immediately run a full reboot with all subsystem tests
• Reboot later (full reboot or faster reboot)</td><td></td><td>Your choice determines whether you complete step 11a or 11b.</td></tr>
<tr><td rowspan="2">12a</td><td>Immediately run a full reboot</td><td>boot system flash <primary | secondary></td><td rowspan="2">The command changes the software from which the switch boots and initiates an immediate boot.</td></tr>
<tr><td>You will be prompted to confirm.</td><td>y</td></tr>
<tr><td rowspan="5">12b</td><td>Reboot later</td><td></td><td></td></tr>
<tr><td>Set the image for the next reboot.</td><td>boot set-default flash <primary | secondary></td><td></td></tr>
<tr><td>You will be prompted to confirm.</td><td>y</td><td></td></tr>
<tr><td>When you are ready, initiate or schedule a reboot.</td><td>Either:
• boot
• reload [after <[dd:]hh:]mm> | at <hh:mm[:ss]> [<mm/dd[/[yy]yy]>]]</td><td>boot initiates a full reboot.
reload initiates a reboot.
reload after sets a delay of a certain length.
reload at schedules the reload for an absolute time.</td></tr>
<tr><td>You will be prompted to confirm.</td><td>y</td><td></td></tr>
<tr><td>13</td><td>Verify the new software version after the reboot.</td><td>show version</td><td></td></tr>
<tr><td>14</td><td>Run acceptance tests and verify system operations. Check the configuration as some defaults might have changed. Backup and revert, as necessary.</td><td></td><td></td></tr>
</table>

Job Aid 4-3: Upgrade ArubaOS switch software (TFTP)

	Step	Command	Notes
1	Complete the pre-work checklist.		See Job Aid 4-1
2	Copy the new software image to the TFTP server.		Note the IP address of the TFTP server. You can use a hostname instead of an IP address if the switch has a DNS server configured.
3	Log into the switch CLI.		Make sure that you are in enable mode.
4	Make sure that the switch has connectivity to the TFTP server.	ping <server IP address>	If the switch cannot reach the server, you should check its IP address and default route.
5	Check the location of the current software image (primary or secondary).	show version	Look for Primary or Secondary next to **Boot Image:**
6	Backup the current configuration and associate this file with the current software.	write memory copy config <startup-config filename> config <backup filename> startup-default <primary | secondary> config <backup filename>	Configurations with the new software might not be backward compatible with the current software, so this step will make it easier for you to revert to the current software if necessary. If you saw Primary, in step 5, enter primary. If you saw Secondary, in step 5, enter secondary.
7	Copy the new image to the software location that is not currently in use.	copy tftp flash <server IP address> <image name> <primary | secondary>	If you saw Primary in step 5, enter secondary. If you saw Secondary in step 5, enter primary.
8	You will be prompted: The Primary | Secondary OS Image will be deleted, continue [y/n]	y	
9	Wait as the software is validated and copied.		The process might take a few minutes. When the process is complete, you will see this message:
10	Choose whether you want to: • Immediately run a full reboot with all subsystem tests • Reboot later (full reboot or faster reboot)		Your choice determines whether you complete step 11a or 11b.

Job Aid 4-3: Continued

<table>
<tr><th></th><th>Step</th><th>Command</th><th>Notes</th></tr>
<tr><td>11a</td><td>Immediately run a full reboot</td><td>boot system flash <primary | secondary></td><td rowspan="2">The command changes the software from which the switch boots and initiates an immediate boot.</td></tr>
<tr><td></td><td>You will be prompted to confirm.</td><td>y</td></tr>
<tr><td>11b</td><td>Reboot later</td><td></td><td></td></tr>
<tr><td></td><td>Set the image for the next reboot.</td><td>boot set-default flash <primary | secondary></td><td></td></tr>
<tr><td></td><td>You will be prompted to confirm.</td><td>y</td><td></td></tr>
<tr><td></td><td>When you are ready, initiate or schedule a reboot.</td><td>Either:
• boot
• reload [after <[dd:]hh:]mm> | at <hh:mm[:ss]> [<mm/dd[/[yy]yy]>]]</td><td>boot initiates a full reboot.
reload initiates a reboot.
reload after sets a delay of a certain length.
reload at schedules the reload for an absolute time.</td></tr>
<tr><td></td><td>You will be prompted to confirm.</td><td>y</td><td></td></tr>
<tr><td>12</td><td>Verify the new software version after the reboot.</td><td>show version</td><td></td></tr>
<tr><td>13</td><td>Run acceptance tests and verify system operations. Check the configuration as some defaults might have changed. Backup and revert, as necessary.</td><td></td><td></td></tr>
</table>

Job Aid 4-4: Upgrade ArubaOS switch software (SFTP)

<table>
<tr><th></th><th>Step</th><th>Command</th><th>Notes</th></tr>
<tr><td>1</td><td>Complete the pre-work checklist.</td><td></td><td>See Job Aid 4-1</td></tr>
<tr><td>2</td><td>Copy the new software image to the SFTP server.</td><td></td><td>Note the IP address of the SFTP server. You can use a hostname instead of an IP address if the switch has a DNS server.
Also note the image name.</td></tr>
<tr><td>3</td><td>Log in to the switch CLI.</td><td></td><td>Make sure that you are in enable mode.</td></tr>
<tr><td>4</td><td>Make sure that the switch has connectivity to the SFTP server.</td><td>ping <IP address | hostname></td><td>If the switch cannot reach the server, you should check its IP address and default route.</td></tr>
</table>

Job Aid 4-4: Continued

<table>
<tr><th></th><th>Step</th><th>Command</th><th>Notes</th></tr>
<tr><td>5</td><td>If the switch does not already support SSH, set up SSH and enable SFTP.</td><td>crypto key generate ssh <rsa | dsa> bits <bits>
ip ssh
ip ssh filetransfer</td><td>The ip ssh command might not be necessary, as SSH is enabled by default.</td></tr>
<tr><td>6</td><td>Check the location of the current software image (primary or secondary).</td><td>show version</td><td>Look for Primary or Secondary next to Boot Image:</td></tr>
<tr><td>7</td><td>Backup the current configuration and associate this file with the current software.</td><td>write memory
copy config <startup-config filename> config <backup filename>
startup-default <primary | secondary> config <backup filename></td><td>Configurations with the new software might not be backward compatible with the current software, so this step will make it easier for you to revert to the current software if necessary.
If you saw Primary, in step 5, enter primary.
If you saw Secondary, in step 5, enter secondary.</td></tr>
<tr><td>8</td><td>Copy the new image to the software location that is not currently in use.</td><td>copy sftp flash {user <name> <IP address | hostname> | <username>@<IP address | hostname> | <IP address | hostname>} <image name> <primary | secondary></td><td>If you saw Primary in step 5, enter secondary.
If you saw Secondary in step 5, enter primary.</td></tr>
<tr><td>9</td><td>If you are prompted for a password, enter it.</td><td></td><td></td></tr>
<tr><td>10</td><td>You will be prompted:
The Primary | Secondary OS Image will be deleted, continue [y/n]</td><td>y</td><td></td></tr>
<tr><td>11</td><td>Wait as the software is validated and copied.</td><td></td><td>The process might take a few minutes. When the process is complete, you will see this message:</td></tr>
<tr><td>12</td><td>Choose whether you want to:
• Immediately run a full reboot with all subsystem tests
• Reboot later (full reboot or faster reboot)</td><td></td><td>Your choice determines whether you complete step 13a or 13b.</td></tr>
<tr><td rowspan="2">13a</td><td>Immediately run a full reboot</td><td>boot system flash <primary | secondary></td><td rowspan="2">The command changes the software from which the switch boots and initiates an immediate boot.</td></tr>
<tr><td>You will be prompted to confirm.</td><td>y</td></tr>
</table>

Job Aid 4-4: Continued

<table>
<tr><th></th><th>Step</th><th>Command</th><th>Notes</th></tr>
<tr><td>13b</td><td>Reboot later</td><td></td><td></td></tr>
<tr><td></td><td>Set the image for the next reboot.</td><td>boot set-default flash <primary | secondary></td><td></td></tr>
<tr><td></td><td>You will be prompted to confirm.</td><td>y</td><td></td></tr>
<tr><td></td><td>When you are ready, initiate or schedule a reboot.</td><td>Either:
• boot
• reload [after <[dd:]hh:]mm> | at <hh:mm[:ss]> [<mm/dd[/[yy]yy]>]]</td><td>boot initiates a full reboot.
reload initiates a reboot.
reload after sets a delay of a certain length.
reload at schedules the reload for an absolute time.</td></tr>
<tr><td></td><td>You will be prompted to confirm.</td><td>y</td><td></td></tr>
<tr><td>14</td><td>Verify the new software version after the reboot.</td><td>show version</td><td></td></tr>
<tr><td>15</td><td>Run acceptance tests and verify system operations. Check the configuration as some defaults might have changed. Backup and revert, as necessary.</td><td></td><td></td></tr>
</table>

Job Aid 4-5: Copy ArubaOS switch files

<table>
<tr><th>Copy from:</th><th>To:</th><th>Command</th></tr>
<tr><td>Config file in flash</td><td>Config file in flash</td><td>copy config <source filename> config <dest filename>
* Copies to the active startup-config are not allowed.</td></tr>
<tr><td></td><td>TFTP server</td><td>copy config <source filename> tftp <IP address | hostname> <dest filename></td></tr>
<tr><td></td><td>SFTP server</td><td>copy config <source filename> sftp {user <name> <IP address | hostname> | <username>@<IP address | hostname> | <IP address | hostname>} <dest filename></td></tr>
<tr><td></td><td>USB</td><td>copy config <source filename> usb <dest filename></td></tr>
<tr><td>TFTP server</td><td>Config file in flash</td><td>copy tftp config <dest filename> <IP address | hostname> <source filename></td></tr>
<tr><td></td><td>Startup-config</td><td>copy tftp startup-config <IP address | hostname> <source filename></td></tr>
<tr><td></td><td>Software image</td><td>copy tftp flash <IP address | hostname> <source filename> <primary | secondary></td></tr>
</table>

Job Aid 4-5: Continued

Copy from:	To:	Command
SFTP server	Config file in flash	copy sftp config <dest filename> {user <name> <IP address | hostname> | <username>@<IP address | hostname> | <IP address | hostname>} <source filename>
	Startup-config	copy sftp startup-config {user <name> <IP address | hostname> | <username>@<IP address | hostname> | <IP address | hostname>} <source filename>
	Software image	copy sftp flash {user <name> <IP address | hostname> | <username>@<IP address | hostname> | <IP address | hostname>} <source filename> <primary | secondary>
USB	Config file in flash	copy usb config <source filename> <dest filename>
	Startup-config	copy usb startup-config <source filename>
	Software image	copy usb flash <source filename> <primary | secondary>
Software	Primary software (Secondary software copies to primary)	copy flash flash primary
	Secondary software (Primary software copies to secondary)	copy flash flash secondary
	TFTP server	copy flash tftp <IP address> <dest filename> <primary | secondary>
	SFTP server	copy flash sftp {user <name> <IP address | hostname> | <username>@<IP address | hostname> | <IP address | hostname>} <dest filename> <primary | secondary>
	USB	copy flash usb <dest filename> <primary | secondary>

Job Aid 4-6: Manage ArubaOS switch config files

Description	Command
Delete a config file.	`delete <filename>`
Rename a config file.	`rename config <filename> <new filename>`
Copy a config file.	`copy config <filename> <destination>` *See Job Aid 5 for more details on copying to various destinations.
View the names of config files; view which files are active and used as startup-config with various software.	`show config files`
View the contents of a config file.	`show config <filename>`
Choose the config file used when the switch boots from primary software, secondary software, or both (no keyword).	`startup-default [primary	secondary] config <filename>` *You must use the **boot** command for the change to take effect.

5 VLANs

EXAM OBJECTIVES

✓ Explain use cases for VLANs

✓ Configure port-based VLANs on ArubaOS switches, using appropriate tagging

✓ Implement basic IP routing between directly connected VLANs or links

ASSUMED KNOWLEDGE

Before reading this chapter, you should have a basic understanding of:

- IP addressing
- Open Systems Interconnection (OSI) model
- Local area networks (LANs)

INTRODUCTION

Virtual LANs (VLANs) help you to control endpoints and their traffic, allowing you to place them in isolated groups according to considerations such as device type, user identity, and location. In this chapter, you learn how to set up VLANs on ArubaOS switches.

Using VLANs to isolate communications

VLANs allow different endpoints connected to the same switch to reside in different networks, protecting the devices in different VLANs from each other's broadcasts and helping you to control communications between them.

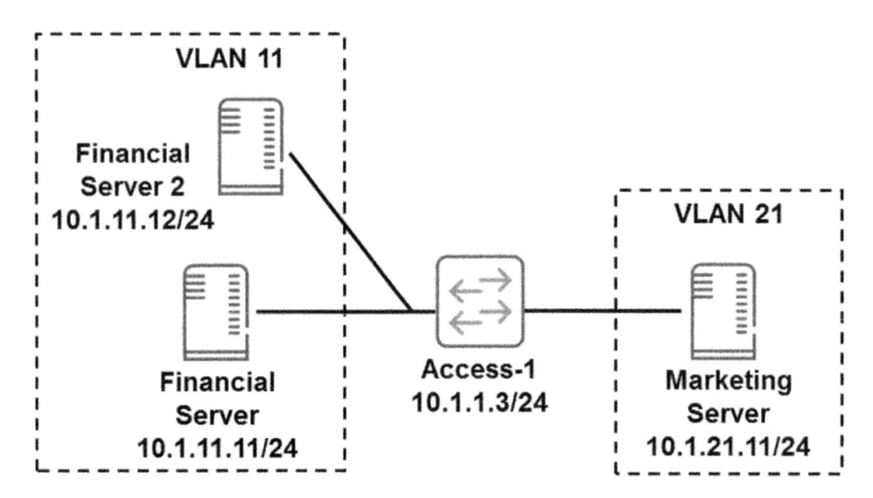

Figure 5-1: Using VLANs to isolate communications

Up to this point, your switches have all been part of a single VLAN, VLAN 1. They have IP addresses on this VLAN, which you can use to access the switches' management interfaces. In the first activity for this chapter, you will place the Windows server in a new VLAN, VLAN 11. The server will then be in a different VLAN from the switches' management addresses. It will also be part of a different subnet, 10.1.11.0/24.

You could add additional VLANs on the same switch to isolate communications between devices used for different purposes. For example, if you had another server that belonged to a different department, you could place that server in VLAN 21.

A VLAN is a Layer 2 concept; it defines the broadcast domain. The switch switches frames that arrive on VLAN 11 only out other ports that belong to VLAN 11. VLAN 11 broadcasts are only flooded out VLAN 11 ports, and so on.

A subnet is a Layer 3 concept. Devices in the same subnet can communicate without the need for routing. Nonetheless, a VLAN and a subnet have a close correspondence. Typically, a VLAN is associated with one subnet, and a subnet is associated with one VLAN, as VLAN 11 is associated with 10.1.11.0/24 in this example.

A switch can support a VLAN without having an IP address on that VLAN. However, if the switch does have an IP address on a VLAN, the subnet for that address becomes linked to that VLAN. For example, you cannot assign IP address 10.1.1.1/24 on VLAN 1 and IP address 10.1.1.2/24 on VLAN 2. The 10.1.1.0/24 subnet belongs to the VLAN 1 interface uniquely. (On the other hand, you can associate more than one subnet with a VLAN through multinetting; however, multinetting is used less typically and is not covered in this guide.)

Later in this chapter, you will explore the implications of moving devices to another VLAN and subnet in more depth.

Assigning an endpoint to a VLAN

When you assign an endpoint to a VLAN, you are actually assigning that endpoint's *traffic* to a particular VLAN. There are several ways to make this assignment. The most common one is a port-based VLAN assignment.

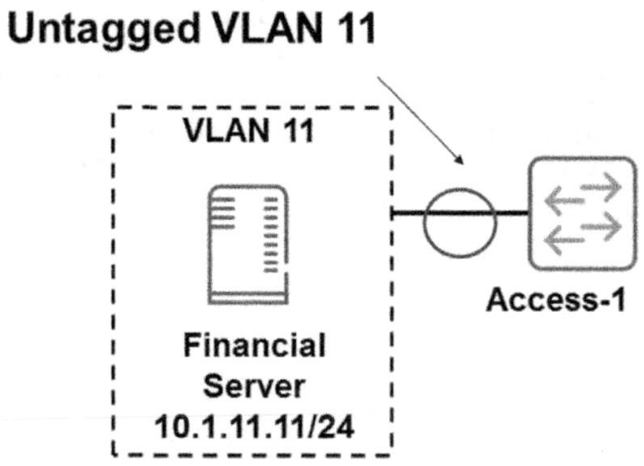

Untagged VLAN 11

VLAN 11

Financial Server 10.1.11.11/24

Access-1

Figure 5-2: Assigning an endpoint to a VLAN

On an ArubaOS switch, you make the port that connects to the endpoint an untagged member of the desired VLAN. Then all traffic that arrives on that port is assigned to that VLAN. Actually, all *untagged* traffic is assigned to this VLAN, and tagged traffic is dropped. You will learn about tagging in a moment. For now, just understand that Ethernet frames are untagged unless a device takes specific steps otherwise.

Endpoints such as desktops and laptops generally send untagged traffic. (Some servers send traffic in multiple VLANs, particularly when they are virtualized servers. You'll learn more about setting up a link to carry multiple VLANs in a moment.)

Extending the VLAN across multiple switches

You might want to place endpoints that connect to different switches in the same VLAN. For example, in Figure 5-3, you can place the server and client in VLAN 11 so that they can communicate with each other at Layer 2. Or a company might have servers and endpoints in the financial department in different locations within the building. Extending the VLAN across multiple switches allows all of those devices to remain in the same subnet regardless of location.

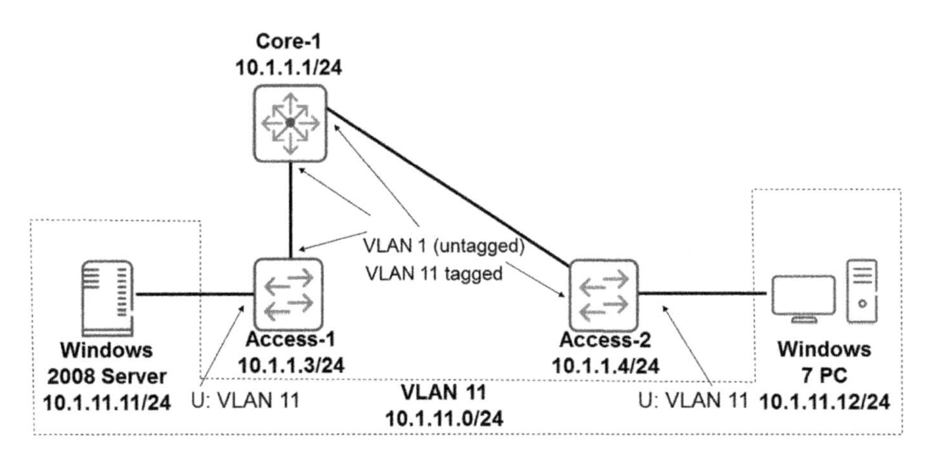

Figure 5-3: Extending the VLAN across multiple switches

You already learned how to assign the server's port to VLAN 11 on Access-1 and the client's port to VLAN 11 on Access-2. But, in order for the server and client to reach each other, you also need to extend the VLAN across the switch topology. The switches do not necessarily need IP addresses in this VLAN, but they need to *support* the VLAN so that the client and server can communicate.

Examine the switch-to-switch links in this topology. They already support VLAN 1, which the switches still use for their management IP addresses. Now they also need to support VLAN 11. In fact, switch-to-switch links often need to support multiple VLANs so that several different types of devices can connect to switches across the LAN.

How can you set up the links to support multiple VLANs? Do you need to add a link for each new VLAN? The cost of that solution would be prohibitively high.

Fortunately, you can set up a single switch-to-switch link to carry multiple VLANs. You simply need to configure the switch ports to distinguish traffic in one VLAN from another. The most common way is through the IEEE 802.1Q standard for VLAN tagging.

This standard defines a 4-byte field that can be inserted into an Ethernet frame. This field allows each Ethernet frame to be identified as part of a particular VLAN. An Ethernet frame can carry a payload of 1500 bytes. An untagged Ethernet frame has a maximum size of 1518 bytes while a tagged frame has a maximum size of 1522 bytes.

Because the VLAN ID component in the tag is 12 bits, valid VLAN IDs range from 1 to 4094. (VLAN 0 and VLAN 4095 are reserved and not permitted as valid VLAN IDs.)

Note

You can enable a switch to carry jumbo frames over 1522 bytes. On ArubaOS switches, jumbo frames are not supported by default, but you can enable them on VLANs.

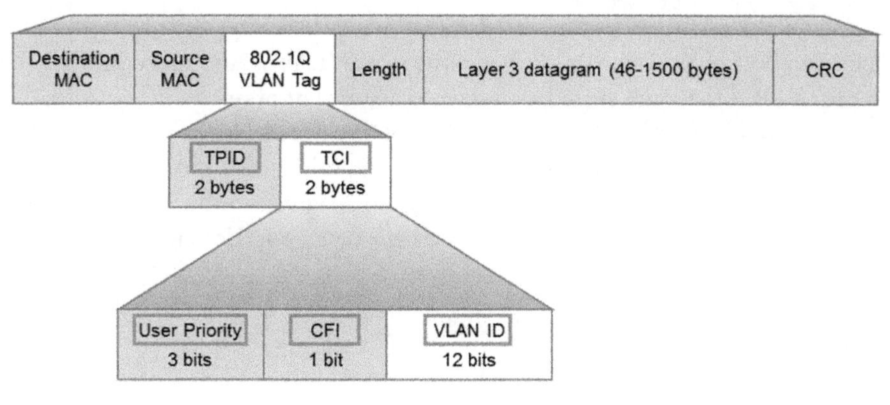

Figure 5-4: 802.1Q field (VLAN tag)

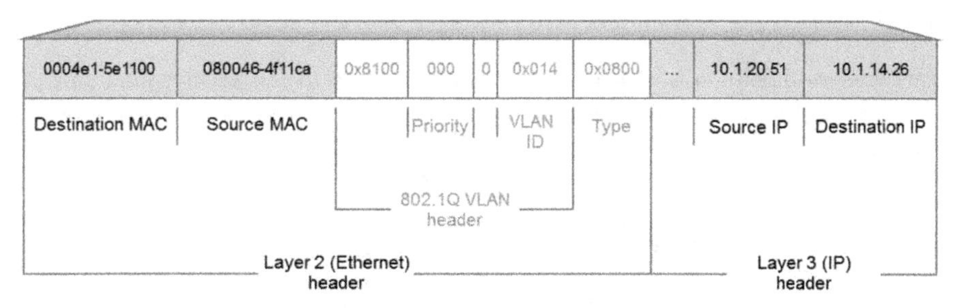

Figure 5-5: Example Ethernet header

The 802.1Q field is often called a VLAN tag, and when the 802.1Q field is inserted into an Ethernet frame, the frame is "tagged." In other words, it has VLAN ID information in it. If an Ethernet frame does not include this field, the frame is "untagged:" it does not give any information about the VLAN ID. The switch must classify the traffic on its own—for example, using a port-based assignment, as you learned earlier.

Using tagging to support multiple VLANs

Setting up a switch port to support VLAN tagging is simple.

On an ArubaOS port, you can simply add tagged VLAN assignments to any port. The port is then allowed to send and receive traffic that has a tag with that specific VLAN ID (remember: tag is another term for the 802.1Q field). The port drops tagged traffic with a VLAN ID that is not tagged on the port.

A port can support multiple tagged VLANs. It can also continue to support a *single untagged* VLAN assignment. You can remove the untagged VLAN assignment if the port has at least one tagged assignment. In that case, the port supports only tagged traffic and drops untagged traffic. If you try to remove the untagged assignment without giving the port a tagged assignment first, you receive an error. A port must be a member of at least one VLAN.

As shown in Figure 5-3, the VLAN assignments **must** match on both sides of a link. Make sure that the switch ports on both sides of the link support the same VLANs. Also make sure that both switch ports assign each VLAN the same tagged or untagged status. For example, if one switch port assigns VLAN 1 as untagged and VLAN 11 as tagged, but the connected switch port assigns VLAN 11 as untagged and VLAN 1 as tagged, connectivity in both VLANs is disrupted between these switches. When one switch sends untagged traffic, the other switch assigns it to the wrong VLAN. When one switch sends tagged traffic, the other switch drops it because its port does not accept that tag.

VLAN uses

Take a moment to consider two questions.

1. What are benefits of placing endpoints in different broadcast domains/subnets (in other words, in VLANs)?
2. What are benefits of using VLANs to create broadcast domains/subnets instead of the traditional method? The traditional method required a router to separate devices in different subnets, as shown below.

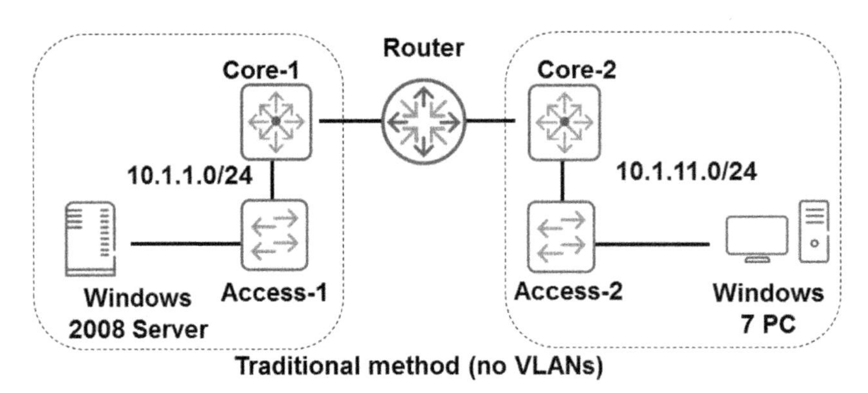

Figure 5-6: Traditional method for dividing subnets

Answers to VLAN uses

Placing endpoints in different VLANs has several benefits:

- You can improve security. For example, you can place endpoints in a separate VLAN from the VLAN on which network infrastructure devices have IP addresses. To reach the devices, they must communicate at Layer 3 through a Layer 3 switch or router. At that point, the traffic can be better controlled.
- You can place devices in different VLANs whenever you want to treat the devices in a different way and set up controls between their communications. By placing the server and the client in

different VLANs, you ensure that they have IP addresses with different ranges and that the clients' traffic must pass through a router to reach the server. Access control lists (ACLs) can control which clients in which VLANs can reach which servers in which VLANs. To help you to impose these controls, you might assign different users to different VLANs based on identity or identity and endpoint type. (You will look at a design in which the PC and server are in different VLANs later in this chapter.)

- Assigning different endpoints to different VLANs also helps to create reasonably sized broadcast domains, which in turn improves performance.
- Broadcast traffic is confined within the VLAN. Isolating the broadcasts can help to reduce congestion and CPU usage on individual devices. In this way, you can use bandwidth more efficiently and improve network performance. VLANs can also help to confine issues (such as excessive broadcasts) to a smaller number of devices. Therefore, you might also assign different endpoints to different VLANs based on their location. Using VLANs rather than routers to create broadcast domains provides additional benefits. You have better flexibility in defining the broadcast domains. For example, in this scenario, you want to place endpoints on ProVision-1 and ProVision-2 in the same VLAN, even though they are in different locations. VLANs let you do so. You'll see how in more detail as the chapter progresses. You could also create more than one broadcast domain on a switch. For example, in a later scenario, you will place servers and clients in different VLANs. You might have a switch that connects to both types of devices, and you can use VLANs to isolate them. Thus network construction and maintenance is much easier and more flexible.

Configure a VLAN

You will now set up the topology shown in the figure to create VLANs on the ArubaOS switches and implement tagged and untagged ports on ArubaOS switches.

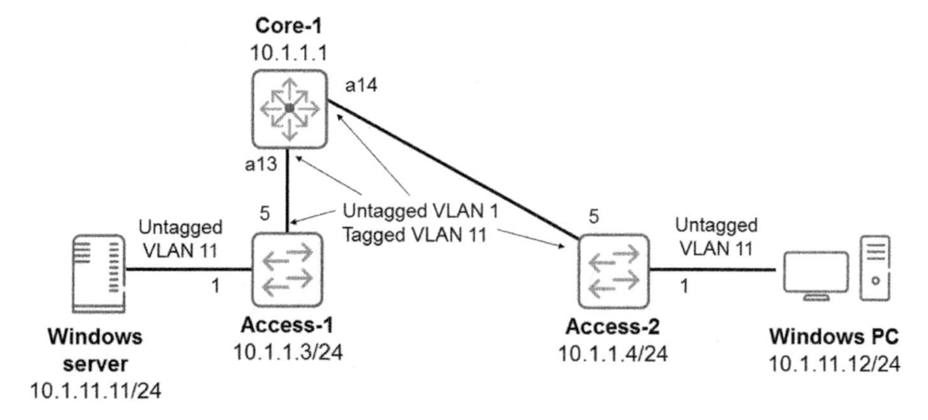

Figure 5-7: Configure a VLAN

Table 5-1 has an updated IP addressing scheme based on the VLANs. You will move the server and Windows 7 PC to VLAN 11. The switches keep their IP addresses in VLAN 1.

Table 5-1: VLAN IP addressing scheme

Switch	VLAN 1	VLAN 11
Core-1	10.1.1.1	
Access-1	10.1.1.3	
Access-2	10.1.1.4	
Windows 2008		10.1.11.11
Windows 7		DHCP Typically 10.1.11.12

Task 1: Assign the server to VLAN 11

In this task, you will assign the server to VLAN 11.

Access-1

1. Access a terminal session with Access-1.
2. Define port 1 (the Windows 2008 server connection) as an untagged member of VLAN 11.

```
Access-1(config)# vlan 11
Access-1(vlan-11)# untagged 1
Access-1(vlan-11)# exit
```

3. Verify that VLAN 11 has the interface as an untagged member.

```
Access-1(config)# show vlan 11
 Status and Counters - VLAN Information - VLAN 11
  VLAN ID : 11
  Name : VLAN11
  Status : Port-based
  Voice : No
  Jumbo : No
  Private VLAN : none
```

```
Associated Primary VID : none

Associated Secondary VIDs : none

Port Information Mode     Unknown VLAN Status
---------------- -------- ------------ ----------
1                Untagged Learn        Up
```

Windows server

4. Access the Windows server desktop and configure its IP address.
   a. Access the Properties window.
   b. Select Internet Protocol version 4 (TCP/IPv4) and click **Properties**.

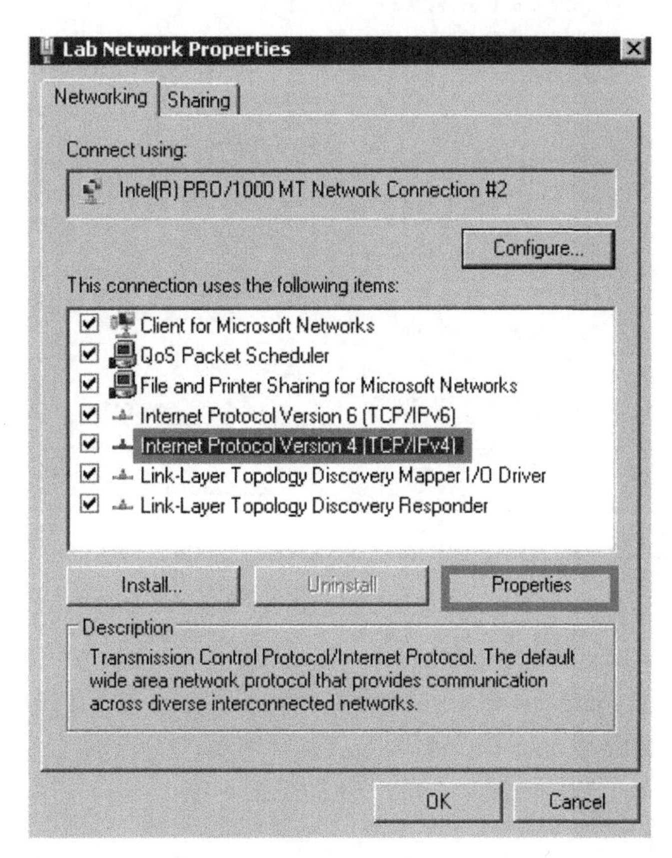

Figure 5-8: Select Internet Protocol version 4 (TCP/IPv4)

c. Enter these settings (the gateway setting will be necessary later):

 – IP address = 10.1.11.11

 – Mask = 255.255.255.0

 – Gateway = 10.1.11.1

d. Make sure that the DNS server address is set to 127.0.0.1.

e. Click **OK**.

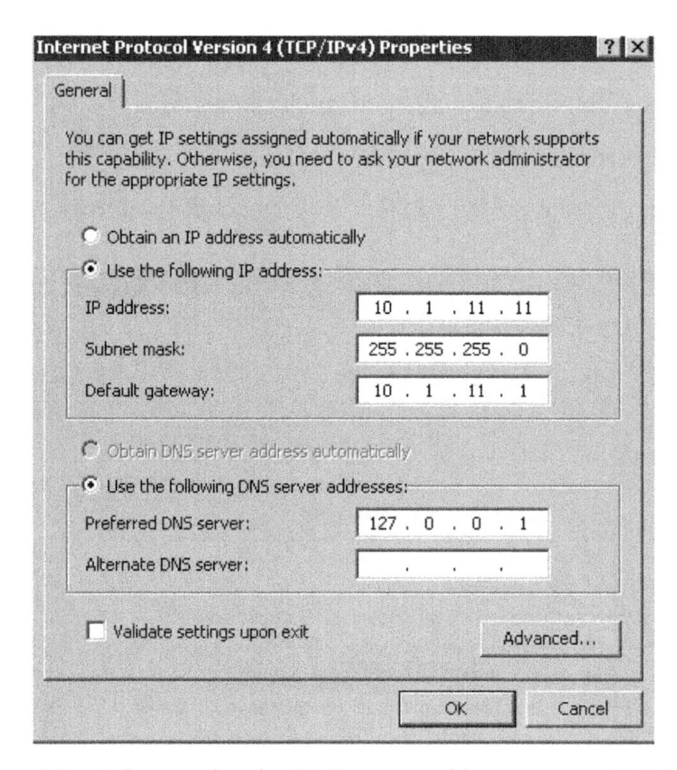

Figure 5-9: Make sure that the DNS server address is set to 127.0.0.1

f. Click **Close**.

5. Open a command prompt on the server. Validate that you have set the IP address correctly.

```
ipconfig
```

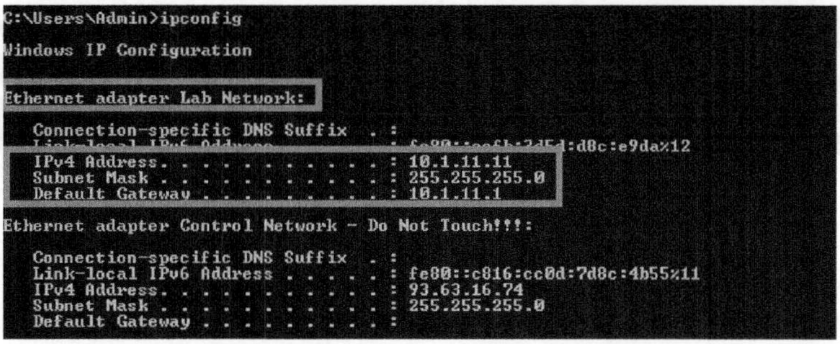

```
C:\Users\Admin>ipconfig

Windows IP Configuration

Ethernet adapter Lab Network:

   Connection-specific DNS Suffix  . :
   Link-local IPv6 Address . . . . . : fe80::[illegible]:d8c:e9da%12
   IPv4 Address. . . . . . . . . . . : 10.1.11.11
   Subnet Mask . . . . . . . . . . . : 255.255.255.0
   Default Gateway . . . . . . . . . : 10.1.11.1

Ethernet adapter Control Network - Do Not Touch!!!:

   Connection-specific DNS Suffix  . :
   Link-local IPv6 Address . . . . . : fe80::c816:cc0d:7d8c:4b55%11
   IPv4 Address. . . . . . . . . . . : 93.63.16.74
   Subnet Mask . . . . . . . . . . . : 255.255.255.0
   Default Gateway . . . . . . . . . :
```

Figure 5-10: Validate that you have set the IP address correctly

Task 2: Assign the Windows PC to VLAN 11

In this task, you will assign the Windows PC to VLAN 11.

Access-2

1. Access a terminal session with Access-2.
2. Define port 1 (which connects to the Windows PC) as an untagged member of VLAN 11.

```
Access-2(config)# vlan 11 untagged 1
```

3. View the VLANs that are configured on interface 1, and verify that the untagged VLAN is 11.

```
Access-2(config)# show vlan port 1 detail

 Status and Counters - VLAN Information - for ports 1

  VLAN ID Name                 | Status     Voice Jumbo Mode
  ------- -------------------- + ---------- ----- ----- --------
  11      VLAN11               | Port-based No    No    Untagged
```

Windows client

4. Access the Windows 7 PC desktop.
5. Open a command prompt on the client. Attempt to renew the DHCP address. (The server is providing DHCP services.)

```
ipconfig/renew
```

```
C:\Users\Admin>ipconfig/renew

Windows IP Configuration

An error occurred while renewing interface Lab Network : unable to contact your
DHCP server. Request has timed out.
An error occurred while releasing interface Loopback Pseudo-Interface 1 : The sy
stem cannot find the file specified.
```

Figure 5-11: Attempt to renew the DHCP address

As you see, the attempt fails because the client cannot reach the server. Adding a VLAN to the edge ports is not enough to establish connectivity. You need to extend the VLAN on the switch-to-switch links.

Task 3: Extending connectivity for VLAN 11

In this task, you will set up VLAN 11 on the switch-to-switch links. By the end of the task, the client and server should be able to reach each other.

Access-1

1. Access the terminal session for the Access-1 switch. Define the interface that connects to Core-1 as a tagged member of VLAN 11. (You should already be in the VLAN 11 context.)

```
Access-1(config)# vlan 11 tagged 5
```

2. Verify the port membership for VLAN 11, including whether each port has a tagged or untagged membership. Notice that port statuses are UP, since the interfaces are currently up.

```
Access-1(config)# show vlan 11

  Status and Counters - VLAN Information - VLAN 11

  VLAN ID : 11

  Name : VLAN11

  Status : Port-based

  Voice : No

  Jumbo : No

  Private VLAN : none

  Associated Primary VID : none

  Associated Secondary VIDs : none
```

```
Port Information Mode     Unknown VLAN Status
---------------- -------- ------------ ----------
1                Untagged Learn        Up
5                Tagged   Learn        Up
```

Access-2

3. Access the terminal session for the Access-2 switch.
4. Define the port that connects to Core-1 as a tagged member of VLAN 11.

```
Access-2((config)# vlan 11 tagged 5
```

5. Verify the VLAN membership for port 5. Use the **detail** option to see the tagging.

```
Access-2(config)# show vlan port 5 detail
 Status and Counters - VLAN Information - for ports 5

  Port name: Core-1
  VLAN ID Name                 | Status       Voice Jumbo Mode
  ------- -------------------- + ---------- ----- ----- --------
  1       DEFAULT_VLAN         | Port-based No    No    Untagged
  11      VLAN11               | Port-based No    No    Tagged
```

Core-1

6. Access the terminal session for the Core-1 switch. Move to the global configuration mode.
7. Allow the ports that connect to Access-1 and Access-2 to carry tagged VLAN 11 traffic.

```
Core-1(config)# vlan 11 tagged A13,A15
```

8. Check the VLAN status using one of the commands that you learned earlier. You can specify multiple ports for the show vlan port command, but you must use the detail option to break out the output per port (otherwise, you see all VLANs that are assigned to any port in the list.)

Windows client

You have now extended VLAN 11 across your topology. The server and client should be able to reach each other. Validate now.

9. Access the client desktop and open a command prompt.
10. Renew the IP address.

ipconfig/release

ipconfig/renew

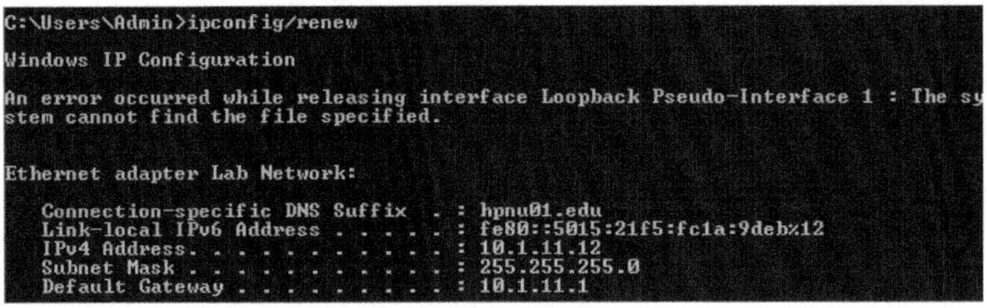

```
C:\Users\Admin>ipconfig/renew

Windows IP Configuration

An error occurred while releasing interface Loopback Pseudo-Interface 1 : The sy
stem cannot find the file specified.

Ethernet adapter Lab Network:

   Connection-specific DNS Suffix  . : hpnu01.edu
   Link-local IPv6 Address . . . . . : fe80::5015:21f5:fc1a:9deb%12
   IPv4 Address. . . . . . . . . . . : 10.1.11.12
   Subnet Mask . . . . . . . . . . . : 255.255.255.0
   Default Gateway . . . . . . . . . : 10.1.11.1
```

Figure 5-12: Attempt to renew the DHCP address

You might need to wait a moment and renew the address again. If the client receives an IP address in VLAN 11, save the configurations. If the process fails again, follow the instructions in the Troubleshooting tip section.

Task 4: Save

Save the configuration on each switch using the **write memory** command.

Troubleshooting tip

You have probably missed adding VLAN 11 as a tagged VLAN on one of the interfaces. You can use this tip to troubleshoot:

1. Assign each switch an IP address in VLAN 11.

Core-1(config)# **vlan 11 ip address 10.1.11.1/24**

Access-1(config)# **vlan 11 ip address 10.1.11.3/24**

Access-2(config)# **vlan 11 ip address 10.1.11.4/24**

2. Access the server desktop and open a command prompt. Ping the IP addresses is the order shown below.

```
ping 10.1.11.3

ping 10.1.11.1

ping 10.1.11.4
```

3. The first ping that fails indicates where the tagging is missing. For example, if the ping to Access-1 succeeds, but the ping to Core-1 fails, the tagging is missing on the link between these switches.
4. Check the tagging on both sides of the link using the **show vlan port <ID>** command.
5. Fix the problem.
6. Check the pings again and verify that you have caught all of the issues.
7. If the server can ping 10.1.11.4, check the VLAN membership on Access-2 port 1 (it should be untagged for VLAN 11).
8. If the VLAN is set up correctly and the client still cannot receive a DHCP address, you can assign the client a static IP address such as 10.1.11.12 255.255.255.0, gateway 10.1.11.1.
9. After the client and server have connectivity, remove the IP addresses from the switches.

```
Core-1(config)# no vlan 11 ip address

Access-1(config)# no vlan 11 ip address

Access-2(config)# no vlan 11 ip address
```

Table 5-2: Troubleshooting table

Failed ping	Link with the issue
10.1.11.3	Link between the server and Access-1
10.1.11.1	Link between the Access-1 and Core-1
10.1.11.4	Link between Core-1 and Access-2

10. After you have fixed the problem and removed the IP addresses, save the configuration on *every* switch (**write memory**).

Learning check

At this point, you should have a working topology in which the server and client have IP addresses on a new VLAN—VLAN 11, subnet 10.1.11.0/24. And they can ping each other on these addresses.

1. Which command can you use to see all interfaces that are part of a VLAN?

2. Which command or commands can you use to see all the VLANs that are carried on an interface?

3. Which option do you need to use to see whether the VLANs are tagged or untagged?

4. Explore how VLAN 11 is a new broadcast domain.
   a. On the server, clear the arp table.

```
netsh interface ip delete arpcache
```

   b. Run a Wireshark capture.
   c. Ping the Windows 7 PC from the server.

```
ping 10.1.11.12
```

   d. In the capture, you can see the ARP request for 10.1.11.12 from the server and the reply from the client.

```
No.  Time      Source            Destination       Protocol Length Info
 1 0.000000   Vmware_97:7c:9f   Broadcast         ARP      42 Who has 10.1.11.12?  Tell 10.1.11.11
 2 0.000684   Vmware_97:7c:77   Vmware_97:7c:9f   ARP      60 10.1.11.12 is at 00:50:56:97:7c:77
 3 0.000703   10.1.11.11        10.1.11.12        ICMP     74 Echo (ping) request  id=0x0001, seq=3/768, ttl=128 (reply in 4)
 4 0.001226   10.1.11.12        10.1.11.11        ICMP     74 Echo (ping) reply    id=0x0001, seq=3/768, ttl=128 (request in 3)
 5 1.005028   10.1.11.11        10.1.11.12        ICMP     74 Echo (ping) request  id=0x0001, seq=4/1024, ttl=128 (reply in 6)
 6 1.006053   10.1.11.12        10.1.11.11        ICMP     74 Echo (ping) reply    id=0x0001, seq=4/1024, ttl=128 (request in 5)
 7 2.019004   10.1.11.11        10.1.11.12        ICMP     74 Echo (ping) request  id=0x0001, seq=5/1280, ttl=128 (reply in 8)
 8 2.020089   10.1.11.12        10.1.11.11        ICMP     74 Echo (ping) reply    id=0x0001, seq=5/1280, ttl=128 (request in 7)
 9 3.033040   10.1.11.11        10.1.11.12        ICMP     74 Echo (ping) request  id=0x0001, seq=6/1536, ttl=128 (reply in 10)
10 3.034080   10.1.11.12        10.1.11.11        ICMP     74 Echo (ping) reply    id=0x0001, seq=6/1536, ttl=128 (request in 9)

Frame 1: 42 bytes on wire (336 bits), 42 bytes captured (336 bits)
Ethernet II, Src: Vmware_97:7c:9f (00:50:56:97:7c:9f), Dst: Broadcast (ff:ff:ff:ff:ff:ff)
Address Resolution Protocol (request)
  Hardware type: Ethernet (1)
  Protocol type: IP (0x0800)
  Hardware size: 6
  Protocol size: 4
  Opcode: request (1)
  Sender MAC address: Vmware_97:7c:9f (00:50:56:97:7c:9f)
  Sender IP address: 10.1.11.11 (10.1.11.11)
  Target MAC address: 00:00:00_00:00:00 (00:00:00:00:00:00)
  Target IP address: 10.1.11.12 (10.1.11.12)
```

Figure 5-13: ARP request

```
No.  Time      Source            Destination       Protocol Length Info
 1 0.000000   Vmware_97:7c:9f   Broadcast         ARP      42 Who has 10.1.11.12?  Tell 10.1.11.11
 2 0.000684   Vmware_97:7c:77   Vmware_97:7c:9f   ARP      60 10.1.11.12 is at 00:50:56:97:7c:77
 3 0.000703   10.1.11.11        10.1.11.12        ICMP     74 Echo (ping) request  id=0x0001, seq=3/768, ttl=128 (reply in 4)
 4 0.001226   10.1.11.12        10.1.11.11        ICMP     74 Echo (ping) reply    id=0x0001, seq=3/768, ttl=128 (request in 3)
 5 1.005028   10.1.11.11        10.1.11.12        ICMP     74 Echo (ping) request  id=0x0001, seq=4/1024, ttl=128 (reply in 6)
 6 1.006053   10.1.11.12        10.1.11.11        ICMP     74 Echo (ping) reply    id=0x0001, seq=4/1024, ttl=128 (request in 5)
 7 2.019004   10.1.11.11        10.1.11.12        ICMP     74 Echo (ping) request  id=0x0001, seq=5/1280, ttl=128 (reply in 8)
 8 2.020089   10.1.11.12        10.1.11.11        ICMP     74 Echo (ping) reply    id=0x0001, seq=5/1280, ttl=128 (request in 7)
 9 3.033040   10.1.11.11        10.1.11.12        ICMP     74 Echo (ping) request  id=0x0001, seq=6/1536, ttl=128 (reply in 10)
10 3.034080   10.1.11.12        10.1.11.11        ICMP     74 Echo (ping) reply    id=0x0001, seq=6/1536, ttl=128 (request in 9)

Frame 2: 60 bytes on wire (480 bits), 60 bytes captured (480 bits)
Ethernet II, Src: Vmware_97:7c:77 (00:50:56:97:7c:77), Dst: Vmware_97:7c:9f (00:50:56:97:7c:9f)
Address Resolution Protocol (reply)
  Hardware type: Ethernet (1)
  Protocol type: IP (0x0800)
  Hardware size: 6
  Protocol size: 4
  Opcode: reply (2)
  Sender MAC address: Vmware_97:7c:77 (00:50:56:97:7c:77)
  Sender IP address: 10.1.11.12 (10.1.11.12)
  Target MAC address: Vmware_97:7c:9f (00:50:56:97:7c:9f)
  Target IP address: 10.1.11.11 (10.1.11.11)
```

Figure 5-14: ARP reply

   e. Also view the ping (ICMP request and reply). Note the source and destination IP addresses and the source and destination MAC addresses, which belong to the server and client.
   f. View the ARP table on the server.

```
arp -a
```

```
C:\Users\Admin>arp -a

Interface: 93.63.16.74 --- 0xb
  Internet Address      Physical Address      Type
  93.63.16.1            d0-7e-28-3e-ff-75     dynamic

Interface: 10.1.11.11 --- 0xc
  Internet Address      Physical Address      Type
  10.1.11.12            00-50-56-97-7c-77     dynamic
```

Figure 5-15: Server ARP table

g. On Access-1, view the VLAN 11 MAC-address table. You will see that the switch has learned the port on which it can reach the server and the client.

```
Access-1# show mac-address vlan 11

Status and Counters - Address Table - VLAN 11

  MAC Address   Port

  ------------- ------

  005056-977c77 5

  005056-977c9f 1
```

h. View the VLAN 1 MAC address table. As you see, the tables for the different VLANs are isolated from each other.

```
Access-1(config)# show mac-address vlan 1

 Status and Counters - Address Table - VLAN 1

  MAC Address   Port

  ------------- ------

  d07e28-cec988 5
```

i. View the ARP table.

```
Access-1# show arp
```

The table is empty. Access-1 is not processing the IP traffic between the client and server. It is simply learning the MAC addresses in the VLAN.

j. You can see that the same is true on the Core-1 switch.

```
Core-1# show mac-address vlan 11

Status and Counters - Address Table - VLAN 11
```

```
  MAC Address    Port
  ------------- ------
  005056-977c77 a15
  005056-977c9f a13
Core-1# show arp
```

As you see, the ARP table is empty.

Answers to Learning check

1. Which command can you use to see all interfaces that are part of a VLAN?

   show vlan <vid>

2. Which command or commands can you use to see all the VLANs that are carried on an interface?

   show vlan port <int-id> [detail]

3. Which option do you need to use to see whether the VLANs are tagged or untagged?

   Use the **detail** option to see whether the VLANs are tagged or untagged.

Trace tagging across the topology

The figure shows a frame that is sent from the Windows 7 PC to the server in your topology. The path is divided into four segments.

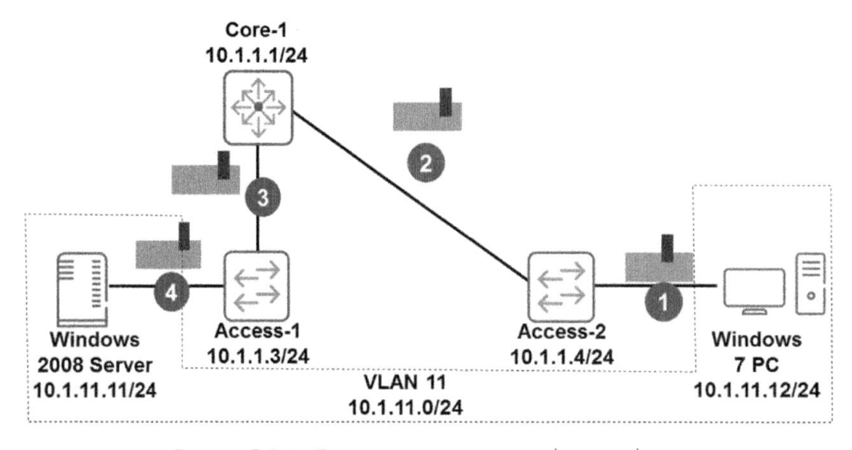

Figure 5-16: Trace tagging across the topology

The table below has a row for each segment. For each segment of the path, decide the VLAN to which the frame belongs and whether the frame is tagged or untagged. Then fill in the second column for the row that corresponds to that second. For example, for the first row, the PC transmits the frame to Access-2:

- To which VLAN does Access-2 assign the frame?
- Is the frame untagged and tagged?

Table 5-3: Tracing a frame across the topology

Segment	VLAN ID Tagged or untagged	Source MAC Address	Destination MAC Address
1—PC to Access-2		Client	Server
2—Access-2 to Core-1		Client	Server
4—Core-1 to Access-1		Client	Server
5—Access-1 to Server		Client	Server

Table 5-4: Answers for Tracing a frame across the topology

Segment	VLAN ID Tagged or untagged	Source MAC Address	Destination MAC Address
1—PC to Access-2	11 Untagged	Client	Server
2—Access-2 to Core-1	11 Tagged	Client	Server
3—Core-1 to Access-1	11 Tagged	Client	Server
4—Access-1 to Server	11 Untagged	Client	Server

Adding another VLAN

Sometimes you do not want all of the endpoints in the same VLAN. In the next activity, you will set up a scenario in which servers and user devices are in different VLANs:

- Servers in VLAN 11
- User devices in VLAN 12

Both Access-1 and Access-2 will support devices in both VLANs, and the VLANs will be extended across the switch topology.

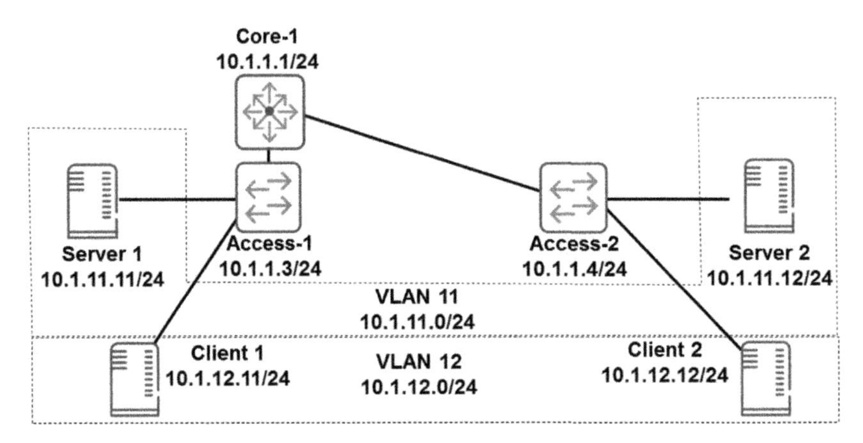

Figure 5-17: Adding another VLAN

Adding another VLAN: Logical topology

Logically, the VLAN 11 and VLAN 12 devices are separated as if by different routers. Physically, though, the devices are located wherever they need to be. And as you learned earlier, the ability to create a solution like this is a benefit of using VLANs.

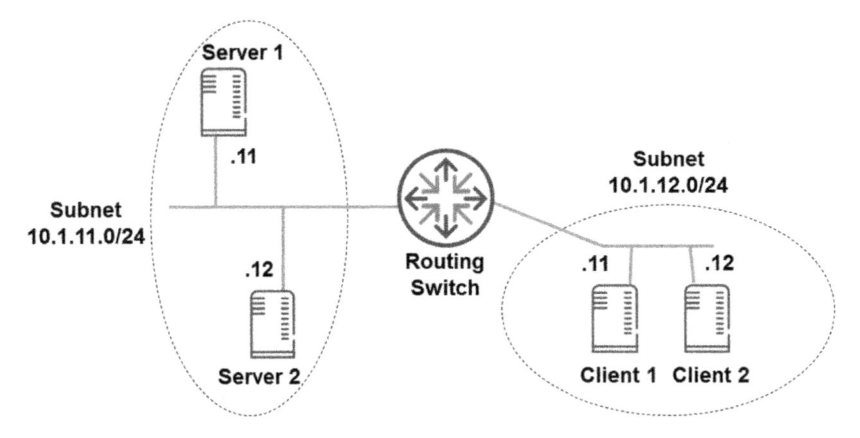

Figure 5-18: Logical topology

The table below summarizes the types of connections in a typical network and the type of VLAN assignments for each one.

Table 5-5: VLAN assignments for different types of connections

Types of connection	Type of VLAN assignment
Switch-to-switch	Untagged in one VLAN (VLAN 1 by default) Tagged in all other VLANs the connection must support
Client-to-switch	Untagged
Server-to-switch	Test network: untagged Servers that support 802.1Q can support multiple VLANs

Adding another VLAN

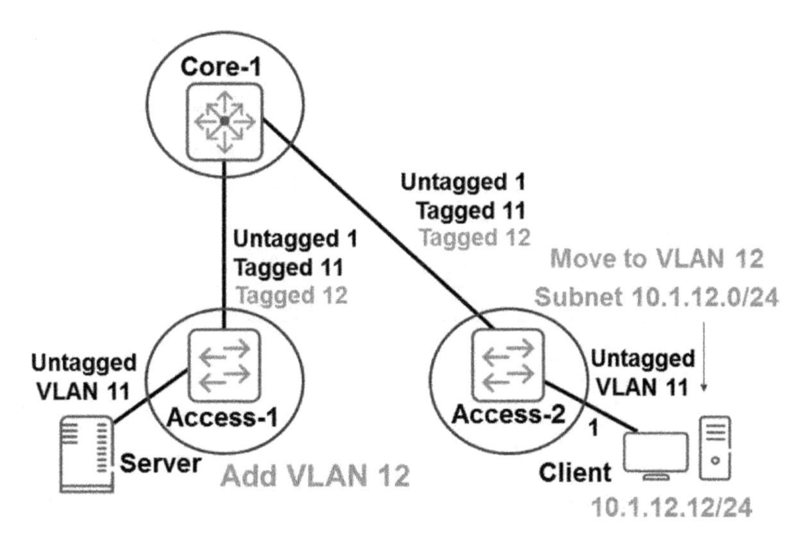

Figure 5-19: Adding another VLAN

In a moment, you will move the client to VLAN 12. You will also extend VLAN 12 across the entire topology. You are setting up a solution in which both Access-1 and Access-2 can connect to clients that should be in VLAN 12.

Make a plan

Begin by making a plan, answering the questions below (refer to the figure above).

1. How do you need to change the configuration for the Access-2 interface that connects to the Windows PC?
2. How do you need to change the configuration for the ArubaOS switch-to-switch interfaces?

3. The Windows PC is receiving its IP address with DHCP, and the server is providing DHCP services. The PC now needs an IP address in the new VLAN and subnet. What issue do you see with the PC receiving a DHCP address?

Answers for Make a plan

1. How do you need to change the configuration for the Access-2 interface that connects to the Windows PC?

   You must make this interface an untagged member of VLAN 12.

2. How do you need to change the configuration for the ArubaOS switch-to-switch interfaces?

   You must add VLAN 12 as a tagged membership. The interfaces continue to have the same untagged membership and to keep the tagged membership in VLAN 11.

3. The Windows PC is receiving its IP address with DHCP, and the server is providing DHCP services. The PC now needs an IP address in the new VLAN and subnet. What issue do you see with the PC receiving a DHCP address?

   DHCP clients such as the Windows PC send broadcasts to discover a DHCP server. However, the client and server are in different VLANs, so the client broadcasts cannot reach the server.

Setting up DHCP relay

Often networks include many VLANs and subnets. Deploying a DHCP server on each of those subnets would be expensive and time-consuming. Companies need a centralized DHCP server in a server subnet, which can provide IP addresses to clients in multiple different subnets. However, DHCP clients use DISCOVER broadcast to find a server from which to request an IP address, and these broadcasts cannot cross the subnet boundaries, so how can clients reach the server with their requests?

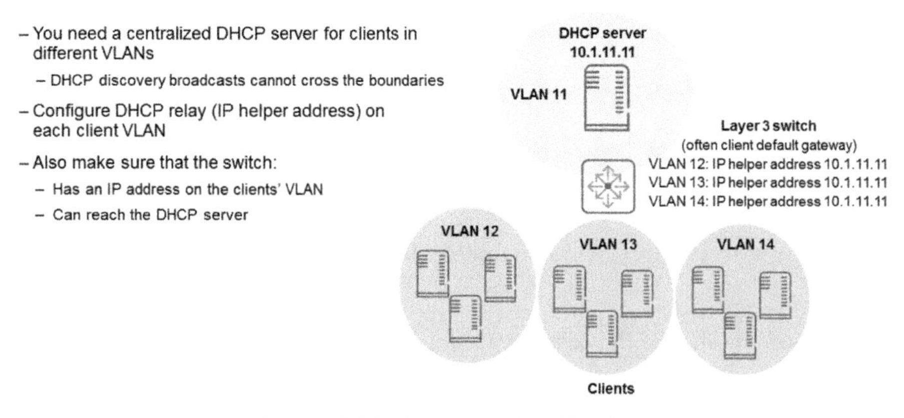

Figure 5-20: Setting up DHCP relay

DHCP relay provides the solution. A router or Layer 3 switch, generally the clients' default gateway, receives the DHCP DISCOVER broadcasts and routes them to the DHCP server as unicasts.

Now the company has a centralized solution for its entire network.

Core-1 will act as the DHCP relay for your solution. You enable DHCP relay by configuring an IP helper address, which specifies the DHCP server IP address, on each client VLAN. The ArubaOS switch must also meet these requirements:

- It has an IP address on the DHCP clients' VLAN.
- It has IP connectivity to the DHCP server (the proper routing is in place).

When the switch receives a client DHCP message on the VLAN, it relays the message to the helper address as a unicast with its own IP address on the VLAN as the source. The server uses the source IP address to select a scope (range of IP addresses) for the IP settings. For example, the switch relays a DHCP message from a client in VLAN 12 with source IP address 10.1.12.1. The server offers an IP address in the 10.1.12.0/24 range to the client. But, when a client in VLAN 13 sends request, the switch relays it with source IP address 10.1.13.1, and the server offers an address in the 10.1.13.0/24 subnet. The relayed requests and server responses include the client MAC address, so the switch can relay the server responses back to the client that sent the message.

Add VLAN 12

You are ready to add VLAN 12 to your network, using the plan that you just created, and as shown in Figure 5-21. The network topology includes VLANs 1 and 11.

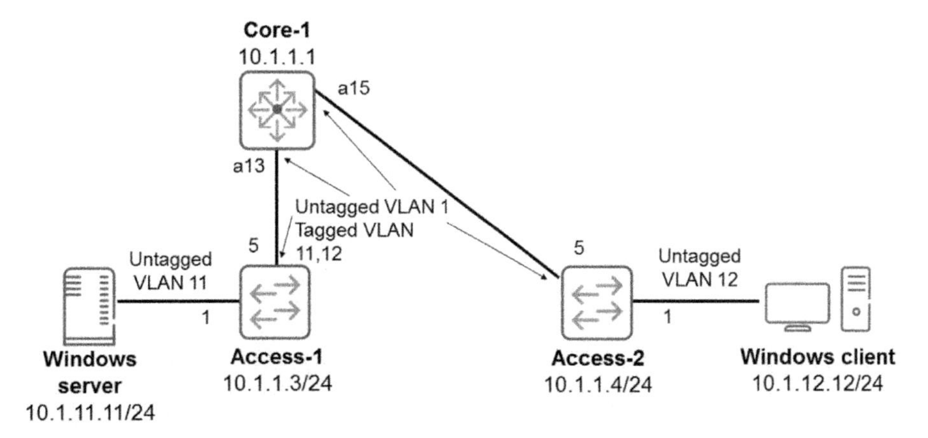

Figure 5-21: Add VLAN 12

The table below shows the VLANs and IP addressing that will be established by the end of the process.

Table 5-6: VLAN IP addressing scheme

Switch	VLAN 1	VLAN 11	VLAN 12
Core-1	10.1.1.1	10.1.11.1	
Access-1	10.1.1.3		
Access-2	10.1.1.4		
Windows 2008		10.1.11.11	
Windows 7			DHCP in 10.1.12.0/24, typically 10.1.12.12

Task 1: Add VLAN 12

You will now add VLAN 12 to the network, extending VLAN 12 across the complete topology, as shown in Figure 5-21.

Try to complete this task by completing each of the steps that you planned, looking up the correct command for the step in Table 5-6.

Access-2

Assign the Windows PC switch interface to VLAN 12 (untagged).

1. Access a terminal session with Access-2. Move to global configuration mode.
2. Create VLAN 12 and move to the VLAN 12 context.

```
Access-2(config)# vlan 12
```

3. Enter this command:

```
Access-2(vlan-12)# untagged 1
```

4. You must also configure the switch-to-switch interface to carry VLAN 12 as a tagged VLAN.

```
Access-2(vlan-12)# tagged 5
```

Core-1

5. Access a terminal session with Core-1. Move to global configuration mode.
6. Add VLAN 12 as a tagged VLAN on the ports that connect to Access-1 and Access-2.

```
Core-1(config)# vlan 12 tagged a13,a15
```

Access-1

7. Access the terminal session with Access-1. Move to global configuration mode.

8. Add VLAN 12 as a tagged VLAN on the port that connects to Core-1.

```
Access-1(config)# vlan 12 tagged 5
```

Task 2: Verify the VLAN topology

In this task, use the show commands that you practiced earlier to verify that the VLAN topology matches the figure below. Remember that you want to extend VLAN 11 and VLAN 12 across the topology even though Access-1 only connects to a VLAN 11 device and Access-2 only connects to a VLAN 12 device.

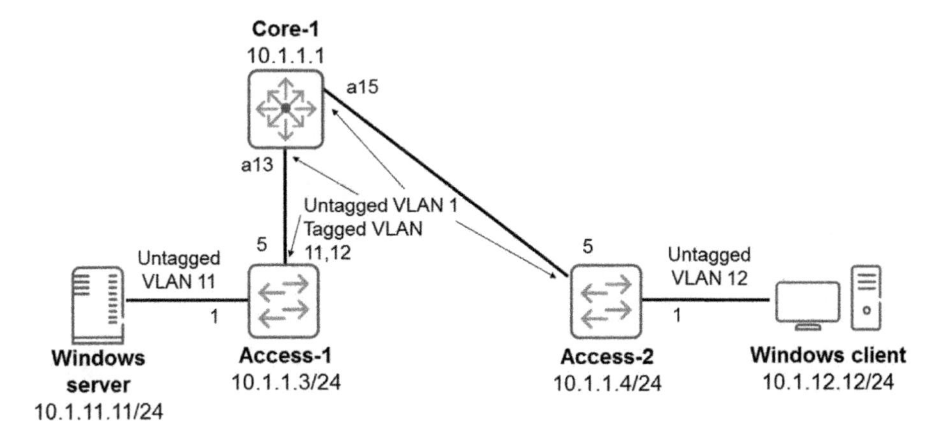

Figure 5-22: Task 2: Verify the VLAN topology

1. Try to remember the show commands or use ? to prompt you.

   The next page gives an example of the commands that you can use.

2. After you have verified the topology, move to Task 3.

```
Access-1# show vlan 12

 Status and Counters - VLAN Information - VLAN 12

  VLAN ID : 12

  <-output omitted->
```

```
  Port Information Mode     Unknown VLAN Status
  ---------------- -------- ------------ ----------
  5                Tagged   Learn        Up
Core-1# show vlan 12
 Status and Counters - VLAN Information - VLAN 12
  VLAN ID : 12
  <-output omitted->
  Port Information Mode     Unknown VLAN Status
  ---------------- -------- ------------ ----------
    A13              Tagged   Learn        Up
    A15              Tagged   Learn        Up

Access-2# show vlan 12
 Status and Counters - VLAN Information - VLAN 12
<-output omitted->
  Port Information Mode     Unknown VLAN Status
  ---------------- -------- ------------ ----------
  1                Untagged Learn        Up
  5                Tagged   Learn        Up
```

Task 3: Set an IP helper address

The client and server are now in different VLANs, so the clients' DHCP broadcasts cannot reach the server. You need to set up Core-1 to relay the DHCP messages between the client and server.

Core-1

1. Access the terminal session with Core-1.
2. Access VLAN 12.

```
Core-1(config)# vlan 12
```

3. The switch requires an IP address on this VLAN to perform DHCP relay. Assign it 10.1.12.1/24.

```
Core-1(vlan-12)# ip address 10.1.12.1/24
```

Note

Typically, the switch performing DHCP also has IP routing enabled on it. However, it can perform DHCP relay with routing disabled.

4. Specify the DHCP server IP address as the IP helper address for this VLAN.

```
Core-1(vlan-12)# ip helper-address 10.1.11.11

Core-1(vlan-12)# exit
```

5. The switch also needs to be able to reach the DHCP server. Give the switch an IP address on VLAN 11.

```
Core-1(config)# vlan 11 ip address 10.1.11.1/24
```

Windows PC

You'll now test the configuration.

6. Access the Windows PC desktop.
7. Open a command prompt on the client. Release the DHCP address and then renew it.

```
ipconfig/release

ipconfig/renew
```

8. Verify the success.

```
C:\Users\Admin>ipconfig/renew

Windows IP Configuration

An error occurred while releasing interface Loopback Pseudo-Interface 1 : The sy
stem cannot find the file specified.

Ethernet adapter Lab Network:

   Connection-specific DNS Suffix  . : hpnu01.edu
   Link-local IPv6 Address . . . . . : fe80::5015:21f5:fc1a:9deb%12
   IPv4 Address. . . . . . . . . . . : 10.1.12.12
   Subnet Mask . . . . . . . . . . . : 255.255.255.0
   Default Gateway . . . . . . . . . : 10.1.12.1
```

Figure 5-23: Verify renewing the DHCP address

Task 4: Save

Save the current configuration on each of the three switches using the **write memory** command.

Learning check

You will now answer some questions about the configuration that you planned and implemented.

1. Can the Windows 7 PC reach the server?
2. When you attempt to ping the Windows 7 PC from the server, for which IP address does the server send an ARP request?

   Here is how you can find this information:

   a. Access the server desktop.

   b. Open Wireshark. Capture traffic on the network interface.

   c. Open a command prompt and ping the PC:

```
ping 10.1.12.12
```

   d. Stop the capture in Wireshark and observe the results.

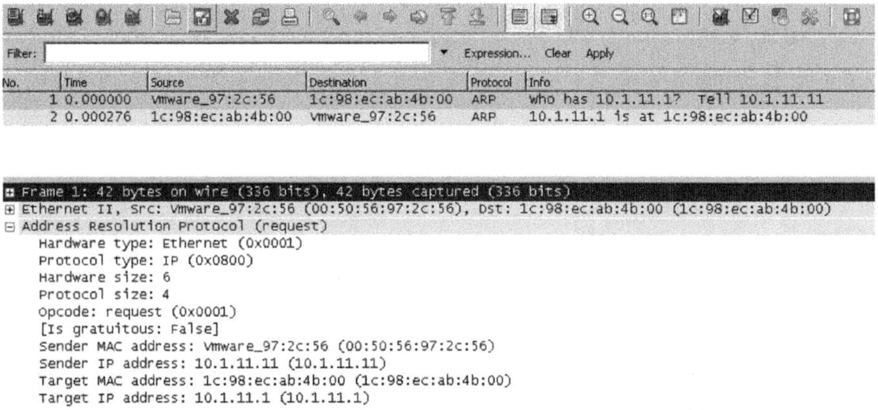

```
Filter:                                   Expression... Clear Apply
No.  Time      Source             Destination        Protocol Info
   1 0.000000  Vmware_97:2c:56    1c:98:ec:ab:4b:00  ARP      Who has 10.1.11.1? Tell 10.1.11.11
   2 0.000276  1c:98:ec:ab:4b:00  Vmware_97:2c:56    ARP      10.1.11.1 is at 1c:98:ec:ab:4b:00

Frame 1: 42 bytes on wire (336 bits), 42 bytes captured (336 bits)
Ethernet II, Src: Vmware_97:2c:56 (00:50:56:97:2c:56), Dst: 1c:98:ec:ab:4b:00 (1c:98:ec:ab:4b:00)
Address Resolution Protocol (request)
    Hardware type: Ethernet (0x0001)
    Protocol type: IP (0x0800)
    Hardware size: 6
    Protocol size: 4
    Opcode: request (0x0001)
    [Is gratuitous: False]
    Sender MAC address: Vmware_97:2c:56 (00:50:56:97:2c:56)
    Sender IP address: 10.1.11.11 (10.1.11.11)
    Target MAC address: 1c:98:ec:ab:4b:00 (1c:98:ec:ab:4b:00)
    Target IP address: 10.1.11.1 (10.1.11.1)
```

Figure 5-24: Server ARP request for an IP address outside of its subnet

Answers to Learning check

1. Can the Windows 7 PC reach the server?

   No. The PC and the server cannot communicate because they are in different subnets, and they need a router to route their traffic.

2. When you attempt to ping the Windows 7 PC from the server, for which IP address does the server send an ARP request?

   The server sent an ARP message, not to the client, but to its default gateway address. It did this because it recognized that the client IP address was not part of its own subnet; therefore, it attempted to find its default gateway, which is within its own subnet. (You set the default gateway address as 10.1.11.1 when you set the server's IP address.)

Currently, the server receives a response because you set this IP address on Core-1. However, Core-1 does not route the traffic because you have not yet enabled that feature.

Routing between VLAN 11 and 12

As you discovered, endpoints in different VLANs can no longer reach each other at Layer 2. Nor can they exchange broadcasts such as ARP messages. When a device needs to reach a device in another subnet, the traffic needs to be routed. Endpoints that cannot route their own traffic send the traffic to a default gateway for routing.

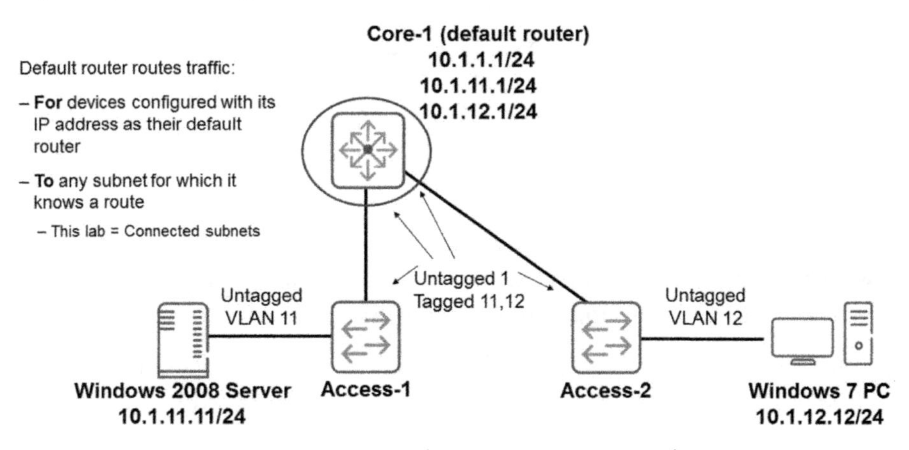

Figure 5-25: Routing between VLAN 11 and 12

When you attempted to ping the PC from the server, the server sent an ARP message, not to the client, but to its default gateway address. It did this because it recognized that the client IP address was not part of its own subnet; therefore, it attempted to find its default gateway, which is within its own subnet. An endpoints learns the gateway IP address for its server from its static or DHCP IP settings.

To allow your endpoints in VLAN 11 and VLAN 12 to communicate, you need to set up the default gateway in each VLAN. Core-1 acts as the default gateway. You've already completed the first requirement: assigning the default gateway an IP address for each subnet on the associated VLAN. The table reminds you of those addresses.

Table 5-7: VLAN IP addressing

VLAN	IP subnet	IP default gateway address
11	10.1.11.0/24	10.1.11.1/24
12	10.1.12.0/24	10.1.12.1/24

Core-1 can then reach devices in every subnet. However, it will **not** route traffic for other devices, such as endpoints configured to use it as their gateway. You must enable IP routing, which is disabled by default.

The switch will then route endpoint traffic to any network for which it knows a route. It automatically knows a route for any subnet on which it has an IP address. This type of subnet is called a direct or connected route. For this task, you simply want to route between VLAN 11, 10.1.11.0/24, and VLAN 12, 10.1.12.0/24. Core-1 will have an IP address on both of those VLANs, so you do not need to set up any additional routing.

Note

You will learn about setting up IP routing for more complicated topologies in "Chapter 9: IP Routing."

Configuring a default route for switches

In this section, you will up a default router for your switches. Access-1 and Access-2 each have only one IP address on one VLAN. They are acting essentially as Layer 2 devices. Just like the client and server, the switches need a default router if they need to communicate with devices outside of their VLAN. Core-1 will be their default router on VLAN 1 using its 10.1.1.1 IP address.

To complete the configuration on the ArubaOS switches, you create a default route with the default router IP address (10.1.1.1, in this scenario) as the next hop, as illustrated in the figure.

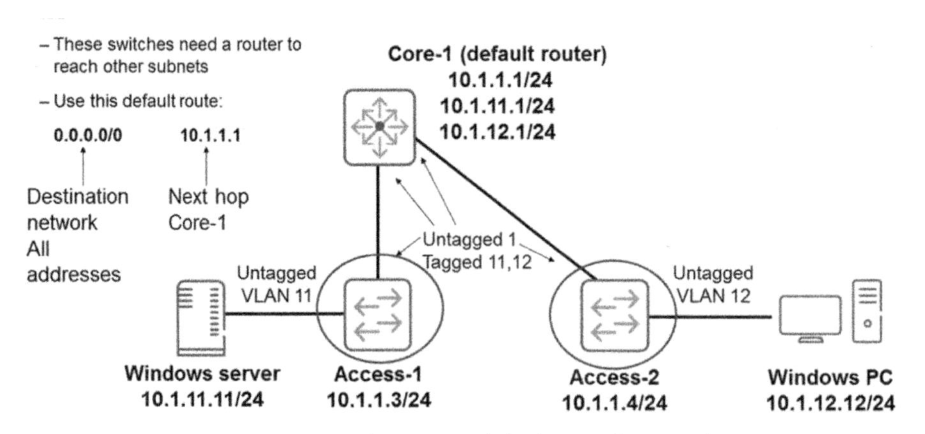

Figure 5-26: Configuring a default route for switches

The default route is a route to network 0.0.0.0/0. The 0 prefix length means that this network address matches *all* IP addresses. In other words, whenever the switch needs to reach any IP address that is not in its same subnet, the switch sends the traffic to the default router.

Note

When an ArubaOS switch has IP routing disabled (the default), you can use an alternative command to set the default router: **ip default-gateway <default router IP address>**

This command only works when IP routing is disabled. The **ip route** command for creating the default route, on the other hand, works whether or not routing is enabled, so you will use it in the task that you perform in the next section.

Set up basic routing

You will now set up Core-1 as the default router and confirm that the server and Windows 7 PC can reach each other. You will also configure default routes on the other switches.

After completing this task, you will be able to:

- Set up an ArubaOS switch as a default router for a VLAN
- Manage VLANs on an ArubaOS switch

The figure below shows the topology. This topology has three VLANs, and you will set up Core-1 to route for them.

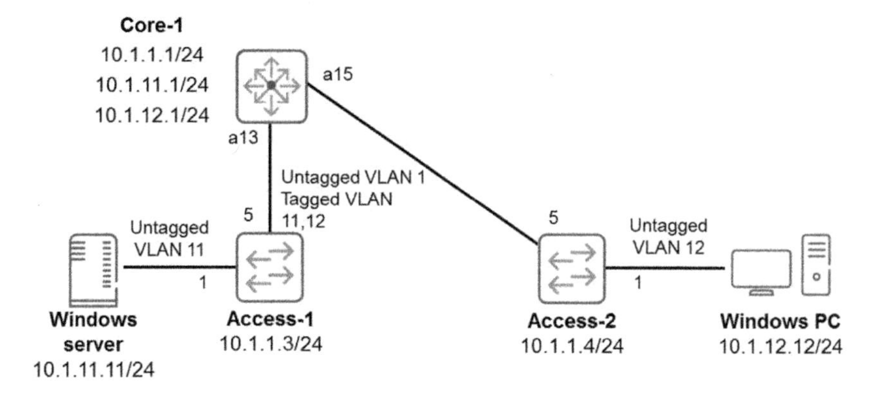

Figure 5-27: Set up basic routing

Task 1: Set up routing between VLANs 11 and 12

You will now configure Core-1 to act as the default router for VLANs 11 and 12.

Core-1

1. You have already assigned Core-1 the default router IP address on VLAN 11 and VLAN 12. On the Core-1 switch, ping the server and verify success.

```
Core-1# ping 10.1.11.11
```

2. Ping the client and verify success. (Note that you should use the DHCP IP address assigned to your client, which is often 10.1.12.12 but can be different. You can check this IP address by entering ipconfig in a command prompt on the Windows PC.)

```
Core-1# ping 10.1.12.12
```

Windows server

3. Access the Windows server desktop and open a command prompt.
4. Ping the VLAN 11 default gateway address and verify success.

```
ping 10.1.11.1
```

5. Ping the client.

```
ping 10.1.12.12
```

6. Why does the ping fail?

   Adding a VLAN to the switches is not enough to establish connectivity. You need to extend the VLAN on the switch-to-switch links.

Core-1

You must enable routing for Core-1 to route traffic that it receives from endpoints.

7. First see that IP routing is not enabled.

```
Core-1# show ip
Internet (IP) Service
 IP Routing : Disabled
<-output omitted->
```

8. Move to the global configuration mode.
9. Enable IP routing.

```
Core-1(config)# ip routing
```

10. Verify that IP routing is enabled.

```
Core-1(config)# show ip

Internet (IP) Service

 IP Routing : Enabled

<-output omitted->
```

Windows server

11. Access the Windows server desktop and ping the client again. The ping should now succeed.

Task 2: Set up default routes on the Layer 2 switches

You want the server to be able to reach all of the switches. Test to determine which switches the server can reach.

Windows server

1. You should still be in the command prompt on the server. Ping the following IP addresses and observe the results. (Some pings will be unsuccessful.)

```
ping 10.1.1.1

ping 10.1.1.3

ping 10.1.1.4
```

2. Why does the ping fail for the access layer switches?

   As you have learned, the switches need a default route to reach devices outside of their subnet. You will now configure the switches to use Core-1 as the next hop in this route; in other words, Core-1 is their default router.

Access-1

3. Access the Access-1 terminal session and move to the global configuration context.

4. Configure the default route.

```
Access-1(config)# ip route 0.0.0.0/0 10.1.1.1
```

Access-2

5. Complete the same steps on Access-2.

```
Access-2(config)# ip route 0.0.0.0/0 10.1.1.1
```

Windows server

6. Access a command prompt on the server. Ping the Access switch IP addresses and verify success.

```
ping 10.1.1.3
ping 10.1.1.4
```

Task 3: Explore routing

You will now perform several experiments to learn more about how the default router works and how to troubleshoot misconfigurations that you might encounter.

Windows server

You will use a number of trace routes throughout this and other tasks. A trace route sends a ping with a TTL of 1. The next routing hop in the path sends a TTL expired message back. The source device then increments the TTL by 1 and sends the ping again. Now the ping reaches the device that is two hops away, which sends a TTL expired message back. In this way, the tracing device can list each routing hop on the path to the destination. (Intervening devices at Layer 2 do not show up in the trace route.)

Instead of using the **tracert** command, you can use the **pathping** command (such as **pathping 10.1.12.12**), which adds the server itself to the beginning of the list.

1. Begin by performing a traceroute to the client at 10.1.12.12 so that you can see what a successful traceroute looks like.

```
tracert 10.1.12.12

tracing a route to client.hpnu01.edu [10.1.12.12]
over a maximum of 30 hops:
1 <1 ms  <1 ms  <1 ms  core-1-vlan11.hpnu01.edu [10.1.11.1]
2 <1 ms  <1 ms  <1 ms  client.hpnu01.edu [10.1.12.12]
```

Note

You can use the -d option (traceroute -d 10.1.12.12) to run the traceroute without looking up hostnames if these aren't available in your environment.

Core-1

As an experiment, you will now change the IP address on VLAN 11 and remove the IP address from VLAN 12. You will then practice using the **traceroute** command to detect the issue.

2. Access the terminal session with Core-1.
3. Remove the IP address from VLAN 11.

```
Core-1(config)# vlan 11
Core-1(vlan-11)# no ip address
```

4. Add a different IP address that is not the server's default gateway address.

```
Core-1(vlan-11)# ip address 10.1.11.2/24
```

Windows server

5. Access a command prompt on the Windows server. Trace a route to the Windows 7 PC.

```
tracert 10.1.12.12
Tracing route to 10.1.12.12 over a maximum of 30 hops
```

6. You might see that the traffic is routed to 10.1.11.2 and then to the client PC, or you might see that the traceroute times out before reaching the first hop. If the former, wait a minute and then perform the traceroute again. You should see that the traceroute times out before reaching the first hop.

Core-1

You will now correct the issue and create a new issue.

7. Return to the terminal session with Core-1.
8. Remove the IP address from VLAN 11 and add the correct default router address.

```
Core-1(vlan-11)# no ip address
Core-1(vlan-11)# ip address 10.1.11.1/24
```

9. Remove the IP address from VLAN 12.

```
Core-1(vlan-11)# exit
Core-1(config)# vlan 12
Core-1(vlan-12)# no ip address
```

Windows server

10. Return to the server command prompt.
11. Trace a route to the Windows PC.

```
tracert 10.1.12.12
Tracing route to pc.hpnu01.edu [10.1.12.12] over a maximum of 30 hops
1 <1 ms    <1 ms    <1 ms  core-1-vlan11.hpnu01.edu [10.1.11.1]
2 core-1-vlan11.hpnu01.edu [10.1.11.1] reports: Destination net
unreachable
```

Core-1

12. Return to the terminal session with Core-1.
13. Add an IP address in 10.1.12.0/24 to VLAN 12 so that Core-1 knows a route to this subnet. However, add the wrong IP address for the default gateway.

```
Core-1(vlan-12)# ip address 10.1.12.2/24
```

14. Ping the PC at 10.1.12.12 and verify that Core-1 can still reach it.

```
Core-1(vlan-12)# ping 10.1.12.12
```

Important

If the ping fails, open a command prompt on the PC and enter ipconfig/renew to renew the address.

Windows server

15. Trace a route to the Windows PC.

```
tracert 10.1.12.12
Tracing route to pc.hpnu01.edu [10.1.12.12] over a maximum of 30 hops
```

```
1       <1 ms    <1 ms    <1 ms  core-1-vlan11.hpnu01.edu [10.1.11.1]
2       *        *        *
```

16. Press <Ctrl+c> to break the traceroute.

Core-1

Fix the problem now.

17. Return to the terminal session with Core-1.

18. Remove the IP address from VLAN 12 and add the correct default router address.

```
Core-1(vlan-12)# no ip address
Core-1(vlan-12)# ip address 10.1.12.1/24
```

Windows server

19. Access a command prompt on the Windows server. Trace a route to the Windows 7 PC and verify success.

```
tracert 10.1.12.12
Tracing route to client.hpnu01.edu [10.1.12.120 over a maximum of 30 hops
1     <1 ms <1 ms <1 ms core-1-vlan11.hpnu01.edu [10.1.11.1]
2     <1 ms <1 ms <1 ms client.hpnu01.edu [10.1.12.12]
```

Task 4: Save

Save the current configuration on each of the switches using the **write memory** command.

Task 5: Practice managing VLANs on ArubaOS switches

The steps in this task will not alter the configuration for your switches, but will give you a chance to practice managing VLANs. You will begin by exploring the role of the primary and default VLAN. VLAN 1 is, by default, the default VLAN and the primary VLAN on ArubaOS switches.

Access-2

1. On Access-2, add VLAN 4000.

```
Access-2(config)# vlan 4000
Access-2(vlan-4000)# exit
```

2. Change the primary VLAN to 4000. VLAN 1 remains the default VLAN.

```
Access-2(config)# primary-vlan 4000
```

3. Define VLAN 13 as the untagged VLAN for an unused interface, such as interface 4.

```
Access-2(config)# vlan 13 untagged 4
```

4. Delete VLAN 13.

```
Access-2(config)# no vlan 13
```

5. Note the warning and enter **y**.
6. Discover the VLAN assigned to the interface.

```
Access-2(config)# show vlan port 4
 Status and Counters - VLAN Information - for ports 4
  VLAN ID Name                             | Status     Voice Jumbo
  ------- -------------------------------- + ---------- ----- -----
  1       DEFAULT_VLAN                     | Port-based No    No
```

7. As you see, the primary VLAN and default VLAN have different roles and changing the primary VLAN does not affect the default VLAN.
8. Assume that you want to change the port 3 VLAN assignment, which is currently untagged for LAN 1, to untagged VLAN 13. What is the correct procedure? Should you remove the current untagged assignment, or simply assign port 3 to VLAN 13?

   Answer: You should assign the port to VLAN 13. If you try to remove the current untagged assignment, you receive an error message that the port would be orphaned.

9. Try removing the current untagged VLAN assignment.

```
Access-2(config)# no vlan 1 untagged 3
```

10. As you see, you receive an error. To change the untagged VLAN assignment, simply enter the **vlan <ID> untagged <port ID>** command.
11. Return VLAN 1 to the primary VLAN.

```
Access-2(config)# primary-vlan 1
```

12. Delete VLAN 4000.

```
Access-2(config)# no vlan 4000
```

Explore the solution

As you discovered, it is very simple to set up a switch as the default gateway for multiple VLANs and enable routing between those VLANs. You will now explore the solution in a bit more depth.

13. Ping the client from the server.

```
ping 10.1.12.12
```

14. View the ARP table on the server.

```
arp -a
```

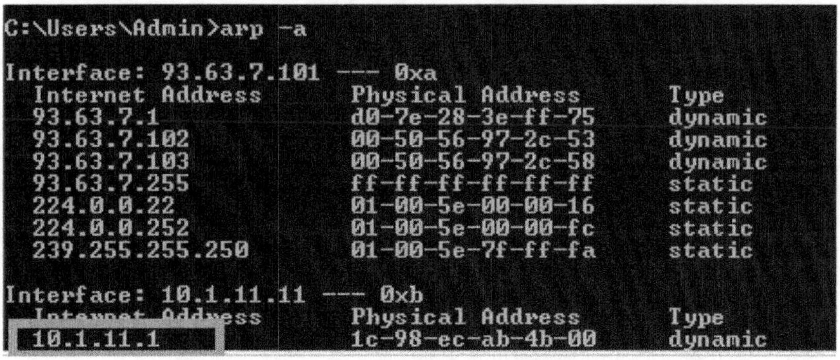

```
C:\Users\Admin>arp -a

Interface: 93.63.7.101 --- 0xa
  Internet Address      Physical Address      Type
  93.63.7.1             d0-7e-28-3e-ff-75     dynamic
  93.63.7.102           00-50-56-97-2c-53     dynamic
  93.63.7.103           00-50-56-97-2c-58     dynamic
  93.63.7.255           ff-ff-ff-ff-ff-ff     static
  224.0.0.22            01-00-5e-00-00-16     static
  224.0.0.252           01-00-5e-00-00-fc     static
  239.255.255.250       01-00-5e-7f-ff-fa     static

Interface: 10.1.11.11 --- 0xb
  Internet Address      Physical Address      Type
  10.1.11.1             1c-98-ec-ab-4b-00     dynamic
```

Figure 5-28: ARP table

As you see, the server only learns the MAC address for its default gateway, not for the client. But it still can reach the client because its default gateway routes the traffic for it.

15. View the ARP table on Core-1.

```
Core-1# show arp

 IP ARP table

  IP Address       MAC Address       Type    Port
  ---------------  ----------------- ------- ----
  10.1.11.11       005056-972c56     dynamic A13
  10.1.12.12       005056-972c54     dynamic A15
```

As you see, Core-1 learns ARP entries for devices in several VLANs because it is routing traffic for those devices.

16. View the IP routing table on Core-1.

```
Core-1(config)# show ip route

                                 IP Route Entries

 Destination      Gateway        VLAN Type       Sub-Type  Metric     Dist.
 ---------------  ------------- ---- -------- --------- ---------- -----
 10.1.1.0/24      DEFAULT_VLAN  1    connected           1          0
 10.1.11.0/24     VLAN11        11   connected           1          0
 10.1.12.0/24     VLAN12        12   connected           1          0
 127.0.0.0/8      reject             static              0          0
 127.0.0.1/32     lo0                connected           1          0
```

Core-1 currently has a simple IP routing table with only connected routes. You will learn more about interpreting a routing table in a later chapter. For now, understand that Core-1 can route traffic to any address that falls within these destination networks.

Learning check

1. What did you discover about the default VLAN?

2. What is the role of the primary VLAN?

3. What did you learn about assigning a port to a new untagged VLAN?

Answers to Learning check

1. What did you discover about the default VLAN?

   The default VLAN is always VLAN 1, and the ID cannot be changed. This is the VLAN to which all interfaces are assigned by default. When you deleted VLAN 13, interface 4 did not have a VLAN assigned to it anymore, so it was moved to the default VLAN, VLAN 1.

2. What is the role of the primary VLAN?

   The primary VLAN is the VLAN that the switch uses to run certain features and management processes. For example, the switch can receive DHCP settings that affect the switch globally on multiple VLANs. Such settings include a default gateway and time server IP address. In this case, the switch uses the global settings only on the primary VLAN. The primary VLAN is VLAN 1 by default, but you can change the primary VLAN ID.

3. What did you learn about assigning a port to a new untagged VLAN?

   You should simply assign the port to the new VLAN without first removing the current untagged assignment. If you try to remove the current untagged assignment, you receive an error message that the port would be orphaned.

Tracing a frame across the routed topology

You have changed the topology so that routing now occurs. You will now trace how a frame is forwarded across the routed topology.

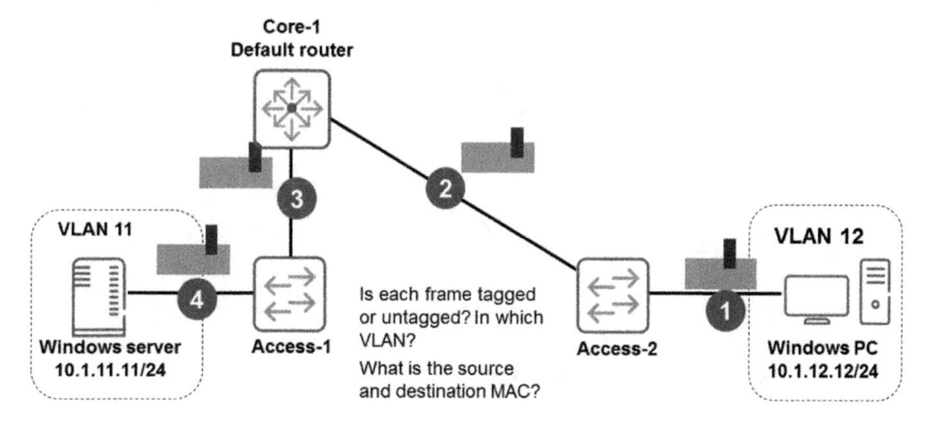

Figure 5-29: Tracing a frame across the routed topology

The figure divides the path from the client to the server into several segments, and the table below has a row for each segment.

In each row of the table, define:

- The frame's VLAN and whether the frame will be tagged using 802.1Q
- The frame's source MAC address
- The frame's destination MAC address

You do not need to write the specific MAC address. Just write: server, PC, VLAN 11 default gateway, or VLAN 12 default gateway.

Table 5-8: Tracing a frame across the routed topology

Segment	VLAN ID Tagged or untagged	Source MAC Address	Destination MAC Address
1—PC to Access-2			
2—Access-2 to Core-1			
3—Core-1 to Access-1			
4—Access-1 to Server			

Table 5-9: Answers to Tracing a frame across the routed topology

Segment	VLAN ID Tagged or untagged	Source MAC Address	Destination MAC Address
1—PC to Access-2	VLAN 12 Untagged	PC	VLAN 12 default gateway
2—Access-2 to Core-1	VLAN 12 Tagged	PC	VLAN 12 default gateway
3—Core-1 to Access-1	VLAN 11 Tagged	VLAN 11 default gateway	server
4—Access-1 to Server	VLAN 11 Untagged	VLAN 11 default gateway	server

Special VLAN types on ArubaOS switches

Sometimes a port needs to receive untagged traffic and sort that traffic into multiple VLANs. How can a port support multiple untagged VLANs? Without a tag, how can the switch determine to which VLAN a frame belongs? It would not be able to if it were only using port-based VLANs—the type of VLANs on which you have focused in this chapter. However, special VLAN types enable the switch to assign untagged traffic received on a port into multiple VLANs based on MAC address or protocol.

A port can receive untagged traffic in multiple VLANs by using non-port-based VLANs:

– MAC-based VLANs (some hardware only)

– Voice VLANs

– Protocol-based VLANs

Figure 5-30: Special VLAN types on ArubaOS switches

These VLAN types are briefly described below. The focus is on the use case; detailed configuration is beyond the scope of this book.

MAC-based VLAN (RADIUS-assigned VLANs for multiple endpoints)

Many ArubaOS switches support MAC-based VLANs. In the typical use case for MAC-based VLANs, a port connects to multiple downstream endpoints—perhaps through a less intelligent switch—and enforces authentication. The endpoints authenticate to a RADIUS server, which assigns them to VLANs based on the user identity, device type, or other criteria. The RADIUS server informs the switch of the assigned VLAN for each endpoint, and the switch port keeps track of all the source-MAC-address-to-VLAN mappings. In this way, the switch and centralized authentication server can work together to provide flexible, secure access for a variety of devices, including PCs, laptops, phones, and so on.

On ArubaOS switches that support it, this feature automatically takes effect when:

- The port enforces port-based authentication in MAC-based mode (has a client limit of one or above applied to it).
- Multiple devices authenticate to the port and receive different VLAN assignments from a RADIUS server.

Aruba 3800 Series and Aruba 3810 Series switches support this feature, as do version 3 and version 2 modules for Aruba 5400R zl2 switches. However, 5400R version 1 modules do not and neither do some older switch series. You can refer to the switch documentation to determine if its hardware supports the feature.

Protocol-based VLANs

ArubaOS switches can also classify traffic into a VLAN based on the traffic's protocol. You simply create a VLAN and then define the protocol for that VLAN. For example, you can define the protocol as IPv6 or IPX. Refer to your switch documentation for a list of supported protocols.

Defining the protocol on a VLAN transforms the VLAN into a protocol-based VLAN. (VLANs are port-based VLANs by default.) You can then assign the protocol-based VLAN to switch ports using the **untagged** and **tagged** commands as usual. However, unlike port-based VLANs, you can assign multiple untagged protocol-based VLANs to a port. The switch examines the network protocol (Ethernet frame type) for incoming frames and uses that to assign them to the correct VLAN. Untagged frames that do not match an untagged protocol-based VLAN on the port are assigned to the untagged port-based VLAN.

Voice VLANs (VoIP phones)

On ArubaOS switches, you can define a VLAN as a voice VLAN. You assign that VLAN to ports as normal, using a tagged or untagged assignment based on the type of traffic sent by the VoIP phone. The voice VLAN definition does *not* play a role in identifying traffic for the VLAN. The port assignment does that. Instead the voice VLAN helps the switch to apply the correct priority to traffic. It can also help the switch to communicate the correct VLAN ID to a VoIP phone using LLDP-MED, a protocol for communicating settings between switches and phones.

Summary

In this chapter, you learned why you would add VLANs to a topology and how to add those VLANs. You now understand how to assign edge ports to the correct VLAN, as well as how to extend a VLAN across switch-to-switch ports. You can use the appropriate ArubaOS tagged or untagged memberships and to complete these configurations.

Finally, you learned how to set up a switch as a default router in multiple VLANs so that it can route traffic between those VLANs.

– Adding VLANs to a topology
– Assigning tagged and untagged memberships appropriate
– Setting up routing between VLANs on a default gateway that supports both VLANs

Figure 5-31: Summary

Learning check

Answer these questions to assess what you have learned in this chapter.

1. What is the maximum number of usable VLANs that 802.1Q supports?
   a. 512
   b. 1,023
   c. 1,024
   d. 2,047
   e. 2,048
   f. 4,094
   g. 4,096
2. What ArubaOS command creates VLAN 10?

3. What ArubaOS command(s) assign interface 11 as an untagged member of VLAN 10?

4. What ArubaOS command(s) configure interface 12 to carry tagged traffic for VLANs 10 and 11?

Answers to Learning check

1. What is the maximum number of usable VLANs that 802.1Q supports?

    a. 512

    b. 1,023

    c. 1,024

    d. 2,047

    e. 2,048

    f. 4,094

    g. 4,096

    The VLAN ID can have up to 4096 values, two of which (0 and 4095) are reserved.

2. What ArubaOS switch command creates VLAN 10?

    vlan 10

3. What ArubaOS switch command(s) assign port 11 as an untagged member of VLAN 10?

    vlan 10 untagged 11

4. What ArubaOS command(s) configure interface 12 to carry tagged traffic for VLANs 10 and 11?

    vlan 10 tagged 12

    vlan 11 tagged 12

6 Spanning Tree

EXAM OBJECTIVES

✓ Understand and configure Rapid Spanning Tree Protocol (RSTP)

✓ Understand how Multiple Spanning Tree Protocol (MSTP) provides load-sharing and implement MSTP to do so

ASSUMED KNOWLEDGE

Before reading this chapter, you should have a basic understanding of:

- Virtual LANs (VLANs)
- Broadcast domains
- IP addressing

INTRODUCTION

This chapter introduces various spanning tree protocol standards, which help switches to manage redundant links without introducing loops and broadcast storms. You will focus on Rapid Spanning Tree Protocol (RSTP) and Multiple Spanning Tree Protocol (MSTP), both of which you will practice configuring via a series of tasks.

Issues adding redundant links to the topology

Networks deliver critical services to users. Failure of a network link may make the network unavailable, resulting in lost time or revenue. To protect a network against these failures, you can install redundant links. Redundant links help to ensure that a path continues to exist across the network even if one link or one switch fails.

For example, in the topology in Figure 6-1, Access-1 could continue to reach Core-1 if its direct link to Core-1 failed. It could also continue to reach the core if Core-1 failed entirely. (Of course, default router services would also have to be made redundant, but this is beyond the scope of this chapter, which focuses on link redundancy.)

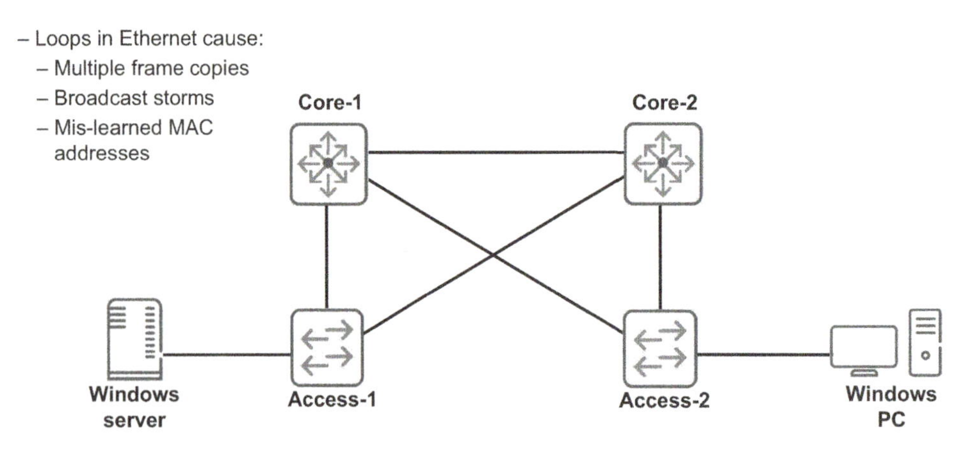

Figure 6-1: Issues adding redundant links to the topology

However, simply adding redundant physical links does not ensure that the switches can use the links correctly.

Adding redundant Layer 2 links without a protocol to manage the links results in network loops. These loops, in turn, create broadcast storms and mis-learned MAC addresses, making the network inaccessible. To function properly, an Ethernet network must have only *one* active pathway between any two devices. The following sections contain more information about the consequences of adding redundant links without properly setting them up.

Multiple frame copies

An Ethernet switch floods three kinds of frames: broadcast, multicast, and unknown destination unicast frames. While flooding is a necessary part of how Ethernet works, it can create serious performance issues in a Layer 2 network that has loops.

Consulting Figure 6-1, suppose that you have cleared the PC's ARP cache and then attempted to ping the server. To complete the ping, the PC would broadcast an Address Resolution Protocol (ARP) request for its default gateway at Core-1. The broadcast would flood to both Core-1 and Core-2, which would then flood the broadcast again to Core-1 and Access-1. Access-1 would flood the frame to Core-1 and Core-2, and so on.

Therefore, although the PC only generated one ARP request, Core-1 (the default gateway) would receive many copies of this request and would therefore need to reply many times.

The same issue holds true for other types of broadcasts and multicasts. The application that receives the duplicate frames might see the duplication as an error and reset its connection to the source of the transmissions, creating connectivity issues.

Broadcast storms

Another issue occurs. As mentioned in the previous section, Access-2 floods the broadcast to Core-1 and Core-2. Core-1 floods the broadcast to Core-2 and Access-1, and Core-2 floods the broadcast to Core-1 and Access-1. Access-1 floods both broadcasts to switches that flood the broadcast again. As you see, the broadcast continues to be flooded between the switches, multiplying again and again, and affecting all devices in the broadcast domain, or VLAN.

The devices are affected because their network interface controllers (NICs) must process all of these broadcasts. Eventually, the network will crash because devices will run out of central processing unit (CPU) cycles to process the broadcasts.

In addition, the broadcasts take more and more bandwidth on switch-to-switch links as they multiply. The congested links begin dropping traffic, which also interferes with connectivity. In fact, the broadcast storms in VLAN 12 and VLAN 1 would affect VLAN 11 as well because the switch-to-switch links that carry VLAN 11 also carry VLAN 1 and VLAN 12.

Note that NICs only process multicasts for which they have registered (typically because an application running on the endpoint uses that address). Therefore, a flood of multicasts does not affect the CPU cycles of all devices to as great a degree. However, floods of multicasts still affect everyone's bandwidth, and a high-speed video stream, for example, could quickly consume all of the available bandwidth.

Mislearned MAC addresses

As you know, Layer 2 loops also cause switches to mis-learn the interface associated with a MAC address.

Remember: Ethernet switches examine the source MAC address for incoming frames and associate that MAC address with the ingress interface for the frame. But in the looped network, Core-1 would receive the ARP request sent by PC, for example, on several different interfaces. It learned the PC's MAC address on interface a19, then a13, then a19, and so on. This switching from interface to interface happens again and again as the broadcasts loop.

The mis-learning can cause a number of problems. For example, Core-1 might currently have learned the PC's MAC address on port a19. It forwards an ARP reply frame destined to the PC to Core-2. But Core-2, due to the changing of the interfaces on which it is learning MAC addresses, might currently have learned the PC's MAC address on its port a19, connected to Core-1. Because switches do not forward traffic back out the port on which they receive it, the switch drops the frame. The PC does not receive its reply, which causes connectivity issues. It might even prompt the PC to send another ARP request, which adds to the number of broadcasts.

In short, a looped Layer 2 network quickly becomes congested and unusable.

Important

A common first indication of a Layer 2 loop is that the computing devices (PCs, servers, printers, switches, routers, and so on) experience very slow performance. Once a large number of broadcasts are built up within the loop, these computing devices must use their CPU (not application specific integrated chips [ASICs]) to process broadcasts. If you look at the CPU usage on a device, you will see that it is running very high.

When you notice a high CPU usage on switches, look at the MAC address table on those switches. If you see flapping MAC addresses (the address constantly changes between two or more interfaces), then you are almost guaranteed to have a loop.

Spanning tree solution

A spanning tree protocol provides the traditional solution for adding redundant links to a Layer 2 network without causing loops. In the first set of tasks for this chapter, you will set up an IEEE 802.1w Rapid Spanning Tree Protocol (RSTP) solution.

You will learn more about how RSTP works as you complete the tasks. For now, just focus on the main ideas.

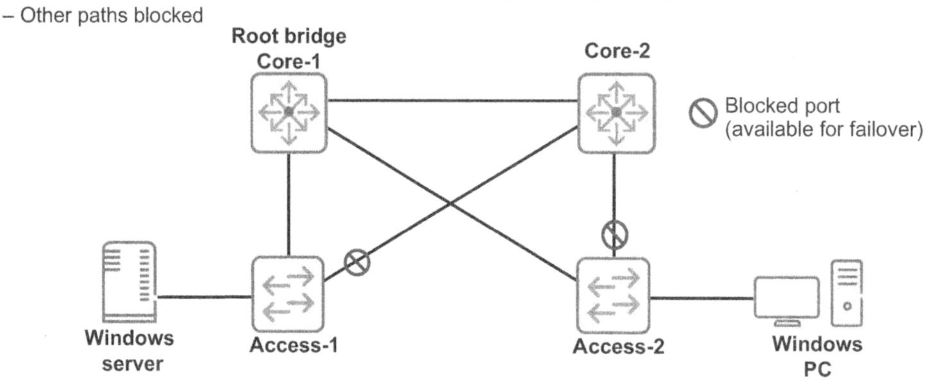

Figure 6-2: Spanning tree solution

RSTP features a single root bridge. (Bridge is the spanning tree terminology for switch; this chapter will use the term switch.) The spanning tree switches elect the root by exchanging Bridge Protocol Data Units (BPDUs) on their links. You will learn more details about the election process after you complete the tasks. For now, you should understand that you can affect the election by configuring certain spanning tree settings.

Every other spanning tree bridge has a lowest-cost path to the root. The links in the lowest cost paths forward traffic. Other links are blocked. As you see in Figure 6-2, the network resembles a tree, rooted at the root bridge, branching to other switches, and finally branching to the endpoints on switch edge ports. Because links that are not part of this tree are blocked, the network has no loops.

The blocked links remain available for use if a change occurs in the topology—for example, a forwarding switch-to-switch link fails. You will learn more about failover later.

Overview of spanning tree protocols

Several spanning tree protocols exist. The original standard, Spanning Tree Protocol (STP), is defined in 802.1D. This standard defined the basic behaviors still used in spanning tree, such as electing a root bridge, calculating the best path to that bridge, and blocking alternate paths. However, STP was designed in a time when a minute or so for convergence was acceptable. It requires ports to move through several port states, each with a relatively long (10 to 20 second) wait time, before the ports can start forwarding. Any change in the topology can force ports to progress slowly through these states again.

STP	RSTP	MSTP
802.1D pre-2004	802.1w merged into 802.1D 2004	802.1s merged into 802.1Q 2005
Original protocol	• Faster convergence • Costs that fit with modern port speeds	RSTP behavior in multiple instances
• Obsolete • ArubaOS switches can be forced to use	Default *operation* between ArubaOS switches (when no MSTP region is set up)	Default *mode* for Aruba switches (but no MSTP region settings)

Tip: Spanning tree is DISABLED on ArubaOS switches by default.

Figure 6-3: Overview of spanning tree protocols

As network connectivity became more crucial, companies could no longer tolerate such long convergence times when topology changes occurred. Rapid Spanning Tree Protocol (RSTP), 802.1w, was developed in 1998 to speed convergence. It also defined port costs that better fit with modern port speeds. As you see in Figure 6-3, in 2004 RSTP was merged into the 802.1D standard, superseding STP. Therefore, this chapter focuses on how RSTP functions over STP. As you work through the tasks and learning checks, you will see how RSTP handles convergence in under a second.

You will also learn about Multiple Spanning Tree Protocol (MSTP), which operates in the same basic way as RSTP but supports multiple instances. Each instance has its own topology, and, as you will see, this behavior allows for load sharing of traffic on redundant paths.

ArubaOS switches use MSTP for their default spanning tree mode. However, the switches do not initially have any MSTP region settings defined. Until you define consistent MSTP region settings, the switches operate as if they were running RSTP. Therefore, you will use RSTP in your first set of tasks.

Important

Although MSTP is the default spanning tree mode, spanning tree as a whole is *disabled* on ArubaOS switches by default. Do *not* simply start establishing redundant links before configuring a spanning tree protocol, or you will cause a broadcast storm.

Spanning tree port roles and states

RSTP switches do not have a complete view of the topology. They maintain the tree by exchanging BPDUs, which indicate how close they are to the root—in other words, the cost of their shortest path to the root, or their root path cost. These BPDUs help switches to calculate the correct port role for each of their ports.

Switches
- Exchange BPDUs about how close to the root they are
- Determine port roles and states accordingly

Port role	What it means	Final state
Designated	I'm closest to the root on this link	Forwarding
Root	Another switch on this link is closer to the root and gives me my best path	Forwarding
Alternate	Another switch on this link is closer to the root, but doesn't give me my best path	Discarding
Backup	I have another port on this link that's closer to the root	Discarding

Figure 6-4: Spanning tree port roles and states

Look first at the links that will become part of the active spanning tree, or, in other words, part of switches' set of lowest cost paths to the root. For each of these links (see Figure 6-4):

- The port on one side is *designated*. This is the port that is closer to the root (or is, in fact, on the root).
- The port on the other side of the link is the root port.

  In other words, the root port is the port that gives a non-root switch its best path to the root. A switch can only have *one* root port because it can have only one best path.

The port role affects the states that the port is allowed to pass through. In normal operation, designated and root ports transition to forwarding state. In this state, ports learn MAC addresses and forward traffic. Because both the designated port and the root port are in a forwarding state, the link is active.

Now look at the links that are *not* part of the spanning tree. For each of these links:

- The port on one side is designated. Again, this is the port that is closer to the root.
- The port one the other side of the link is either:
  - Alternate (it is connected to a different switch)
  - Backup (it is connected to the same switch)

  Typically, you will see alternate ports. A backup port occurs if a switch port is accidently looped to a port on the same switch. It also occurs if a switch port is connected to downstream switches that do not support spanning tree and that introduce a loop. From the point of view of spanning tree, those switches are transparent, so the spanning tree switch appears to be connected to itself.

Alternate and backup ports remain in the discarding state. The discarding state is sometimes called the blocking state, and you might see this in switch CLI output. In this state, ports drop incoming traffic, do not learn MAC addresses from incoming traffic, and do not forward traffic. They continue to receive and process BPDUs from the designated port; as you learned, these BPDUs communicate information about and maintain the spanning tree.

In short, links that are not part of the spanning tree have a designated port on one side and an alternate or backup port on the other. These links are effectively blocked because the downstream side of the link is in a discarding state.

Note

These rules properly apply to full-duplex Ethernet links, which are between two switches. The rules are a bit different for half-duplex links, which might connect several switches through a hub. In the latter case, one switch port on the link would be the designated port and all other ports would be alternate, backup, or root ports, depending on their lowest cost path to the root. However, this guide focuses on the full-duplex behavior.

Note that ports that do not connect to spanning tree devices—such as the ArubaOS switches' connections to the server and PC—are still part of the spanning tree. These switch ports have a designated role so that they can forward traffic to the endpoints. See Table 6-1.

Table 6-1: RSTP port roles

Port role	Description	Potential port state
Designated	Port is closer to the root than any other port on this link	Learning (when the port first comes up) Discarding (when certain events occur) Forwarding (final state during normal operation)
Root	• Port is not the closest port to the root on this link. • This link offers the switch its best path to the root.	Forwarding
Alternate	• Port is not the closest port to the root on this link. • This link does not offers the switch its best path to the root. It offers a second (or third, etc.) best path. • Port is not connected to the same switch.	Discarding
Backup	• Port is not the closest port to the root on this link. • This link does not offers the switch its best path to the root. • Port is connected to the same switch. The switch has a looped connection to itself. Or it has a loop through downstream devices that do not support spanning tree.	Discarding

Configuring Rapid Spanning Tree Protocol

You are now ready to complete a set of tasks in which you set up RSTP, add Core-2 and redundant links to the topology, and observe the results. As you complete the set of tasks, you will find the port roles and states for each port on the switches in Figure 6-5.

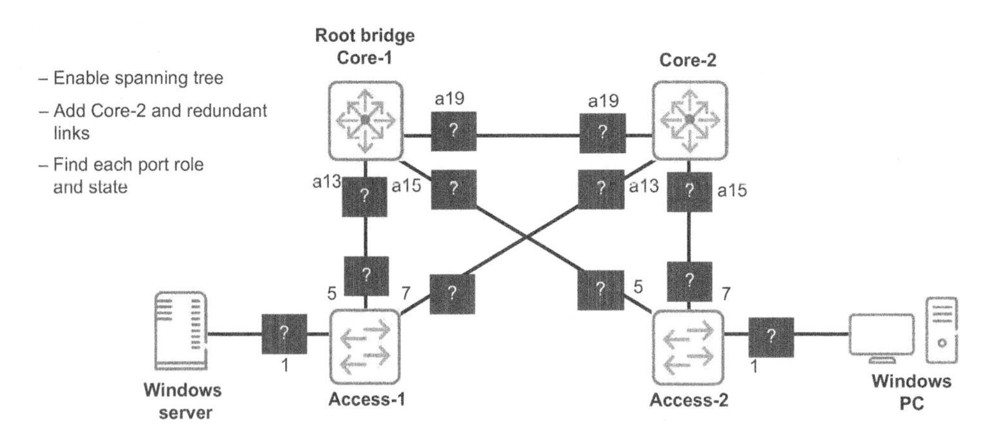

Figure 6-5: Configuring Rapid Spanning Tree Protocol

Task 1: Configure Core-1 as the root

Important

Before beginning this task, make sure that your switches are running the configuration that you established at the end of Chapter 5.

You learned that RSTP sets up the spanning tree based on the root bridge. In this task, you will configure Core-1 as the primary root bridge, which ensures that it wins the root bridge election.

1. Access a terminal session with Core-1 and move to the global configuration context.
2. Configure Core-1 to be elected the root bridge by giving it the lowest priority value.

```
Core-1(config)# spanning-tree priority 0
```

Task 2: Enable spanning tree on each switch

In this task you will enable spanning tree, which is disabled by default, on ArubaOS switches.

Core-1

1. Enable spanning tree on Core-1.

```
Core-1(config)# spanning-tree enable
```

Access-1

2. Access the Access-1 CLI and move to global configuration mode.
3. Enable spanning tree. You can also simply enter this command:

```
Access-1(config)# spanning-tree
```

Access-2

4. Access the Access-1 CLI and move to global configuration mode.
5. Enable spanning tree.

```
Access-2(config)# spanning-tree
```

Task 3: Add the redundant core switch and redundant links

You will now add Core-2 to the topology. Figure 6-6 shows the new redundant links that this switch will add. It is very important to enable spanning tree, which you did in Task 1, before adding the redundant links.

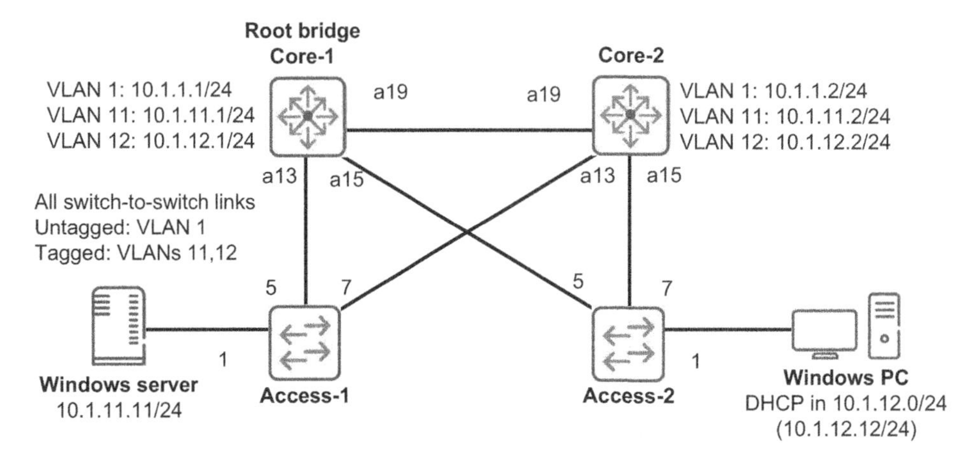

Figure 6-6: Configuring Rapid Spanning Tree Protocol

The Core-2 configuration will be very similar to the Core-1 configuration. You will create the configuration by copying and editing the Core-1 configuration.

Core-1

1. Core-1 will connect to Core-2 on one link, which will be untagged for VLAN 1 and tagged for VLANs 11 and 12 like other switch-to-switch links. Configure the VLAN settings.

```
Core-1(config)# vlan 11 tagged a19
Core-1(config)# vlan 12 tagged a19
```

2. Enable the interface.

```
Core-1(config)# interface a19
Core-1(eth-A19)# enable
```

3. Give the interface an intuitive name.

```
Core-1(eth-A19)# name Core-2
```

4. Establish the physical connection if applicable.

5. It is easier to copy the Core-1 configuration, if you can scroll through the entire running-config at once. Disable page breaks.

```
Core-1(eth-A19)# end
```

```
Core-1# no page
```

6. View the running-config.

```
Core-1# show running-config
```

7. Select the complete output up through the hostname line. Copy the output (<Alt+c>).

Notepad

8. Open Notepad and paste the output into the document.

Core-1

9. Re-enable paging on Core-1.

```
Core-1# page
```

Notepad

10. Edit the output in the Notepad document as follows:
    - Change the hostname to Core-2.
    - Change the name of interface A19 to "Core-1."
    - Change the host portion of each IP address to 2:
      - 10.1.1.2/24
      - 10.1.11.2/24
      - 10.1.12.2/24
    - Change the spanning-tree root command to **spanning-tree priority 1**.
11. Refer to the following configuration to ensure that yours is correct.
12. Select and copy all the text in Notepad.

```
hostname "Core-2"

module A type j9990a

ip routing
```

```
interface A1
   disable
   exit
interface A2
   disable
   exit
interface A3
   disable
   exit
interface A4
   disable
   exit
interface A5
   disable
   exit
interface A6
   disable
   exit
interface A7
   disable
   exit
interface A8
   disable
   exit
interface A9
   disable
   exit
interface A10
```

```
   disable
   exit
interface A11
   disable
   exit
interface A12
   disable
   exit
interface A13
   name "Access-1"
   exit
interface A14
   disable
   exit
interface A15
   name "Access-2"
   exit
interface A16
   disable
   exit
interface A17
   disable
   exit
interface A18
   disable
   exit
interface A19
   name "Core-1"
```

```
   exit
interface A20
   disable
   exit
interface A21
   disable
   exit
interface A22
   disable
   exit
interface A23
   disable
   exit
interface A24
   disable
   exit
snmp-server community "public" unrestricted
oobm
   ip address dhcp-bootp
   exit
vlan 1
   name "DEFAULT_VLAN"
   untagged A1-A24
   ip address 10.1.1.2 255.255.255.0
   exit
vlan 11
   name "VLAN11"
   tagged A13,A15,A19
```

```
   ip address 10.1.11.2 255.255.255.0
   exit
vlan 12
   name "VLAN12"
   tagged A13,A15,A19
   ip address 10.1.12.2 255.255.255.0
   ip helper-address 10.1.11.11
   exit
spanning-tree
spanning-tree priority 1
```

Core-2

13. Open a terminal session with Core-2.
14. Move to global configuration mode.

```
HP-Switch-5406Rzl2# config
```

15. Paste the commands. Verify that they execute successfully.
16. This switch was using a configuration that disabled all ports. Therefore, you must enable the interfaces that connect to Core-1, Access-1, and Access-2 now.

```
Core-2(config)# interface a13,a15,a19 enable
```

Access-1

You need to add the redundant links to the access layer switches and configure these links to carry VLANs 11 and 12, ensuring that connectivity for these VLANs isn't disrupted during failover situations.

17. Access the terminal session with Access-1.
18. Configure the interface that connects to Core-2 as a tagged member of VLANs 11 and 12.

```
Access-1(config)# vlan 11 tagged 7
Access-1(config)# vlan 12 tagged 7
```

19. Enable the interface.

```
Access-1(config)# interface 7 enable
```

20. If you want, give the interface an intuitive name.

```
Access-1(config)# interface 7 name Core-2
```

21. Establish the physical connection if applicable.

Access-2

22. Access the terminal session with Access-2.

23. Follow the same steps to set up the redundant connection to Core-2.

```
Access-2(config)# vlan 11 tagged 7

Access-2(config)# vlan 12 tagged 7

Access-2(config)# interface 7 enable

Access-2(config)# interface 7 name Core-2
```

24. Establish the physical connection if applicable.

Task 4: Verify the root bridge

You will now verify that Core-1 has been elected the root bridge, as desired, and that spanning tree is blocking the redundant links.

Core-1

1. View spanning tree. For now focus on the CST section of the output. (CST refers to Common Spanning Tree [CST], which is the spanning tree used by all switches running spanning tree and the only tree that you need to consider when your switches are using RSTP.) The CST root Port should indicate that this switch is root.

```
Core-1# show spanning-tree

Multiple Spanning Tree (MST) Information

  STP Enabled    : Yes

  Force Version : MSTP-operation

  <-output omitted->

  CST Root MAC Address : 1c98ec-ab4b00

  CST Root Priority    : 0
```

```
  CST Root Path Cost    : 0
  CST Root Port         : This switch is root
  <-output omitted->
```

2. The CST Root Priority field indicates the priority value that the root (which, in this case, is this switch) is using. This priority was set when you specified the spanning-tree priority command on Core-1. What is it? (You will find the answer in the following section.)
3. Record the CST root MAC address, which is Core-1's bridge ID. It probably will not match the ID shown in the output on the previous page because this ID depends on the switch hardware. (You will find the answer in the following section.)

Access-1

4. Access the terminal session with Access-1.
5. Use the same spanning tree command to verify that Core-1 is the root on this switch. Match the CST Root MAC Address to the address that you recorded above.

```
Access-1(config)# show spanning-tree
 Multiple Spanning Tree (MST) Information
  STP Enabled   : Yes
  Force Version : MSTP-operation
<-output omitted->
  CST Root MAC Address : 1c98ec-ab4b00
  CST Root Priority    : 0
  CST Root Path Cost   : 20000
  CST Root Port        : 5
<-output omitted->
```

Access-2

6. Access the terminal session with Access-2.
7. You can also view the history for the CST's root, which shows the current root and the number of times that the root has changed. This command can be useful for troubleshooting instability in the spanning tree.

```
Access-2(config)# show spanning-tree root-history cst
```

```
 Status and Counters - CST Root Changes History
  MST Instance ID         : 0
  Root Changes Counter    : 2
  Current Root Bridge ID : 0:1c98ec-ab4b00
  Root Bridge ID        Date      Time
  ------------------- -------- --------
      0:1c98ec-ab4b00 06/06/16 02:47:59
  32768:843497-0223c0 06/06/16 02:47:59
```

8. Verify that the current root bridge ID matches the Core-1 bridge ID that you recorded above.

Core-2

9. Access the terminal session with Core-2.
10. Use either command that you prefer to verify that Core-2 recognizes Core-1 as the root.

Task 5: Check CPU

Verify that the CPU usage on *each* switch is normal. A high CPU usage could indicate a loop, but RSTP should have removed the loops.

```
Core-1(config)# show cpu
1 percent busy, from 300 sec ago
1 sec ave: 1 percent busy
5 sec ave: 1 percent busy
1 min ave: 1 percent busy
Core-2(config)# show cpu
1 percent busy, from 300 sec ago
1 sec ave: 1 percent busy
5 sec ave: 1 percent busy
1 min ave: 1 percent busy
Access-1(config)# show cpu
5 percent busy, from 300 sec ago
```

```
1 sec ave: 1 percent busy

5 sec ave: 1 percent busy

1 min ave: 1 percent busy

Access-2(config)# show cpu

4 percent busy, from 300 sec ago

1 sec ave: 4 percent busy

5 sec ave: 1 percent busy

1 min ave: 1 percent busy
```

Learning check

You will now answer questions and complete activities to expand on what you learned in the preceding tasks.

1. Which commands did you use to discover this information:

   The root switch:

   The port role and port state for each interface:

2. The topology that you created when you performed the preceding tasks should resemble the one shown in Figure 6-7. How was this topology established?

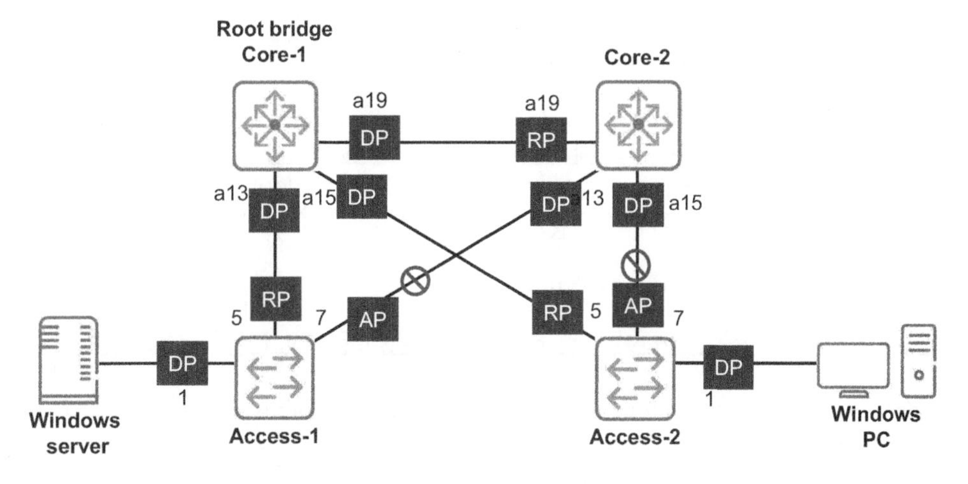

Figure 6-7: RSTP topology with the port states filled in

3. Which switch has only designated ports?

4. How many root ports does each switch have?

5. The **show spanning-tree** commands also showed the port cost on each port. What was the cost on each link?

Answers to Learning check

1. Which commands did you use to discover this information:

   The root switch:

   show spanning-tree root-history cst

   The port role and port state for each interface:

   show spanning-tree [*interface ID list*] instance ist

2. The topology that you created as you completed the preceding tasks should resemble the one shown in Figure 6-7. How was this topology established?

   Core-1 was elected the root. Switches sent BPDUs that indicated how close they were to the root. Access-1 and Access-2 both have two uplinks that offer paths to the root. In this topology, each switch's root port connects directly to Core-1. Redundant links between Access-1 and Core-2 and between Access-2 and Core-2 were blocked. As you see in Figure 6-7, Core-2 has the designated port on these links, and the access switches have alternate ports.

3. Which switch has only designated ports?

   The root, which is Core-1, has only designated ports. Ports on the root are always closer to the root than connected ports, by definition. Therefore, the root bridge has only designated ports and no root port. The one exception is if the root had a looped connection to itself (directly or through non-STP switches). In this case, it could also have a backup port.

4. How many root ports does each switch have?

   Each switch has *one* root port except Core-1, which is the root, and has no root ports.

5. The **show spanning-tree** commands also showed the port cost on each port. What was the cost on each link?

   20,000

Root election

You will now examine in more detail how a root bridge is elected. In the preceding set of tasks, you set Core-1's spanning tree priority value to 0. A *lower* priority value gives a *higher* priority for being elected root, so Core-1 was elected root. (See Figure 6-8.)

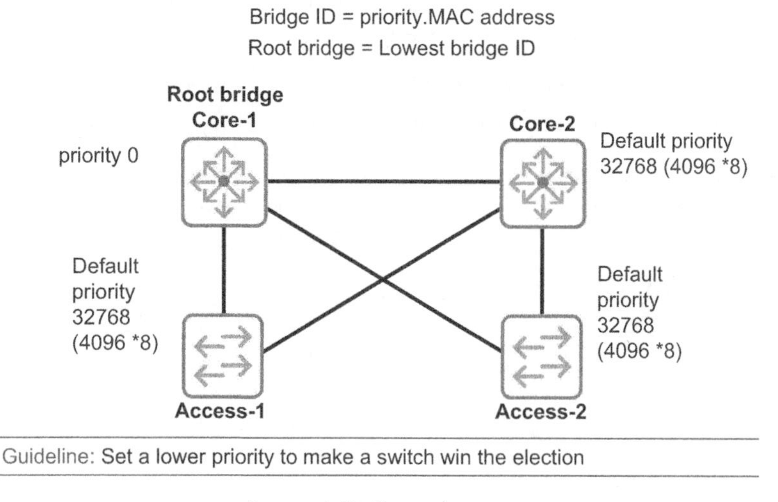

Figure 6-8: Root election

Next consider what happens if multiple switches have the same lowest priority value, creating a tie. This situation occurs if you do not set a lower priority on any switches. In this case, all switches use the default priority value, which is 32,768 or 8 (priorities are set in steps of 4096; 8 * 4096 = 32,768). See Figure 6-9. This situation could occur if Core-1 failed, leaving only active switches with the default priority.

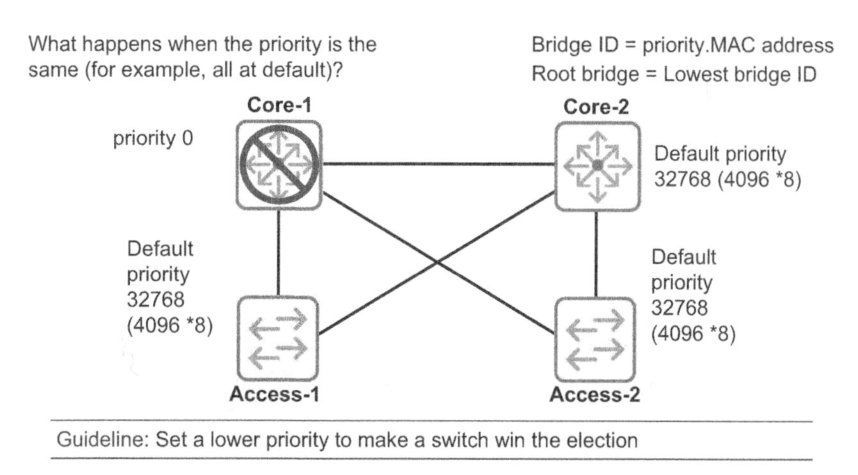

Figure 6-9: Situation in which all active switches are using the default priority

It is actually difficult to predict which switch becomes root in this case.

To understand why, you need to understand that switches do not use the priority alone to elect the root. They actually use their bridge ID for the election. The bridge ID is:

Priority + system ID (MAC address)

The priority value is more significant, so it affects the election most. However, when multiple switches have the same priority, the switch with the lowest MAC address becomes root.

To prevent an unexpected root, you should always manually configure a priority on the desired root *and* the desired backup (or secondary) root bridge, as indicated in Figure 6-10.

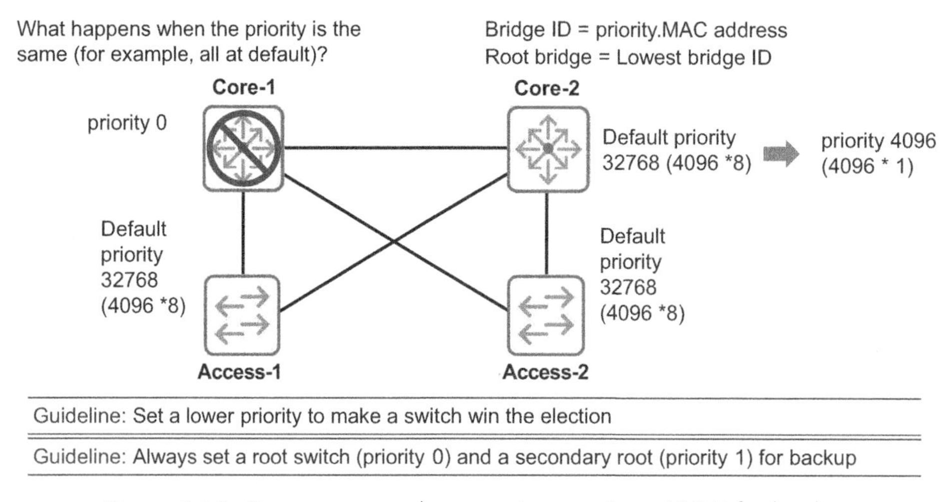

Figure 6-10: Setting a secondary root (priority 1 or 4096) for backup

Table 6-2 shows RSTP bridge priority values and the associated step value, which you use in the **spanning-tree priority** command.

Table 6-2: Priority

Priority value	Step
0	0
4096	1
8092	2
12288	3
16384	4
20480	5
24576	6
28672	7
32768	8
36864	9
40960	10
45056	11
49152	12
53248	13
57344	14
61440	15

Port costs

Up to this point, you have been looking at the topology as if every link has the same cost, and you can simply count links to determine the best path to the root. This is not always true, however. Sometimes links have different bandwidths, so the ports have different costs. You might also have a heterogeneous network with switches that set port costs in different ways. See Figure 6-11.

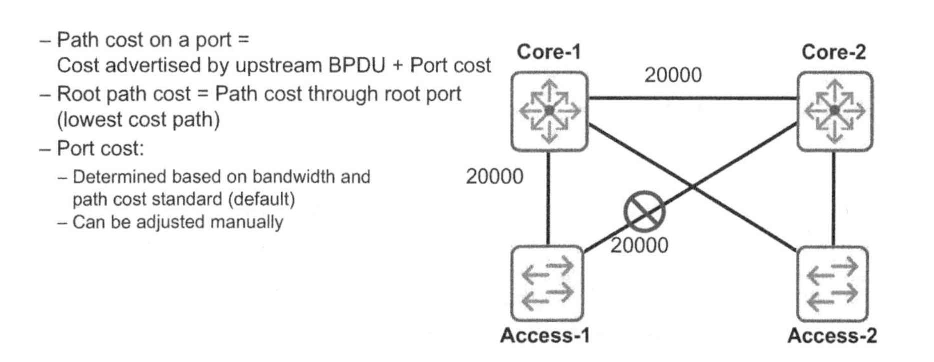

Figure 6-11: Port costs

You will now learn a bit more about how the root path cost is calculated so that you can follow best practices and understand how topologies are formed.

The cost for reaching the root on any port is:

The cost advertised in the BPDU + the port cost

The port cost derives from the bandwidth on the port, as defined by the spanning tree standard. See Table 6-3.

Table 6-3 Default port costs

Connection Type	RSTP/MSTP (ArubaOS default)	STP (802.1D pre-2004)
10 Gbps	2,000	-
1 Gbps	20,000	4
100 Mbps	200,000	10
10 Mbps	2,000,000	100

In the preceding set of tasks, you observed that ArubaOS switches assign a cost of 20,000 to GbE links. You'll now look at what this means for the path costs.

First, look at the cost for reaching the root on Core-2's root port:

The cost advertised in the BPDU is 0. Core-1 is the root, so it does not have a cost for reaching the root. The port cost is 20,000, which is the RSTP standard cost for a GbE link. So the path cost is:

0 + 20,000 = 20,000

This is Core-2's best path to the root, so this becomes its *root path cost*. Core-2 advertises this cost in BPDUs that it sends on any designated port.

Next examine Access-1's cost for reaching the root on its root port. Again Core-1 advertises a root path cost of 0 because it is root. So the path cost to the root on this ArubaOS switch is:

0 + 20,000 = 20,000

Now look at the path cost on Access-1's alternate port. Core-2, as you saw earlier, advertises a root path cost of 20,000. The port cost is 20,000. Therefore, the path cost to the root is:

20,000 + 20,000 = 40,000

This cost is higher than the cost on the root port, which is exactly why the first port is the root port and the second one is the alternate port.

Access-1 has a root path cost of 20,000 (through the interface connected directly to Core-1), which is what it would advertise to any downstream devices.

Note that Core-2 and Access-1 have equal root path costs. How do the switches decide which has the designated port on the link between them? They look at other criteria to break the tie. The criteria are:

- Bridge ID (lower value wins; remember: the bridge ID is the priority plus MAC address)
- If Bridge ID is the same, port priority (lower value wins)
- If port priority is the same, port ID (lower ID wins)

In this example, Core-2 has a lower priority value, so it has a lower Bridge ID, and its port is designated.

Consistent port costs on heterogeneous networks

If you have a heterogeneous network with different types of switches, you should be aware that some switches use different port costs from those suggested by the standard. The discrepancy could cause an issue. For example, consider the topology on Figure 6-12.

Access-1 is a switch using proprietary costs. Core-1 and Core-2, though, are ArubaOS switches. Access-1 has a GbE link to Core-1 and a 10GbE link to Core-2. Core-2 also has a 10GbE link to Core-1. Figure 6-12 shows the new port costs in this situation.

You see that, in this case, Access-1's link to Core-1 has a root path cost of:

0 + 20 = 20

Access-1's link to Core-2 has a root path cost of:

2 + 2000 = 2002

Therefore, Access-1 chooses the slower link. As a best practice, you should set consistent costs on all switches—generally, you should use the standard RSTP costs that ArubaOS switches use by default.

When the switches all use the same costs, Access-1 chooses the 10GbE link to Core-2 as the root port (path cost 4000 as opposed to 20,000). See Figure 6-13.

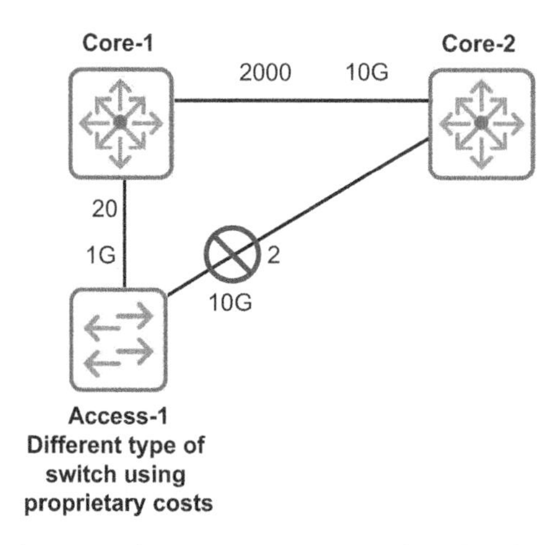

Figure 6-12: Topology in a heterogeneous network with GbE and 10GbE links

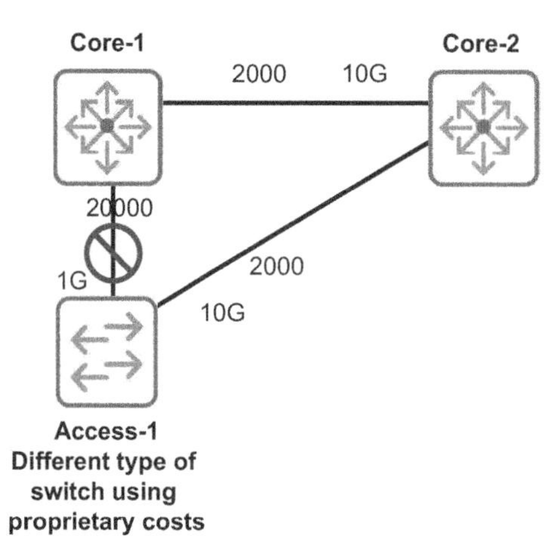

Figure 6-13: Topology in a heterogeneous network with GbE and 10GbE links and consistent port costs

Failing over from a root to an alternate port

You will now examine how failover works in an RSTP environment, as well as how other topology changes are handled. First you will examine a scenario in which Access-1 loses its active link to Core-1 and must failover to its alternate link through Core-1.

Figure 6-14 shows the instant at which the Access-1's link to Core-1 is lost.

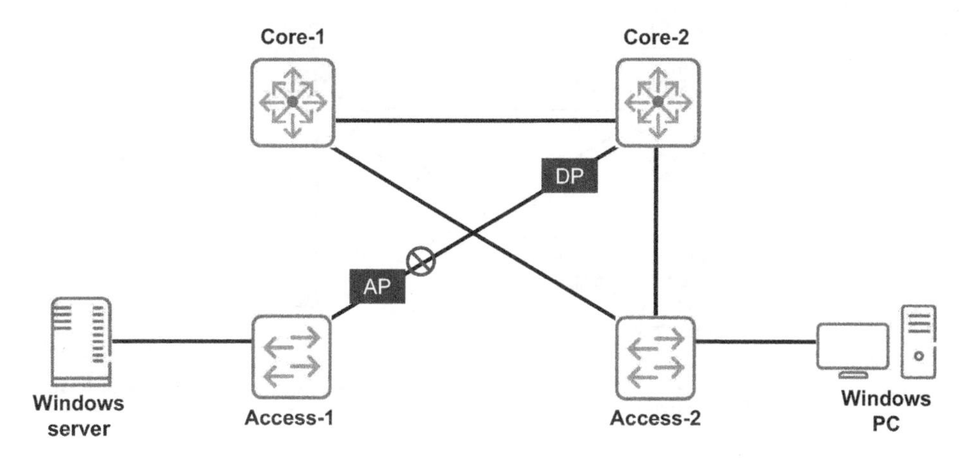

Figure 6-14: Root port lost on a switch with an alternate port

As you see in Figure 6-15, as soon as Access-1 detects that its root port is down, it changes its alternate port to a root port, which is immediately set to the forwarding state. Other ports on Access-1 and on other switches are unaffected. The convergence takes much less than a second.

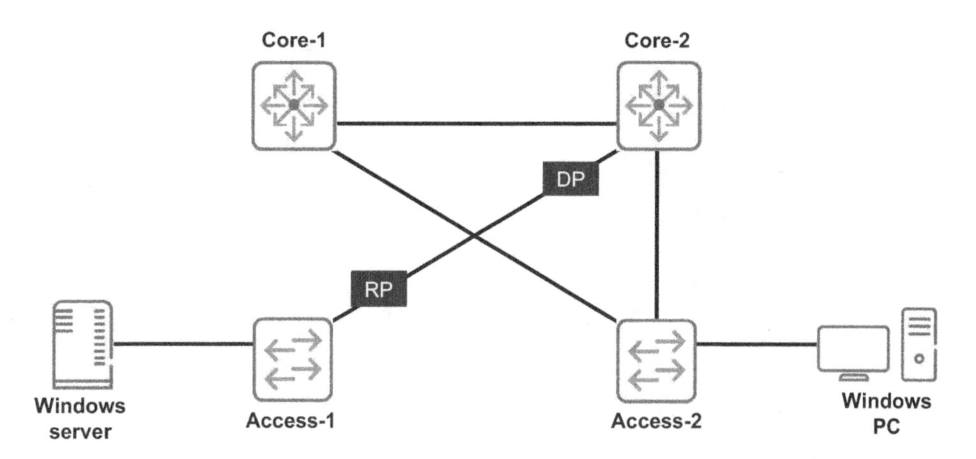

Figure 6-15: Alternate port becomes root port

Event messages on the other switches also indicate that Access-1 sends a topology change (TC) BPDU on all forwarding ports. The TC BPDU lets other switches know that they should flush their learned MAC addresses on other ports because they might now be learning the addresses on a new port. For example, Core-2 will now learn the MAC address for the server on the port that connects to Access-1 instead of on the port that connects to Core-1. (See Figure 6-16.)

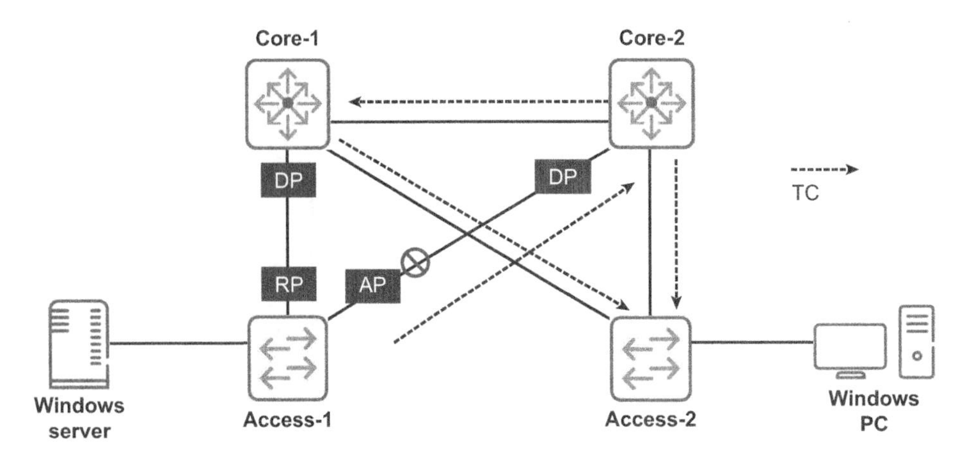

Figure 6-16: TC BPDUs

Failing over from a root to a designated port

Next you will examine a scenario in which Core-2 loses its active link to Core-1. (Figures 6-17, 6-18, and 6-19 do not show the link between Core-1 and Core-2 because this link has failed.)

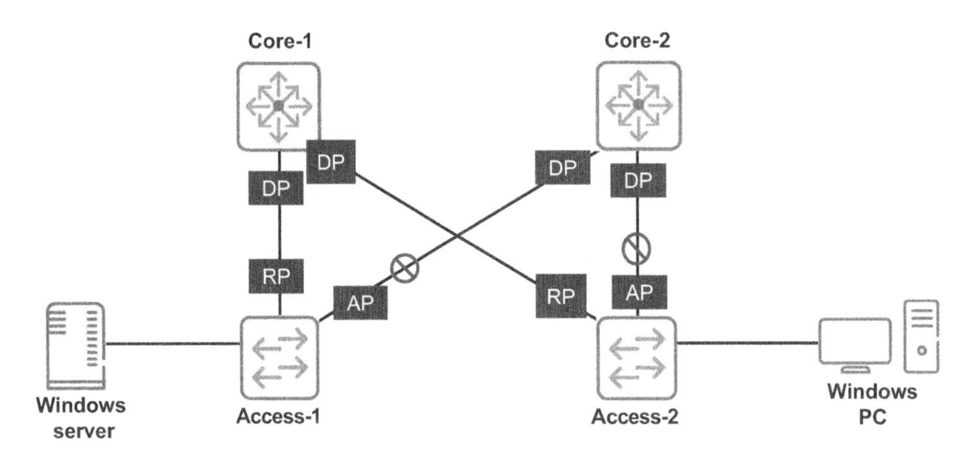

Figure 6-17: Root port lost on a switch without an alternate port

The convergence in this case is a bit more complicated. Core-2 must transition one of its designated ports (which used to offer the connected switch a better path to the root) to a root port (which now receives a better path to the root from the connected switch—see Figure 6-17). It must do so without introducing temporary loops due to the convergence process.

Several events occur when Core-2 detects that it has lost its root port:

1. It sends a BPDU on each of its designated ports. (See Figure 18.)
    a. This BPDU indicates that Core-2 is now the root. (It has lost contact with the root, and it believes that it has the best priority.)
    b. In legacy STP, the BPDU from Core-2—called an inferior BPDU because Core-2 has an inferior switch ID to Core-1—would have caused the access switch ports to undergo a lengthy reconvergence process. But with RSTP, the process is much faster. The access switches keep using the best root—in this case, Core-1—about which they have current information. (They trust that they will soon receive another BPDU from Core-1. If they do not—if Core-1 had actually failed, then the convergence process would proceed differently.)

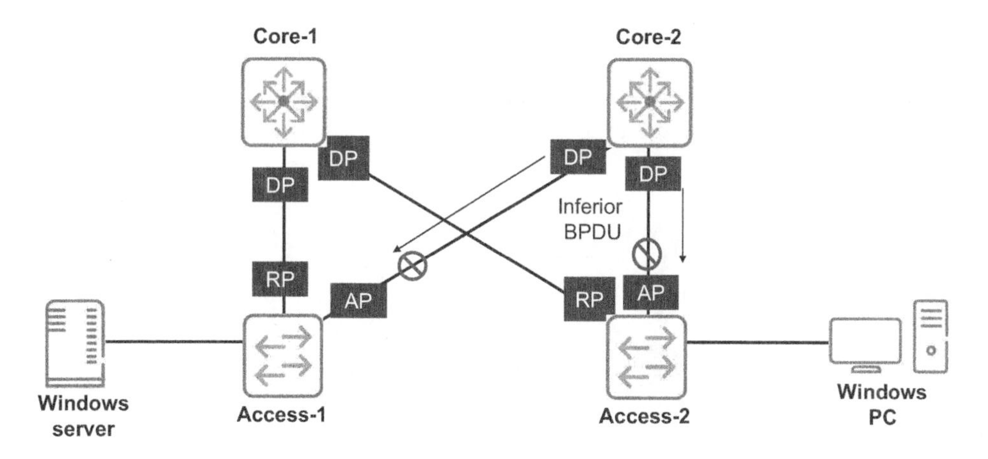

Figure 6-18: Step 1a and b in convergence

2. Each access switch knows that it has a better path to the root than Core-2 is advertising in its inferior BPDU. It undergoes a quick exchange with the connected Core-2 port.
    a. The access switch port asserts itself as now offering a better connection to the root. See Figure 6-19. The port becomes a designated port, but it is in discarding state to prevent temporary loops during convergence.

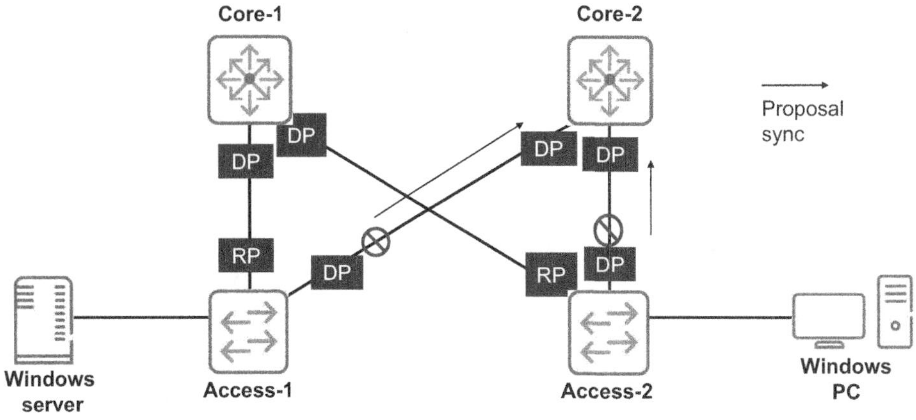

Figure 6-19: Step 2a in convergence

b. Core-2 chooses the best BPDU—the lowest cost path—and sets the root port. This port can immediately transition to forwarding. The other port is an alternate port, which is discarding. See Figure 6-20.

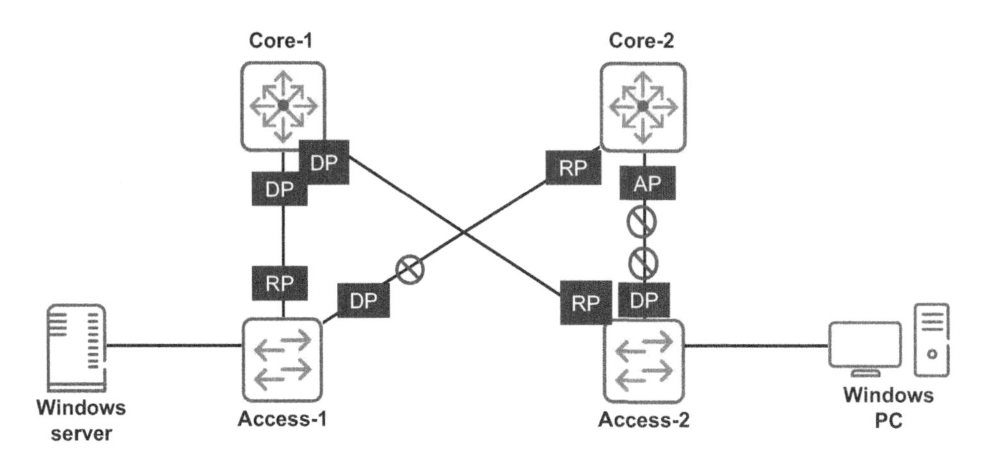

Figure 6-20: Step 2b in convergence

c. Core-2 informs each connected port that it has set a single root port and other ports are discarding—ensuring that it will not cause loops. The access switches then let their designated ports transition to forwarding. See Figure 6-21.

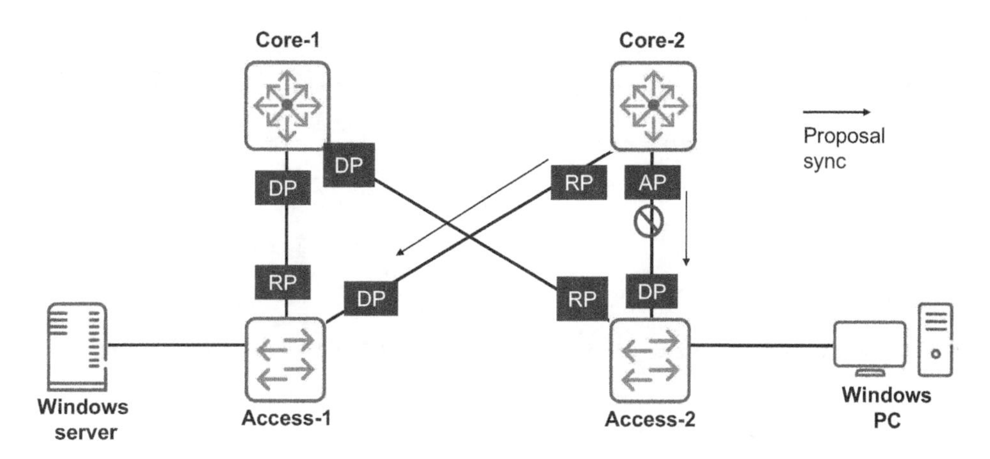

Figure 6-21: Step 2c in convergence

3. At the same time that the access switches are transitioning their alternative ports to designated, these switches send BPDUs with the topology change (TC) bit set on all other ports. These TC BDPUs notify other switches that a new link is forwarding. Then those switches can flush the MAC addresses that they have learned on other ports, which might no longer be part of the active topology. See Figure 6-22.

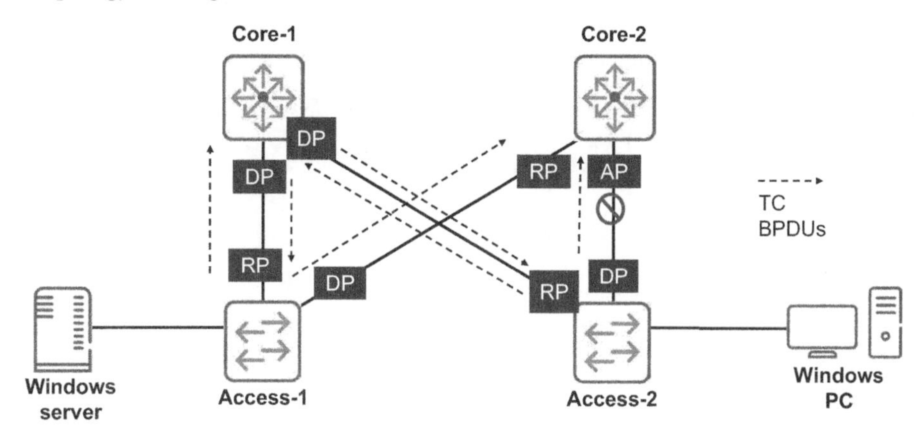

Figure 6-22: Step 3 in convergence

Although this process involves more steps than the previous failover scenario, it occurs in about a second or less.

These events would cause the following command outputs on the Core-2 switch:

```
Core-2# show spanning-tree interface a13,a15,a19 instance ist

<-output omitted->
```

```
                                                                 Designated
  Port  Type       Cost   Priority Role        State      Bridge
  ----- --------- ------ -------- ----------- ---------- -------------
  A13   100/1000T 20000  128      Designated  Forwarding 70106f-0d2100
  A15   100/1000T 20000  128      Designated  Forwarding 70106f-0d2100
  A19   100/1000T 20000  128      Root        Forwarding 1c98ec-ab4b00
Core-2# show spanning-tree interface a13,a15,a19 instance ist
<-output omitted->
                                                                Designated
  Port  Type       Cost   Priority Role        State      Bridge
  ----- --------- -----  -------- ---------   ---------- --------
  A13   100/1000T 20000  128      Root        Forwarding 6c3be5-6208c0
  A15   100/1000T 20000  128      Alternate   Blocking   843496-0223c0
  A19   100/1000T Auto   128      Disabled    Disabled
```

Access-1 and Access-2 both generate TC BPDUs.

```
Access-X(config)# show spanning-tree topo-change-history originated
 Status and Counters - CST Regional Topology Changes History
 MST Instance Id  : CST

      Current       Previous
 Port Role          Role          Date        Time
 ---- ------------- ------------- ----------- ------------
 7    Designated    Alternate     07/07/2016  04:14:59
```

Reconvergence when a better path is added

You will now examine what happens when the link between Core-2 and Core-1 is repaired so that Core-2 now has a better path to the root. As you see in Figure 6-23, the link comes up blocking to prevent loops.

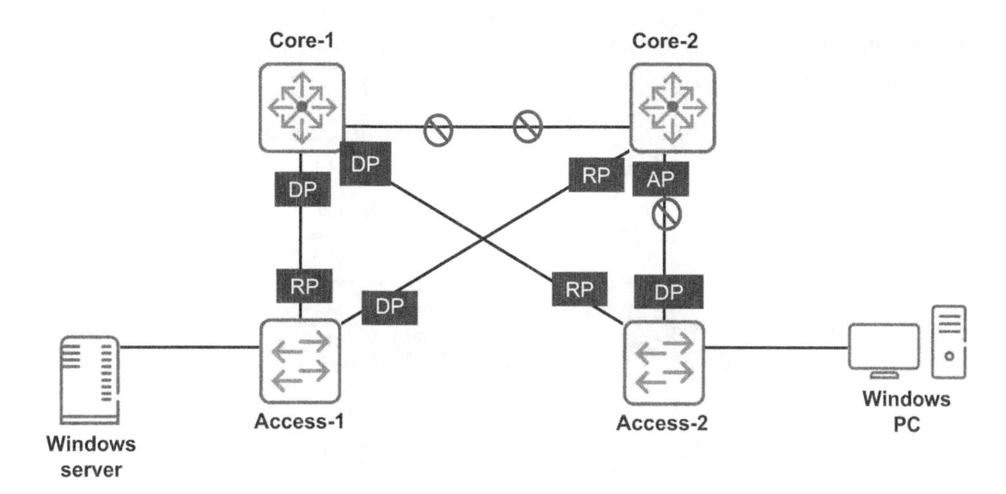

Figure 6-23: Better path to the root restored

Several events now occur:

1. Core-1 asserts that it has a superior path to the root, and Core-2 sees that this is the case.
   a. The new port on Core-2 becomes its root port, which transitions immediately to forwarding. The link is still blocked on the Core-1 side, so loops do not occur. See Figure 6-24.

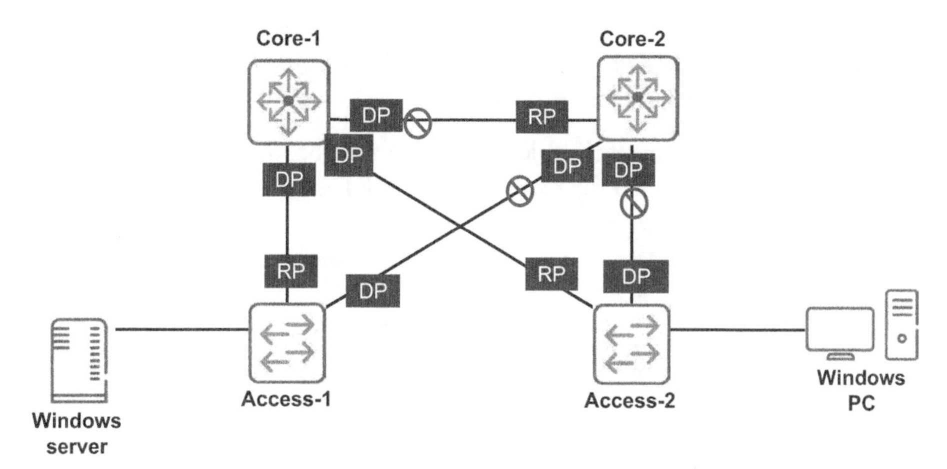

Figure 6-24: Convergence step 1a

   b. At the same time, Core-2 sets any other port that is currently forwarding to discarding. This action guarantees that loops will not occur during convergence. (Core-2 now believes that it has a better path to the root than each connected access switch does, so the ports are designated.)

c. Core-2 informs Core-1 that it has a single root port and other ports are discarding. With this assurance, Core-1 sets its designated port to forwarding.

2. Core-1 and Core-2 send TC BPDUs indicating that a new link is active. These BPDUs propagate throughout the tree. See Figure 6-25.

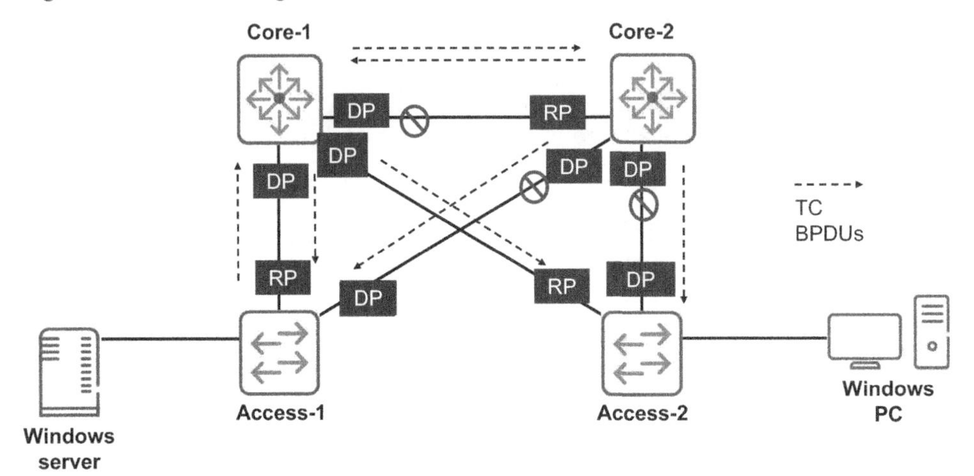

Figure 6-25: Convergence step 2

3. The new link is now up, but Core-2 needs to move any discarding designated ports to a forwarding state. To do this, it completes an agreement proposal handshake mechanism with the access switches similar to the one that it completed with Core-1.

   a. In this exchange, Core-2 takes the role that Core-1 took because Core-2 believes that it offers a better path to the root than the correct switch. See Figure 6-26.

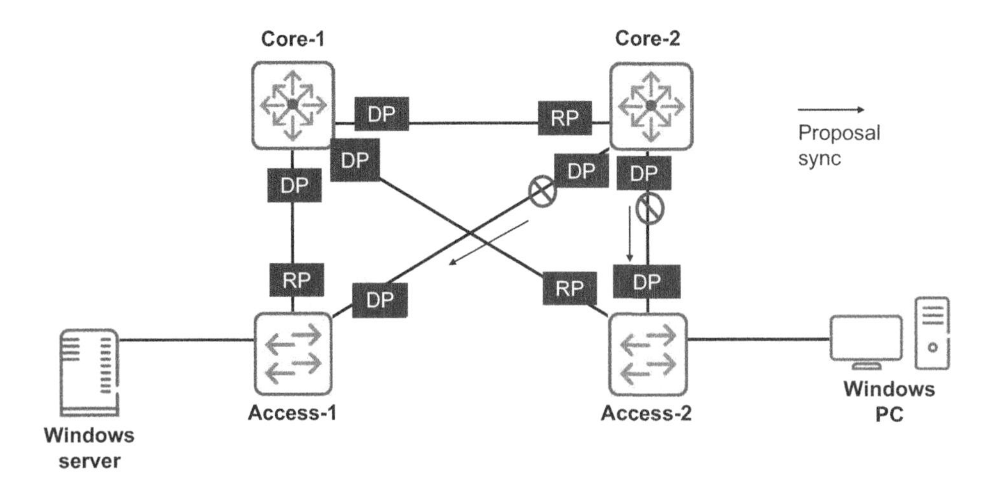

Figure 6-26: Convergence step 3a

b. Each access switch recognizes that Core-2 is using the same root as it is, but Core-2 does now offer a better root path (superior BPDU) than it can. So each access switch changes the port role accordingly. See Figure 6-27.

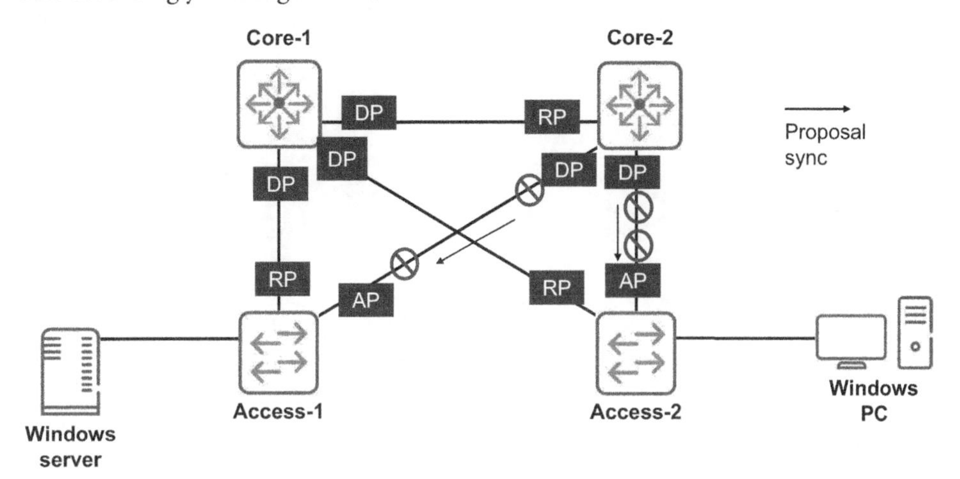

Figure 6-27: Convergence step 3b

c. Each access switch informs the connected Core-2 port that it has one root port and other ports are blocking. (If the switch had other designated ports, it would need to block them, and the convergence would ripple downstream.) Therefore, Core-2 sets its designated port to forwarding. See Figure 6-28.

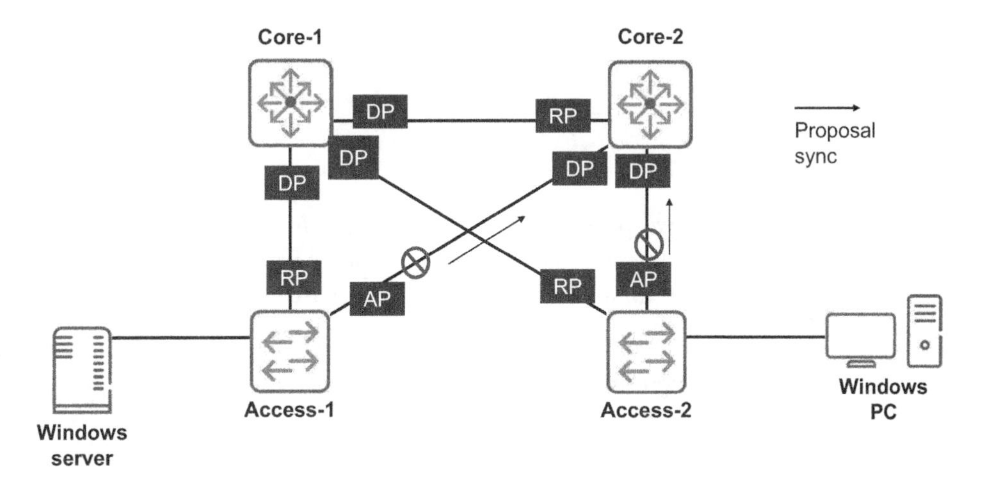

Figure 6-28: Convergence step 3c

d. Core-2 also sends TC BPDUs about the alternate port becoming a designated port.

```
Core-2# show spanning-tree a13,a15,a19 instance ist
<-output omitted->
                                                         Designated
  Port  Type      Cost  Priority Role       State      Bridge
  ----- --------- ----- -------- ---------- ---------- -------------
  A13   100/1000T 20000 128      Designated Forwarding 70106f-0d2100
  A15   100/1000T 20000 128      Designated Forwarding 70106f-0d2100
  A19   100/1000T 20000 128      Root       Forwarding 1c98ec-ab4b00
Core-2(config)# show spanning-tree topo-change originate

 Status and Counters - CST Regional Topology Changes History

 MST Instance Id  : CST

      Current    Previous
 Port Role       Role      Date       Time
 ---- ---------- --------- ---------- --------
 A15  Designated Alternate 07/07/2016 21:58:20
 A19  Designated Disabled  07/07/2016 21:58:18
```

Spanning tree edge ports

Suppose that your network has two ports that connect to endpoints, one on Access-1 and one on Access-2. Of course, a real network would have many more such ports. Because these ports connect to endpoints, with a single link to the spanning tree topology, they can never introduce loops.

RSTP lets you define such ports as "edge ports." The edge ports have the designated role, and they are allowed to transition to forwarding immediately. See Figure 6-29.

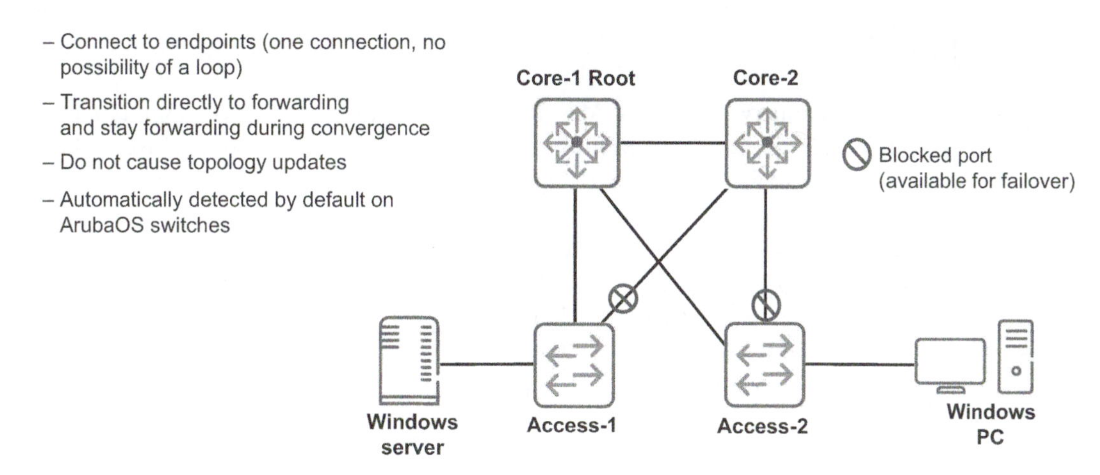

– Connect to endpoints (one connection, no possibility of a loop)
– Transition directly to forwarding and stay forwarding during convergence
– Do not cause topology updates
– Automatically detected by default on ArubaOS switches
Core-1 Root
Core-2
Blocked port (available for failover)
Windows server
Access-1
Access-2
Windows PC

Figure 6-29: Spanning tree edge ports

On the previous pages, you learned about various failover situations. Several of these situations cause a switch to temporarily set all of its designated ports to a discarding state to prevent loops during convergence. If one of those designated ports connects to an endpoint, that endpoint would lose connectivity while the switch waits to receive a BPDU or for a timeout. The temporary loss of connectivity could be damaging, particularly for a server. However, a switch keeps edge ports forwarding during these situations—because it knows that the edge ports cannot introduce loops—helping endpoints and servers to maintain connectivity.

In addition, changes in edge port status do not affect the RSTP topology. If an edge port goes down, the switch does not send a topology update to the root. It does not need to recalculate its port roles, and neither do other switches. In this way, properly defined edge ports make the overall RSTP topology more stable.

ArubaOS switch ports are enabled for auto edge port detection by default. If the port does not receive a BPDU within three seconds, the switch defines the port as an edge port. If it later receives a BPDU, the port is reclassified as non-edge. You can enable and disable this feature with this command:

```
ArubaOS(config)# [no] spanning-tree <int-id-list> auto-edge-port
```

You can also configure a port as an edge port manually, which makes the port always act as an edge port immediately. However, if the switch detects a BPDU on the port, it disables the edge port function.

```
ArubaOS(config)# [no] spanning-tree <int-id-list> admin-edge-port
```

Issues with RSTP

RSTP guarantees a loop-free network, but it does not guarantee optimization. For example, in this network, Access-1 and Access-2 each have two uplinks. However, only one uplink is active, so half of the uplink bandwidth is essentially blocked. See Figure 6-30.

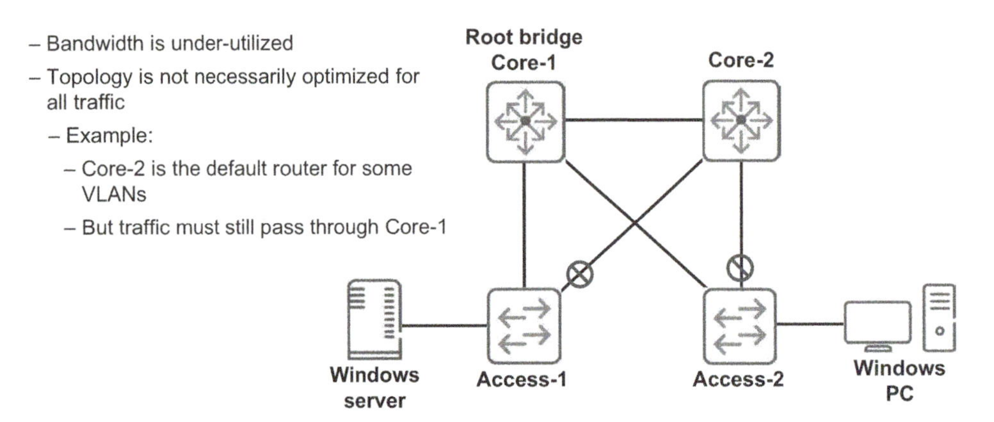

Figure 6-30: Issues with RSTP

In addition, the spanning tree always sets up shortest paths to the root, which might not be the best path for specific traffic. You set Core-1 as the root because it is the default gateway, so most traffic is directed to it. However, what if you wanted to set up Core-2 as the default gateway for some VLANs? Traffic would still need to pass through Core-1 because Core-1 is the root.

MSTP solution

Multiple Spanning Tree Protocol (MSTP), IEEE standard 802.1s, provided an extension to STP and RSTP r to deal with the issues we just discussed. 802.1s was later merged into 802.1Q-2005, so MSTP is now part of the Ethernet standard.

MSTP allows switches to set up multiple spanning trees, called spanning tree instances. Each spanning tree instance is associated with a different group of VLANs. Different spanning tree instances can have different root bridges, leading to different topologies that use different links. Traffic for some VLANs can use some links while traffic for other VLANs use other links. This significantly improves network resource use while maintaining a loop-free environment in each VLAN broadcast domain. See Figure 6-31.

– Multiple spanning tree instances
– A different root and topology for each instance (set of VLANs)
– Load-sharing on redundant links (VLAN-based)

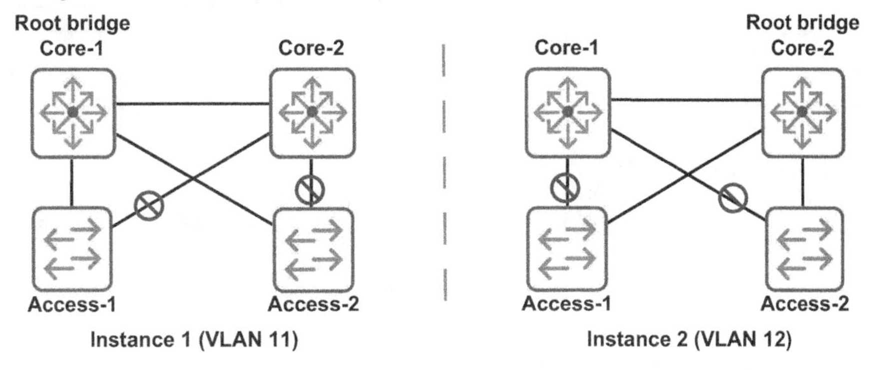

Figure 6-31: MSTP solution

In upcoming tasks, you will create two MSTP instances in addition to the default instance 0. Instance 0 is always required; it has some special roles, about which you will learn in a moment. Table 6-4 indicates the root bridge for each instance, as well as the VLANs that are associated with each instance.

Table 6-4: MSTP instances

Instance	VLANs	Root (priority 0)	Secondary root (priority 1 or 4096)
0	1 (and all unused)	Core-1	Core-2
1	11	Core-1	Core-2
2	12	Core-2	Core-1

As you see in Figure 6-31, this plan creates two spanning tree topologies, one used by instance 1 (and 0) and one used by instance 2. The first topology uses the uplinks between the access layer switches and Core-1, and the second topology uses the uplinks between the access layer switches and Core-2.

MSTP region

Before you set up MSTP in the following set of tasks, you should understand a bit about the necessary configuration steps, including how to set up a region. See Figure 6-32.

- Switches in an MSTP region must match region settings exactly
- Ports inside a region can use all instances

Parameter	Setting for your lab
Name	hpe
Revision number	1
Instance-to-VLAN mapping	0 = VLAN1 and all un used 1 = VLAN 11 2 = VLAN 12

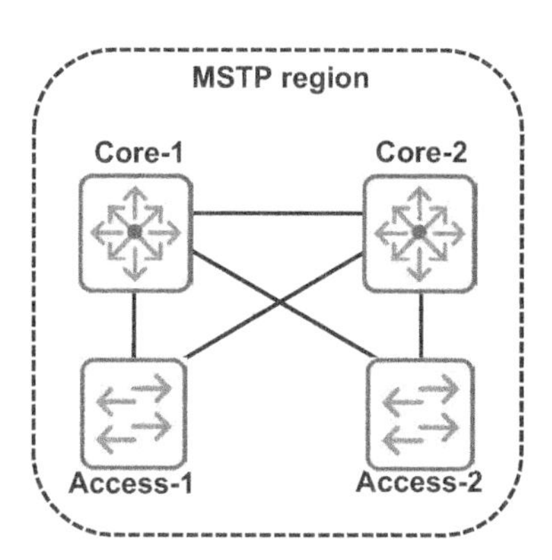

Figure 6-32: MSTP region

The MSTP region lets MSTP switches guarantee that they are using the same VLAN-to-instance mappings. It consists of these settings:

- Region name
- Revision number
- Instance-to-VLAN mapping

All of the settings must match for switches to become part of the same MSTP region. (See Table 6-5.)

Table 6-5: MSTP region settings

Parameter	Value
Name	hpe
Revision	1
Instance-to-VLAN mapping	0: 1-10,13-4094 1: 11 2: 12

MSTP region incompatibility

When two connected switches run MSTP but do not have the same settings, they have an MSTP region incompatibility. Switches in different MSTP regions cannot use their instances because they do not know that they map the same VLANs to these instances. Therefore, the switches use RSTP between each other. The ports between the regions, which are called boundary ports have a forwarding or blocking state that applies to *all* instances configured on the switch. As far as each switch is concerned, the different MSTP region is an RSTP switch. (See Figure 6-33.)

– Switches in different MSTP regions use RSTP between each other

– IST (instance 0) joins each region to the complete CIST

Common spanning tree (CST) or
Common internal spanning tree (CIST)

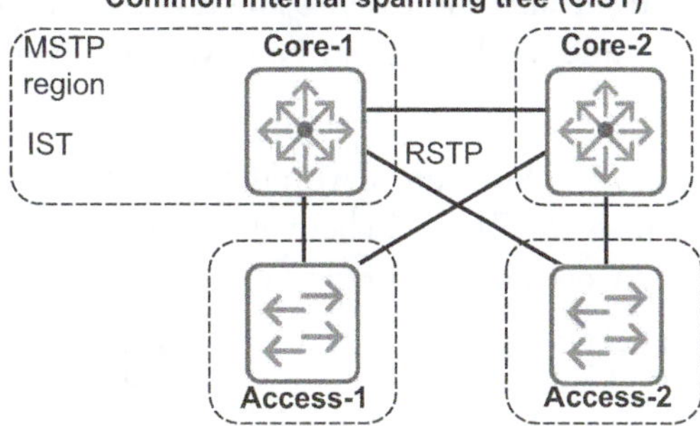

Figure 6-33: MSTP region incompatibility

The ArubaOS switches operate in MSTP mode, so you cannot truly configure RSTP on them. Instead, when you want to use RSTP, you simply do not establish any MSTP region settings. Because the switches have incompatible regions, each switch is in its own MSTP region and looks look an RSTP switch to its neighbors.

All of the switches across the different regions, as well as any actual RSTP switches, establish a common internal spanning tree (CIST). The CIST elects a single root. MSTP switches use their instance 0 settings in the CIST election. (The CIST is sometimes called just the common spanning tree [CST] when the connections between regions are emphasized.)

Instance 0 is also called the internal spanning tree (IST). The IST connects the MSTP region to the rest of the CIST. The region selects one boundary port—the one closest to the CIST root—as the region master port. In switch command outputs, you will see this port designated as a root port in the IST and CIST.

If the region has other boundary ports—such as Access-1's second uplink in Figure 6-33—those ports become designated, alternate, or backup ports in the CIST. You can determine the selected role by thinking of each MSTP region as a single RSTP switch and using the rules that you learned for RSTP. Remember: boundary ports function like RSTP ports, so the boundary port role is the *single* role used by *all* VLANs on the port, regardless of MSTP instance settings.

Note that, for ArubaOS switches, any settings that you configure globally for spanning tree apply to instance 0.

Configure Multiple Spanning Tree Protocol

In the following set of tasks, you will configure switches to support MSTP. The switches actually operate in MSTP mode by default. They were using RSTP in past tasks because you did not configure them to operate in the same MSTP region. Now you will configure them with matching MSTP region settings.

You will also configure Core-1 and Core-2 with the correct priorities to become elected the root and secondary root in each instance. The switches will alternate the roles so that Core-1 is primary root in instances 0 and 1, and Core-2 is the primary root in instance 2. (See Figure 6-34.)

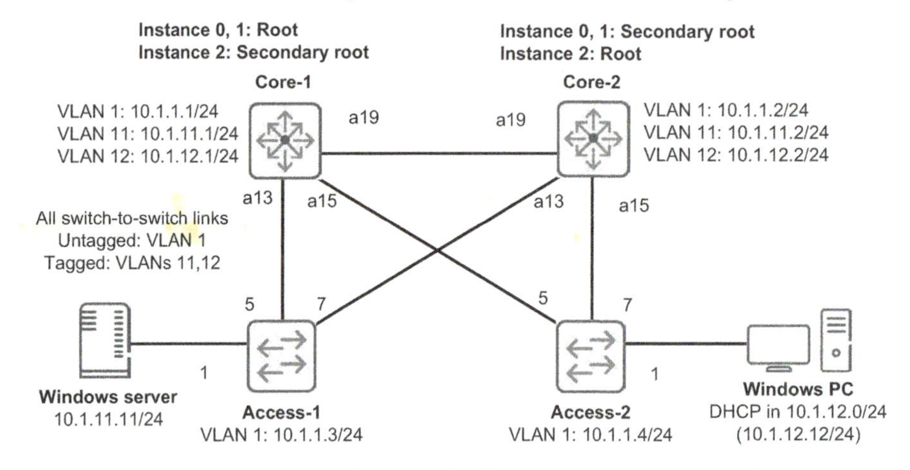

Figure 6-34: Configure Multiple Spanning Tree Protocol

Task 1: Configure the MSTP region

Important

Your switches should be running the configuration established at the end of the previous set of tasks.

In this task, you will configure a consistent MSTP region on all of your switches. Table 6-6 indicates the settings.

Table 6-6: MSTP region settings

Parameter	Value
Name	hpe
Revision	1
Instance-to-VLAN mapping	0: 1-10,13-4094 1: 11 2: 12

Core-1

1. Access the terminal session with Core-1 and move to the global configuration context.
2. Establish the settings indicated in the table.

```
Core-1(config)# spanning-tree config-name hpe
Core-1(config)# spanning-tree config-revision 1
Core-1(config)# spanning-tree instance 1 vlan 11
Core-1(config)# spanning-tree instance 2 vlan 12
```

Note

You do not need to configure the instance 0 mapping; by default, instance 0 includes all VLANs that have not been mapped to a different instance. If you did need to move VLANs back to instance 0, you would use this command: **no spanning-tree instance <current instance> vlan <list>**.

3. Verify the settings. Check the configuration name, revision, and digest. The digest is calculated from the VLAN mapping.

```
Core-1(config)# show spanning-tree mst-config
 MST Configuration Identifier Information
  MST Configuration Name : hpe
  MST Configuration Revision : 1
  MST Configuration Digest : 0xBE0284D20F4D46A8DA89C5D9B3B4F78A
  IST Mapped VLANs : 1-10,13-4094
  Instance ID Mapped VLANs
  ---------- --------------------------------------------------------
  1          11
  2          12
```

Core-2

4. Access a terminal session with Core-2.
5. Move to the global configuration context.

6. Configure the MSTP region using the same commands. Try to remember the commands yourself. (If you need help, refer to the previous page.)
7. Verify the MST configuration. Check the configuration name, configuration revision, and configuration digest or VLAN mappings, which must **all** exactly match these settings on Core-1.

```
Core-2(config)# show spanning-tree mst-config

 MST Configuration Identifier Information

  MST Configuration Name : hpe

  MST Configuration Revision : 1

  MST Configuration Digest : 0xBE0284D20F4D46A8DA89C5D9B3B4F78A

  IST Mapped VLANs : 1-10,13-4094

  Instance ID Mapped VLANs

  ----------- ------------------------------------------------------

  1           11

  2           12
```

8. The commands for setting up the MSTP region are exactly the same for the other three switches. What strategy could you use to configure the settings more quickly, as well as ensure that they match on all switches?

   Answer: Use a copy and paste approach.
9. Follow your strategy to configure the settings on Access-1 and Access-2. Do not move on until you have verified that the configuration name, revision, and digest or VLAN mappings match on all switches.

Important
Do not move on until you have configured identical region settings on all switches.

Task 2: Configure the instance root settings

Now that the switches are in the same MSTP region, they will elect a root for each instance. As discussed, you might want to select different roots for different instances to load-balance traffic. Table 6-7 shows the plan.

Table 6-7: MSTP instances

Instance	VLANs	Root (priority 0)	Secondary root (priority 4096)
0	1 (and all unused)	Core-1	Core-2
1	11	Core-1	Core-2
2	12	Core-2	Core-1

Core-1

1. Access the Core-1 terminal session and make sure that you are in the global configuration context.
2. In the previous set of tasks, you configured Core-1 with priority 0 without specifying a particular instance. This configured the setting on instance 0 (the IST). Use the same command but add the instance number to ensure that Core-1 becomes the root for instance 1.

```
Core-1(config)# spanning-tree instance 1 priority 0
```

3. The typical priority for the switch that you want to act as secondary root is 4096. You will specify 1 for the priority, which is then multiplied by 4096.

```
Core-1(config)# spanning-tree instance 2 priority 1
```

Core-2

4. Access the Core-2 switch terminal session.
5. Core-2 is already configured as the secondary root for instance 0 (IST).
6. Configure Core-2 as the secondary root for instance 1 by setting its priority to 1 (4096).

```
Core-2(config)# spanning-tree instance 1 priority 1
```

7. Configure Core-2 as the primary root for instance 2 by setting its priority to 0.

```
Core-2(config)# spanning-tree instance 2 priority 0
```

Task 3: Verify the configuration

If all of the switches are operating in the same MSTP region, they should have elected the same switches as root in each instance:

- Core-1 in instance 0 and 1
- Core-2 in instance 2

You will now verify that this is the case.

Core-1

1. Access the terminal session with Core-1. Verify that it is the root for instance 0.

```
Core-1(config)# show spanning-tree
 Multiple Spanning Tree (MST) Information
  STP Enabled   : Yes
  Force Version : MSTP-operation
  IST Mapped VLANs : 1-10,13-4094
  Switch MAC Address : 1c98ec-ab4b00
  Switch Priority    : 0
<-output omitted->
  IST Regional Root MAC Address : 1c98ec-ab4b00
  IST Regional Root Priority    : 0
  IST Regional Root Path Cost   : 0
  IST Remaining Hops            : 20
```

2. Also verify that it is root for instance 1 by adding the instance to the command.

```
Core-1(config)# show spanning-tree instance 1
 MST Instance Information
  Instance ID : 1
  Mapped VLANs : 11
  Switch Priority          : 0

  Topology Change Count    : 9
  Time Since Last Change   : 10 hours

  Regional Root MAC Address : 1c98ec-ab4b00
  Regional Root Priority    : 0
  Regional Root Path Cost   : 0
  Regional Root Port        : This switch is root
  <-output omitted->
```

3. Record the instance 0 and 1 root bridge ID (or at least the last four characters):

Core-2

4. Access the terminal session for Core-2. Verify that it is the root for instance 2.

```
Core-2(config)# show spanning-tree instance 2

 MST Instance Information

  Instance ID : 2

  Mapped VLANs : 12

  Switch Priority          : 0

  Topology Change Count    : 39

  Time Since Last Change   : 10 hours

  Regional Root MAC Address : 70106f-0d2100

  Regional Root Priority    : 0

  Regional Root Path Cost   : 0

  Regional Root Port        : This switch is root

  Remaining Hops            : 20
```

5. Record the instance 2 root bridge ID (or at least the last four characters):

6. Verify that Core-2 is using Core-1 for the instance 0 and 1 root bridge (ID recorded in step 4).

```
Core-2(config)# show spanning-tree instance ist

 IST Instance Information

  Instance ID : 0

  Mapped VLANs : 1-10,13-4094

  Switch Priority          : 4096

  Topology Change Count    : 0

  Time Since Last Change   : 10 hours

  Regional Root MAC Address : 1c98ec-ab4b00

  Regional Root Priority    : 0

  Regional Root Path Cost   : 20000

  Regional Root Port        : A19
```

```
  Remaining Hops              : 19
<-output omitted->
Core-2(config)# show spanning-tree instance 1
 MST Instance Information
  Instance ID : 1
  Mapped VLANs : 11
  Switch Priority           : 32768

  Topology Change Count     : 48
  Time Since Last Change    : 10 hours

  Regional Root MAC Address : 1c98ec-ab4b00
  Regional Root Priority      : 0
  Regional Root Path Cost     : 20000
  Regional Root Port          : A19
  Remaining Hops              : 19
<-output omitted->
```

Core-1

7. Check back in the command output on Core-1. Verify that it is using Core-2 for the instance 2 root bridge (ID recorded in 4).

```
Core-1(config)# show spanning-tree instance 2

 MST Instance Information

  Instance ID : 2
  Mapped VLANs : 12
  Switch Priority           : 4096

  Topology Change Count     : 9
  Time Since Last Change    : 10 hours

  Regional Root MAC Address : 70106f-0d2100
```

```
  Regional Root Priority     : 0
  Regional Root Path Cost    : 20000
  Regional Root Port         : A19
  Remaining Hops             : 19
```

Access-1

8. Access the Access-1 terminal session and verify that its root for each instance matches the root on the Core switches. On this switch, you'll practice using the **show spanning-tree root-history** command. You must enter the command for each instance separately.

Note

You can also use the **show spanning-tree** command; however, this command produces more output. The **show spanning-tree root-history** command makes it easier for you to focus on the root ID. It also helps you to troubleshoot because you can see changes to the root.

```
Access-1(config)# show spanning-tree root-history ist
 Status and Counters - IST Regional Root Changes History
  MST Instance ID         : 0
  Root Changes Counter    : 2
  Current Root Bridge ID : 0:1c98ec-ab4b00
<-output omitted->

Access-1(config)# show spanning-tree root-history msti 1
 Status and Counters - MST Instance Regional Root Changes History
  MST Instance ID         : 1
  Root Changes Counter    : 3
  Current Root Bridge ID : 0:1c98ec-ab4b00
<-output omitted->

Access-1(config)# show spanning-tree root-history msti 2
 Status and Counters - MST Instance Regional Root Changes History
  MST Instance ID         : 2
  Root Changes Counter    : 4
```

```
  Current Root Bridge ID : 0:70106f-0d2100
<-output omitted->
```

Access-2

9. Complete the same step on Access-2. The output should be identical.

Task 4: Map the topology

Based on what you learned in the chapter and the previous set of tasks, you should be able to predict the port roles for each switch-to-switch port in each instance. Do so using Figure 6-35, Figure 6-36, and Figure 6-37.

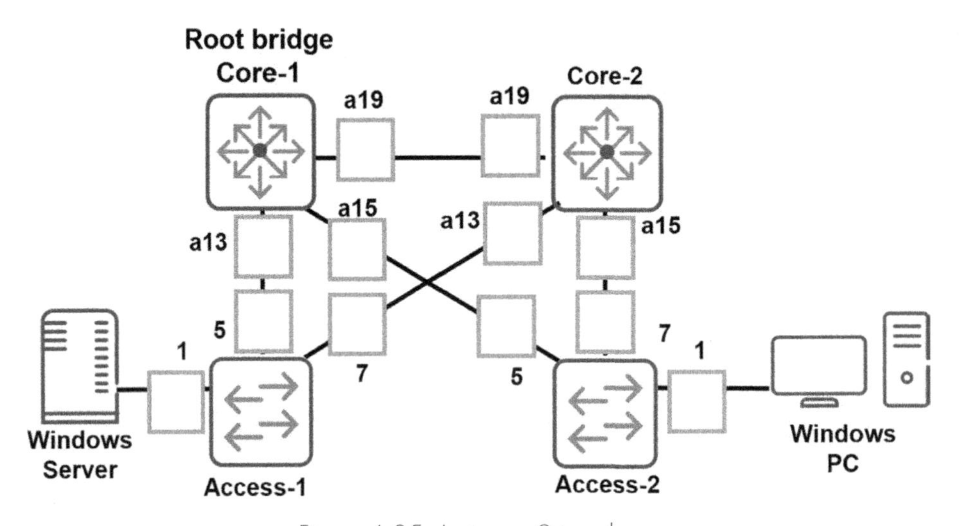

Figure 6-35: Instance 0 topology

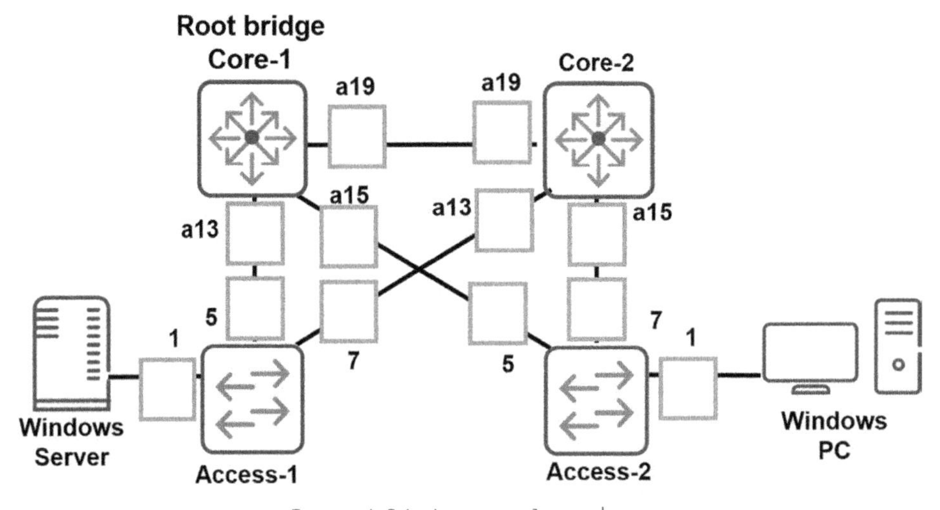

Figure 6-36: Instance 1 topology

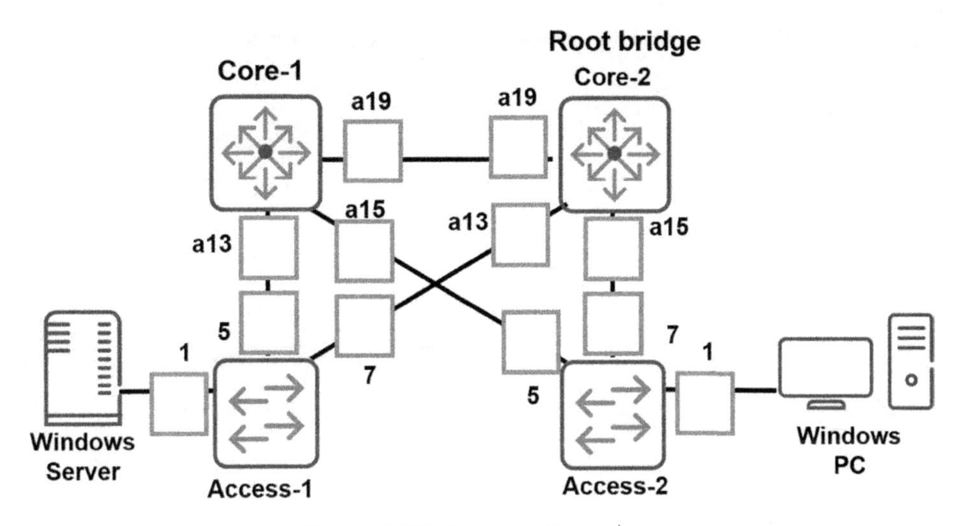

Figure 6-37: Instance 2 topology

You can verify your predictions in the following sections.

Map instance 0

Begin with instance 0.

Core-1

1. Access the terminal session with Core-1. View the port roles for instance 0. Check them against your predictions and adjust if necessary.

```
Core-1(config)# show spanning-tree a13,a15,a19 instance ist

<-output omitted->

                                                              Designated
 Port  Type      Cost   Priority Role        State       Bridge
 ----- --------- ------ -------- ----------  ----------  ----------
 A13   100/1000T 20000  128      Designated  Forwarding  c98ec-ab4b00
 A15   100/1000T 20000  128      Designated  Forwarding  1c98ec-ab4b00
 A19   100/1000T 20000  128      Designated  Forwarding  1c98ec-ab4b00
```

Core-2

2. Access the terminal session with Core-2. View the port roles for instance 0. Check them against your predictions and adjust if necessary.

```
Core-2(config)# show spanning-tree a13,a15,a19 instance ist

 IST Instance Information

<-output omitted->

                                                           Designated
  Port  Type      Cost    Priority Role       State      Bridge
  ----- --------- -----   -------- ---------- ---------- -------------
  A13   100/1000T 20000   128      Designated Forwarding 1c98ec-ab4b00
  A15   100/1000T 20000   128      Designated Forwarding 1c98ec-ab4b00
  A19   100/1000T 20000   128      Root       Forwarding 1c98ec-ab4b00
```

Access-1

3. Access the terminal session with Access-1. View the port roles for instance 0 (also called the IST). Check them against your predictions and adjust if necessary. Remember that you will find the port roles and states at the bottom of the output.

```
Access-1(config)# show spanning-tree 1,5,7 instance ist

 <-output omitted->

                                                           Designated
  Port  Type       Cost    Priority Role       State      Bridge
  ----- ---------  -----   -------- ---------- ---------- -------------
  1     100/1000T  Auto    128      Designated Forwarding 1c98ec-ab4b00
  5     100/1000T  20000   128      Root       Forwarding 1c98ec-ab4b00
  7     100/1000T  20000   128      Alternate  Blocking   1c98ec-ab4b00
```

Access-2

4. Access the terminal session with Access-2. View the port roles for instance 0. Check them against your predictions and adjust if necessary.

```
Access-2(config)# show spanning-tree 1,5,7 instance ist

  <-output omitted->
```

```
                                                        Designated
Port   Type      Cost  Priority Role       State      Bridge
-----  --------- ----- -------- --------   ---------- -------------
1      100/1000T Auto  128      Designated Forwarding 1c98ec-ab4b00
5      100/1000T 20000 128      Root       Forwarding 1c98ec-ab4b00
7      100/1000T 20000 128      Alternate  Blocking   1c98ec-ab4b00
```

Map the topology in instance 1

5. Follow the same steps to check your predictions for instance 1. Example command output is provided for your reference.

```
Core-1(config)# show spanning-tree a13,a15,a19 instance 1
<-output omitted->
                                                        Designated
Port   Type      Cost  Priority Role       State      Bridge
-----  --------- ----- -------- ---------- ---------- -------------
A13    100/1000T 20000 128      Designated Forwarding 1c98ec-ab4b00
A15    100/1000T 20000 128      Designated Forwarding 1c98ec-ab4b00
A19    100/1000T 20000 128      Designated Forwarding 1c98ec-ab4b00
```

```
Core-2(config)# show spanning-tree a13,a15,a19 instance 1
                                                        Designated
Port   Type      Cost  Priority Role       State      Bridge
-----  --------- ----- -------- ---------- ---------- -------------
A13    100/1000T 20000 128      Designated Forwarding 70106f-0d2100
A15    100/1000T 20000 128      Designated Forwarding 70106f-0d2100
A19    100/1000T 20000 128      Root       Forwarding 1c98ec-ab4b00
```

```
Access-1(config)# show spanning-tree 1,5,7 instance 1
 <-output omitted->
                                                     Designated
  Port  Type      Cost  Priority Role       State      Bridge
  ----- --------- ----- -------- ---------- ---------- -------------
  1     100/1000T 20000 128      Designated Forwarding 6c3be5-6208c0
  5     100/1000T 20000 128      Root       Forwarding 1c98ec-ab4b00
  7     100/1000T 20000 128      Alternate  Blocking   70106f-0d2100

Access-2(config)# show spanning-tree 1,5,7 instance 1
  <-output omitted->
                                                     Designated
  Port  Type      Cost  Priority Role       State      Bridge
  ----- --------- ----- -------- ---------- ---------- -------------
  1     100/1000T 20000 128      Designated Forwarding 843497-0223c0
  5     100/1000T 20000 128      Root       Forwarding 1c98ec-ab4b00
  7     100/1000T 20000 128      Alternate  Blocking   70106f-0d210
```

Map the topology in instance 2

6. Follow the same steps to check your predictions for instance 2. Example command output is provided for your reference.

```
Core-1(config)# show spanning-tree a13,a15,a19 instance 2
<-output omitted->
                                                     Designated
  Port  Type      Cost  Priority Role       State      Bridge
  ----- --------- ----- -------- ---------- ---------- -------------
  A13   100/1000T 20000 128      Designated Forwarding 1c98ec-ab4b00
  A15   100/1000T 20000 128      Designated Forwarding 1c98ec-ab4b00
  A19   100/1000T 20000 128      Root       Forwarding 70106f-0d2100
```

```
Core-2(config)# show spanning-tree a13,a15,a19 instance 2

                                                          Designated
  Port  Type       Cost   Priority Role       State       Bridge
  ----- ---------- -----  -------- ---------- ----------  -------------
  A13   100/1000T  20000  128      Designated Forwarding  70106f-0d2100
  A15   100/1000T  20000  128      Designated Forwarding  70106f-0d2100
  A19   100/1000T  20000  128      Designated Forwarding  70106f-0d2100

Access-1(config)# show spanning-tree 1,5,7 instance 2
 <-output omitted->
                                                          Designated
  Port  Type       Cost   Priority Role       State       Bridge
  ----  ---------- -----  -------- ---------- ----------  -------------
  1     100/1000T  20000  128      Designated Forwarding  6c3be5-6208c0
  5     100/1000T  20000  128      Alternate  Blocking    1c98ec-ab4b00
  7     100/1000T  20000  128      Root       Forwarding  70106f-0d210

Access-2(config)# show spanning-tree 1,5,7 instance 2
  <-output omitted->
                                                          Designated
  Port  Type       Cost   Priority Role       State       Bridge
  ----  ---------- -----  -------- ---------- ----------  -------------
  1     100/1000T  20000  128      Designated Forwarding  843497-0223c0
  19    100/1000T  20000  128      Alternate  Blocking    1c98ec-ab4b00
  21    100/1000T  20000  128      Root       Forwarding  70106f-0d2100
```

Task 5: Save your configurations

Save the current configuration on each of the four switches (**`write memory`**).

Task 6: Exploration activity

Task A and Task B guide you through making two common errors with spanning tree configurations. You then explore the effect and correct the issue.

Task A: Add a VLAN to a switch and map it to an instance

Follow these steps:

1. Make sure that you have saved the configuration on all of your switches. Do **not** save your configuration at any point during this task. You will reboot at the end to return to the correct configuration.

Core-1

2. Access the terminal session with Core-1 and move to global configuration mode.
3. Configure the interface that connects to Core-2 as a tagged member of VLAN 13.

```
Core-1(config)# vlan 13 tagged a19
```

Core-2

4. Follow the same steps on Core-2.

```
Core-2(config)# vlan 13 tagged a19
```

Access-1

5. On Access-1, check the switch-to-switch interfaces' port roles for each MSTP instance using the **show spanning-tree <int-id-list> instance <id>** command. You should not see any change from before.

Core-1

6. On Core-1, add VLAN 13 to instance 1.

```
Core-1(config)# spanning-tree instance 1 vlan 13
```

Core-2

7. Do the same on Core-2.

```
Core-2(config)# spanning-tree instance 1 vlan 13
```

Access-1

Although you have not made any changes to the access layer switches, you will see that this configuration has actually affected them.

8. On Access-1, check the switch-to-switch interfaces' port roles for each MSTP instance.

```
Access-1(config)# show spanning-tree 5,7 instance ist

<-output omitted->

  Port Type      Cost  Priority Role       State      Bridge
  ---- --------- ----- -------- ---------- ---------- -------------
  5    100/1000T 20000 128      Root       Forwarding 1c98ec-ab4b00
  7    100/1000T 20000 128      Alternate  Blocking   70106f-0d2100

Access-1(config)# show spanning-tree 19,21 instance 1

<-output omitted->

  Port  Type      Cost  Priority Role      State      Bridge
  ----- --------- ----- -------- --------- ---------- -------------
  5     100/1000T 20000 128      Master    Forwarding 6c3be5-6208c0
  7     100/1000T 20000 128      Alternate Blocking   6c3be5-6208c0

Access-1(config)# show spanning-tree 5,7 instance 2

<-output omitted->

                                                      Designated
  Port Type      Cost  Priority Role      State      Bridge
  ---- --------- ----- -------- --------- ---------- -------------
  5    100/1000T 20000 128      Master    Forwarding 6c3be5-6208c0
  7    100/1000T 20000 128      Alternate Blocking   6c3be5-6208c0
```

9. As you see, Access-1 now is forwarding only on the link to Core-1 no matter what the instance or VLAN.

10. Compare the MSTP region settings on Core-1, Core-2, and Access-1 for a clue as to what has happened.

```
Access-1# show spanning-tree mst-config

 MST Configuration Identifier Information
```

```
  MST Configuration Name : hp
  MST Configuration Revision : 1
  MST Configuration Digest : 0xBE0284D20F4D46A8DA89C5D9B3B4F78A

  IST Mapped VLANs : 1-10,13-4094
  Instance ID Mapped VLANs
  ----------- --------------------------------------------------------
  1           11
  2           12

Core-1(config)# show spanning-tree mst-config
 MST Configuration Identifier Information

  MST Configuration Name : hpe
  MST Configuration Revision : 1
  MST Configuration Digest : 0xF843355B493955BCD42BEE4C4E2FFB00
  IST Mapped VLANs : 1-10,14-4094

  Instance ID Mapped VLANs
  ----------- --------------------------------------------------------
  1           11,13
  2           12

Core-2(config)# show spanning-tree mst-config
 MST Configuration Identifier Information

  MST Configuration Name : hpe
  MST Configuration Revision : 1
  MST Configuration Digest : 0xF843355B493955BCD42BEE4C4E2FFB00

  IST Mapped VLANs : 1-10,14-4094
```

```
Instance ID Mapped VLANs
----------- ------------------------------------------------
1           11,13
2           12
```

11. As you see, Access-1 has different MST region settings. This discrepancy explains why Access-1 is interacting with Core-1 and Core-2 as if it was running RSTP. It is using the CIST rather than its individual MSTP instances.

12. Why are Access-1's MST region settings now different from the Core switches' settings? See the following section for an explanation.

Core-1 and Core-2

13. Reboot the Core-1 and Core-2 switches using the **reload** command. (Also reboot Access-1 if you made any changes to its configuration.) Do **not** save the configuration.

Explanation of Task A

You can add a new VLAN to a switch without affecting the MSTP topology.

But any difference in VLAN-to-instance *mappings* will place switches in different regions even if some switches are not actively using the VLAN. As you learned earlier, when switches are in different regions, they interact using RSTP, and the boundary ports between regions are part of a CIST. See Figure 6-38.

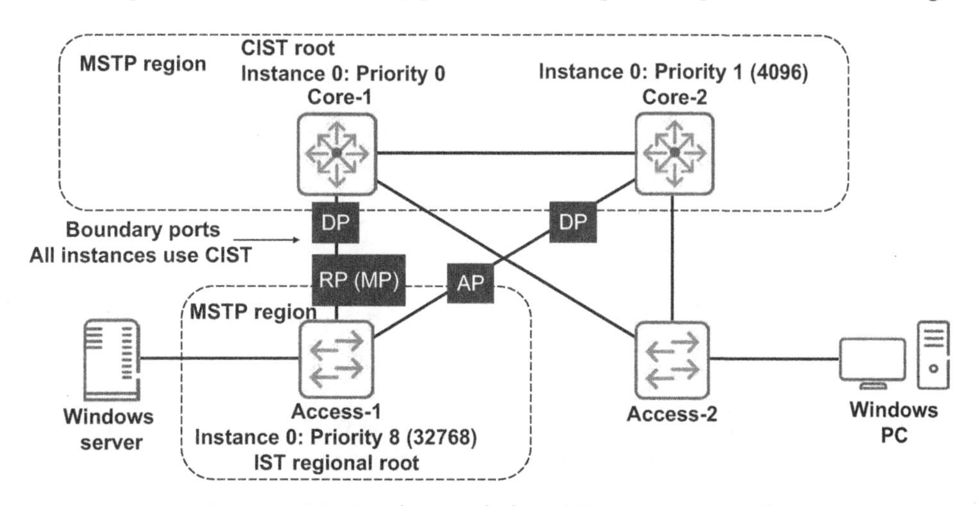

Figure 6-38: Topology with the MST region mismatch

On an MSTP boundary port, all instances refer to the IST port role and state. Here, Access-1's link to Core-1 is its root port in the CIST. So instance 1 and instance 2 also forward on that port. (Master

Port is another name for a region's root port in the CIST; this is the port that offers the MSTP region its link to the rest of the CST). Access-1's link to Core-2 is its alternate port in the CIST, so the other instances also block on this port.

In short, no load-balancing occurs on the MSTP boundary ports. Therefore, adding a VLAN and moving that VLAN from instance 0 to another instance in an active network can cause disruptions and non-optimal link usage.

Recommendation

Plan in advance and place all VLANs that might be used in the future in the desired instance on all switches.

More details

As you learned earlier, Core-1 is elected the CIST root because it has the lowest instance 0 priority.

Switches in other MSTP regions elect a regional root for the CIST based on which switch offers the best path to the CIST root. The regional root becomes the IST root. In this case, Access-1 is the only switch in its region, so it is the regional root. Its boundary port with the lowest cost to the CIST root becomes the CIST root port (IST master port), and its other boundary port becomes an alternate port.

Note an important implication: an MSTP region could have an IST root that does not have the lowest priority value. This can occur when the region is part of a CIST and does not contain the CIST root. The switch with the best path to the CIST root might not be the switch with the lowest instance 0 priority value, but it still becomes the IST root.

Recommendation

It is important to set priorities for instance 0 even if you are not using that instance for carrying data traffic. You want to control which switch becomes CIST root in case of misconfigurations.

Task B: Remove a VLAN from a link

Follow these steps:

1. Make sure that you have saved the configuration on all of your switches. Do **not** save your configuration at any point during this task. You will reboot at the end to return to the correct configuration.

Windows client

2. Access the client desktop and open a command prompt.

3. Establish a continuous ping to the server.

```
ping 10.1.11.11 -t
```

Access-1

4. Access the terminal session with Access-1 and move to the global configuration context.
5. Remove VLAN 12 from the links to Core-1 and Core-2.

```
Access-1(config)# no vlan 12 tagged 5,7
```

6. Assign Access-1 priority multiplier 4 (priority 16384) in each instance.

   The issue that you want to observe might or might not occur depending on which access layer switch is selected as being closer to the root, which often depends on MAC address in a real network. For this experiment, you want to make sure that Access-1 is selected so that you can be sure to see the potential issue.

```
Access-1(config)# spanning-tree priority 4
Access-1(config)# spanning-tree instance 1 priority 4
Access-1(config)# spanning-tree instance 2 priority 4
```

Access-2

7. Access the terminal session with Access-2 and move to the global configuration mode.
8. Remove VLAN 11 from the links to Core-1 and Core-2.

```
Access-2(config)# no vlan 11 tagged 5,7
```

Windows client

9. Check the ping.

   As you see, at this point, the ping continues normally. During normal operation for this topology, Access-1 does not need to be part of VLAN 12 and Access-2 does not need to be part of VLAN 11.

Core-1

Now observe what happens if the link between Core-1 and Core-2 fails. Access the terminal session with Core-1.

10. Shut down the port that connects to Core-2.

```
Core-1(config)# interface a19 disable
```

Windows client

11. View the ping on the client. After a moment, the ping begins failing. Why did the failover not occur properly? (If you do not know, do not worry. The following steps help you to discover the problem.)

Core-1

12. The client needs to reach Core-1, its default gateway in VLAN 12. VLAN 12 is associated with instance 2. Which link is active on Core-1 in instance 2? View the port roles on Core-1 in instance 2 to find out.

    Enter this command:

```
Core-1(config)# show spanning-tree a13,a15 instance 2

<-output omitted->

                                                          Designated
  Port Type      Cost  Priority Role       State      Bridge
  ---- --------- ----- -------- ---------- ---------- -------------
  A13  100/1000T 20000 128      Root       Forwarding 6c3be5-6208c0
  A15  100/1000T 20000 128      Alternate  Blocking   843497-0223c0
```

Access-1

13. The root port on Core-1 connects to Access-1. Does the link between Core-1 and Access-1 carry VLAN 12? To remind yourself, view VLANs on port 5 on Access-1.

```
Access-1(config)# show vlan port 5

 Status and Counters - VLAN Information - for ports 5

  VLAN ID Name                             | Status     Voice Jumbo
  ------- -------------------------------- + ---------- ----- -----
  1       DEFAULT_VLAN                     | Port-based No    No
  11      VLAN11                           | Port-based No    No
```

As you see, the active link does not carry VLAN 12. Therefore, the client cannot reach its default gateway, causing a disruption in connectivity.

Core-1

14. Re-enable the port on Core-1.

```
Core-1(config)# interface a19 enable
```

Access-1 and Access-2

15. Reboot the Access-1 and Access-2 switches using the **reload** command. Do **not** save the configuration.

Windows client

End the continuous ping on the PC with <Ctrl+c>.

Explanation of Task B

MSTP uses a *single* BPDU, which is backward compatible with RSTP and also includes MSTP instance information within it. This approach has the advantage of causing minimal impact on the switch's CPU; only one BPDU must be processed. However, it also introduces some important implications, which you must understand.

An MSTP BPDU is carried on the Ethernet link regardless of the VLANs carried on that link, so the switches are essentially using a single physical Layer 2 topology for all instances. The switches then use the specific instance settings to calculate the port roles and states for each spanning tree instance. But, again, this process is not VLAN aware. The VLAN mappings only come into play when the switch decides how to treat traffic in a specific VLAN on a specific port. For example, instance 2 is forwarding on a port, VLAN 12 is part of instance 2, so VLAN 12 traffic is allowed to be forwarded on that port.

Note that the link between Core-1 and Core-2 is not shown in Figure 6-39 below because this link is down.

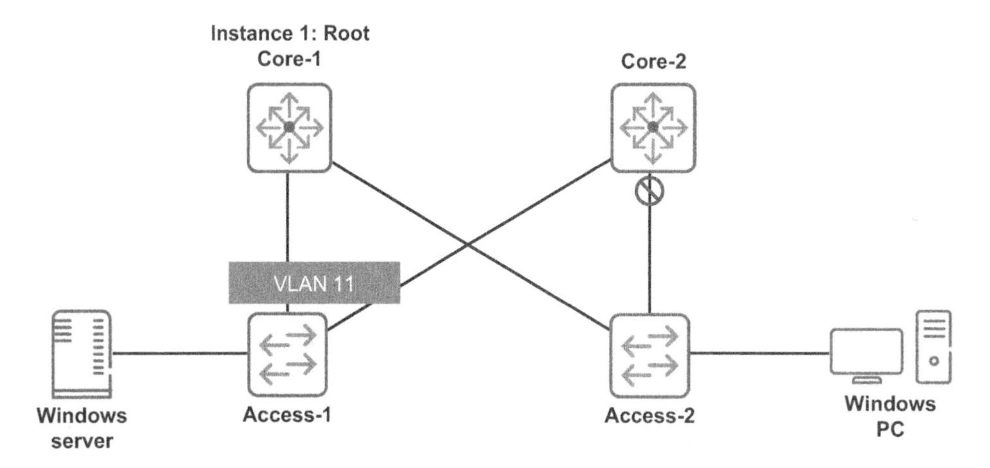

Figure 6-39: Instance 1 topology and VLAN 11 assignments

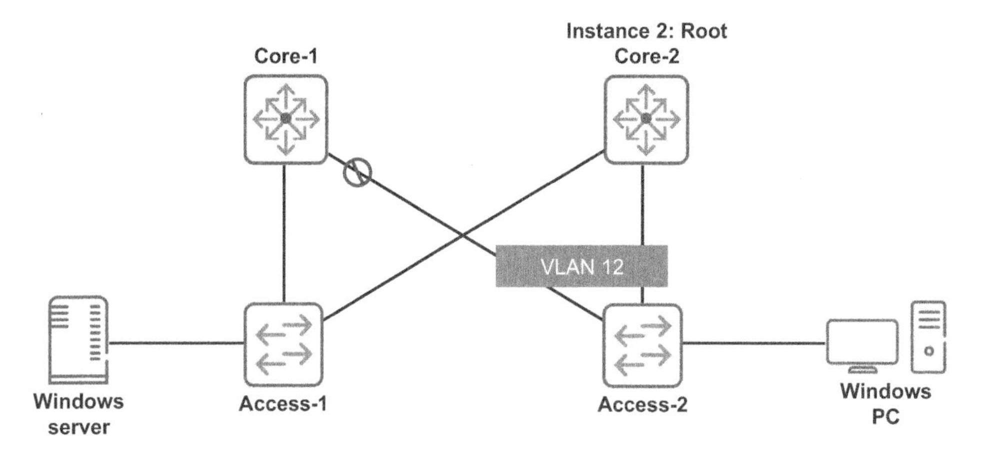

Figure 6-40: Instance 2 topology and VLAN 12 assignments

Because the spanning tree instances are set up without regard to which VLANs are actually carried on a link, it is possible that an instance could select a root port that does not carry one of the instance's VLANs and an alternate port that does—as occurred in this scenario. See Figure 6-40. In other words, even though *spanning tree* allows the forwarding of VLAN 12 on a port, VLAN 12 might not be allowed by the tagging on the port. And even though another port is available that *does* have the proper VLAN 12 tagging, that port is still blocked in instance 2.

You need to be aware of the VLAN topology on your switches and make sure that it works with the spanning tree instance topology *both in normal operation and failover situations.*

Recommendation

To prevent misconfigurations, configure all switch-to-switch links to support all VLANs used in the network segment.

Note

This explanation helps you to understand more completely why MSTP region settings must match on switches. Because the spanning trees are set up without regard to VLAN topology, it is very important that every switch assigns the same VLANs to the same instance. Otherwise, one switch might use one topology to flood a broadcast in a VLAN while a different switches uses a different topology: loops could occur. If switches have different settings, they must use RSTP to prevent loops—which is, as you have learned, what they do.

Learning check

You will now review what you learned in this set of tasks.

1. Did your predictions for the port roles and states match the actual roles and states in each spanning tree instance? If your predictions were incorrect in any way, think about why this was the case. Explain what you now understand about how MSTP works.

2. Describe the differences between the topology in instance 1 and in instance 2.

3. When you changed the region settings in this set of tasks, the settings took effect immediately. Instead of configuring the new settings directly, though, you can configure pending settings with the following commands. You can then apply or revert the pending configurations. What are the differences between these approaches?

   Configure pending settings with these commands:

```
ArubaOS(config)# spanning-tree pending config-name <name>
ArubaOS(config)# spanning-tree pending config-revision <number>
ArubaOS(config)# spanning-tree pending instance <ID> vlan <vid range>
```

   When you are ready, activate the pending region settings with this command:

```
ArubaOS(config)# spanning-tree pending apply
```

   If, before you apply the pending settings, you decide that you want to revert them to match the currently active settings instead, use this command:

```
ArubaOS(config)# spanning-tree pending reset
```

Answers to Learning check

1. Did your predictions for the port roles and states match the actual roles and states in each spanning tree instance? If your predictions were incorrect in any way, think about why this was the case. Explain what you now understand about how MSTP works.

   The answer to this question depends on your experience.

2. Describe the differences between the topology in instance 1 and in instance 2.

   Core-1 is the root in instance 1, so the access layer switches use their direct link to Core-1 and block their other link. Core-2 is the root in instance 2, so the access layer switches use their direct Core-2 and block their other link. In this way, each access layer switch has two links, one of which is used in each instance.

3. When you changed the region settings in the task set, the settings took effect immediately. Instead of configuring the new settings directly, though, you can configure pending settings with the commands shown at the bottom of this page. You can then apply or revert the pending configurations. What are the differences between these approaches?

   Configure pending settings with these commands:

```
ArubaOS(config)# spanning-tree pending config-name <name>
ArubaOS(config)# spanning-tree pending config-revision <number>
ArubaOS(config)# spanning-tree pending instance <ID> vlan <vid range>
```

   When you are ready, activate the pending region settings with this command:

```
ArubaOS(config)# spanning-tree pending apply
```

   If, before you apply the pending settings, you decide that you want to revert them to match the currently active settings instead, use this command:

```
ArubaOS(config)# spanning-tree pending reset
```

   It can be useful to configure the region settings in advance without applying them. Then you can set up the new configuration on many switches in advance. When you are ready, you apply the changes to all the switches at once, minimizing the disruption. (Changing the region settings always causes a brief disruption as the topology reconverges.)

Plan instances for load-sharing

Figure 6-41 shows an expanded network with more VLANs.

1. How many MSTP instances, besides the IST (instance 0), do you need to have most effective load sharing?

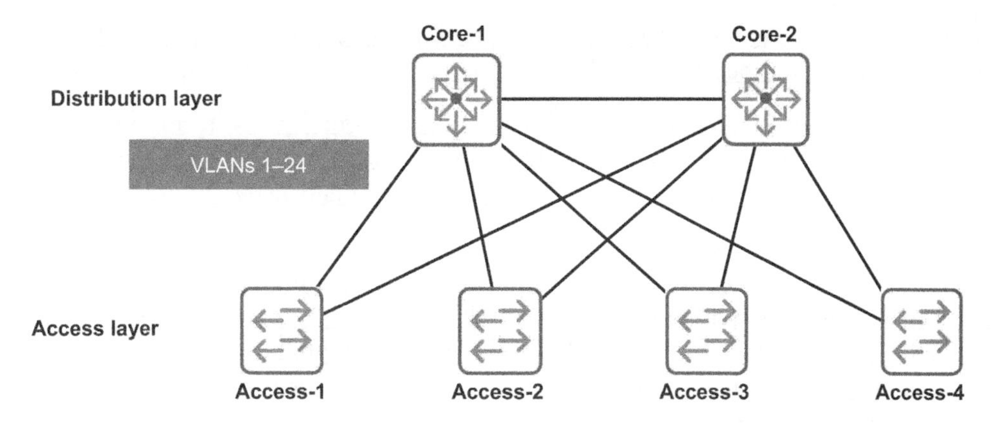

Figure 6-41: Activity (question 1)

Answer: To implement the most effective load sharing, you need a separate instance for each distribution/aggregation switch (Core-1 and Core-2). Then each of these switches can act as root in one instance and uplinks to that switch can be active in that instance. Therefore, in the first scenario, two instances (in addition to instance 0) work.

The number of VLANs does not affect the number of instances that you need. You can assign half of the VLANs to one instance and half to the other. Or you might take a more nuanced approach to load-balancing. For example, you can look at the typical traffic load for each VLAN and combine the VLANs such that each instance has about half the load (but not necessarily half the VLAN IDs).

You also can choose not to load-balance traffic evenly. For example, maybe you want to reserve a path for a critical VLAN, but you do not care that other VLANs have to share bandwidth. You would then give the critical VLAN its own instance and assign other VLANs to the other instance.

The choice is yours and depends on the company's needs.

2. Consider a network that has expanded even more, as shown in Figure 6-42.

   How many MSTP instances, besides the IST (instance 0), do you need to have most effective load sharing?

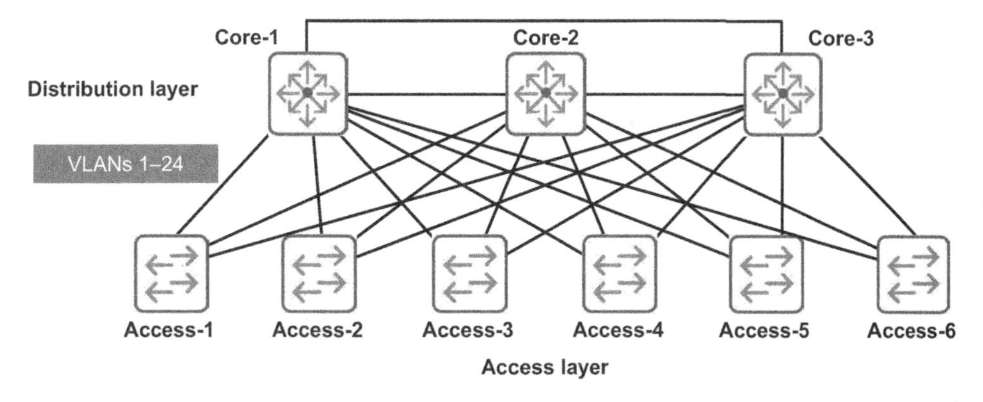

Figure 6-42: Activity (question 2)

Answer: You now have three switches at the distribution layer, so you should use three instances. Again, you should distribute the VLANs among these instances evenly or evenly based on traffic load.

As you can see from this second scenario, the more distribution layer switches you have, the more configuration you need to perform in order to use the bandwidth on the access layer switches' uplinks. The design becomes more complex, which is why many networks use a pair of distribution switches.

Note finally that you do not *have* to load share traffic. You should think about why you do or do not want to. Maybe you can provision enough bandwidth so that an access layer switch can handle all traffic on a single uplink. If so, you can use RSTP and avoid the complexities of MSTP configuration. If the backup link has the same bandwidth as the main link, you also know that the backup link will have enough bandwidth to handle all traffic during failover.

Summary

In this chapter, you learned about how to use RSTP and MSTP to add Layer 2 link redundancy without introducing dangerous loops and broadcast storms. You can set the appropriate priority values to select a primary and secondary root. You understand how an RSTP topology is created and the port roles and states used in the topology. You also understand how failover occurs.

- RSTP to support redundant links without loops
- MSTP to utilize bandwidth more efficiently
- Best practices

Figure 6-43: Summary

You learned about how you can use MSTP to use your network bandwidth more efficiently. MSTP operates much like RSTP, but supports multiple instances, each with its own root, topology, and port roles and states. You also explored some best practices for implementing MSTP and learned how to avoid potential issues.

Learning check

1. The root switch in STP is elected using which of the following criteria?
   a. Lowest MAC address
   b. Highest MAC address
   c. Lowest priority
   d. Highest priority
   e. Lowest bridge ID
   f. Highest bridge ID

2. Which RSTP port role designates a port that offers a second (or third, etc.) best path to the root? In other words, the port connects to a switch that:
   a. Offers a better path to the root than this switch does
   b. But does not offer this switch its best path to the root
   c. Root port
   d. Backup port
   e. Designated port
   f. Alternate port

3. What ArubaOS switch command enables a spanning tree protocol? Also specify the default mode for the protocol.

4. You want to configure MSTP region settings on your switches in advance and then apply them all at the same time. How do you achieve this goal?

Answers to Learning check

1. The root switch in STP is elected using which of the following criteria?
    a. Lowest MAC address
    b. Highest MAC address
    c. Lowest priority
    d. Highest priority
    e. Lowest bridge ID
    f. Highest bridge ID

2. Which RSTP port role designates a port that offers a second (or third, etc.) best path to the root? In other words, the port connects to a switch that:
    a. Offers a better path to the root than this switch does
    b. But does not offer this switch its best path to the root
    c. Root port
    d. Backup port
    e. Designated port
    f. Alternate port

3. What ArubaOS switch command enables a spanning tree protocol? Also specify the default mode for the protocol.

    The command is **spanning-tree**. The default mode is MSTP.

4. You want to configure MSTP region settings on your switches in advance and then apply them all at the same time. How do you achieve this goal?

    Use the **pending** options for the spanning tree region configuration command. Then apply the pending configurations after all switches are configured.

7 Link Aggregation

EXAM OBJECTIVES

✓ Differentiate between different types of link aggregation and understand the benefits of Link Aggregation Control Protocol (LACP)

✓ Configure and troubleshoot link aggregation on ArubaOS switches

ASSUMED KNOWLEDGE

Before reading this chapter, you should have a basic understanding of:

- Virtual LANs (VLANs)
- Spanning Tree Protocol (STP)

INTRODUCTION

Link aggregation helps you to add redundant links and more bandwidth between switches more easily and efficiently.

Adding redundant links between the same two switches

In the previous chapter, you learned how to create a redundant Layer 2 topology using spanning tree. Now think about a situation in which you want to increase bandwidth between two switches—for example, Core-1 and Core-2 in the figure. You decide to add another link between the switches.

If you use RSTP, how will this protocol deal with the new link? If you use MSTP, how could the protocol deal with this new link, and what would you need to configure to obtain the desired results? What are drawbacks of allowing RSTP/MSTP to handle the redundant links in this situation?

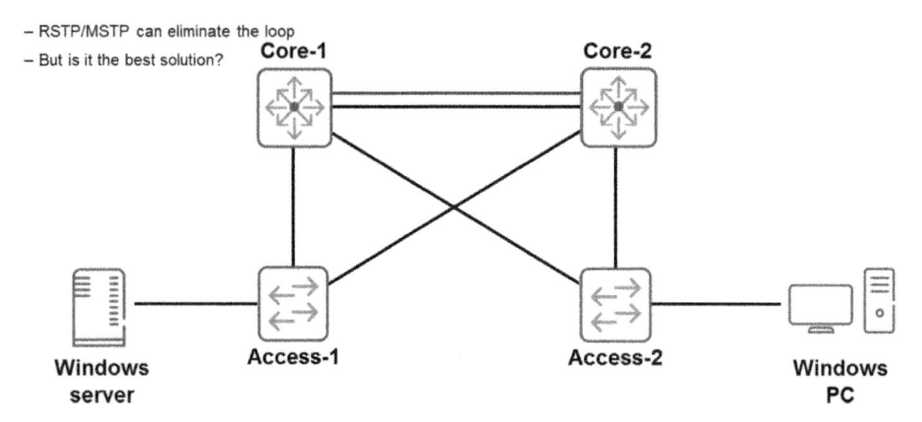

Figure 7-1: Adding redundant links between the same two switches

Using MSTP for Redundant Links between Two Switches

You will now find out the answers to the questions posed on the previous page. You will add a redundant link between Core-1 and Core-2 and see how MSTP handles the new topology.

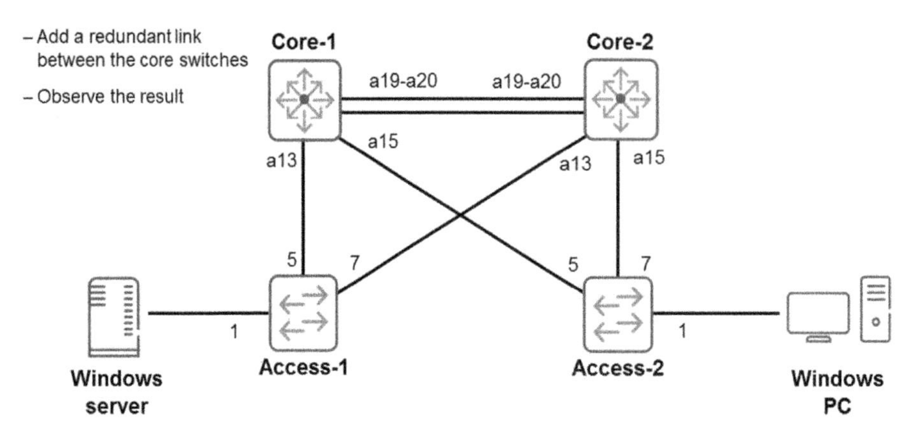

Figure 7-2: Using MSTP for Redundant Links between Two Switches

The figure below shows the final topology.

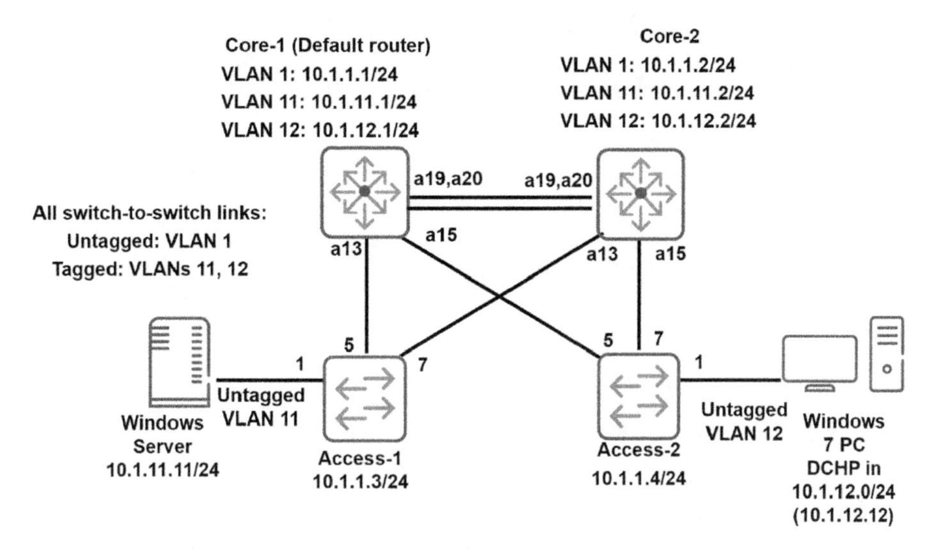

Figure 7-3: Final topology—Using MSTP for Redundant Links between Two Switches

Task 1: Verify the MSTP configuration

In this task you will verify the existing MSTP configuration.

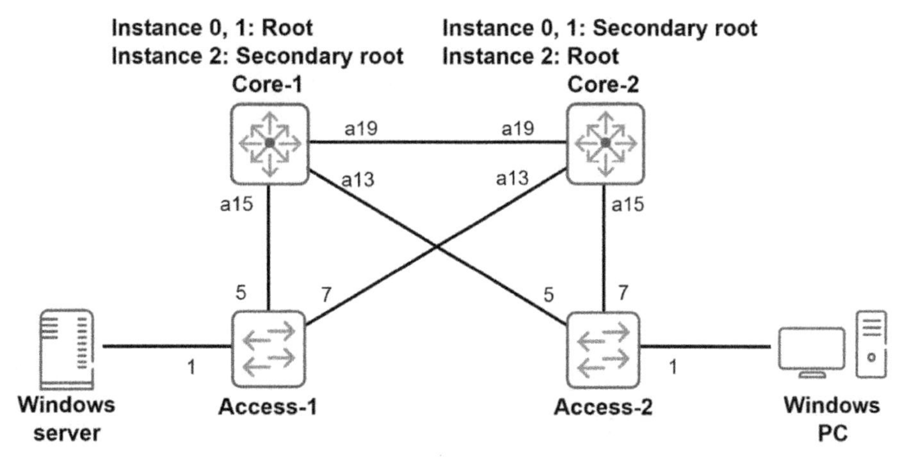

Figure 7-4: Task 1: Verify the MSTP configuration

Core-1

1. Access a terminal session with Core-1.
2. Verify that Core-1 is root for the CST, as well as instance 0 (IST) and instance 1.

```
Core-1# show spanning-tree
<-output omitted->

  CST Root MAC Address : 70106f-0d4100
  CST Root Priority    : 0
  CST Root Path Cost   : 0
  CST Root Port        : This switch is root

  IST Regional Root MAC Address : 70106f-0d4100
<-output omitted->

Core-1(config)# show spanning-tree instance 1
 MST Instance Information

  Instance ID : 1
  Mapped VLANs : 11
  Switch Priority          : 0
  Topology Change Count    : 12
  Time Since Last Change   : 32 secs
  Regional Root MAC Address : 70106f-0d4100
  Regional Root Priority    : 0
  Regional Root Path Cost   : 0
  Regional Root Port        : This switch is root
<-output omitted->
```

3. Also check instance 2. If Core-2 is root, the interface that connects to Core-2 should be the root port.

```
Core-1(config)# show spanning-tree instance 2
 MST Instance Information
  Instance ID : 2
```

```
Mapped VLANs : 12

Switch Priority          : 4096

Topology Change Count    : 9
Time Since Last Change   : 2 mins

Regional Root MAC Address : 70106f-0e2f00
Regional Root Priority    : 0
Regional Root Path Cost   : 20000
Regional Root Port        : A19
```

Task 2: Add a redundant link

You will now add a redundant link between Core-1 and Core-2.

Core-1

1. On Core-1, permit the second physical interface that connects to Core-2 to carry tagged traffic in VLANs 11 and 12. The port is an untagged member of VLAN 1 by default.

```
Core-1# config
Core-1(config)# vlan 11 tagged a20
Core-1(config)# vlan 12 tagged a20
```

2. Enable the interface.

```
Core-1(config)# interface a20 enable
```

3. Verify the configuration.

```
Core-1(config)# show vlan port a20 detail
```

Core-2

4. Access Core-2. Follow similar steps to configure the other side of the link.

```
Core-2# config
Core-2(config)# vlan 11 tagged a20
Core-2(config)# vlan 12 tagged a20
Core-2(config)# interface a20 enable
```

5. Verify the configuration.

```
Core-2(config)# show vlan port a20 detail
```

6. Verify that Core-2 detects Core-1 as an LLDP neighbor on both interfaces.

```
Core-2(config)# show lldp info remote-device

 LLDP Remote Devices Information

  LocalPort | ChassisId                 PortId PortDescr SysName
  --------- + ------------------------- ------ --------- -------------
  A13       | 6c 3b e5 62 08 c0         7      7         Access-1
  A15       | 84 34 97 02 23 c0         7      7         Access-2
  A19       | 1c 98 ec ab 4b 00         19     A19       Core-1
  A20       | 1c 98 ec ab 4b 00         20     A20       Core-1
```

Task 3: Observe MSTP with the new link

You will now examine how MSTP handles the two links between Core-1 and Core-2. You should be familiar with the ArubaOS **display** and **show** commands for STP from Chapter 6.

Core-1

1. Core-1 is root in instance 0 and 1. Core-2 is root in instance 2. What do you know about the role and state for all interfaces on Core-1 in instance 0 and 1 and on Core-2 in instance 2?
2. But you don't know the status in instance 2 for the two Core-1 interfaces that connect to Core-2. Check this status now.

```
Core-1(config)# show spanning-tree a19,a20 instance 2

 MST Instance Information

  Instance ID : 2

<-output omitted->

                                                                Designated
  Port Type       Cost    Priority Role      State        Bridge
  ---- ---------- ------- -------- --------- ------------ -------------
  A19  100/1000T  20000   128      Root      Forwarding   70106f-0d2100
  A20  100/1000T  20000   128      AlternateBlocking      70106f-0d2100
```

Core-2

3. On Core-2, closely examine the status for the two ports that connect to Core-1 in instance 0 and in instance 1.

```
Core-2(config)# show spanning-tree a19,a20 instance ist

 IST Instance Information

<-output omitted->

                                                 Designated
  Port  Type       Cost   Priority Role      State      Bridge
  ----- ---------- ------ -------- -------   ---------- ------------
  A19   100/1000T  20000  128      Root      Forwarding 1c98ec-ab4b00
  A20   100/1000T  20000  128      Alternate Blocking   1c98ec-ab4b00

Core-2(config)# show spanning-tree a19,a20 instance 1

 MST Instance Information

  Instance ID : 1

<-output omitted->

                                                           Designated
  Port  Type       Cost    Priority Role      State      Bridge
  ----- ---------- ------- -------- -------   ---------- ------------
  A19   100/1000T  20000   128      Root      Forwarding 1c98ec-ab4b00
  A20   100/1000T  20000   128      Alternate Blocking   1c98ec-ab4b00
```

Table 7-1: Criteria for selecting the root port

Criteria
Port with the lowest cost path to the root (receives a BPDU with the lowest cost)
If a tie, port that connects to the neighbor with the lowest switch ID (priority + MAC address)
If a tie, port that connects to the port with the lowest port priority value
If a tie, port with the lowest ID

As an optional step, you can try adjusting spanning tree port priorities to achieve a degree of load balancing.

4. Set the Core-2 A20 port priority lower than the default in instance 2.

```
Core-2(config)# spanning-tree instance 2 a20 priority 0
```

Core-1

5. A lower value is preferred so you should see that Core-1 changes its root port to a20. Now this link is forwarding in instance 2.

```
Core-1(config)# show spanning-tree a19,a20 instance 2

 MST Instance Information

  Instance ID : 2

<-output omitted->

                                                                Designated
  Port Type       Cost    Priority Role      State      Bridge
  ---- ---------- ------- -------- -------   ---------  -------------
  A19  100/1000T  20000   128      Alternate Blocking   70106f-0d2100
  A20  100/1000T  20000   128      Root      Forwarding 70106f-0d2100
```

Task 4: Save

Save the current configuration on the core switches using the **write memory** command.

Learning check

1. How did MSTP handle the two links between Core-1 and Core-2?

2. What are the implications? Does this new link add bandwidth to the connection between Core-1 and Core-2?

3. What would happen if you simply disabled spanning tree?

Answers to Learning check

1. How did MSTP handle the two links between Core-1 and Core-2?

   On Core-1, both ports are designated ports and in a forwarding state in instances 1 and 0. In instance 2, for which Core-2 is the root, interface a9 is the root port and interface a20 is the alternate port (placed in a discarding/blocking state). Interface a19 and interface a20 have the same root cost, the same bridge ID for the upstream (Core-2), and the same port priority. Interface a19 was selected as the root port because it has a lower ID.

   On Core-2, both ports are alternate ports and in a blocking state for instances 1 and 0. They are designated in instance 2.

2. What are the implications? Does this new link add bandwidth to the connection between Core-1 and Core-2?

   In other words, the new link is blocked in all instances. The new link is not adding any bandwidth to the connection. It adds some redundancy, but it does not increase the throughput between Core-1 and Core-2.

   You could try to manipulate the configuration to enable some load-balancing across the links on a per-instance basis. For example, you could assign a higher port priority to one port in instance 1 and a higher port priority to a different port in instance 2. However, the planning is more complex, and the load-balancing is not granular. If you wanted to add more than links than two, they would be blocked.

3. What would happen if you simply disabled spanning tree?

   On the other hand, you cannot simply disable spanning tree. That would cause a loop and broadcast storm.

Link aggregation

As you observed, when you simply add a link between two switches, RSTP/MSTP sees it as another redundant link. It blocks the link as an alternate path to the root for Core-2. The new link adds a bit of resiliency, but it does *not* add any bandwidth to the connection.

Link aggregation solves this issue.

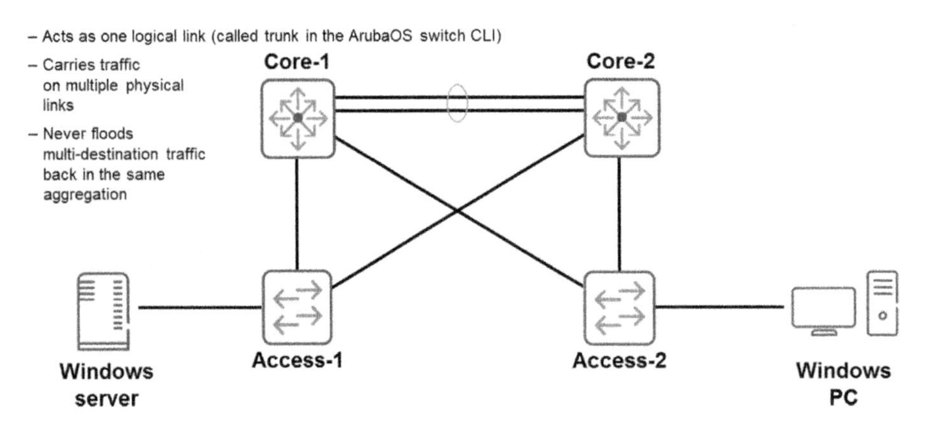

Figure 7-5: Link aggregation

You define a link aggregation on the switch and add physical interfaces to that aggregation. Now, from the point of view of the switch, the *link aggregation* is the logical link—not the individual physical interfaces assigned to it:

- RSTP/MSTP sees the link aggregation as an interface, and the link aggregation as a whole has a forwarding or blocking status.
- The switch learns MAC addresses on the link aggregation not on an individual interface.

The switch can forward traffic across any of the links in the aggregation. In effect, the link aggregation is a link with the combined bandwidth of the physical interfaces assigned to it. (At a lower level, the switch is assigning specific traffic to specific links within the aggregation, and load balancing is not always perfect. You will learn about the load-balancing mechanism in more depth a bit later.)

Link aggregations can handle unicast and multi-destination traffic (multi-destination traffic includes multicasts, broadcasts, and unknown unicasts). When a switch receives multi-destination traffic on a link aggregation, it never floods that traffic on another link in the same link aggregation. When it needs to flood this traffic over a link aggregation, it chooses just one link for the transmission. This behavior ensures that the link aggregation does not create a loop.

Important

The ArubaOS switch CLI refers to link aggregations as "trunks." However, on some other types of switches, the term "trunk" refers to a link that uses VLAN tagging to carry multiple VLANs. To avoid confusion, this guide will use the term link aggregation except when referring to the actual syntax for setting up a link aggregation on ArubaOS switches.

Configure a manual link aggregation

You will now configure both links between the core switches as part of a manual link aggregation. A manual link aggregation does not use any protocol to establish the aggregation. You simply select the physical interfaces that you want for the aggregation and assign them to the link aggregation interface (called in trunk in the CLI).

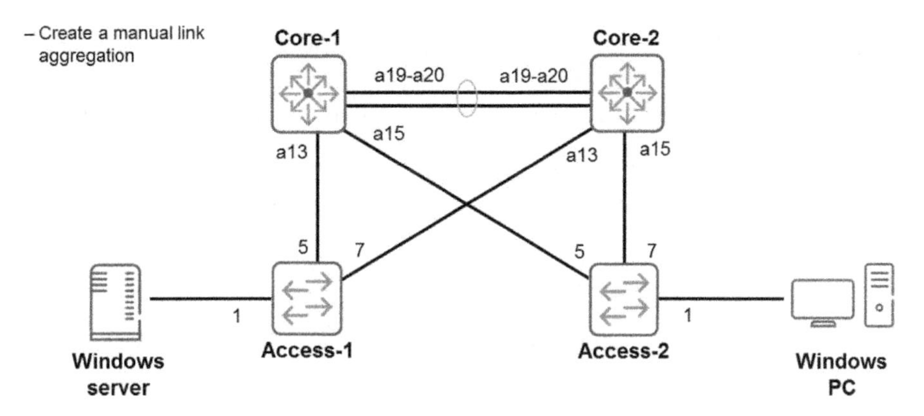

Figure 7-6: Configure a manual link aggregation

The figure below shows the final topology with the new link aggregation.

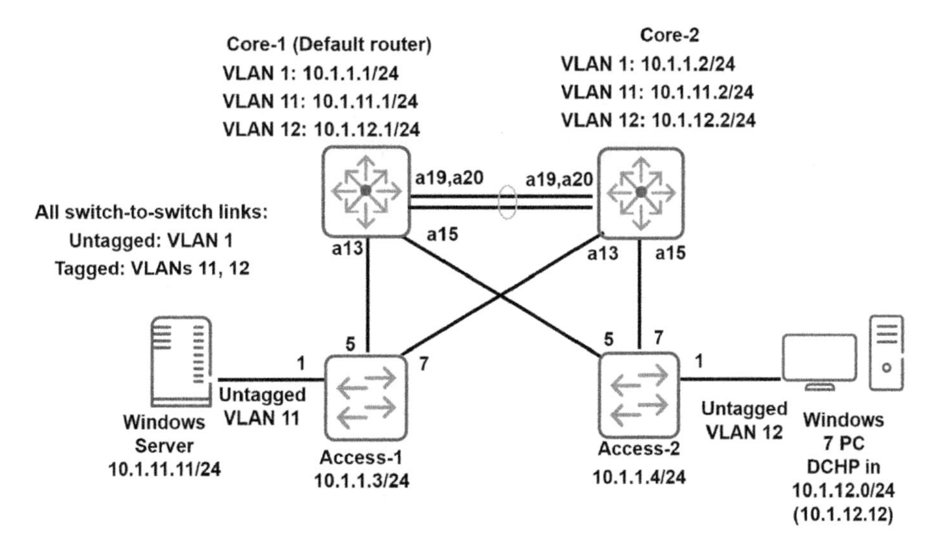

Figure 7-7: Final topology—Configure a manual link aggregation

Task 1: Configure a link aggregation between the Core switches

In this task you create a manual link aggregation on Core-1 and Core-2. A link aggregation is a logical interface, and on Core switches, you create a bridge-aggregation interface as the logical interface for each link aggregation. You then assign physical interfaces to the logical interface.

Core-1

1. Access a terminal session with the Core-1 switch. Move to the global configuration context.
2. It is best practice to wait to connect interfaces until after you have configured the link aggregation on both sides. If your test network has permanent physical links, disable the interfaces that connect to Core-2.

```
Core-1(config)# interface a19,a20 disable
```

3. Create a link aggregation (trunk) that includes both interfaces that connect to Core-2. The trunk option indicates that this is a manual link aggregation.

```
Core-1(config)# trunk a19-a20 trk1 trunk
```

4. You must define the VLAN settings on the trunk interface.

```
Core-1(config)# vlan 11 tag trk1
Core-1(config)# vlan 12 tag trk1
```

Core-2

5. Repeat the same steps on the Core-2 interfaces. If you don't specify a particular option, trunk (static) is used by default.

```
Core-2(config)# trunk a19-a20 trk1
Core-2(config)# vlan 11 tag trk1
Core-2(config)# vlan 12 tag trk1
```

Core-1

6. Return to the Core-1 terminal session and enable the interfaces.

```
Core-1(config)# interface a19,a20 enable
```

Task 2: Observe the link aggregation

You will now view the link aggregation. You will also observe the effect on MSTP.

Core-2

1. Return to the Core-2 terminal session.
2. View the link aggregation.

```
Core-2(config)# show trunk

 Load Balancing Method: L3-based (default)

  Port       | Name                Type         | Group  Type
  ---------- + ---------------     ----------   + ------ --------
  A19        | Core-1              100/1000T    | Trk1   Trunk
  A20        |                     100/1000T    | Trk1   Trunk
```

3. Also verify the VLAN memberships. Notice that the ***trk1*** interface appears as a VLAN member instead of the physical interfaces of a19 and a20.

```
Core-2(config)# show vlan 11
Status and Counters - VLAN Information - VLAN 11
  VLAN ID : 11
  Name : VLAN11
  <-output omitted->

  Port Information Mode     Unknown VLAN Status
  ---------------- -------- ------------ ----------
  A13              Tagged   Learn        Up
  A15              Tagged   Learn        Up
  Trk1             Tagged   Learn        Up

Core-2(config)# show vlan 12
 Status and Counters - VLAN Information - VLAN 12
  VLAN ID : 12
```

```
<-output omitted->
Port Information Mode     Unknown VLAN Status
---------------- -------- ------------ ----------
A13              Tagged   Learn        Up
A15              Tagged   Learn        Up
Trk1             Tagged   Learn        Up
```

4. Examine the MSTP topology for instance 0. Notice that you no longer see ports a19 and a20, but the trunk interface instead. Both interfaces in the trunk are allowed to forward. The same holds true for other instances.

```
Core-2(config)# show spanning-tree instance ist
 IST Instance Information
 <-output omitted->
                                                               Designated
 Port Type       Cost  Priority Role       State       Bridge
 ---- ---------- ----- -------- ---------- ----------- -------------
 A1   100/1000T  Auto  128      Disabled   Disabled
 <-output omitted->
 A13  100/1000T  20000 128      Designated Forwarding  1c98ec-ab4b00
 A14  100/1000T  Auto  128      Disabled   Disabled
 A15  100/1000T  20000 128      Designated Forwarding  1c98ec-ab4b00
 <-output omitted->
 Trk1 20000            64       Root       Forwarding  1c98ec-ab4b00
```

Task 3: Save

Save the current configuration on the Core switches using the **write memory** command.

Learning check

1. What step did you have to take after configuring the manual link aggregation to ensure that the aggregation could carry the necessary traffic?

2. What command did you use to view the manual link aggregation on ArubaOS switches? Take note of the information revealed by the command.

Answers to Learning check

1. What step did you have to take after configuring the manual link aggregation to ensure that the aggregation could carry the necessary traffic?

   You had to configure the VLAN settings that used to be on the physical interface on the link aggregation (trk1).

2. What command did you use to view the manual link aggregation on ArubaOS switches? Take note of the information revealed by the command.

   The command is **display trunks**.

   This command shows the interfaces that are assigned to trunk interfaces and the type of link aggregation.

Requirements for links

Are you allowed to add any link that you want to a link aggregation, or does the link have to meet any criteria?

You actually need to follow several guidelines. Links that are in the same link aggregation should have matching settings for:

- Duplex mode (full or half)
- Link speed
- Media (copper or fiber)

- Ensure compatibility to prevent undesirable behavior (no check on manual link aggregation):
  - Full-duplex or half-duplex
  - Speed and media type
- Check maximum links allowed per-aggregation

Guideline: Remember to configure the VLAN settings on the link aggregation (trk interface).

Figure 7-8: Requirements for links

ArubaOS: Behavior under incompatibility

ArubaOS switches do not check for compatibility on links in a *manual* link aggregation. You could, theoretically, combine links of different speeds and duplex modes, and the links could become part of the link aggregation. However, you *should not* do so.

You must be very careful to check the compatibility yourself for:

- Duplex mode
- Link speed
- Media

The link aggregation will not work as well as expected if, for example, you combine links of different speeds. You could even disrupt network connectivity if you aggregate incompatible links between an ArubaOS switch and different type of switch that does check for compatibility. The ArubaOS switch will use the incompatible links, but the other switch might not accept traffic on the incompatible links, disrupting connectivity.

VLAN settings

You do not need to configure matching VLAN settings on physical interfaces in order to add them to a link aggregation. Instead ArubaOS switch ensure that all interfaces in an aggregated link have compatible VLAN settings by having you configure these settings on the link aggregation itself (trk interface).

When you add an interface to a link aggregation it loses any VLAN settings that it had and takes the settings from the trk interface. You can no longer configure VLAN assignments on physical interfaces that are part of a link aggregation. Instead the settings automatically apply to interfaces in the aggregation, including links that you add to the aggregation at a later point.

The trk interface is untagged for VLAN 1, by default. If you want the link aggregation to carry tagged traffic or to be in a different untagged VLAN, you must remember to configure these settings on the trk interface.

Requirements for maximum number of links

A link aggregation can only support a certain number of links per link aggregation. The number of links allowed in an aggregation depend on the switch model. For example, many switches have a maximum of eight links. Refer to the documentation for your switch for more information.

Potential issue with manual link aggregations

You just considered the settings that you need ensure match when you add interfaces to a link aggregation. You also need to be very careful that all interfaces actually connect to the same switch.

Here you see a situation in which the administrator wanted to create two link aggregations on Access-1: one to Core-1 and one to Core-2. However, the admin accidentally misconfigured the two link aggregations on Access-1. Each link aggregation has one link to Core-1 and one link to Core-2. The link aggregations are configured correctly on Core-1 and Core-2.

This mismatch can cause serious consequences.

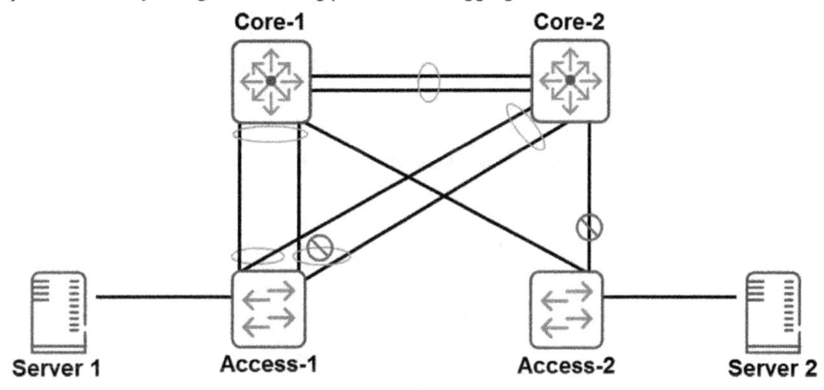

Figure 7-9: Potential issue with manual link aggregations

The figure below shows two servers in VLAN 11, one connected to Access-1 and one connected to Access-2. What happens when server 2 tries to contact server 1?

The frame reaches Core-1, which forwards the frame on the link aggregation to Access-1. The switch happens to select the link that is misconfigured on Access-1. This link is part of a link aggregation that is blocked by MSTP. The communication fails.

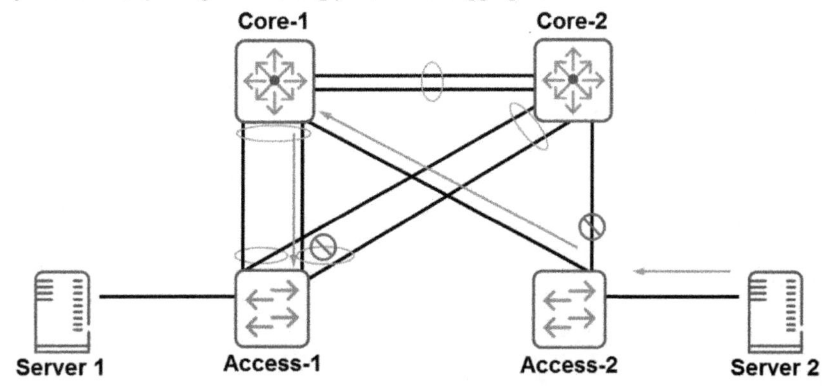

Figure 7-10: Traffic flow across the misconfigured link aggregations (1)

What can happen when server 1 sends a broadcast? Access-1 should flood the broadcast on its link aggregation to Core-1 (its MSTP root port) but not on its link aggregation to Core-2 (its MSTP alternate port). But it happens to choose the misconfigured link and sends the broadcast to Core-2. Core-2 accepts the broadcast because it does not block the link aggregation; on its side, the aggregation is an MSTP designated port. Core-2 floods the broadcast back to Core-1, which will forward the broadcast to Access-2 *and* back to Access-1.

If Core-1 selects the interface that correctly connects to Access-1. Access-1 will start the process again, leading to a broadcast storm.

As more conversations like this occur, connectivity is disrupted as far as the VLAN extends.

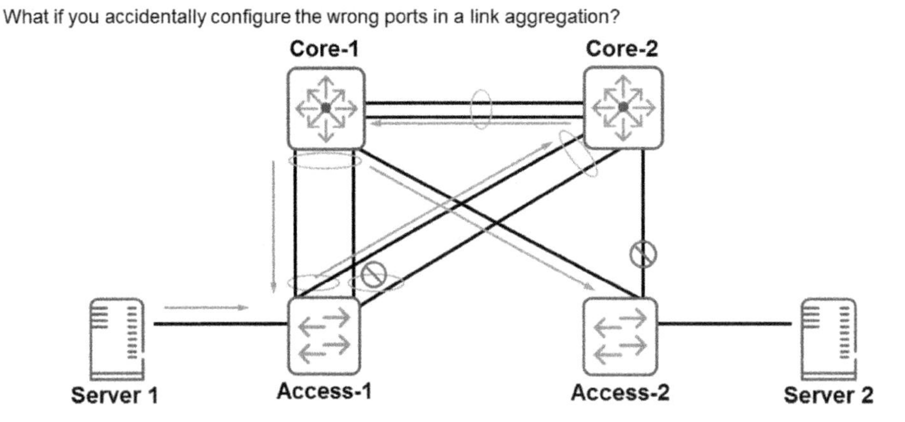

Figure 7-11: Traffic flow across the misconfigured link aggregations (2)

LACP

IEEE 802.3ad, LACP, provides a solution to the issues described on the previous pages. This protocol requires switches to exchange messages, called LACP data units (DUs), to establish the link aggregation.

These messages include a system ID, which uniquely identifies the switch. Each LACP peer checks the messages that arrive on each link in the aggregation. A link aggregation establishes correctly only if the incoming system ID and operational key matches on all the links. The system ID identifies the switch and the operational key identifies the link aggregation group. In this way, each peer ensures that all links are connected to the same other peer and also that all links are connected to the same link aggregation on that peer.

- Uses messages to establish the link aggregation
- Ensures that all links are connected to the same peer (system ID) and link aggregation (operational key)
- Ensures compatibility for other settings
- Manages adding and removing links

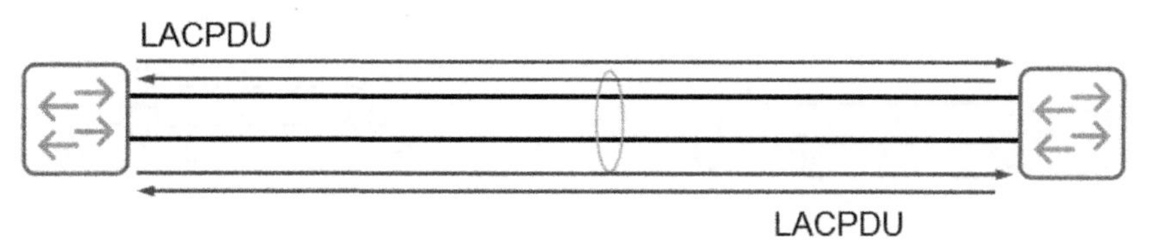

Figure 7-12: LACP

Consider what would have happened if you had used LACP for the misconfigured link aggregations in the example on previous page. For link aggregation 1 on Access-1, LACP would have detected that the system ID for the peer on the two links was different. One of the links would not have been allowed to become active.

LACP allows the peers to exchange other information about the link aggregation as well. Using LACP, ArubaOS switches check compatibility for:

- Media type and speed
- Duplex mode—LACP requires full-duplex operation.

Then they only allow compatible links to join the aggregation.

A reference port determines which port's settings are used for the aggregation; other ports must match or be excluded. LACP determines which switch has the reference port by system priority (smaller value has higher priority; the priority is derived from a priority value and MAC address). That switch then selects the reference port by port priority (smaller value has higher priority; the priority is derived from a priority value and port ID).

LACPDUs allow the devices to manage the logical aggregated link. LACP manages adding or removing physical links. It also handles failovers, ensuring that all frames are delivered in order even when traffic needs to move to a new link.

LACP operational modes

LACP has two operational modes.

- Static
  - LACP is specified when trk interface created
  - Uses Active/Active mode—Both sides actively send LACP messages
  - Typically recommended mode

- Dynamic
  - No trk interface is created
  - Instead LACP is enabled on physical interfaces
  - Uses one of two modes:
    - Active/Active
    - Active/Passive—Passive side waits
  - Allows extra links to be on standby
  - Does not support non-default VLAN assignments or various other settings on the aggregation

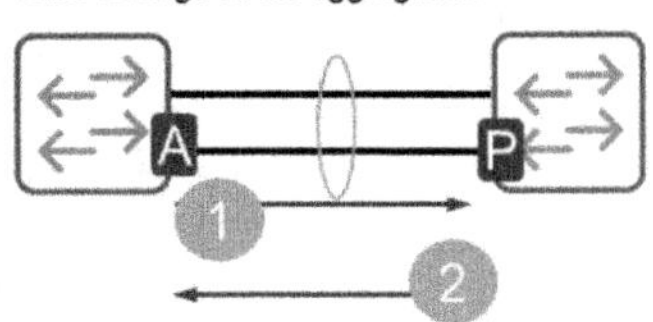

Figure 7-13: LACP operational modes

Static mode

You set up a static mode LACP link aggregation much as you do a manual link aggregation. However, when you create the link aggregation (trk) interface, you specify LACP for that interface.

The physical interfaces that are assigned to that link aggregation then operate in active LACP state. That is, both sides of the link send LACPDUs to each other and set up the link as you learned on the previous page.

- Dynamic mode

ArubaOS switches also support dynamic mode LACP.

To set up a dynamic link aggregation, you enable LACP on the switch interfaces that you want to function as part of a link aggregation. But you do not actually create a link aggregation (trk) interface. You do the same on the switch on the other side of the link.

When you enable LACP on a physical interface, you choose one of these states:

- Active—Transmits LACPDUs to advertise that it can create an aggregated links
- Passive—Listens for LACPDUs and responds with LACPDUs only after it receives one from an active port

As long as one or both sides of each link are in active mode, the link aggregation automatically establishes. (If both sides are in a passive state, then neither will initiate the exchange process.) LACP automatically determines which links are between the same two devices and places those links in the same link aggregation.

When you use dynamic LACP, ArubaOS switches can have standby links beyond the maximum allowed in the aggregation. For example, if the switch supports only eight links, you might be able to add 12. Links over the maximum are standby links, which cannot be used to forward traffic. However, if a link fails, LACP automatically adds one of the standby links to the link aggregation.

A dynamic LACP link aggregation does not have a logical interface in the switch configuration. Therefore, you cannot assign non-default STP, VLAN, and other such settings to it. (You can use a GVRP, a protocol for dynamically learning VLANs, to add VLANs to the dynamic link aggregation. However, this protocol is not covered in this book.) For this reason, static LACP is typically recommended.

Configure an LACP link aggregation

You will now add LACP link aggregations between Core-1 and Access-1 and between Core-1 and Access-2. You will also convert the manual link aggregation between the core switches into an LACP aggregation.

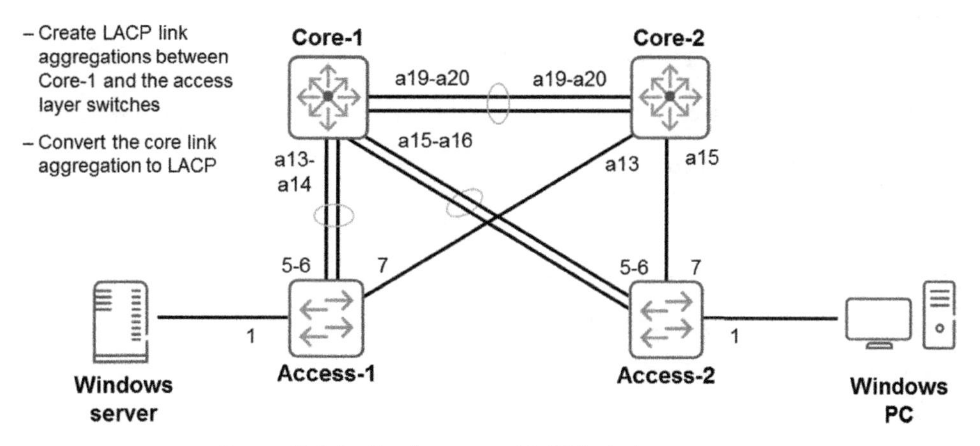

Figure 7-14: Configure an LACP Link Aggregation

The figure below shows the final topology.

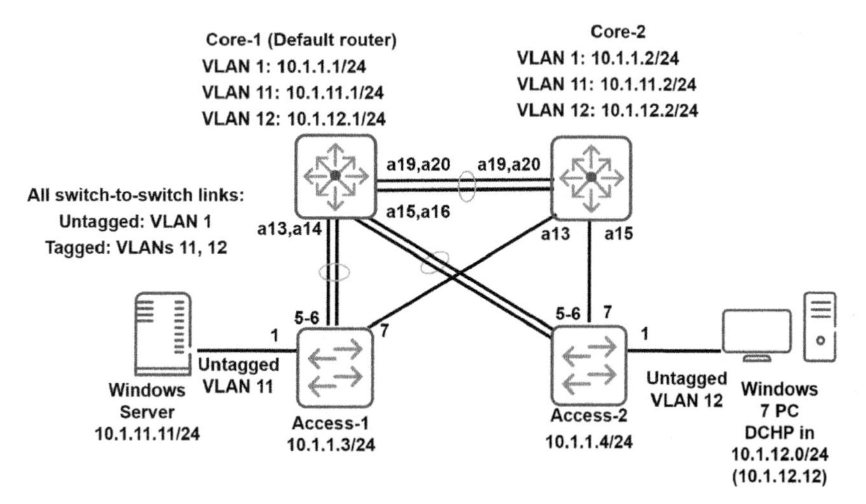

Figure 7-15: Configure an LACP link aggregation

Task 1: Configure LACP link aggregations

You will begin by configuring an LACP link aggregation between Core-1 and the access switches.

Core-1

1. Access a terminal session with Core-1. Move to the global configuration mode.
2. Create a trunk interface that includes two interfaces that connect to Access-1. This is the second link aggregation on this switch, so you will use trk2 for the link aggregation name. Specify the **lacp** option to use static LACP.

```
Core-1(config)# trunk a13,a14 trk2 lacp
```

3. Repeat the command, but specify the two interfaces that connect to Access-2 and a new trk interface ID.

```
Core-1(config)# trunk a15,a16 trk3 lacp
```

4. Configure the VLAN settings on the trunk interfaces.

```
Core-1(config)# vlan 11 tagged trk2,trk3
Core-1(config)# vlan 12 tagged trk2,trk3
```

Access-1

5. Access a terminal session with Access-1. Move to the global configuration context.
6. Create a static LACP link aggregation on the two ports that connect to Core-1.

```
Access-1(config)# trunk 5,6 trk1 lacp
```

7. Tag VLAN 11 and 12 on the link aggregation (trk1).

```
Access-1(config)# vlan 11 tagged trk1
Access-1(config)# vlan 12 tagged trk1
```

8. Enable the second interface in the trunk.

```
Access-1(config)# interface 6 enable
```

Core-1

9. Enable the connected interface on the Core-1 side as well. It is best practice to enable these interfaces after you've added them to the link aggregations to prevent loops.

```
Core-1(config)# interface a14 enable
```

10. If you have not already, establish the physical connection (interface 6 on Access-1 connects to interface a14 on Core-1).

Access-2

11. Open a terminal session with Access-2.

12. Follow the same steps. Try to remember the commands on your own, but you can refer back to the previous section if you need help.

Important

Remember to return to Core-1 and enable interface a16.

13. If you have not already, establish the physical connection (interface 6 on Access-2 connects to interface a16 on Core-1).

Task 2: View the link aggregation

You will now examine the link aggregation.

Access-1

1. View the trunk.

```
Access-1(config)# show trunks

 Load Balancing Method:  L3-based (default)

  Port    | Name                             Type       | Group Type
  ------- + -------------------------------- ---------- + ----- ------
  5       | Core-1                           100/1000T  | Trk1  LACP
  6       |                                  100/1000T  | Trk1  LACP
```

2. LACP provides more information about the link aggregation. View LACP information. Verify that the Partner is Yes and LACP status is Success on both links.

```
Access-1(config)# show lacp
```

```
                         LACP
        LACP     Trunk  Port             LACP     Admin Oper
Port    Enabled  Group  Status Partner  Status   Key   Key
------  -------- ------ ------ -------- -------- ----  ----
5       Active   Trk1   Up     Yes      Success  0     562
6       Active   Trk1   Up     Yes      Success  0     562
```

3. You can also view information specific to the local side of the link aggregation. For example, you can see this switch's system ID and that the links are aggregated.

```
Access-1(config)# show lacp local
LACP Local Information.
            LACP                  Tx     Rx Timer
Port Trunk  Mode      Aggregated  Timer  Expired   System ID
---- ------ --------  ----------- ------ --------  -------------
5    Trk1   Active    Yes         Slow   No        6c3be5-6208c0
6    Trk1   Active    Yes         Slow   No        6c3be5-6208c0
```

4. And you can view information about the peer. Note that both peer interfaces have the same system ID and operational key.

```
Access-1(config)# show lacp peer
LACP Peer Information.
System ID: 6c3be5-6208c0
Local  Local                         Port       Oper     LACP      Tx
Port   Trunk   System ID      Port   Priority   Key      Mode      Timer
------ ------ -------------- -----  ---------  -------  --------  -----
5      Trk1   1c98ec-ab4b00  13     0          963      Active    Slow
6      Trk1   1c98ec-ab4b00  14     0          963      Active    Slow
```

5. View log messages on Access-1.

```
Access-1(config)# show log -r
```

The -r option shows the most recent logs first, so the events occurred chronologically from the bottom of the list.

You can see that the interfaces were blocked by LACP, added to the link aggregation interface (Trk1), blocked by STP, and then brought online as a part of the link aggregation.

```
Keys:   W=Warning    I=Information

        M=Major      D=Debug E=Error

----  Reverse event Log listing: Events Since Boot  ----

I 06/08/16 20:19:33 00076 ports: port 6 in Trk1 is now on-line

I 06/08/16 20:19:33 00435 ports: port 6 is Blocked by STP

I 06/08/16 20:19:33 00435 ports: port 6 is Blocked by LACP

I 06/08/16 20:19:31 04743 ports: Port 6 recovery occurred.

I 06/08/16 20:19:07 00076 ports: port 5 in Trk1 is now on-line

I 06/08/16 20:19:06 00435 ports: port 5 is Blocked by STP

I 06/08/16 20:19:06 00078 ports: trunk Trk1 is now active

I 06/08/16 20:19:06 00435 ports: port 5 is Blocked by LACP

I 06/08/16 20:19:03 04743 ports: Port 5 recovery occurred.

I 06/08/16 20:19:03 00077 ports: port 5 is now off-line

I 06/08/16 20:19:01 05101 dhcp: ZTP is disabled due to first
configuration

I 06/08/16 20:18:34 00076 ports: port 5 is now on-line

I 06/08/16 20:18:31 00435 ports: port 5 is Blocked by STP

I 06/08/16 20:18:29 00001 vlan: DEFAULT_VLAN virtual LAN enabled

I 06/08/16 20:18:29 00002 vlan: DEFAULT_VLAN virtual LAN disabled

I 06/08/16 20:18:29 00077 ports: port 5 is now off-line
```

You could use the same commands to examine the link aggregation from the Core-1 side, if you want.

Access-2

6. Also verify that the link aggregation has established successfully on Access-2.

```
Access-2(config)# show lacp
```

```
                              LACP
       LACP     Trunk Port            LACP     Admin Oper
  Port Enabled  Group Status Partner Status    Key   Key
  ----- -------- ----- ------ ------- ------- ---   ----
  5     Active   Trk1  Up     Yes     Success 0     562
  6     Active   Trk1  Up     Yes     Success 0     562
```

7. Note that the link aggregation is the root port in spanning tree instance 0.

```
Access-2(config)# show spanning-tree instance ist
 IST Instance Information
<-output omitted->
                                                                Designated
  Port     Type     Cost      Priority Role State           Bridge
  -------- -------- --------  -------- ---- -----------     -------------
  <-output omitted->
  Trk1              20000     64       Root Forwarding      1c98ec-ab4b00
```

Task 3: Observe load sharing

You have confirmed that Access-1 and Access-2 is using their link aggregations in instance 0, which includes VLAN 1. You will now observe load balancing across both links on Access-2.

Access-2

1. Verify connectivity in VLAN 1. From Access-2, ping Core-1 (10.1.1.1), Core-2 (10.1.1.2) and Access-1 (10.1.1.3).

```
Access-2(config)# ping 10.1.1.1
10.1.1.1 is alive, time = 2 ms
Access-2(config)# ping 10.1.1.2
10.1.1.2 is alive, time = 5 ms
Access-2(config)# ping 10.1.1.3
10.1.1.3 is alive, time = 2 ms
```

2. View the ARP table and the MAC forwarding table for VLAN 1 on Access-2.

   Notice that Access-2 has learned the Core-1 and Core-2 MAC addresses on Trk1 because this is the logical interface that is forwarding in VLAN 1 (instance 0). MAC learning also occurs on the link aggregation rather than individual link.

```
Access-2(config)# show arp

 IP ARP table

  IP Address       MAC Address       Type    Port

  --------------- ----------------- ------- ----

  10.1.1.1         1c98ec-ab4b00     dynamic Trk1

  10.1.1.2         70106f-0d2100     dynamic Trk1

  10.1.1.3         6c3be5-6208c0     dynamic Trk1

Access-2(config)# show mac-address vlan 1

 Status and Counters - Address Table - VLAN 1

  MAC Address   Port

  ------------- ------

  1c98ec-ab4b00 Trk1

  1c98ec-ab5bf0 Trk1

  1c98ec-ab5bf1 Trk1

  6c3be5-6208c0 Trk1

  70106f-0d2100 Trk1
```

Now you will observe how traffic flows over the physical interfaces within the link aggregation.

3. Clear the interface statistics and start up the interface menu display.

```
Access-2(config)# clear statistics global

Access-2(config)# show interface display
```

4. Use the <DOWN> arrow key until you can see the Trk 1interfaces displayed.

   You should already see traffic traversing both links in a trunk from things like STP BPDUs and miscellaneous traffic from the Windows computers (like NetBIOS and IPv6 management traffic if enabled). Keep this console window open.

```
                              Status and Counters - Port Counters

                                                                    Flow Bca*
  Port    Total Bytes Total Frames  Errors Rx Drops Tx Ctrl Lim*
  ------ ----------- ------------ --------- -------- ---- ---*
  1          450,640         2926         0        0 off  0
  2                0            0         0        0 off  0
  3                0            0         0        0 off  0
  4                0            0         0        0 off  0
  5-Trk1     158,505         1066         0        0 off  0
  6-Trk1      65,665          379         0        0 off  0
  <-output omitted->
 Actions-> Back  Show details Reset Help
Use up/down arrow keys to scroll to other entries, left/right arrow keys to
change action selection, and <Enter> to execute action.
```

Core-1

5. Access the terminal session with Core-1.
6. Send 500 pings to Access-2.

```
Core-1(config)# ping repetitions 500 10.1.1.4
```

Access-2

7. On Access-2, look at the statistics for the interfaces in Trk1. You should see statistics for both ports. The statistics for one of the ports will probably increase more rapidly depending on how your switches choose links for the conversation.

```
                    Status and Counters - Port Counters

                                                          Flow Bca*
  Port    Total Bytes Total Frames Errors Rx Drops Tx Ctrl Lim*
  ------ ----------- ------------ --------- -------- ---- ---*
  1          480,060         3145         0        0 off  0
  2                0            0         0        0 off  0
  3                0            0         0        0 off  0
  4                0            0         0        0 off  0
  5-Trk1     184,403         1253         0        0 off  0
  6-Trk1      76,044          485         0        0 off  0
<-output omitted->
 Actions->       Back        Show details       Reset        Help
```

Core-2

8. Now start another conversation. Access a terminal session with Core-2.
9. Ping Access-2 500 times.

```
Core-2# ping repetitions 500 10.1.1.4
```

Access-2

10. On Access-2, look at the statistics for the interfaces in Trk1. With another conversation, you are more likely to see traffic on both links.

```
Status and Counters - Port Counters
                      Status and Counters - Port Counters

                                                                Flow Bca*
  Port        Total Bytes  Total Frames  Errors Rx   Drops Tx Ctrl Lim*
  --------    -----------  ------------  ----------  -------- ---- ---*
  1               523,199          3358           0         0 off  0
  2                     0             0           0         0 off  0
  3                     0             0           0         0 off  0
  4                     0             0           0         0 off  0
```

```
                                                                   Flow Bca*
 Port        Total Bytes  Total Frames  Errors Rx   Drops Tx  Ctrl Lim*
 --------  -----------  ------------  ----------  --------  ---- ---*
 5-Trk1        216,268          1463           0          0 off   0
 6-Trk1         91,749           545           0          0 off   0
Actions->     Back      Show details      Reset      Help
```

11. Press **<Enter>** on the Back option to back out of the display. Then press **<Ctrl+c>**.
12. On Core-1 and on Core-2, press **<Ctrl+c>** to end the pings.

Task 4: Convert a manual link aggregation to an LACP aggregation

In this task, you will see how easy it is to convert a manual link aggregation to a static LACP one.

Core-1

1. Core-1 currently has a manual link aggregation to Core-2. To change the aggregation to an LACP one, simply re-enter the trunk command with the **lacp** option.

```
Core-1(config)# trunk a19,a20 trk1 lacp
```

2. At this point, Core-1 doesn't detect any LACP partner *on any link in the aggregation*. LACP allows the link aggregation to form in this case, so connectivity isn't disrupted if the other side doesn't support LACP. As you can confirm with this command, the LACP status remains Success. (If Core-1 detected LACP messages on one of the links, it would block the links that didn't send messages.)

```
Core-1(config)# show lacp
```

```
                                   LACP

        LACP      Trunk    Port              LACP      Admin   Oper
 Port   Enabled   Group    Status   Partner  Status    Key     Key
 -----  -------   -------  -------  -------  -------   ------  ------
 A13    Active    Trk2     Up       Yes      Success   0       963
 A14    Active    Trk2     Up       Yes      Success   0       963
 A15    Active    Trk3     Up       Yes      Success   0       964
 A16    Active    Trk3     Up       Yes      Success   0       964
 A19    Active    Trk1     Up       No       Success   0       962
 A20    Active    Trk1     Up       No       Success   0       962
```

Core-2

3. Repeat the same step on Core-2.

```
Core-2(config)# trunk a19,a20 trk1 lacp
```

4. Verify that LACP has succeeded and has a partner.

```
Core-2(config)# show lacp

                                 LACP

        LACP    Trunk Port                LACP    Admin Oper
   Port Enabled Group Status Partner Status  Key   Key
   ---- ------- ----- ------ ------- ------  ---   ---
   A19  Active  Trk1  Up     Yes     Success 0     962
   A20  Active  Trk1  Up     Yes     Success 0     962
```

Task 5: Save

Save the current configuration on each of the four switches using the **write memory** command.

Learning check

Answer the following questions to check your understanding of LACP link aggregation.

1. What is the difference between setting up an LACP link aggregation and a manual link aggregation on ArubaOS switches?

2. What ArubaOS switch command shows whether an interface has successfully been added to a link aggregation and can pass traffic?

3. You checked Access-2's MAC forwarding table. What forwarding interface did this switch learn for other switches?

4. When you pinged Access-2 from other switches, did you see traffic on both physical interfaces? Did you see exactly the same amount on both interfaces?

Answers to Learning check

1. What is the difference between setting up an LACP link aggregation and a manual link aggregation on ArubaOS switches?

   The process is very similar for both types of aggregation. For LACP, you simply use the **lacp** option with the **trunk** command.

2. What ArubaOS switch command shows whether an interface has successfully been added to a link aggregation and can pass traffic?

   The **show lacp [local | peer]** command shows this information. You should verify that the LACP status is Success. If you use the **local** or **peer** option, you can see more information about the status on the local interfaces or the peer interfaces.

3. You checked Access-2's MAC forwarding table. What forwarding interface did this switch learn for other switches?

   The MAC address forwarding table learns the link aggregation interface (trk1) as the forwarding interface for traffic, not individual physical interfaces.

4. When you pinged Access-2 from other switches, did you see traffic on both physical interfaces? Did you see exactly the same amount on both interfaces?

   You should have seen traffic on both interfaces, but the amount might not have been equal. Link aggregations load-share traffic over all active links in the aggregation, but, as you will learn in a moment, the sharing tends to become more even when more devices are communicating (as in a real network).

Load-sharing traffic over a link aggregation

You will now explore how the link aggregation decides which link to use for specific traffic.

The link aggregation selects a link for each *conversation*, or one-way communication between a source and a destination. For example, when a workstation sends an Ethernet frame to a server, a conversation begins. The switch hashes the source and destination address and, based on the hash, assigns the conversation to a link within the aggregation. All subsequent frames from the same workstation to the same server are part of that conversation and assigned to the same physical link, avoiding out-of-order packet delivery. The link aggregation will only select a new link for the conversation if the current one fails. If you are using LACP, LACP ensures that all packets are delivered in order as the link fails over.

The link aggregation uses a hash to select a link for each *conversation*:

- Conversation defined by the load-sharing mode
  - IP source and destination address (L3-based) = Default option
  - Source and destination UDP/TCP ports (L4-based) = Option for situations in which few IP addresses are sent on the link
- Link for a conversation stays constant
- Traffic tends to be balanced more evenly as conversations increase

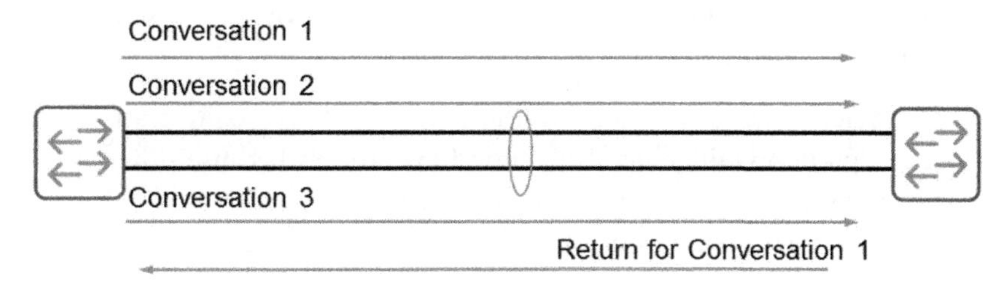

Figure 7-16: Load-sharing traffic over a link aggregation

Which addresses are used to define the source and destination for a conversation depends on the link aggregation's load-balancing mode, set with the **trunk-load-balance** command. The default mode uses source IP address and destination IP address for IP (or IPv6) traffic. In some cases, you might want to set L4-based mode instead. For example, the link aggregation might support traffic between a limited number of devices or, if network address translation (NAT) is involved, IP addresses. In this case, balancing by TCP or UDP ports could create more conversations.

Table 7-2: Options for load-balance mode

Mode	Addresses
L3-based (default)	• Source and destination IP address when available • When Layer 3 information is not available (non-IP or IPv6 traffic), source and destination MAC address
L4-based	• Source and destination UDP or TCP port when available • When Layer 4 information is not available (non-UDP and TCP traffic), source and destination IP address when available • When Layer 3 information is not available (non-IP traffic), source and destination MAC address

Load-sharing multiple conversations

The hash assigns conversations to links arbitrarily, not based on round-robin. In other words, a conversation between client 1 and server 1 begins and is assigned to link 2 in the aggregation. When another conversation between client 2 and server 2 begins, it *could* be is assigned to the same link. However, when a link aggregation carries many conversations, the conversations tend to spread out over the links relatively evenly.

Even when all physical links carry precisely the same *number* of conversations, though, the links might experience significant differences in traffic flow. This is because different conversations will have different bandwidth requirements, different levels of burstiness, and so on. In a test environment with a relatively small number of source/destination pairs, you might see dramatic differences in traffic distribution.

In a real world network, which has larger numbers of conversations each carrying variable bandwidth, the links within an aggregation will tend to carry roughly even amounts of traffic. Although, distribution will never be perfectly equal, it will be good enough for practical purposes—while still allowing the switch to quickly assign traffic to a link at line rate.

Considering the other sides of the conversation

Each switch is responsible for assigning a conversation to a link locally. The return traffic is another conversation, and the connected switch might assign it to a different link even when that switch is using the same load-sharing mode. However, this behavior does not typically cause issues. Each switch considers the link aggregation as a whole as the logical interface, learns MAC addresses on the link aggregation interface, and accepts traffic on any active link.

Other options for link aggregations

In this chapter you have focused on link aggregations between switches, but other use cases exist.

For example, you can establish a link aggregation between a switch and a server with multiple NICs. You can use a manual link aggregation or an LACP link aggregation, but to choose the correct option, you need to understand the server settings and capabilities. When the server is a virtual host for virtual machines (VMs), the virtual switch configuration determines the mode for the link aggregation—some VMware and Citrix XenServer versions do not support LACP—and even whether redundant links should be part of a link aggregation at all

Similarly, you can configure link aggregations between a switch and a storage device that uses Ethernet.

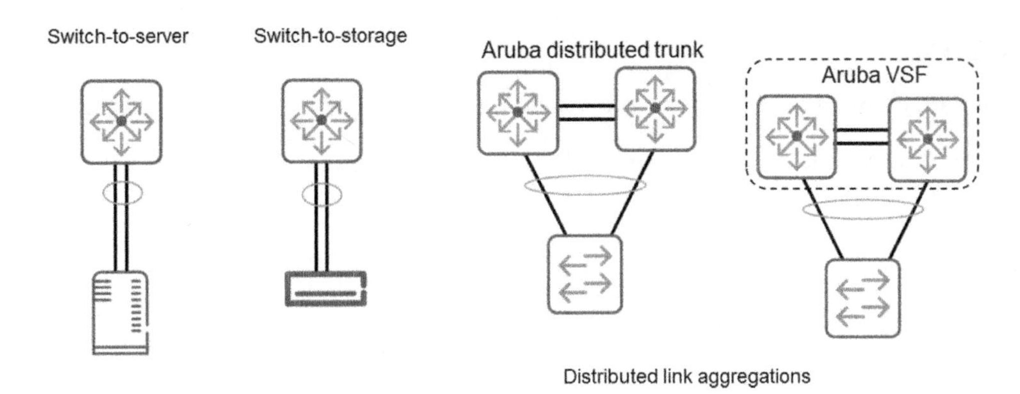

Figure 7-17: Other options for link aggregations

You should attend other HPE training for more details on these topics or refer to the appropriate documentation. When you search for information, keep in mind that the feature might be called bonding or NIC teaming rather than link aggregation.

Finally, note that you can break the rule about how aggregated links must connect the same switches in special circumstances. You can create distributed trunks between a downstream device and two ArubaOS switches. This feature is covered in other HPE training.

As you will learn in Chapter 9, you can also create a link aggregation between a switch and two members of a Virtual Switching Framework (VSF) fabric. The VSF fabric actually operates as a single logical switch, so, unlike with a distributed trunk, you can up this type of link aggregation just like you learned in this chapter.

Summary

This chapter has introduced you to link aggregations, which combine multiple physical interfaces into a single logical interface that load-shares traffic across it.

You learned how to create manual, or static, link aggregations on ArubaOS switches. You then learned about the advantages of LACP for managing the link aggregation, facilitating failover, and ensuring that links are compatible and connected to the same peer. You then practiced setting up an LACP link aggregation.

- Manual link aggregations
- LACP link aggregation

Figure 7-18: Summary

Learning check

Answer these questions to check your understanding of what you have learned in this chapter.

1. What is the ArubaOS switch command for adding interfaces to an aggregated static link?
    a. trunk <int-id-list> trk<ID>
    b. interface <int-id-list> trk<ID>
    c. interface bridge-aggregation <number>
    d. port trunk

2. You want to configure a static LACP link aggregation on an ArubaOS switch. The trunk will use ID 1. The aggregation should include interfaces A1 and A2. What is the command?

3. What is the default load sharing mode for link aggregation on ArubaOS switches?

Answers to Learning check

1. What is the ArubaOS switch command for adding interfaces to an aggregated static link?

   a. trunk <int-id-list> trk<ID>

   b. interface <int-id-list> trk<ID>

   c. interface bridge-aggregation <number>

   d. port trunk

2. You want to configure a static LACP link aggregation on an ArubaOS switch. The trunk will use ID 1. The aggregation should include interfaces A1 and A2. What is the command?

   trunk a1,a2 trk1 lacp

3. What is the default load sharing mode for link aggregation on ArubaOS switches?

   The default mode is source IP address and destination IP address for IP traffic and source MAC address and destination MAC address for non-IP traffic.

8 IP Routing

EXAM OBJECTIVES

✓ Configure static routes on HPE switches

✓ Interpret IP routing tables

✓ Configure a basic OSPF solution

ASSUMED KNOWLEDGE

Before reading this chapter, you should have a basic understanding of:

- Layer 2 and 3 in the Open Systems Interconnection (OSI)
- IP addressing
- VLANs

INTRODUCTION

"Chapter 5: VLANs" introduced you to how routing enables communications between devices in different subnets (VLANs). Many ArubaOS switches support routing, even though it is disabled by default. This chapter teaches you more about IP routing.

IP routes

Core-1 is already using its routing table to route traffic for your network. You will now look at what Core-1 is doing in more detail.

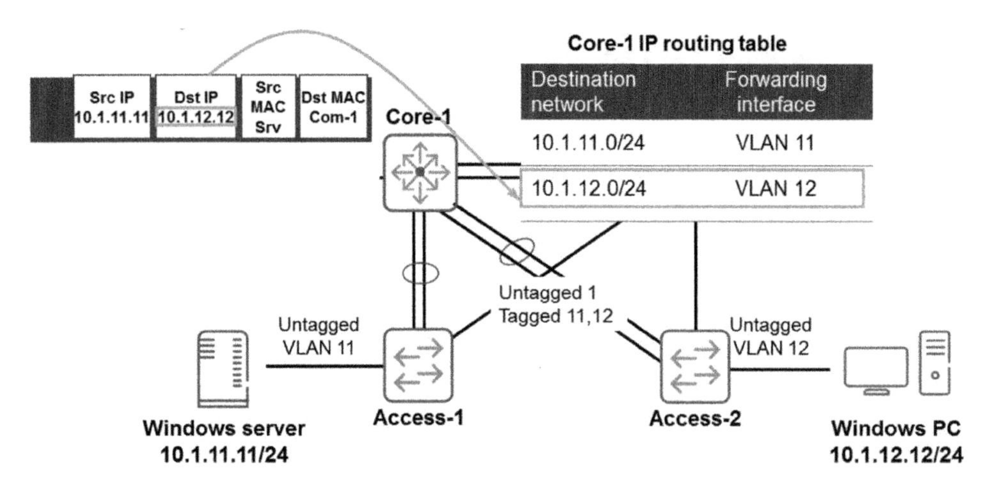

Figure 8-1: IP routes

Switches acting at Layer 2 make switching decisions based on destination MAC addresses. When a Layer 3 routing switch receives IP traffic that is destined to its own MAC address, though, it looks at the destination IP address inside the Ethernet frame. If the destination IP address does not belong to it, the routing switch knows that it needs to route the traffic.

The routing switch uses the destination IP address to decide how to forward the traffic. It looks in its IP routing table and finds the route with the most specific match to the packet's destination IP address—in other words, the route with a matching destination network that has the longest subnet mask.

For example, the server needs to send a packet to the client. It directs the frame to Core-1's MAC address because Core-1 is its default gateway. Core-1 receives a frame and decapsulates it. It matches the destination IP address, 10.1.12.12, to IP route 10.1.12.0/24 in its routing table.

Direct IP routes

Direct routes are for local networks on which the routing device has an IP address itself. They are associated with a Layer 3 forwarding interface such as a VLAN. The direct route is automatically added to the routing table when you configure the IP address on the device.

ArubaOS switches, however, have routing disabled by default. If you want the switch to use its direct routes (or any other routes) to route traffic for other devices, you must enable IP routing.

- Routes to local networks
- No configuration necessary

Core-1 IP addresses

VLAN1: 10.1.1.1/24 → 10.1.1.0/24
VLAN 11: 10.1.11.1/24 → 10.1.11.0/24
VLAN 12: 10.1.12.1/24 → 10.1.12.0/24

IP routing table

Destination network	Forwarding interface
10.1.1.0/24	VLAN 1
10.1.11.0/24	VLAN 11
10.1.12.0/24	VLAN 12

Note: Remember to enable IP routing.

Figure 8-2: Direct IP routes

Indirect IP routes and default routes

An indirect route is a route to a remote network, which does not exist on the routing device. You must configure this route on the device manually, or the device must learn it dynamically with a routing protocol. (This will be covered later.)

As well as a destination network address, an indirect route also includes the next hop—the next device that will forward the packet to its final destination. This next hop is sometimes called the gateway. The next hop IP address should be on a network connected to the routing switch. The routing switch automatically determines the forwarding interface based on which interface connects to the next hop.

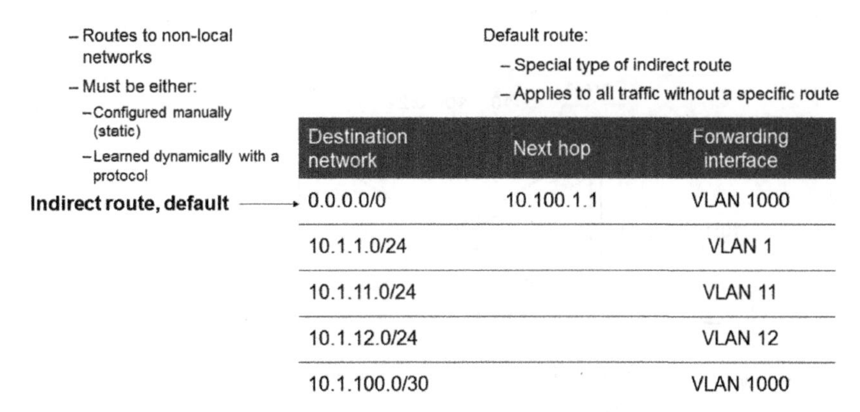

Figure 8-3: Indirect IP routes and default routes

You have already created one special type of indirect route: a default route. As you recall, a default route is a route to 0.0.0.0/0. It applies to all IP traffic for which the device does not have a more specific route.

It is very important to remember that the switch matches routed traffic to the most specific route. For example, in the figure below, the switch has an indirect route to 10.1.0.0/16 and 10.1.12.0/24. A packet that is destined to 10.1.12.12 matches both routes. However, /24 is more specific than /16 (the prefix length is longer), so the switch forwards this traffic to 10.1.102.1. It would forward a packet destined to 10.1.13.13, on the other hand, to 10.1.100.1.

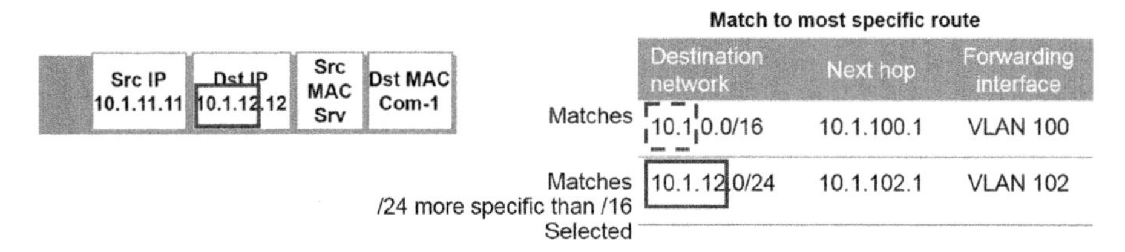

Figure 8-4: Matching to the most specific route

Topology that requires indirect routes

In the topology that you used in previous chapters, direct routes with a few default routes worked well. All the VLANs in your network used the same default router, which routed traffic between its connected networks. You will now set up a different logical topology, which introduces a Layer 3 hop between each switch. Now each switch must act as a routing switch, and it requires indirect routes. For example, Access-1 will now be routing traffic for VLAN 11, subnet 10.1.11.0/24. Subnet 10.1.12.0/24 is not local on this switch, so Access-1 requires an indirect (remote) route to it.

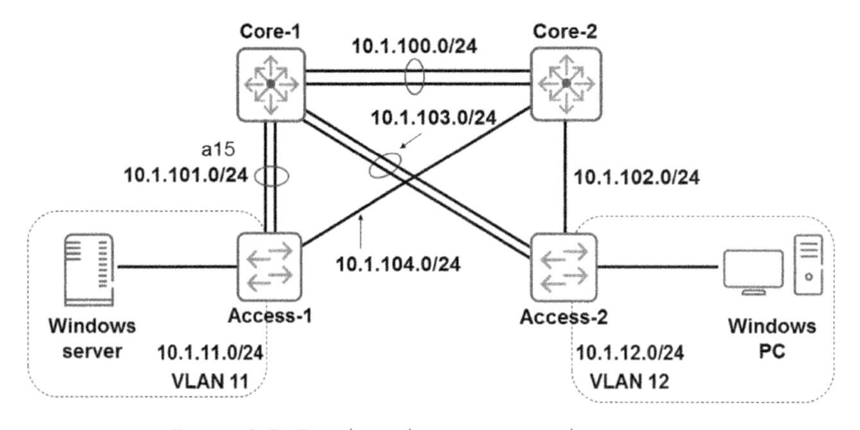

Figure 8-5: Topology that requires indirect routes

As shown in the figure, each switch-to-switch link is associated with a different IP subnet. You should create a VLAN that is specific to the switch-to-switch link—for example, VLAN 102 for the link between Access-2 and Core-2. On both sides of the link, you assign only this VLAN to the switch port. You do not assign any other ports to this VLAN. You also assign an IP address to this VLAN, making it a Layer 3 interface.

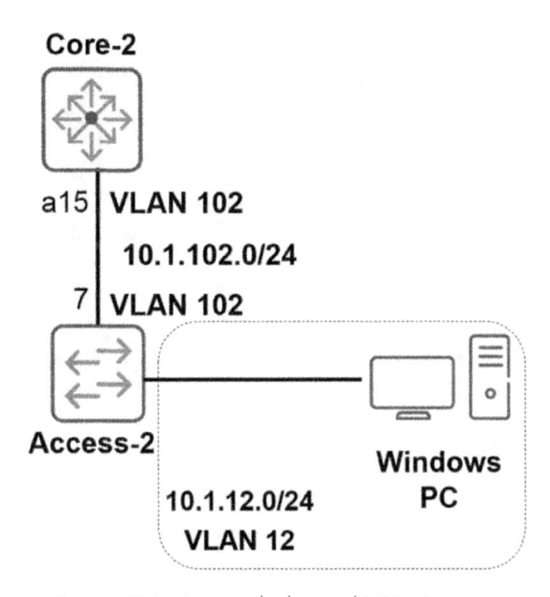

Figure 8-6: Using dedicated VLANs

You might choose to use a topology like this for several reasons:

- You want to eliminate RSTP/MSTP on the switch-to-switch links. In this new topology, VLANs assigned to edge ports are terminated at each switch and not carried on the switch-to-switch links. Therefore, Layer 2 loops cannot be introduced. You have seen how MSTP provides some load-sharing of traffic. But using routing to determine traffic's path instead can result in more optimal paths.

Note

This topology eliminates the need for spanning tree on the switch-to-switch links. But you would still run spanning tree on the switches to prevent accidental loops downstream. You will learn how to disable spanning tree on individual ports in the next section.

- You want to configure the default router role on different switches. In this topology, Access-1 is default router for devices connected to it, and Access-2 is default router for devices connected to it. The burden is distributed, and the negative effect of one switch failing is minimized.

Configure a base topology for routing

You will now set up this topology. After you have the topology in place, you will learn how to configure the correct routes for it.

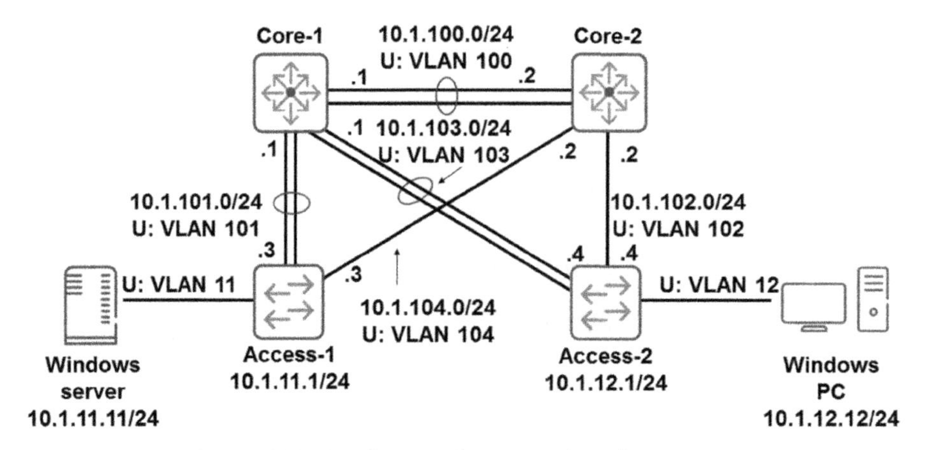

Figure 8-7: Configure a base topology for routing

Task 1: Set up the topology for routing between switches

You will now configure the VLAN and IP topology.

Core-1

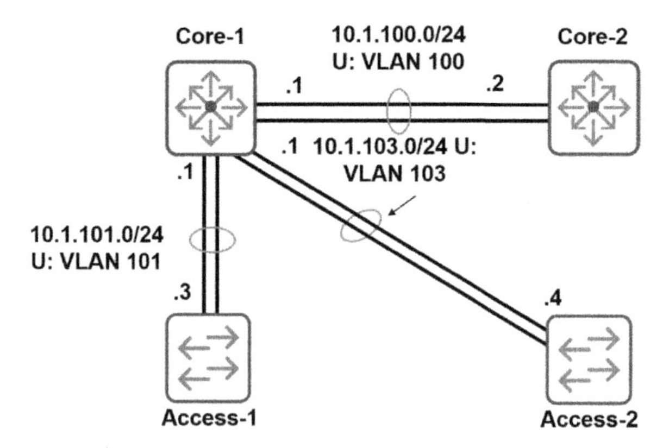

Figure 8-8: Task 1: Set up the topology for routing between switches

1. Access a terminal session with Core-1.
2. Move to global configuration mode.

3. Remove VLANs 11 and 12.

```
Core-1(config)# no vlan 11
Core-1(config)# no vlan 12
```

4. Create VLAN 100 and assign it to the link aggregation that connects to Core-2. Assign it IP address 10.1.100.1/24.

```
Core-1(config)# vlan 100
Core-1(vlan-100)# untagged trk1
Core-1(vlan-100)# ip address 10.1.100.1/24
Core-1(vlan-100)# exit
```

5. Create VLAN 101. Assign it to the link aggregation that connects to Access-1, and assign it IP address 10.1.101.1/24.

```
Core-1(config)# vlan 101
Core-1(vlan-101)# untagged trk2
Core-1(vlan-101)# ip address 10.1.101.1/24
Core-1(vlan-101)# exit
```

6. Create VLAN 103. Assign it to the link aggregation that connects to Access-2, and assign it IP address 10.1.103.1/24.

```
Core-1(config)# vlan 103
Core-1(vlan-103)# untagged trk3
Core-1(vlan-103)# ip address 10.1.103.1/24
Core-1(vlan-103)# exit
```

Access-1

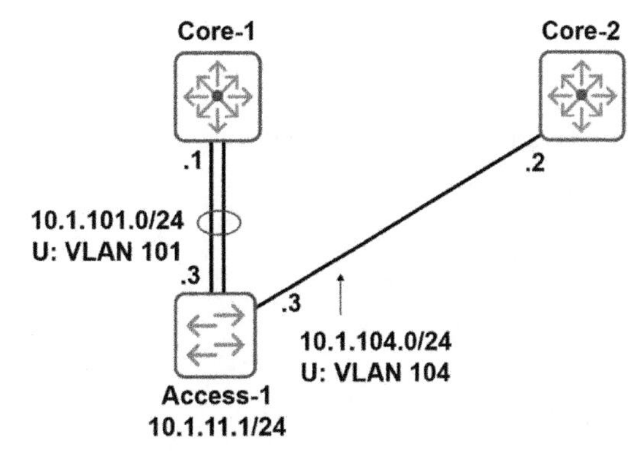

Figure 8-9: Topology showing only Access-1 links

7. Access a terminal session with Access-1.
8. Either paste in the commands from the Task1-Access-1.cfg file and then move to step 15 (last step in this section) or follow the steps below.
9. Move to global configuration mode.
10. Remove VLAN 12.

```
Access-1(config)# no vlan 12
```

11. Access-1 will now be the router for VLAN 11, which will terminate at this switch. Remove VLAN 11 from the switch-to-switch links and configure IP address 10.1.11.1/24 on it.

```
Access-1(config)# vlan 11
Access-1(vlan-11)# no tagged trk1,7
Access-1(vlan-11)# ip address 10.1.11.1/24
```

12. Create VLAN 101. Assign it to the link aggregation that connects to Access-1, and assign it IP address 10.1.101.3/24.

```
Access-1(config)# vlan 101
Access-1(vlan-101)# untagged trk1
Access-1(vlan-101)# ip address 10.1.101.3/24
Access-1(vlan-101)# exit
```

13. Create VLAN 104. Assign it to the interface that connects to Core-2, and assign it IP address 10.1.104.3/24.

```
Access-1(config)# vlan 104
Access-1(vlan-104)# untagged 7
Access-1(vlan-104)# ip address 10.1.104.3/24
Access-1(vlan-104)# exit
```

14. You should now be able to reach Core-1 at 10.1.101.1 since you have already set the IP address in this VLAN on the other Core-1.

    If the ping fails, check the IP addressing on Core-1 and Access-1.

```
Access-1(config)# ping 10.1.101.1
10.1.101.1 is alive, time = 1 ms
```

Access-2

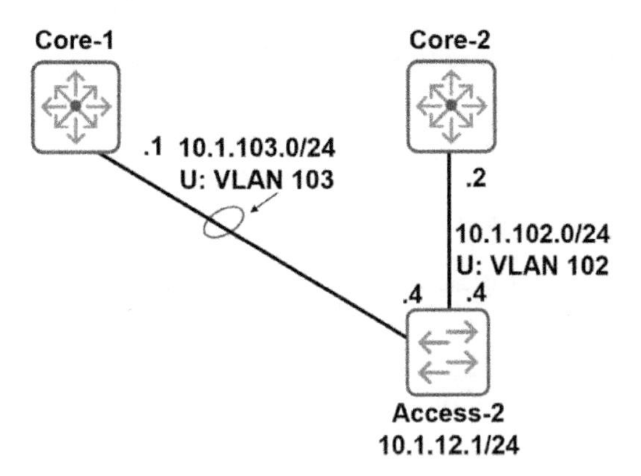

Figure 8-10: Topology showing Access-2 links

15. Access a terminal session with Access-2.
16. Either paste in the commands from the Task1-Access-2.cfg file and then move to step 23 (last step in this section) or follow the steps below.
17. Move to global configuration mode.
18. Remove VLAN 11.

```
Access-2(config)# no vlan 11
```

19. Access-2 will now be the router for VLAN 12, which will terminate at this switch. Remove VLAN 12 from the switch-to-switch links and configure IP address 10.1.12.1/24 on it.

```
Access-2(config)# vlan 12
Access-2(vlan-12)# no tagged trk1,7
Access-2(vlan-12)# ip address 10.1.12.1/24
```

20. Remember that VLAN 12 requires an IP helper address so that clients can receive DHCP addresses from the server in VLAN 11.

```
Access-2(vlan-12)# ip helper-address 10.1.11.11
Access-2(vlan-12)# exit
```

21. Create VLAN 102. Assign it to interface that connects to Core-2, and assign it IP address 10.1.102.4/24.

```
Access-2(config)# vlan 102
Access-2(vlan-102)# untagged 7
```

```
Access-2(vlan-102)# ip address 10.1.102.4/24
Access-2(vlan-102)# exit
```

22. Create VLAN 103. Assign it to the link aggregation that connects to Core-1, and assign it IP address 10.1.103.4/24.

```
Access-2(config)# vlan 103
Access-2(vlan-103)# untagged trk1
Access-2(vlan-103)# ip address 10.1.103.4/24
Access-2(vlan-103)# exit
```

23. You should now be able to reach Core-1 at 10.1.103.1 since you have already set the IP address in this VLAN on the other Core-1.

    If the ping fails, check the IP addressing on Core-1 and Access-2.

```
Access-2(config)# ping 10.1.103.1
10.1.103.1 is alive, time = 1 ms
```

Core-2

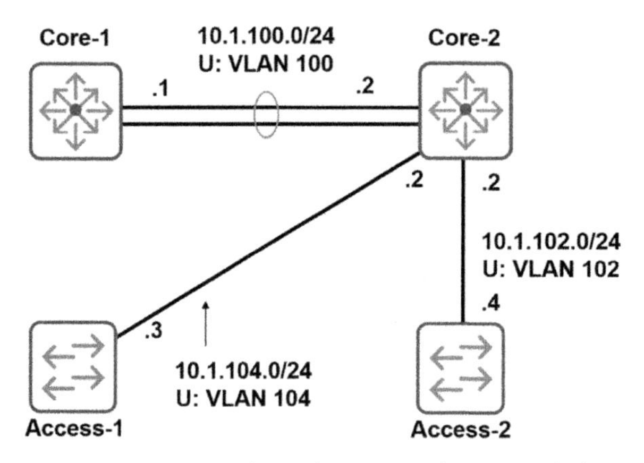

Figure 8-11: Topology showing only Core-2 links

24. Access a terminal session with Core-2.
25. Either paste in the commands from the Task1-Core-2.cfg file and then move to step 31 (second to last step in this section) or follow the steps below.
26. Move to global configuration mode.

27. Remove VLANs 11 and 12.

```
Core-2(config)# no vlan 11
Core-2(config)# no vlan 12
```

28. Create VLAN 100. Assign it to the link aggregation that connects to Core-1, and assign it IP address 10.1.100.2/24.

```
Core-2(config)# vlan 100
Core-2(vlan-100)# untagged trk1
Core-2(vlan-100)# ip address 10.1.100.2/24
Core-2(vlan-100)# exit
```

29. Create VLAN 102. Assign it to the interface that connects to Access-2, and assign it IP address 10.1.102.2/24.

```
Core-2(config)# vlan 102
Core-2(vlan-102)# untagged a15
Core-2(vlan-102)# ip address 10.1.102.2/24
Core-2(vlan-102)# exit
```

30. Create VLAN 104. Assign it to the interface that connects to Access-1, and assign it IP address 10.1.104.2/24.

```
Core-2(config)# vlan 104
Core-2(vlan-102)# untagged a13
Core-2(vlan-104)# ip address 10.1.104.2/24
Core-2(vlan-104)# exit
```

31. You have set up VLAN 100, subnet 10.1.100.0/24, on both sides of the link. You should be able to ping Core-1 from Core-2.

    If the ping fails, check the IP addressing on Core-1 and Core-.

```
Core-2(config)# ping 10.1.100.1
10.1.100.1 is alive, time = 1 ms
```

32. Now try to ping Access-1 on its IP address on its Core-2 link. Also try to ping Access-2.

    The pings will fail.

```
Core-2(config)# ping 10.1.104.3
Request timed out.

Core-2(config)# ping 10.1.102.4
Request timed out.
```

Note

At the end of this task, you still have IP addresses on VLAN 1 on your switches.

Task 2: Explore why some links are unavailable

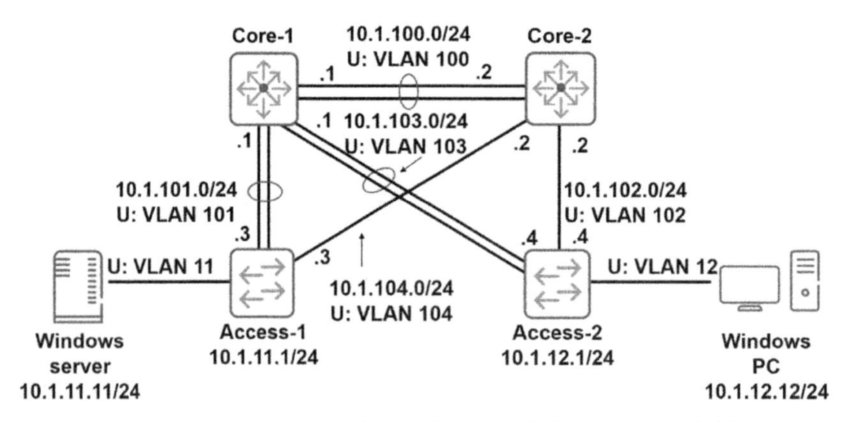

Figure 8-12: Task 2: Explore why some links are unavailable

You will now explore why some links are unavailable.

1. Refer to the topology above. Why do you think that Core-2 cannot ping Access-1 and Access-2?

Access-1

Explore this question by viewing settings on Access-1.

2. Access the terminal session with Access-1.
3. View the IST roles for the connections to Core-1 and Core-2. (The new VLANs are part of the IST.)

```
Access-1# show spanning-tree trk1,7 instance ist
<-output omitted->
                                                    Designated
  Port Type      Cost  Priority Role       State      Bridge
  ---- --------- ----- -------- ---------- ---------- -------------
  7    100/1000T 20000 128      Alternate  Blocking   1c98ec-ab4b00
  Trk1           20000 64       Root       Forwarding 1c98ec-ab4b00
```

Task 3: Resolve the spanning tree issues with filters

As you see, MSTP blocks the link to Core-2 on Access-1.You would find the same result on Access-2. Every link in the new topology has its own VLAN, so loops will not form within VLANs. But MSTP builds the topology based on link regardless of the VLAN. You need to disable spanning tree on the links between the switches to allow routing over all links. However, you want to keep spanning tree enabled to prevent against accidental loops on edge ports. You will disable spanning tree specifically on the switch-to-switch ports.

Access-1

1. On Access-1, verify that the switch-to-switch links are correctly in different, unique VLANs.

```
Access-1(config)# show vlan port trk1,7 detail
 Status and Counters - VLAN Information - for ports 7
  Port name: Core-2
  VLAN ID Name                 | Status     Voice Jumbo Mode
  ------- -------------------- + ---------- ----- ----- --------
  104     VLAN104              | Port-based No    No    Untagged
 Status and Counters - VLAN Information - for ports Trk1
  VLAN ID Name                 | Status     Voice Jumbo Mode
  ------- -------------------- + ---------- ----- ----- --------
  101     VLAN101              | Port-based No    No    Untagged
```

2. Move to global configuration mode.
3. Implement BPDU filters to disable spanning-tree on trk1 and on the interface that connects to Core-2.

```
Access-1(config)# spanning-tree trk1,7 bpdu-filter
```

Note

You will see a warning because the BPDU filter could cause loops to occur. You already checked your topology to ensure that you will not introduce a loop.

4. Optional: To clean up your configuration, you can also remove the MSTP instances. They are no longer needed for load sharing because you are only using MSTP to protect against accidental loops. Note that you do **not** need to complete this step if you initialized the configuration at the beginning of this task.

```
Access-1(config)# no spanning-tree instance 1

Access-1(config)# no spanning-tree instance 2
```

Core-1, Core-2, and Access-2

5. Follow similar steps on the other switches. The next page provides the commands for your reference.

```
Core-1(config)# show vlan port trk1,trk2,trk3 detail

 Status and Counters - VLAN Information - for ports Trk1

  VLAN ID Name                 | Status     Voice Jumbo Mode

  ------- -------------------- + ---------- ----- ----- --------

  100     VLAN100              | Port-based No    No    Untagged

 Status and Counters - VLAN Information - for ports Trk2

  VLAN ID Name                 | Status     Voice Jumbo Mode

  ------- -------------------- + ---------- ----- ----- --------

  101     VLAN101              | Port-based No    No    Untagged

 Status and Counters - VLAN Information - for ports Trk3

  VLAN ID Name                 | Status     Voice Jumbo Mode

  ------- -------------------- + ---------- ----- ----- --------

  103     VLAN103              | Port-based No    No    Untagged
```

```
Core-1(config)# spanning-tree trk1,trk2,trk3 bpdu-filter
Core-1(config)# no spanning-tree instance 1
Core-1(config)# no spanning-tree instance 1

Core-2(config)# show vlan port trk1,a13,a15 detail

 Status and Counters - VLAN Information - for ports A13
  Port name: Access-1
  VLAN ID Name                 | Status     Voice Jumbo Mode
  ------- -------------------- + ---------- ----- ----- --------
  104     VLAN104              | Port-based No    No    Untagged
 Status and Counters - VLAN Information - for ports A15
  Port name: Access-2
  VLAN ID Name                 | Status     Voice Jumbo Mode
  ------- -------------------- + ---------- ----- ----- --------
  102     VLAN102              | Port-based No    No    Untagged
 Status and Counters - VLAN Information - for ports Trk1
  VLAN ID Name                 | Status     Voice Jumbo Mode
  ------- -------------------- + ---------- ----- ----- --------
  100     VLAN100              | Port-based No    No    Untagged
Core-2(config)# spanning-tree trk1,a13,a15 bpdu-filter
Core-2(config)# no spanning-tree instance 1
Core-2(config)# no spanning-tree instance 1

Access-2(config)# show vlan port trk1,7 detail
 Status and Counters - VLAN Information - for ports 7
  Port name: Core-2
  VLAN ID Name                 | Status     Voice Jumbo Mode
  ------- -------------------- + ---------- ----- ----- --------
  102     VLAN102              | Port-based No    No    Untagged
```

```
 Status and Counters - VLAN Information - for ports Trk1
  VLAN ID Name                 | Status     Voice Jumbo Mode
  ------- -------------------- + ---------- ----- ----- --------
  103     VLAN103              | Port-based No    No    Untagged
Access-2(config)# spanning-tree trk1,7 bpdu-filter
Access-2(config)# no spanning-tree instance 1
Access-2(config)# no spanning-tree instance 1
```

Core-2

6. Return to the terminal session with Core-2.
7. Verify that Core-2 can reach Access-1 and Access-2 on the directly connected links. The pings should now be successful because the BPDU filters ensure that all routed links remain up.

```
Core-2(config)# ping 10.1.104.3
10.1.104.3 is alive, time = 3 ms

Core-2(config)# ping 10.1.102.4
10.1.102.4 is alive, time = 3 ms
```

Think about what you have learned about RSTP/MSTP.

Task 4: Explore the need for routing

You have now tested connectivity on each individual switch-to-switch link. Try testing connectivity between other subnets.

Access-2

1. Ping 10.1.11.1 (Access-1) from Access-2.

```
Access-2(config)# ping 10.1.11.1
The destination address is unreachable.
```

2. Now trying pinging 10.1.100.1 (Core-1's link to Core-2) from Access-2.

```
Access-2(config)# ping 10.1.100.1
The destination address is unreachable.
```

3. Why did the pings fail?

   Answer: The switches can only communicate on their direct links because they do not know how to reach the other subnets. You need to set up routing to establish full connectivity.

Windows PC

4. Open a command prompt on the Windows PC.
5. Try to obtain an IP address (ipconfig/release and ipconfig/renew).
6. You set an IP helper address on VLAN 12 on Access-2. Why does DHCP relay fail?

   Answer: Access-2 must have a route to reach the server, and the server's default gateway must have a route to 10.1.12.0/24 for DHCP relay to succeed.

Task 5: Save

Save the current configuration on each of the four switches using the **write memory** command.

Learning check

1. How were you able to disable spanning tree on switch-to-switch links (implement BPDU filters) without introducing broadcast storms? Why did you need to disable spanning tree on these links in your initial setup?

2. Which IP addresses can Access-1 ping? Which IP addresses can it not ping, and why not?

3. View the IP routing table on Access-1 (**show ip route**) and note what you see.

Answers to Learning check

1. How were you able to disable spanning tree on switch-to-switch links (implement BPDU filters) without introducing broadcast storms? Why did you need to disable spanning tree on these links in your initial setup?

   In your new topology, VLANs are isolated on one switch only. Because the broadcast domain does not extend over the redundant switch-to-switch links, broadcast storms cannot occur.

   You needed to disable spanning tree on these links (filter BPDUs) so that the redundant switch-to-switch links are available for routing. Otherwise, MSTP puts some of the interfaces in a blocking state.

2. Which IP addresses can Access-1 ping? Which IP addresses can it not ping, and why not?

   Access-1 can ping IP addresses on the networks that are directly connected to it:

   - Core-1 at 10.1.101.1
   - Core-2 at 10.1.104.2
   - The server at 10.1.11.11

   Access-1 cannot ping any other IP addresses, including other IP addresses on Core-1 and Core-2.

3. View the IP routing table on Access-1 (**show ip route**) and note what you see.

   You can see direct routes to the directly connected networks, so the switch can reach IP addresses on those networks. You do not see routes for networks such as 10.1.12.0/24 and 10.1.100.0/24, so the switch cannot reach IP addresses in those networks.

Indirect IP routes

The figure shows a simplified network with just Access-1, Access-2, and Core-1 (no redundant connections through Core-2). You will now learn how to set up routes in this topology.

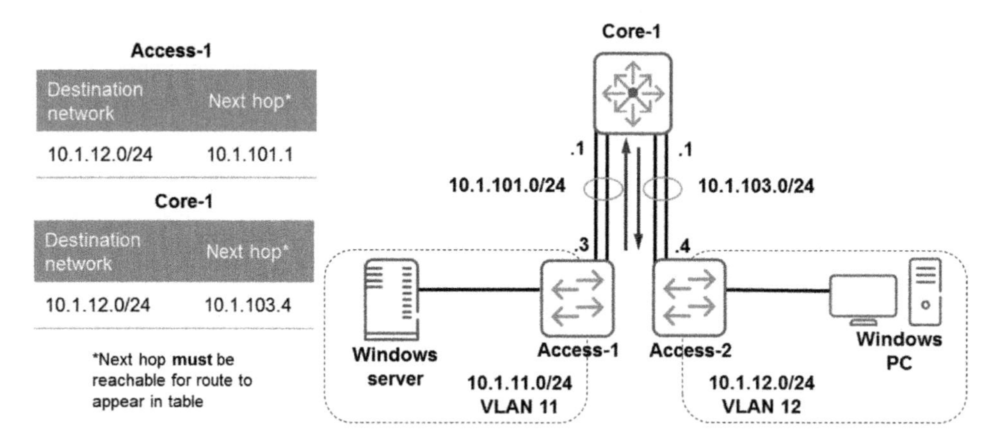

Figure 8-13: Indirect IP routes

You need to create specific indirect routes (as opposed to just direct and default routes). For example, you might create a route to 10.1.12.0/24 on Access-1 so that this switch can route server traffic to clients.

In addition to the destination network address, the route specifies the next hop. It specifies *only* the next hop; as far as the local switch's routing table is concerned, the rest of the path does not exist.

The next hop for Access-1's route to 10.1.12.0/24 is Core-1. Core-1 has several IP addresses. For the next hop address, you should specify the IP address on the subnet that connects the two switches:

10.1.101.1. If you specified a different IP address on Core-1, Access-1 would not know how to reach that address. A route with an unreachable next hop is not added to the active IP routing table, and the switch cannot use it.

To finish the path to 10.1.12.0/24, you must also create a route to 10.1.12.0/24 on Core-1. Again you specify the next hop in the path. This next hop is Access-2 and, specifically, IP address 10.1.103.1 on Access-2.

Now Access-1 can route traffic to 10.1.12.0/24 successfully.

Of course, you would need to create a return route for traffic to enable full connectivity.

For example, you want to set up connectivity between servers in subnet 10.1.11.0/24 and clients in subnet 10.1.12.0/24. You have already configured a route to 10.1.12.0/24 on Access-1 and Core-1. You would need to configure these routes for the return traffic:

- Access-2: A route to 10.1.11.0/24 through next hop 10.1.103.1
- Core-1: A route to 10.1.11.0/24 through next hop 10.1.104.1

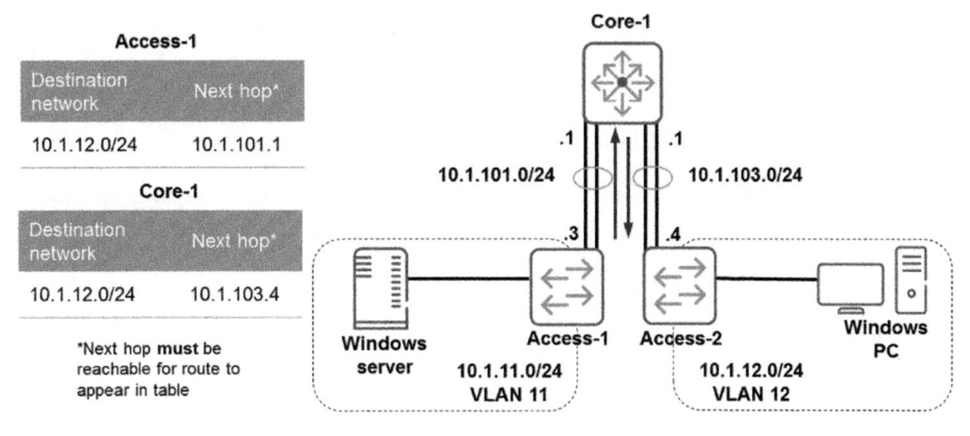

Figure 8-14: Indirect routes for return traffic to 10.1.11.0/24

Static IP routing

Switches can learn indirect routes statically or dynamically through a routing protocol. For now, you are focusing on static routes, which are configured manually by the administrator. When you use static routing, you must configure the routes manually on each router or routing switch in the path. You must take care to specify the next hops correctly. And you must take care to create all the necessary paths.

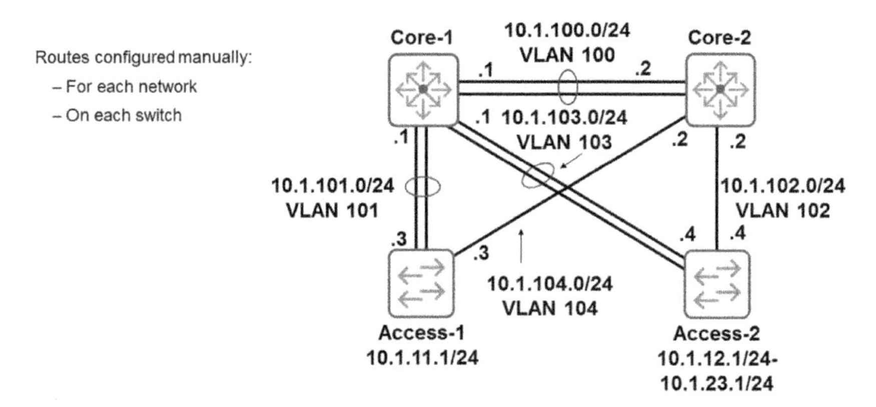

Figure 8-15: Static IP routing

For example, the routes that you examined establish connectivity between 10.1.11.0/24 and 10.1.12.0/24. However, Access-1 still cannot route traffic to the 10.1.102.0/24 subnet on Access-2, for example. You might not care about routing traffic to a subnet devoted to a switch-to-switch link.

But what if Access-2 supported several other subnets with devices in them? For example, it might support 10.1.13.0/24, 10.1.14.0/24, and so on up to 10.1.23.1/24. You would need to create a route for each on each routing device in the pathway. (Or, if subnets are contiguous, you could create a route to a larger destination subnet that includes the other subnets within it.)

As you can see, the process can be laborious.

As a final note, your network has redundancy, but you are not currently taking advantage of the redundant links. You will look at managing redundant static routes a bit later.

Create static IP routes

You will now set up static routes on Access-1, Access-2, and Core-1 to enable connectivity between the server in 10.1.11.0/24 and the PC in 10.1.12.0/24.

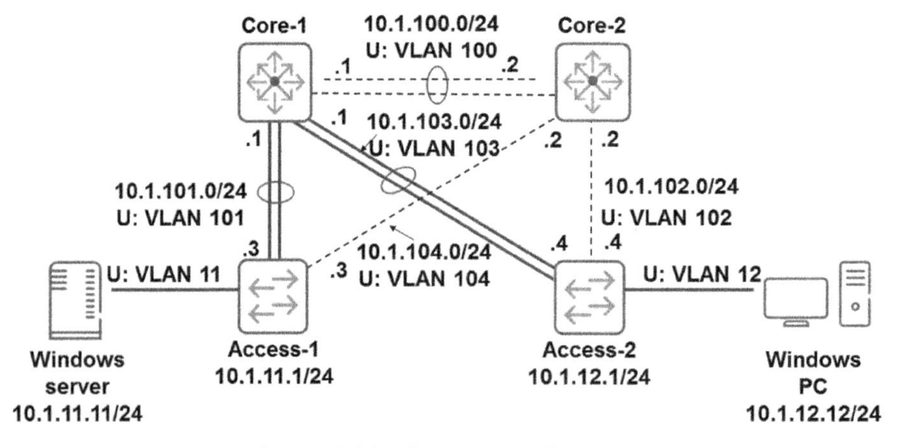

Figure 8-16: Create static IP routes

You will use default routes on the access layer switches, which are using Core-1 as their single next hop for all other destinations. You will create routes to specific destination networks on Core-1, which needs to route some traffic to Access-1 and some traffic to Access-2.

Task 1: Configure the static routes

Table 8-1 shows the routes required to establish the path between the server and Windows PC.

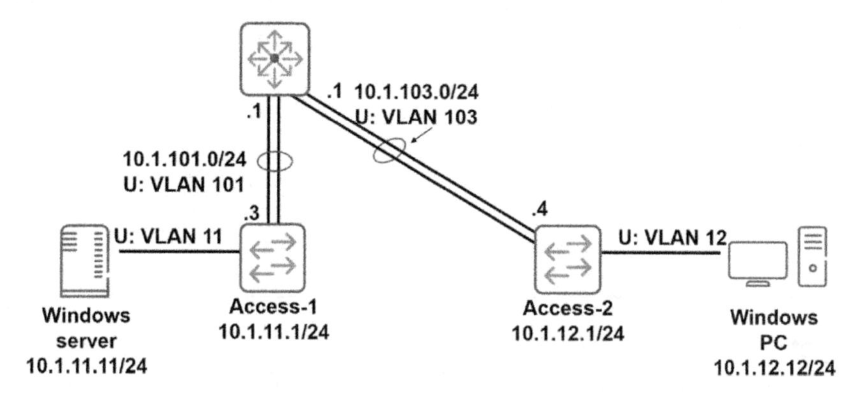

Figure 8-17: Task 1: Configure the static routes

Table 8-1: Static routes

Switch	Destination network	Next hop
Access-1	10.1.12.0/24 or 0.0.0.0/0	10.1.101.1
Core-1	10.1.12.0/24	10.1.103.4
	10.1.11.0/24	10.1.101.3
Access-2	10.1.11.0/24 or 0.0.0.0/0	10.1.103.1

Access-1

1. Access a terminal session with Access-1 and move to the global configuration mode context.
2. Enable IP routing.

```
Access-1(config)# ip routing
```

3. Create a default route through Core-1 at 10.1.101.1.

```
Access-1(config)# ip route 0.0.0.0/0 10.1.101.1
```

4. View the IP routing table.

```
Access-1(config)# show ip route

                          IP Route Entries

  Destination    Gateway    VLAN Type      Sub-Type Metric Dist.
  -----------    ---------- ---- --------- -------- ------ -----
  0.0.0.0/0      10.1.101.1 101  static             1      1
  10.1.11.0/24   VLAN11     11   connected          1      0
  10.1.101.0/24  VLAN101    101  connected          1      0
  10.1.104.0/24  VLAN104    104  connected          1      0
  127.0.0.0/8    reject          static             0      0
  127.0.0.1/32   lo0             connected          1      0
```

Important

Access-1 still has the IP route to 0.0.0.0/0 through 10.1.1.1 that you configured in Chapter 5. However, this route is not in the IP routing table because the next hop is not available. In the real world, you would remove this legacy route (and if you check the config files on the Windows PC, they do not have this route).

However, if you are planning to do all the test network tasks, you must keep the route because you will need it again later. The same applies to Access-2.

Windows server

You will use a number of trace routes throughout this and other tasks. A trace route lists each routing hop between the device that executes the trace route and the destination. (Intervening devices at Layer 2 do not show up in the trace route.)

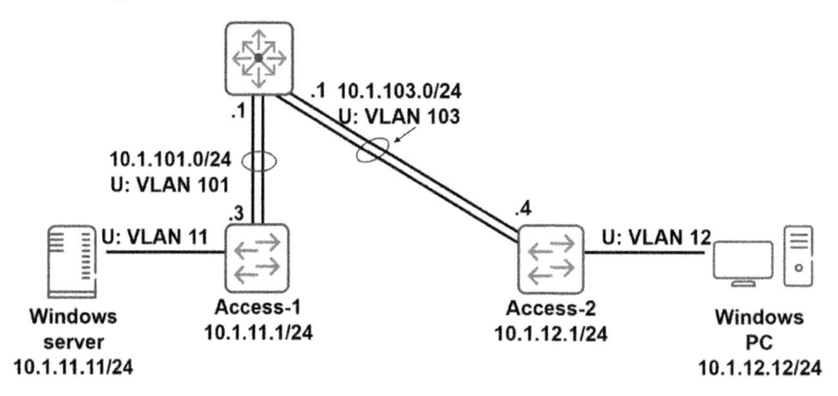

Figure 8-18: Topology showing only the links for the route

5. Access the Windows server desktop and open a command prompt.

The traceroutes will include the intuitive hostnames for the devices because the test network environment is set up with DNS services, which map the IP addresses to the hostnames. You can use the -d option (traceroute -d 10.1.12.12) to run the traceroute without looking up hostnames if these aren't available in your environment.

6. Execute a trace route to the Windows PC (10.1.12.12).

```
tracert 10.1.12.12

tracing a route to pc.hpnu01.edu [10.1.12.12] over a maximum of 30 hops:

  1    <1 ms    <1 ms     1 ms  vlan11_defaultrouter.hpnu01.edu [10.1.11.1]

  2    *        *         *     Request timed out.

  3    *        *         *     Request timed out.
```

7. Notice that the trace fails after one hop. Press <Ctrl+C> to end it.

8. Why did the routing fail?

   Answer: Access-1 routes the ping to Core-1, but Core-1 does not yet know how to reach the destination. Core-1 also does not know a route to 10.1.11.0/24, so it can't send an ICMP unreachable message back.

Core-1

9. Access a terminal session with Core-1 and move to global configuration mode.

10. Examine the routing table to validate that Core-1 has no route to 10.1.11.0/24 or 10.1.12.0/24.

```
Core-1(config)# show ip route

  IP Route Entries

  Destination   Gateway VLAN Type      Sub-Type Metric  Dist.

  ------------- ------- ---- --------- -------- ------- -----

  10.1.100.0/24 VLAN100 100  connected           1       0

  10.1.101.0/24 VLAN101 101  connected           1       0

  10.1.103.0/24 VLAN103 103  connected           1       0

  127.0.0.0/8   reject       static              0       0

  127.0.0.1/32  lo0          connected           1       0
```

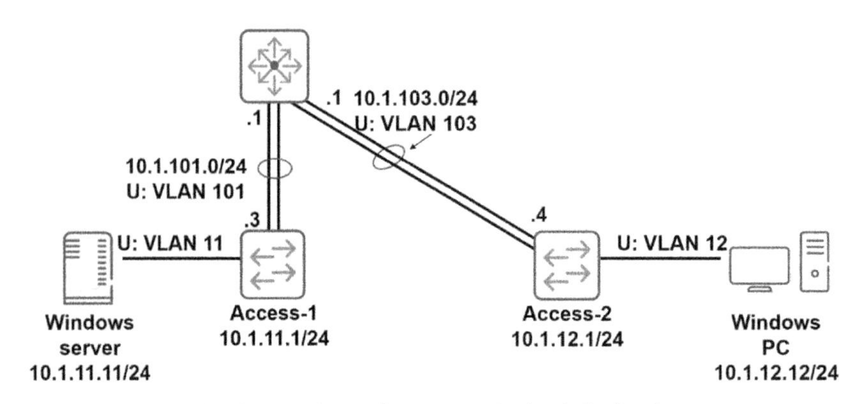

Figure 8-19: Topology showing only the links for the route

11. IP routing is already enabled on Core-1. Create the static route to 10.1.12.0/24 through Access-2 at 10.1.103.4.

```
Core-1(config)# ip route 10.1.12.0/24 10.1.103.4
```

12. Also create the static route for return traffic to 10.1.11.0/24. The next hop is Access-1 at 10.1.101.3.

```
Core-1(config)# ip route 10.1.11.0/24 10.1.101.3
```

13. View the new routes in the IP routing table.

```
Core-1(config)# show ip route

                         IP Route Entries

  Destination    Gateway     VLAN  Type      Sub-Type Metric   Dist.
  -------------- ----------  ----- --------- -------- -------- -----
  10.1.11.0/24   10.1.101.3  101   static             1        1
  10.1.12.0/24   10.1.103.4  103   static             1        1
  10.1.100.0/24  VLAN100     100   connected          1        0
  10.1.101.0/24  VLAN101     101   connected          1        0
  10.1.103.0/24  VLAN103     103   connected          1        0
  127.0.0.0/8    reject            static             0        0
  127.0.0.1/32   lo0               connected          1        0
```

Windows server

14. Access the Windows server and re-perform the trace route.

```
tracert 10.1.12.12

tracing a route to pc.hpnu01.edu [10.1.12.12] over a maximum of 30 hops:

  1     <1 ms     <1 ms     <1 ms   core-1-vlan11.hpnu01.edu [10.1.11.1]
  2      1 ms      1 ms      1 ms   core-1_access-1_link [10.1.101.1]
  3      *         *         *      Request timed out.
  4      *         *         *      Request timed out.
```

15. Now after two hops, the trace route begins to time out. Press <Ctrl+c> to end it.

    Why does the trace route fail?

    Answer: The trace route ICMP reaches Access-2, but Access-2 does not know how to route return traffic.

Access-2

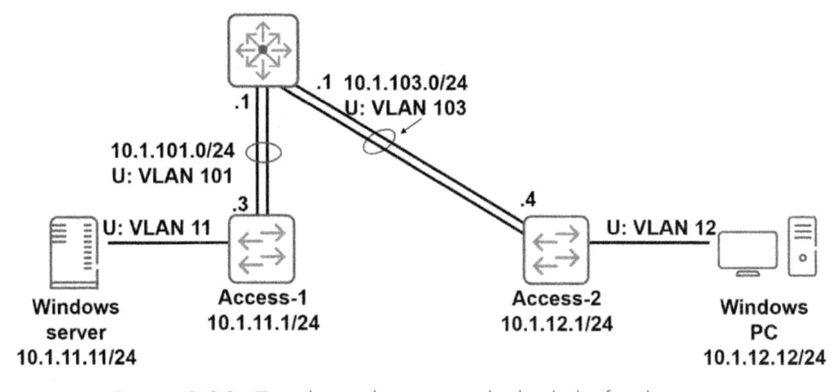

Figure 8-20: Topology showing only the links for the route

16. Access a terminal session with Access-2 and move to the global configuration mode context.
17. Enable IP routing.

```
Access-2(config)# ip routing
```

18. Create a default route through Core-1 at 10.1.103.1.

```
Access-2(config)# ip route 0.0.0.0/0 10.1.103.1
```

19. View the IP routing table

```
Access-2(config)# show ip route

                          IP Route Entries

  Destination     Gateway      VLAN Type       Sub-Type Metric Dist.
  -----------     ----------   ---- ---------  -------- ------ -----
  0.0.0.0/0       10.1.103.1   103  static                1      1
  10.1.12.0/24    VLAN12       12   connected             1      0
  10.1.102.0/24   VLAN102      102  connected             1      0
  10.1.103.0/24   VLAN103      103  connected             1      0
  127.0.0.0/8     reject            static                0      0
  127.0.0.1/32    lo0               connected             1      0
```

Windows PC

20. Access the Windows PC desktop.
21. Open a command prompt.
22. Renew the DHCP address by entering **ipconfig/release** and **ipconfig/renew**.
23. Record the IP address that the client receives, which should be in the 10.1.12.0/24 subnet. If it is different from 10.1.12.12, use that IP address instead in the next trace route command.

 **Note**

If the client does not receive an IP address, go to the server and trace the route to 10.1.12.1. If you do not see the 10.1.103.4 hop, check the route on Access-2. If this traceroute succeeds, make sure that you have deleted VLAN 12 entirely on Core-1. (If Core-1 still has IP address 10.1.12.1, it will prevent the DHCP offer from returning to Access-2.)

After fixing these issues, try to obtain the DHCP address on the client again.

If DHCP does not work, simply assign your client a static IP address such as 10.1.12.32 255.255.255.0, default gateway 10.1.12.1.

Windows server

24. Access the Windows server and re-perform the trace route. It should be successful. Remember to use the correct IP address for your client.

```
tracert 10.1.12.12

tracing a route to pc.hpnu01.edu [10.1.12.12] over a maximum of 30 hops:

  1   <1 ms  <1 ms  <1 ms  vlan11_default_router.hpnu01.edu [10.1.11.1]
  2    1 ms   1 ms   1 ms  core-1_access-1_link.hpnu01.edu [10.1.101.1]
  3   <1 ms  <1 ms  <1 ms  access-2_core-1_link.hpnu01.edu [10.1.103.4]
  4    4 ms  <1 ms  <1 ms  pc.hpnu01.edu [10.1.12.12]

Trace complete.
```

Note

If you do not see the 10.1.12.12 hop, check that the Windows client has the correct IP address and default gateway (use **ipconfig**).

Task 2: Save

Save the current configuration on each of the switches using the **write memory** command.

Learning check

Review what you have learned by answering these questions.

1. To enable a ping between the server and the client, on which switches did you need to configure static routes?

2. What would happen if Access-1 loses its connection to Core-1? Will the server and PC still be able to reach each other?

3. View an IP routing table and take notes on the output.

Answers to Learning check

1. To enable a ping between the server and the client, on which switches did you need to configure static routes?

   You needed to configure a static route on Access-1, Core-1, and Access-2. Every routing hop in the pathway requires a route, so both Access-1 and Core-1 required a route to 10.1.12.0/24 to enable the server traffic to reach the client. Because the client return traffic also required routing, you had to set up routes to 10.1.11.0/24 on Access-2 and Core-1.

2. What would happen if Access-1 loses its connection to Core-1? Will the server and PC still be able to reach each other?

   No. When Access-1 loses its connection to Core-1, its 10.1.101.0/24 subnet becomes unavailable. Access-1 can no longer reach the next hop in its static route, so the route is no longer available. Even though Access-1 still has a physical pathway to Access-2 through Core-2, it cannot use this pathway because it has no route.

3. View an IP routing table and take notes on the output.

   In addition to the static routes that you created, you see direct routes. You also see a route through the loopback interface.

   For each route, you see the destination network IP address, the gateway (or next hop), and the forwarding interface. You might notice other parameters as well, including:

   - The type, which indicates the route type such as static or direct.
   - Metric
   - Dist, which stands for administrative distance

   Metric and administrative distance help the routing switch prefer the right route when it has more than one route to the same destination, as you will learn in a moment.

Managing redundant static routes

As you learned, the static routing topology that you set up does not provide redundancy. You can add to the resiliency of your network by setting up redundant routes. The redundant routes can also introduce load-sharing.

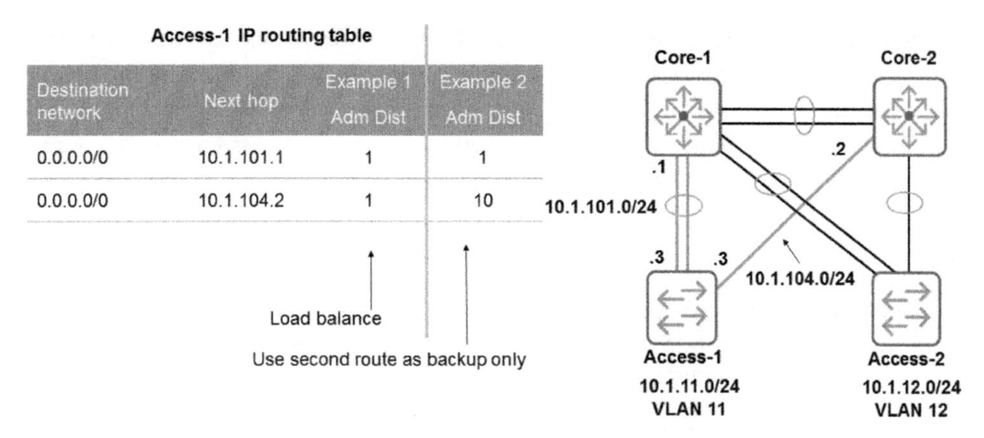

Access-1 IP routing table

Destination network	Next hop	Example 1 Adm Dist	Example 2 Adm Dist
0.0.0.0/0	10.1.101.1	1	1
0.0.0.0/0	10.1.104.2	1	10

Figure 8-21: Managing redundant static routes

For example, you can create a second default route on Access-1 through Core-2 at IP address 10.1.104.2. (Of course, Core-2 needs to have the correct routes so that it can route the traffic successfully.)

Now Access-1 has two routes to the same destination network. It needs to decide whether it will add the first route, the second, or both to its routing table. Assuming that both routes have a valid next hop, the switch makes this decision based on these route properties:

- Administrative distance—Routing devices use administrative distance to distinguish between routes that are learned in different ways such as through different routing protocols. You will examine administrative distance in more detail when you learn about dynamic routing protocols.

  Note, though, that you can set a static route's administrative distance when you create the route. The default administrative distance for a static route is 1 on ArubaOS switches.

  Because you can set different administrative distances for different static routes, you can use administrative distance to distinguish between static routes. The route with the lowest administrative distance is preferred.

- Metric—Routing devices use metrics to distinguish between routes that are learned in the same way such as statically or with the same routing protocol. Generally, these routes have the same administrative distance. When two identical routes have the same administrative distance, the metric breaks the tie and permits the routing device to choose a route for the routing table. The route with the lower metric is preferred.

  The default metric for a static route is 1 on ArubaOS switches, but you can specify a different metric when you create the route.

  Metric is sometimes referred to as cost.

Consider some examples. You could add a second default route on Access-1 that uses the default metric and administrative distance just like the first default route. Because the default routes tie for administrative distance and metric, Access-1 adds both routes to its routing table and uses both routes. The switch uses a hashing mechanism to decide which traffic to assign to each route. Thus, it load-shares the traffic over the links to Core-1 and Core-2.

Note

The switch must support Equal Cost Multi-Pathing (ECMP) routes in order to add both routes. By default, ArubaOS switches support this feature. Refer to your switch documentation for details on the number of ECMP routes permitted. If the number of ECMP routes exceeds the limit, the switch uses a factor such as next hop address to break the tie.

In your topology, Access-1 has a link aggregation to Core-1 (2 Gbps), but just a single Gbps link to Core-2. You might decide, therefore, that you do not want to load share the traffic but instead prefer the route through Core-1. In that case, you set the administrative distance on the route through Core-2 higher than the distance on the route through Core-1.

ArubaOS switches let you adjust both the metric and the administrative distance. As you learned, the ArubaOS switch first compares the static routes' administrative distance and chooses the route with the lower distance. But if the routes have the same distance, it breaks the tie with metric and chooses the route with the lower metric. So you could set a higher metric on the route through Core-2 rather than a higher administrative distance.

In either case, Access-1 now adds only the route through Core-1 to its routing table. The other route still exists, but it is not active. If, however, the link to Core-1 fails and that route is removed from the IP routing table, Access-1 starts using the route through Core-2.

It is also important for you to understand that planning redundancy involves examining the entire path. For example, the redundant route on Access-1 lets the server continue to reach the client if Access-1's link to Core-1 fails. But Core-1 also needs a redundant route for 10.1.11.0/24 so that it can continue to return traffic from the client.

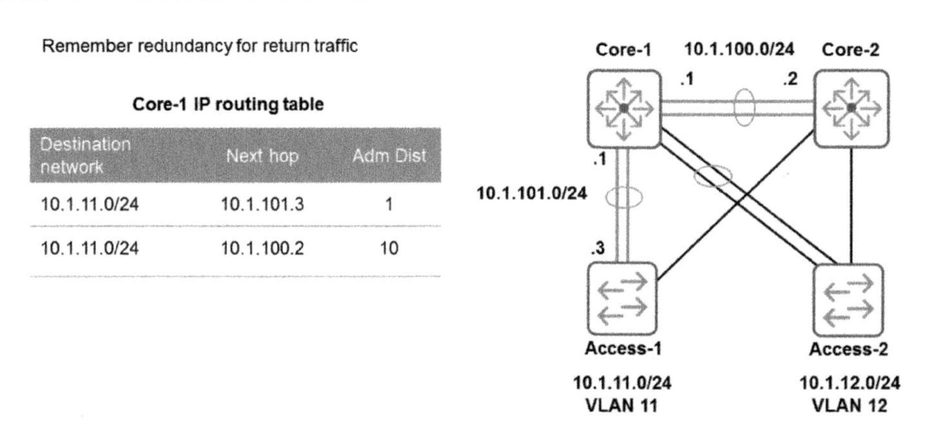

Core-1 IP routing table

Destination network	Next hop	Adm Dist
10.1.11.0/24	10.1.101.3	1
10.1.11.0/24	10.1.100.2	10

Figure 8-22: Additional redundant routes

Create redundant static IP routes

You will now set up the routes on your test network equipment to enable redundancy for the Access-1 links.

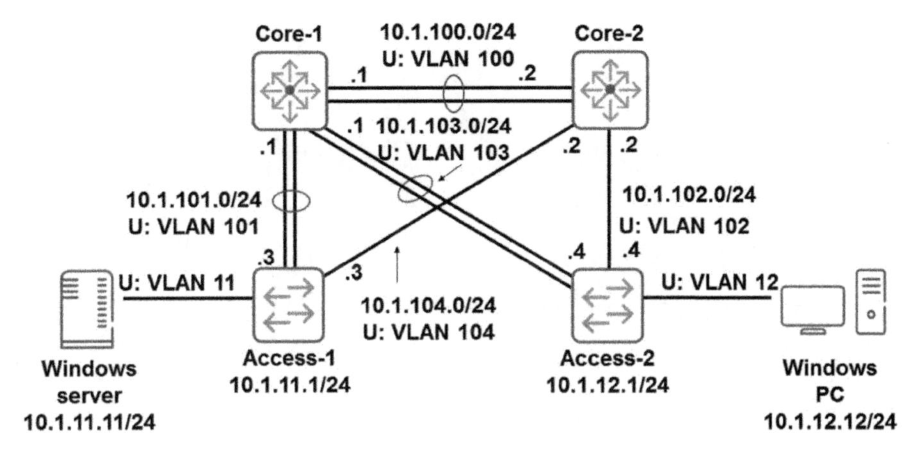

Figure 8-23: Create redundant static IP routes

Task 1: Set up static routes on Core-2

You want Access-1 to be able to route traffic through Core-2, so you must set up routing on Core-2.

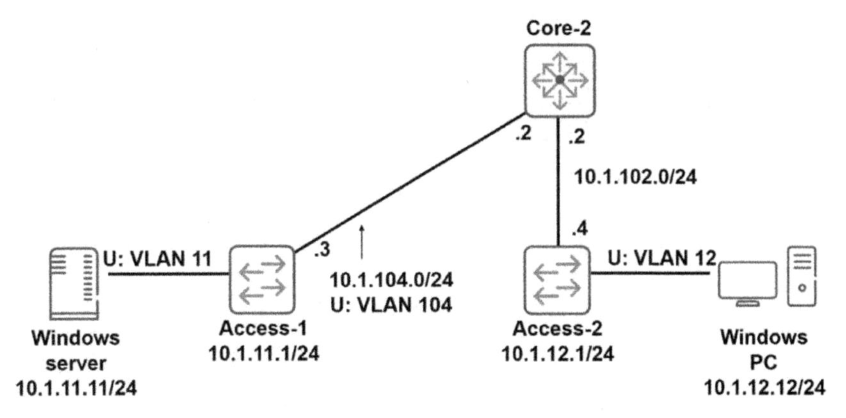

Figure 8-24: Test network topology showing the links for the Core-2 routes

1. Access a terminal session with Core-2 and move to global configuration mode.
2. Create a static route to 10.1.12.0/24 through Access-2 at 10.1.102.4.

```
Core-2(config)# ip route 10.1.12.0/24 10.1.102.4
```

3. Create a static route to 10.1.11.0/24 through Access-1 at 10.1.104.3.

```
Core-2(config)# ip route 10.1.11.0/24 10.1.104.3
```

4. Check the routes.

```
Core-2(config)# show ip route

                            IP Route Entries

  Destination     Gateway     VLAN  Type       Sub-Type  Metric  Dist.
  ------------    ----------  ----  ---------  --------  ------- -----
  10.1.11.0/24    10.1.104.3  104   static               1       1
  10.1.12.0/24    10.1.102.4  102   static               1       1
  10.1.100.0/24   VLAN100     100   connected            1       0
  10.1.102.0/24   VLAN102     102   connected            1       0
  10.1.104.0/24   VLAN104     104   connected            1       0
  127.0.0.0/8     reject            static               0       0
  127.0.0.1/32    lo0               connected            1       0
```

Task 2: Create a second default route on Access-1

You will now create a default route on Access-1 through Core-2. You will first assign the route the default metric and administrative distance and observe the behavior. You will then assign the route a higher metric and observe the behavior.

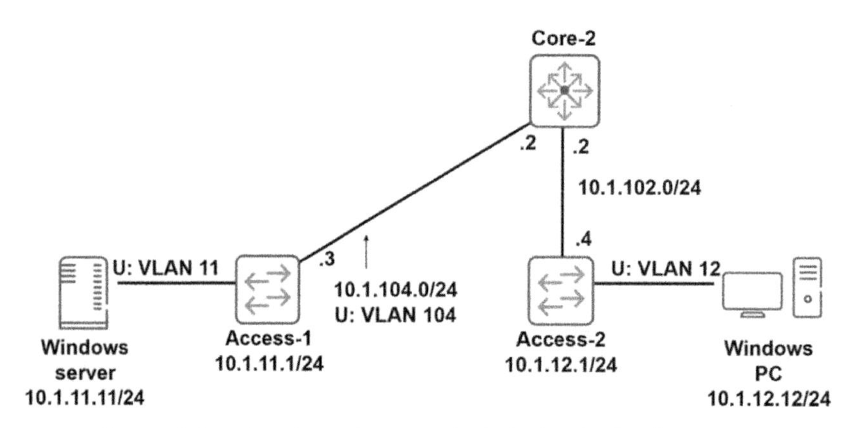

Figure 8-25: Topology showing the links for the backup route

Access-1

1. Access a terminal session with Access-1 and move to the global configuration mode context.
2. Create a default route through Core-2 at 10.1.104.2.

```
Access-1(config)# ip route 0.0.0.0/0 10.1.104.2
```

3. View the IP routing table and verify that you see both routes. As you see, they have the same metric and administrative distance.

```
Access-1(config)# show ip route

                              IP Route Entries

  Destination    Gateway      VLAN  Type        Sub-Type Metric   Dist.
  ------------   ----------   ----  ---------   -------- -------  -----
  0.0.0.0/0      10.1.101.1   101   static               1        1
  0.0.0.0/0      10.1.104.2   104   static               1        1
  10.1.11.0/24   VLAN11       11    connected            1        0
  10.1.101.0/24  VLAN101      101   connected            1        0
  10.1.104.0/24  VLAN104      104   connected            1        0
  127.0.0.0/8    reject             static               0        0
  127.0.0.1/32   lo0                connected            1        0
```

Windows server

4. Access the Windows server desktop and open a command prompt.

Note

The traceroutes will include the intuitive hostnames for the devices because the test network environment is set up with DNS services, which map the IP addresses to the hostnames. You can use the -d option (traceroute -d 10.1.12.12) to run the traceroute without looking up hostnames if these aren't available in your environment.

5. Execute a trace route to the Windows 7 PC (10.1.12.12).

```
tracert 10.1.12.12

tracing a route to pc.hpnu01.edu [10.1.12.12] over a maximum of 30 hops:

  1   <1 ms <1 ms 1 ms  vlan11_defaultrouter.hpnu01.edu [10.1.11.1]
```

```
  2    <1 ms <1 ms 1 ms   core-2_access-1_link.hpnu01.edu [10.1.104.2]
  3    <1 ms <1 ms 1 ms   access-2_core-1_link.hpnu01.edu [10.1.103.4]
  4    <1 ms <1 ms 1 ms   pc.hpnu01.edu [10.1.12.12]
```

6. Execute a trace route to 10.1.12.1.

```
tracert 10.1.12.1
Tracing route to pc.hpnu01.edu [10.1.12.12] over a maximum of 30 hops
  1    <1 ms <1 ms 1 ms   vlan11_defaultrouter.hpnu01.edu [10.1.11.1]
  2    <1 ms <1 ms 1 ms   core-1_access-1_link.hpnu01.edu [10.1.101.1]
  3    <1 ms <1 ms 1 ms   vlan12_defaultrouter.hpnu01.edu [10.1.12.1]
```

Access-1

As you see, Access-1 is using both routes, assigning some traffic to one and other traffic to the other. Access-1 has a higher bandwidth connection to Core-1, so now you will make Access-1 prefer the route through Core-1.

7. Return to the terminal session with Access-1.
8. Delete the current route through Core-2. Then add the route with a metric of 10.

```
Access-1(config)# no ip route 0.0.0.0/0 10.1.104.2
Access-1(config)# ip route 0.0.0.0/0 10.1.104.2 metric 10
```

9. View the IP routing table and see that only the route through Core-1 is present. The other route does not display in the active routing table because it has a higher metric and is not preferred.

```
Access-1(config)# show ip route

                        IP Route Entries

  Destination    Gateway    VLAN  Type       Sub-Type Metric  Dist.
  ------------   ---------- ----  ---------  -------- ------  -----
  0.0.0.0/0      10.1.101.1 101   static              1       1
  10.1.11.0/24   VLAN11     11    connected           1       0
  10.1.101.0/24  VLAN101    101   connected           1       0
  10.1.104.0/24  VLAN104    104   connected           1       0
  127.0.0.0/8    reject           static              0       0
  127.0.0.1/32   lo0              connected           1       0
```

10. You will now observe failover to the higher metric route. Disable the links to Core-1.

```
Access-1(config)# interface 5,6 disable
```

11. View the IP routing table and see that it now contains the route through Core-2.

```
Access-1(config)# show ip route

                        IP Route Entries

  Destination    Gateway    VLAN Type       Sub-Type Metric Dist.
  ------------ ---------- ---- --------- -------- ------ -----
  0.0.0.0/0      10.1.104.2 104  static              10      1
  10.1.11.0/24   VLAN11     11   connected           1       0
  10.1.104.0/24 VLAN104     104  connected           1       0
  127.0.0.0/8    reject          static              0       0
  127.0.0.1/32   lo0             connected           1       0
```

Windows server

12. Access the Windows server desktop and open a command prompt.

13. Execute a trace route to the Windows 7 PC (10.1.12.12).

```
tracert 10.1.12.12

tracing a route to pc.hpnu01.edu [10.1.12.12] over a maximum of 30 hops:
  1   <1 ms <1 ms 1 ms   vlan11_defaultrouter.hpnu01.edu [10.1.11.1]
  2   <1 ms <1 ms 1 ms   core-2_access-1_link.hpnu01.edu [10.1.104.2]
  3   *      *      *    Request timed out
```

14. As you see the trace route fails. Press **<Ctrl+c>** to end it.

You already saw that Access-1 started using the redundant route. Why is connectivity still disrupted?

Task 3: Create a redundant route on Core-1

Connectivity is disrupted because Access-2 routes return traffic to Core-1, and Core-1 cannot route traffic back to 10.1.11.0/24. It needs a redundant route through Core-2, as well.

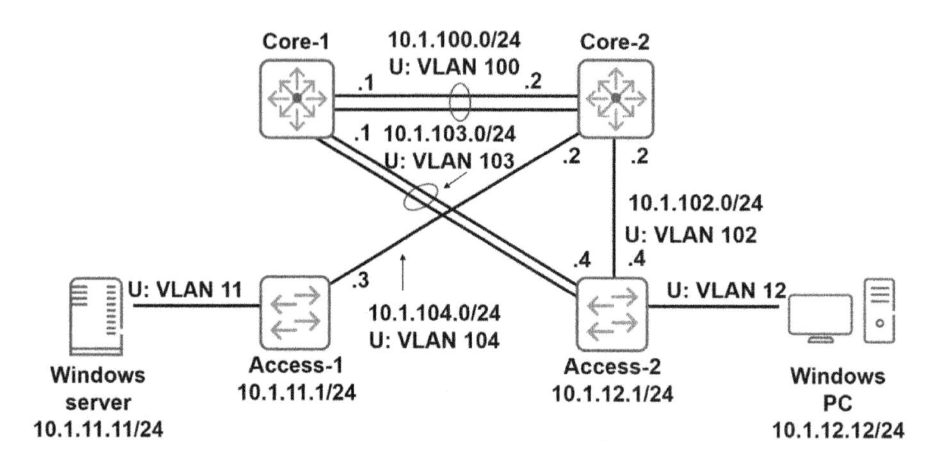

Figure 8-26: Topology with one link down

1. Access a terminal session with Core-1. Move to global configuration mode.
2. Examine the routing table to validate that Core-1 has no active route to 10.1.11.0/24.

```
Core-1(config)# show ip route

                       IP Route Entries

 Destination     Gateway       VLAN  Type        Sub-Type  Metric  Dist.
 ------------    ----------    ----  ---------   --------  ------  -----
 10.1.12.0/24    10.1.103.4    103   static                1       1
 10.1.100.0/24   VLAN100       100   connected             1       0
 10.1.103.0/24   VLAN103       103   connected             1       0
 127.0.0.0/8     reject              static                0       0
 127.0.0.1/32    lo0                 connected             1       0
```

3. Create the redundant static route to 10.1.11.0/24 through Core-2 at 10.1.100.2. Set the metric to 10, which is higher than the default.

```
Core-1(config)# ip route 10.1.11.0/24 10.1.100.2 metric 10
```

4. View the new route in the IP routing table.

```
Core-1(config)# show ip route

                         IP Route Entries

  Destination    Gateway     VLAN  Type       Sub-Type Metric Dist.
  ------------   ----------  ----  ---------  -------- ------ -----
  10.1.11.0/24   10.1.100.2  100   static              10     1
  10.1.12.0/24   10.1.103.4  103   static              1      1
  10.1.100.0/24  VLAN100     100   connected           1      0
  10.1.103.0/24  VLAN103     103   connected           1      0
  127.0.0.0/8    reject            static              0      0
  127.0.0.1/32   lo0               connected           1      0
```

Windows server

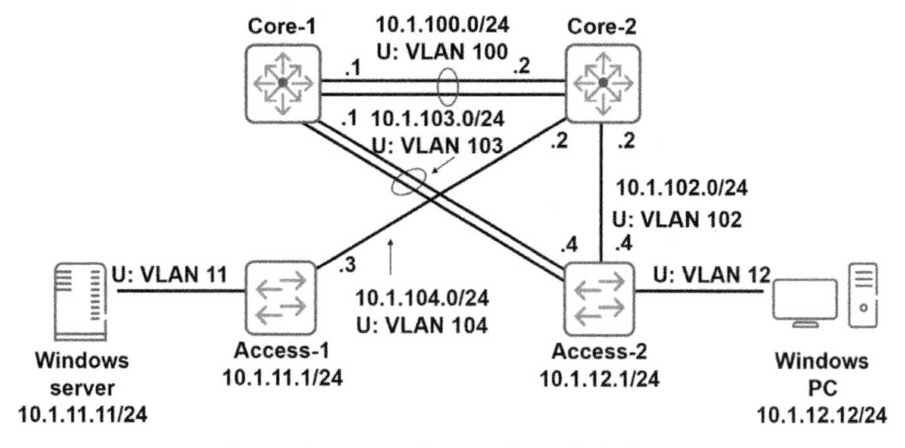

Figure 8-27: Topology with one link down

5. Access the Windows server and re-perform the trace route. It is now successful.

```
tracert 10.1.12.12

tracing a route to pc.hpnu01.edu [10.1.12.12] over a maximum of 30 hops:

  1   <1 ms <1 ms 1 ms  vlan11_defaultrouter.hpnu01.edu [10.1.11.1]

  2   <1 ms <1 ms 1 ms  core-2_access-1_link.hpnu01.edu [10.1.104.2]
```

```
3    <1 ms <1 ms 1 ms   access-2_core-1_link.hpnu01.edu [10.1.103.4]

4    <1 ms <1 ms 1 ms   pc.hpnu01.edu [10.1.12.12]
```

Access-1

6. Return to the terminal session with Access-1. Restore the failed link.

```
Access-1(config)# interface 5,6 enable
```

Core-1

7. Observe that Core-1 is using its preferred route through Access-1 again.

```
Core-1(config)# show ip route

                              IP Route Entries

  Destination     Gateway      VLAN   Type          Sub-Type  Metric  Dist.
  ------------    ----------   ----   ---------     --------  ------  -----
  10.1.11.0/24    10.1.101.3   101    static                  1       1
  10.1.12.0/24    10.1.103.4   103    static                  1       1
  10.1.100.0/24   VLAN100      100    connected               1       0
  10.1.101.0/24   VLAN101      101    connected               1       0
  10.1.103.0/24   VLAN103      103    connected               1       0
  127.0.0.0/8     reject              static                  0       0
  127.0.0.1/32    lo0                 connected               1       0
```

Task 4: Save

Save the current configuration on each of the switches using the **write memory** command.

Learning check

1. You have set up routes such that Access-1 can use its link to Core-1 during normal operation and its link to Core-2 as a backup. Now assume that you want to do the same on Access-2. What steps must you complete?

2. The network illustrated below is using the routes that you established when you created redundant static IP routes. Now consider a situation in which the link between Access-1 and

Core-1 fails. (The figure below illustrates this situation by not showing the link.) Trace the traffic flow for traffic from the server to the PC. Then trace the flow from the PC back to the server.

If you need help, you can trace routes and use the **show ip route** and **display ip routing** commands on your test network equipment.

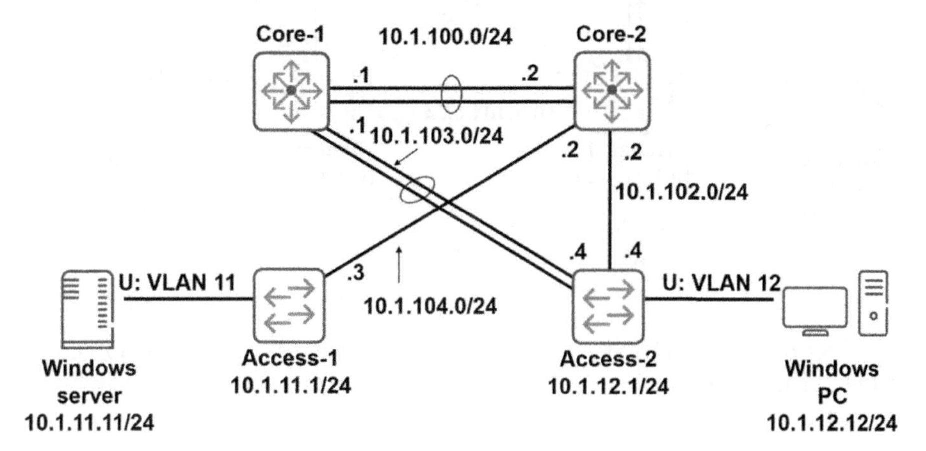

Figure 8-28: Traffic flow from the server to the PC

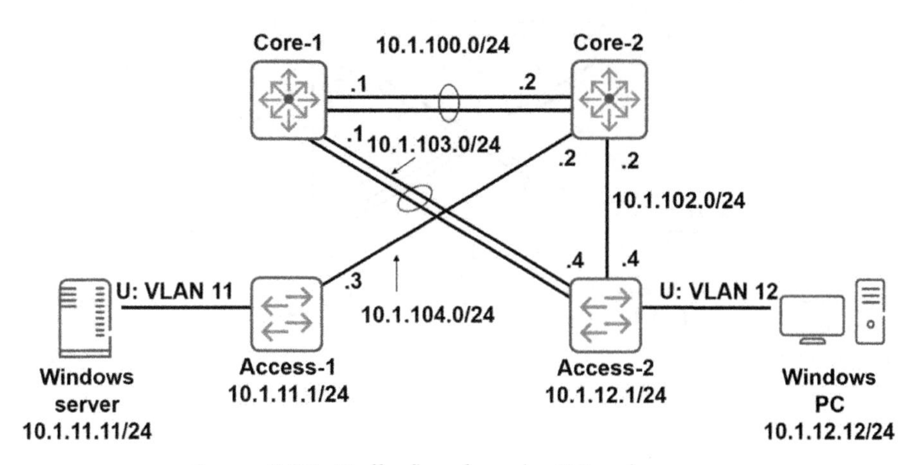

Figure 8-29: Traffic flow from the PC to the server

3. Now consider a network in which the bandwidth for all switch-to-switch connections is the same. You set up the redundant routes on Access-1 and Access-2 to have the same metric and administrative distance. What is the traffic flow from the PC to the server during normal operation? Use dashed lines to indicate load-sharing.

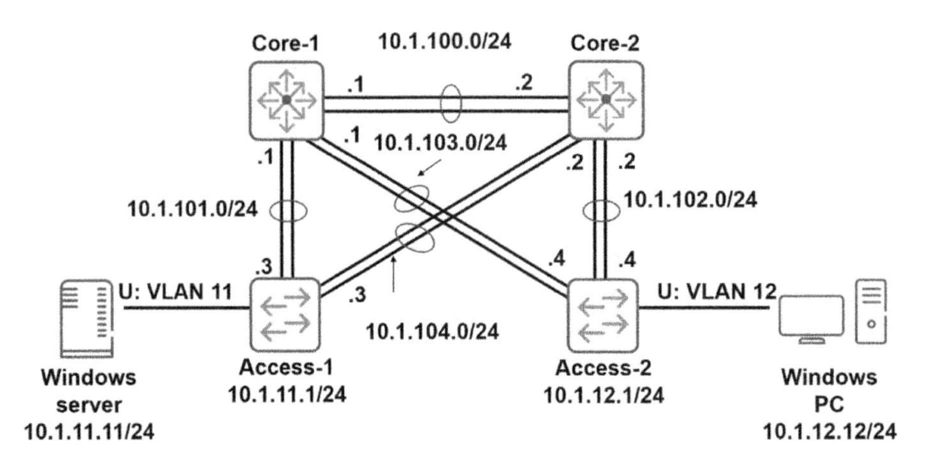

Figure 8-30: Traffic flow from the PC to the server

4. Now consider a scenario like that in question 3; however, the link between Access-1 and Core-1 fails. (The figure below does not show the link to indicate that it has failed.) What is the traffic flow from the PC to the server now?

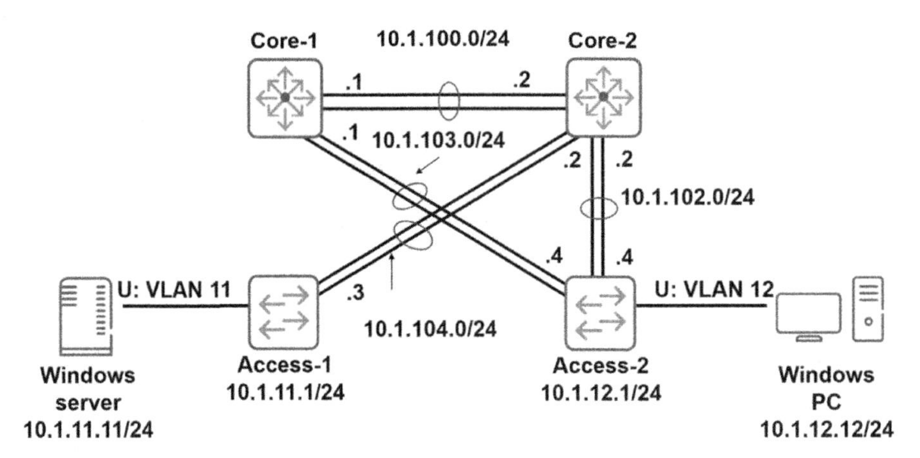

Figure 8-31: Traffic flow from the PC to the server

5. In either scenario, is the traffic flow during failover situations ideal?

Answers to Learning check

1. You have set up routes such that Access-1 can use its link to Core-1 during normal operation and its link to Core-2 as a backup. Now assume that you want to do the same on Access-2. What steps must you complete?

   You need to create a second default route through Core-2 on Access-1. You might set the administrative distance higher than the route through Core-1 to favor the Core-1 route. You must make sure that Core-2 has the proper routes, which it should from the previous setup. You must also create a route on Core-1 to 10.1.11.0/24 through Core-2, again setting the administrative distance (preference) higher than the existing route.

   You might want to provide failover in case the Core-1 link on one access layer switch fails and the Core-2 link on the other access layer switch fails. To achieve this redundancy, you must set up redundant routes on Core-2 to 10.1.11.0/24 and 10.1.12.0/24 through Core-1.

2. The network illustrated below is using the routes that you established when you created redundant static IP routes. The link between Access-1 and Core-1 fails. Trace the traffic flow for traffic from the server to the PC. Then trace the flow from the PC back to the server.

   If you need help, you can trace routes and use the **show ip route** and **display ip routing** commands on your test network equipment.

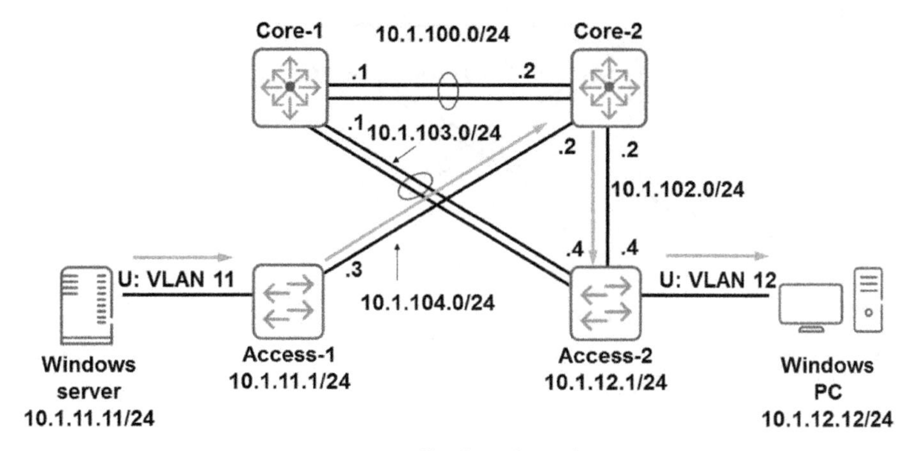

Figure 8-32 (answer): Traffic flow from the server to the PC

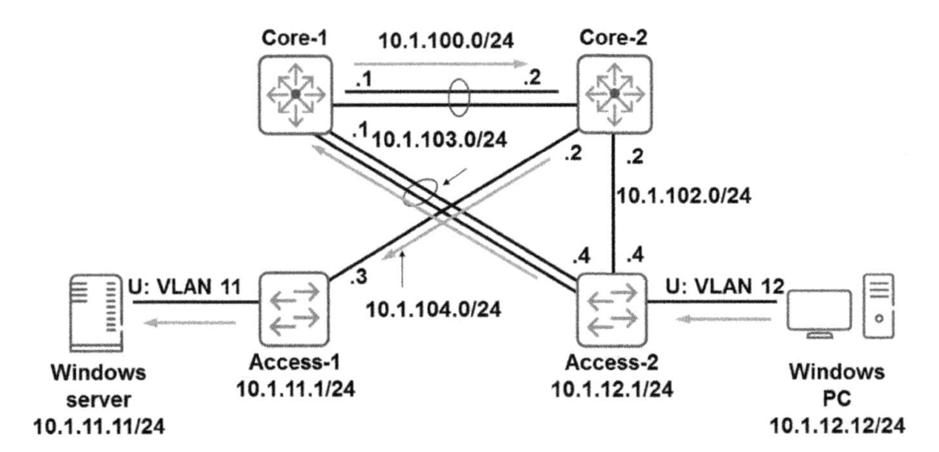

Figure 8-33 (answer): Traffic flow from the PC to the server

3. Now consider a network in which the bandwidth for all switch-to-switch connections is the same. You set up the redundant routes on Access-1 and Access-2 to have the same metric and administrative distance. What is the traffic flow from the PC to the server during normal operation? Use dashed lines to indicate load-sharing.

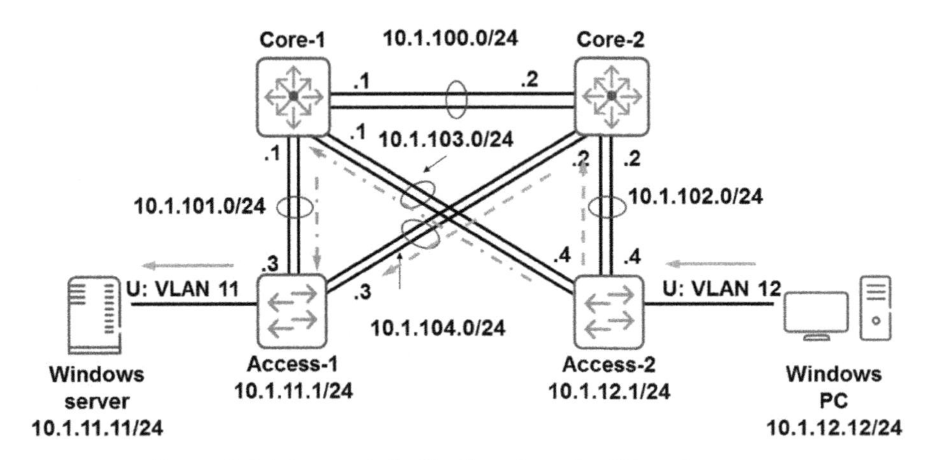

Figure 8-34 (answer): Traffic flow from the PC to the server

4. In the scenario for question 3, the link between Access-1 and Core-1 fails. What is the traffic flow from the PC to the server now?

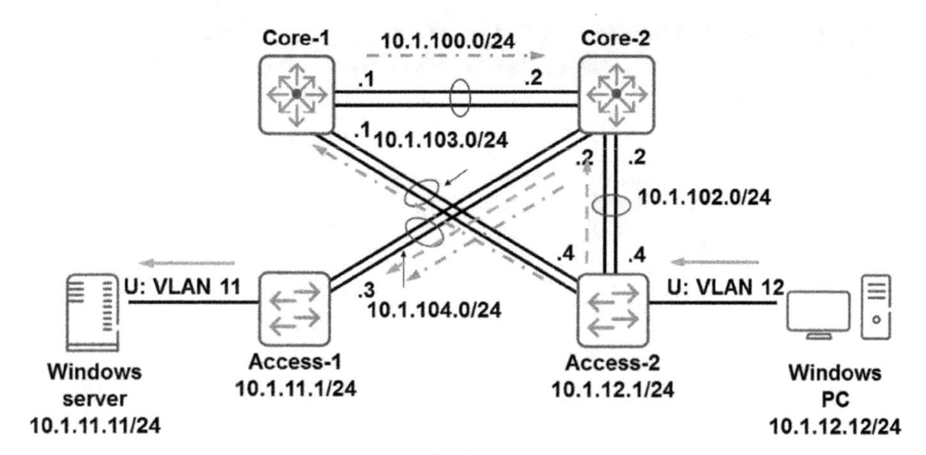

Figure 8-35 (answer): Traffic flow from the PC to the server

5. In either scenario, is the traffic flow during failover situations ideal?

   The flow in the failover situations is not ideal. It would be better if Access-2 used only the route to Core-2 because Core-1 has to send the traffic back to Core-2 anyway. Sending all traffic through Core-2 would also eliminate the asymmetric routing in which the PC-to-server traffic takes a different path from the server-to-PC traffic. (One reason that asymmetric routing is undesirable is that it can sometimes, although not often, lead to the flooding of unknown unicasts. In any case, it is better to use symmetric routing.) However, static routing does not let Access-1 change its routes in response to topology changes elsewhere.

Dynamic routing protocols

Static routes have a role to play in routing design, but they tend not to work well in a redundant network with multiple paths. As the tasks that you completed in the last section demonstrated, you might find it difficult to plan all of the routes that are necessary for all failover situations. Also the static routes on one device cannot adjust based on changes in the topology in other areas of the network, interfering with the solution's resiliency. You could also accidentally introduce routing loops, which can disrupt connectivity.

For these reasons, you should typically use a dynamic routing protocol for more complex network topologies. In addition, as networks grow, configuring static routes on each routing device becomes infeasible (as well as complex and error prone).

Routing protocols eliminate the need to create routes manually. Instead routers communicate with each other to discover available routes and the best paths to destinations automatically.

- Work well for complex, redundant topologies
- Allow routers to:
  - Communicate routing and topology information
  - Learn routes automatically
  - Respond to changes automatically

Type	Operation	Example
Distance vector	Advertise routes Choose route based on hop count	RIP
Link state	Advertise link information Calculate lowest cost route based on topology Converge quickly	OSPF

Class	Operates	Example
IGP	Within an organization (AS)	OSPF
EGP	Between organizations (ASes)	BGP

Figure 8-36: Dynamic routing protocols

Depending on the protocol, routers exchange different types of information and learn routes differently:

- Distance vector—Routers advertise routes that they know, and the advertisements propagate across all routers participating in the protocol. Routers select routes with the hop count. Routing Information Protocol (RIP) is an example of a distance vector protocol.
- Link state—Routers advertise information about their connected Layer 3 interfaces, also called links, and the cost (or metric) associated with that interface. These link state advertisements propagate across participating routers. Routers use the advertisements to construct a topology of the network, consisting of all the known links, the routers connected to the links, and the associated costs. Using the topology, routers calculate their own lowest cost route to each destination subnet. Open Shortest Path First (OSPF) is a link state protocol.

Important

Link state protocols such as OSPF tend to choose better routing paths and to converge faster than distance vector protocols such as RIP.

You can also classify dynamic routing protocols based on how they are used:

- Interior Gateway Protocol (IGP)—Facilitates the exchange of routing information among routers under the same organizational control, or in other words within the same autonomous system (AS). A common IGP is OSPF.
- Exterior Gateway Protocol (EGP)—Facilitates exchange of routing information among routers in different ASes. Border Gateway Protocol v4 (BGP4), a common example, is used by ISPs to route traffic on the Internet backbone.

This guide focuses on routing within a private network, so you will look at OSPF, an IGP.

Basic OSPF setup

As mentioned earlier, OSPF is a link state protocol. OSPF routers advertise their links (connected subnets) to each other using link state advertisements (LSAs), build a topology based on the advertisements, and calculate best routes based on this topology.

In this guide, you will just look at the steps required to set up a basic, functional OPSF solution. You should attend the appropriate HPE training course if you want to learn more about topics such as these:

- How OSPF routers discover their neighbors
- How OSPF routers advertise topology information, build topologies, and choose routes
- How to configure router IDs and use loopback interfaces to enhance stability
- How to configure metrics (costs) to help routing switches choose the best paths
- How to import external routes into OSPF
- How to set up multi-area solutions

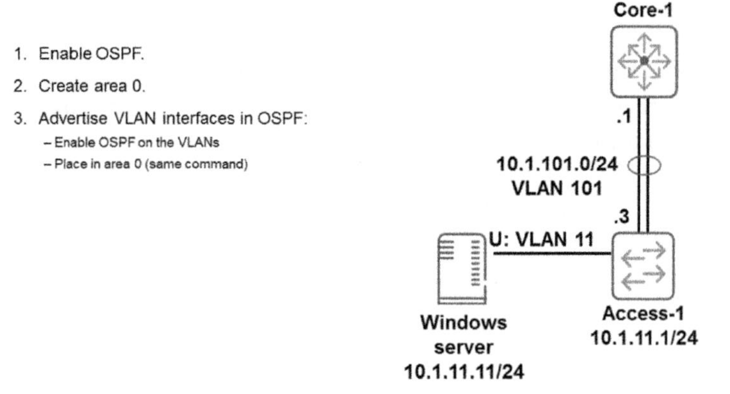

Figure 8-37: Basic OSPF setup

To set up a basic OSPF solution, follow these steps:

1. Enable OSPF.
2. Create area 0.

   Area 0 is a special area called the backbone. You can learn about the purpose of the backbone area in multi-area topologies in other HPE training. For a single area topology like yours, you should usually use area 0 (in case you want to move to a multi-area topology later).

3. Enable OSPF on the VLAN interfaces. Enabling OSPF on an interface has two effects:
   - The routing switch sends OSPF messages on the interface and can form neighbor relationships with other routers on the same interface. Neighbors that fully exchange their topology information are called fully adjacent.
   - The routing switch advertises this interface in OSPF. For example, if the interface is associated with 10.1.11.0/24, the routing switch advertises that it has a link to 10.1.11.0/24.

   VLANs that need to be advertised in OSPF include:
   - Those that support endpoints (like VLAN 11 on Access-1)
   - Those that connect routing switches (like VLAN 101 on Access-1)

   When you enable OSPF on a VLAN, you also specify the VLAN's OSPF area. In this simple topology, all VLANs are in the same area (area 0).

An optional step, but one that you should typically perform is setting a router ID. The router ID does not have to be a valid IP address on the switch, but you often should use an IP address on a loopback interface. On your test network, you will just assign a unique router ID to each switch that is an IP address not used elsewhere. (Other training explains in more detail about router IDs and loopback interfaces.)

After you complete these steps, your routing switches will automatically create a topology of your network and learn routes to every subnet within it. If a link fails, the routing switches inform each other of the change and learn new routes.

Table 8-2: Settings that must match for OSPF neighbors to achieve adjacency

Setting
Network address and mask
Area ID
Area type (OSPF defines different types of areas covered in other training)
Authentication method and key if authentication is used
Network type
Hello timer

Table 8-3: LSA types used in single area configurations

Type	Name	Information advertised	Advertised by	Scope
1	Router	The OSPF links on a routing device (transit and stub networks)	Each OSPF routing device	Area
2	Network	The routers on broadcast or NBMA networks	Designated Router (DR) for the network	Area

Table 8-4: Roles for routers on broadcast networks

Role	Description
Designated Router (DR)	Achieves adjacency with all other routers on the network Advertises Type 2 (Network) LSAs Receives Link State Updates (LSUs) from other routers Advertises LSUs on a multicast address 224.0.0.6
Backup Designated Router (BDR)	Achieves adjacency with all other routers on the network Receives LSUs from other routers (and sends its own LSUs to the DR) Takes over other DR roles if the DR fails
Other	Achieves adjacency with the DR and BDR Sends LSUs to both DR and BDR

Note

The DR for a subnet running OSPF (called an OSPF link) is elected based on priority (higher priority value wins) and, in the case of a tie, higher router ID.
When a link has many routers (for example, if you extended VLAN 12 across all of your routing switches and ran OSPF on that VLAN), the DR and BDR roles reduce the number of adjacencies on routers and allow routers to listen for LSUs from a single source. When a link has only two routers, as in the design in your topology, which devices take the DR and BDR role do not usually matter.

Choosing the preferred route with static and dynamic routing

You were introduced to the concepts of metric and administrative distance earlier. You will now return to these concepts so that you can understand what happens if your routing switch has a static route and dynamic route to the same destination.

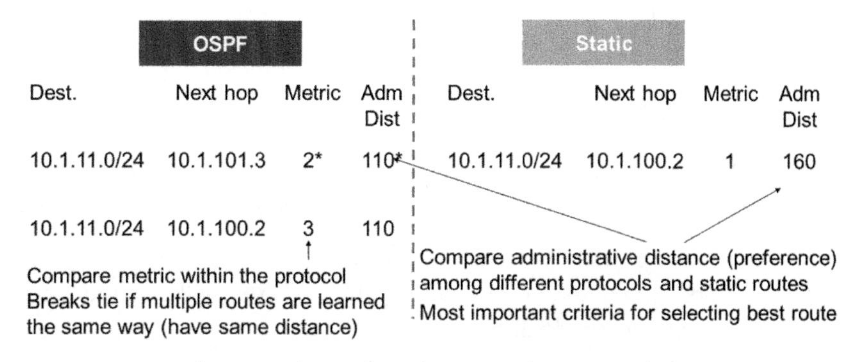

Figure 8-38: Choosing the preferred route with static and dynamic routing

- Administrative distance—It is the administrative distance that allows the switch to choose between a static route and an OSPF (or other IGP) route to the same destination. You can set a static route's administrative distance when you create the route. Each routing protocol is associated with an administrative distance, which applies to all routes learned with that protocol.
- Metric—Routing devices use metrics to distinguish between routes that are learned in the same way. Dynamic routing protocols each have their own way of defining a route's metric. An OSPF routing switch might automatically determine the metric for a link based on the link speed, or you can set a metric (OSPF cost) on a VLAN interface. The switch advertises that metric when it advertises the link. When a switch calculates a route to a destination network, it adds up the metric on all the links on the way to that network. If OSPF learns two routes to the same destination, it chooses the route with the lowest metric to propose to the IP routing table (or multiple routes with the same lowest metric).

  Remember: metrics are locally significant to static routes or to a particular routing protocol. In other words, the switch will compare the metric of two OSPF routes, but it will not compare the metric of an OSPF route to the metric of a static route (unless the static route and OSPF route have the same administrative distance).

Remember: administrative distance is the most important criteria for selecting the route. The route with the *lowest administrative distance is always chosen* even if it has a higher metric than a route with a higher administrative distance. Metrics only matter if the switch has multiple routes with the same administrative distance.

The table shows the default administrative distances for static routes and OSPF routes on ArubaOS switches. ArubaOS switches prefer static routes to OSPF routes when the static routes use the default distance. The figure shows an example in which the static route was configured with a higher distance.

Table 8-5: Default administrative distances

Routing method	ArubaOS
Static	1
OSPF	110

Configure basic OSPF

You will now set up OSPF routing on your switches. You will see how the switches learn routes automatically. You will also observe how the switches choose between static and OSPF routes.

The tasks in this section build on the tasks you completed in the previous section. The next two figures show the test network topology that you have established. You will continue to use this same topology as you set up Open Shortest Path First (OSPF) routing.

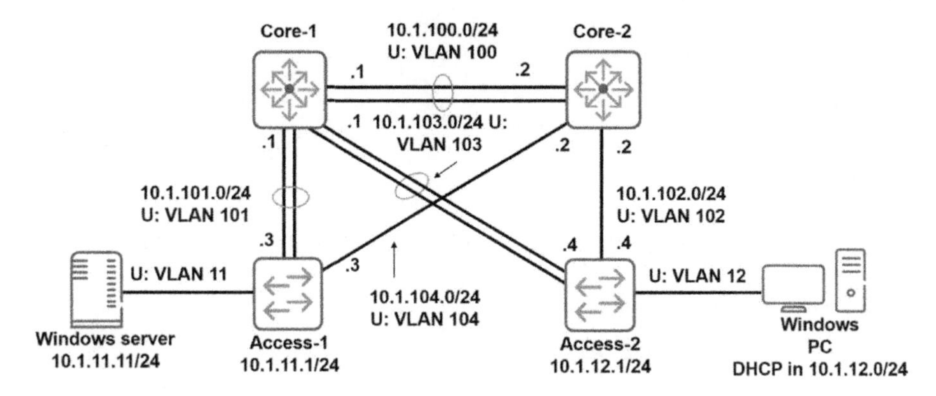

Figure 8-39: Final test network topology (logical)

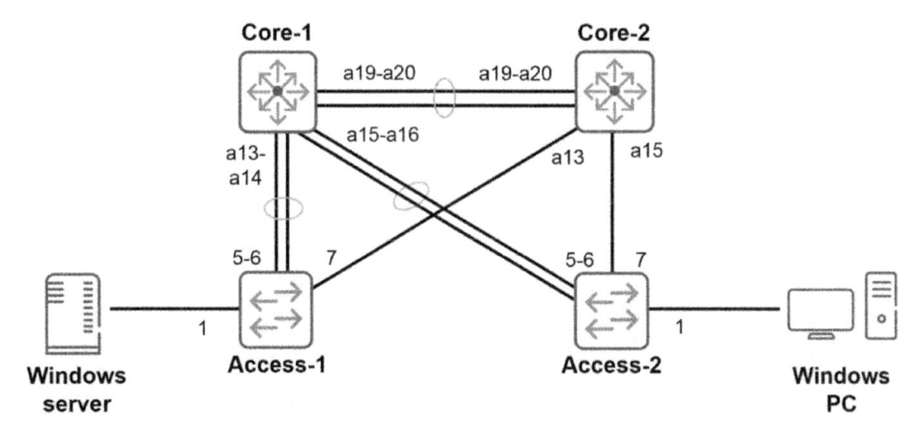

Figure 8-40: Final test network topology (physical)

Task 1: Configure OSPF on Access-1

You will now configure OSPF on Access-1.

1. Access a terminal session with Access-1. Move to global configuration mode.
2. Set the router ID to 10.0.0.3. In this case, the router ID simply identifies the Access-1 to other OSPF routers. It is not an IP address. Other HPE training gives guidelines for configuring the router ID on a loopback interface so that you can ping it.

```
Access-1(config)# ip router-id 10.0.0.3
```

3. Move to the OSPF configuration mode.

```
Access-1(config)# router ospf
```

4. Create area 0.

```
Access-1(ospf)# area 0
```

5. Enable OSPF.

```
Access-1(ospf)# enable
```

6. Access each of the switch's active VLANs. Enable OSPF and place the VLAN in area 0.

```
Access-1(ospf)# vlan 11
Access-1(vlan-11)# ip ospf area 0
Access-1(vlan-11)# exit
Access-1(config)# vlan 101
Access-1(vlan-101)# ip ospf area 0
Access-1(vlan-101)# exit
Access-1(config)# vlan 104
Access-1(vlan-104)# ip ospf area 0
Access-1(vlan-104)# exit
```

7. If you want, you can make VLAN 11 a passive interface, which prevents the switch making neighbors on the interface. Access-1(config)# **vlan 11 ip ospf passive**

8. Verify that OSPF is enabled on the VLAN interfaces. Note that area 0 is also called the backbone area.

```
Access-1(config)# show ip ospf interface

 OSPF Interface Status

 IP Address  Status    AreaID    State  Auth-type  Cost  Pri  Passive
 ----------  -------   --------  -----  ---------  ---   ---  -------
 10.1.11.1   enabled   backbone  WAIT   none       1     1    yes
 10.1.101.3  enabled   backbone  WAIT   none       1     1    no
 10.1.104.3  enabled   backbone  WAIT   none       1     1    no
```

Task 2: Configure OSPF on Core-1

1. Table 8-6 shows the OSPF settings for Core-1. Try to remember the steps and commands for configuring these settings on your own and using the help keys.

   The commands are provided on the next page for your reference.

Table 8-6: Core-1 OSPF settings

Setting	Parameter
Router ID	10.0.0.1
OSPF area ID	0
VLAN interfaces in area 0	100 101 103
Passive interfaces	None

2. When you are finished verify that OSPF is enabled on the correct interfaces.

```
Core-1(config)# show ip ospf interface

OSPF Interface Status

 IP Address Status   Area ID  State Auth-type Cost Pri Passive
 ---------- ------- -------- ----- --------- --- --- -------
 10.1.100.1 enabled backbone WAIT  none      1   1   no
 10.1.101.1 enabled backbone BDR   none      1   1   no
 10.1.103.1 enabled backbone WAIT  none      1   1   no
```

3. Verify that Core-1 and Access-1 are neighbors. Check for the FULL state, which indicates the routing switches have exchanged all link state information and achieved adjacency.

```
Core-1(config)# show ip ospf neighbor

OSPF Neighbor Information

 Router ID Pri IP Address NbIfState State QLen Events Status
 --------- --- ---------- --------- ----- ---  ------ ------
 10.0.0.3  1   10.1.101.3 DR        FULL  0    7      None

Core-1(config)# ip router-id 10.0.0.1

Core-1(config)# router ospf

Core-1(ospf)# area 0

Core-1(ospf)# enable

Core-1(ospf)# vlan 100

Core-1(vlan-100)# ip ospf area 0
```

```
Core-1(vlan-100)# vlan 101
Core-1(vlan-101)# ip ospf area 0
Core-1(vlan-101)# vlan 103
Core-1(vlan-103)# ip ospf area 0
Core-1(vlan-103)# exit
```

Task 3: Configure OSPF on Core-2

1. Table 8-7 shows the OSPF settings for Core-2. Again, try to remember the steps and commands for configuring these settings on your own and using the help keys.

   You can also use one of the strategies that you've learned before such as copying the command history into Notepad, adjusting the commands according to the table, and pasting them in the Core-2 CLI.

   The commands are provided on the next page for your reference.

Table 8-7: Core-2 OSPF settings

Setting	Parameter
Router ID	10.0.0.2
OSPF area ID	0
VLAN interfaces in area 0	100 102 104
Passive interfaces	None

2. When you are finished verify that OSPF is enabled on the correct interfaces.

```
Core-2(config)# show ip ospf interface
 OSPF Interface Status
  IP Address   Status   Area ID   State  Auth-type  Cost  Pri  Passive
  ----------   -------  --------  -----  ---------  ----  ---  -------
  10.1.100.2   enabled  backbone  BDR    none       1     1    no
  10.1.102.2   enabled  backbone  WAIT   none       1     1    no
  10.1.104.2   enabled  backbone  BDR    none       1     1    no
```

3. Core-2 should now be OSPF neighbors with Core-1 and Access-1, which are already running OSPF on the same networks. Verify and check for the Full state.

   If you do not see both peers, wait about 30 seconds or a minute and try again.

```
Core-2(config)# show ip ospf neighbor

 OSPF Neighbor Information

  Router ID  Pri  IP Address  NbIfState  State  QLen  Events  Status
  ---------  ---  ----------  ---------  -----  ----  ------  ------
  10.0.0.1   1    10.1.100.1  DR         FULL   0     6       None
  10.0.0.3   1    10.1.104.3  DR         FULL   0     6       None

Core-2(config)# ip router-id 10.0.0.2
Core-2(config)# router ospf
Core-2(ospf)# area 0
Core-2(ospf)# enable
Core-2(ospf)# vlan 100
Core-2(vlan-100)# ip ospf area 0
Core-2(vlan-100)# vlan 102
Core-2(vlan-102)# ip ospf area 0
Core-2(vlan-102)# vlan 104
Core-2(vlan-104)# ip ospf area 0
Core-2(vlan-104)# exit
```

Task 4: Configure OSPF on Access-2

You will now configure OSPF on Access-2.

1. Table 8-8 shows the OSPF settings for Access-2. Again, try to remember the steps and commands for configuring these settings on your own and using the help keys.

   You can also use one of the strategies that you've learned before such as copying the command history into Notepad, adjusting the commands according to the table, and pasting them in the Access-2 CLI. Remember to make VLAN 12 a passive interface.

   The commands are provided on the next page for your reference.

Table 8-8: Access-2 OSPF settings

Setting	Parameter
Router ID	10.0.0.4
OSPF area ID	0
VLAN interfaces in area 0	102 103 12
Passive interfaces	12

2. When you are finished verify that OSPF is enabled on the correct interfaces.

```
Access-2(config)# show ip ospf interface

 OSPF Interface Status

  IP Address   Status   Area ID  State Auth-type  Cost Pri Passive
  ----------   -------  -------- ----- ---------  ---  --- -------
  10.1.12.1    enabled  backbone WAIT  none       1    1   yes
  10.1.102.4   enabled  backbone BDR   none       1    1   no
  10.1.103.4   enabled  backbone BDR   none       1    1   no
```

3. Verify that Access-2 is OSPF neighbors with the two connected switches, Core-1 and Core-2. Also verify that the state is Full for each. If you do not see both peers, wait about 30 seconds or a minute and try again.

```
Access-2(config)# show ip ospf neighbor

 OSPF Neighbor Information

  Router ID Pri IP Address NbIfState State QLen Events Status
  --------- --- ---------- --------- ----- ---  ------ ------
  10.0.0.2  1   10.1.102.2 DR        FULL  0    6      None
  10.0.0.1  1   10.1.103.1 DR        FULL  0    6      None

Access-2(config)# ip router-id 10.0.0.4
Access-2(config)# router ospf
Access-2(ospf)# area 0
Access-2(ospf)# enable
Access-2(ospf)# vlan 102
```

```
Access-2(vlan-102)# ip ospf area 0
Access-2(vlan-102)# vlan 103
Access-2(vlan-103)# ip ospf area 0
Access-2(vlan-103)# vlan 12
Access-2(vlan-12)# ip ospf area 0
Access-2(vlan-12)# ip ospf passive
Access-2(vlan-12)# exit
```

Task 5: Verify the routes

You can now verify that the switches have learned the routes.

1. View routes on Access-1. You should see that the switch has learned a route to every subnet in your topology. You should also see that the switch has multiple routes to some destinations such as 10.1.12.0/24.

   Also notice the "ospf" route type and the administrative distance associated with OSPF routes.

```
Access-1(config)# show ip route

                          IP Route Entries

  Destination    Gateway     VLAN  Type       Sub-Type   Metric Dist.
  -----------    ----------  ----  ---------  ---------- ------ -----
  0.0.0.0/0      10.1.101.1  101   static                1      1
  10.1.11.0/24   VLAN11      11    connected             1      0
  10.1.12.0/24   10.1.101.1  101   ospf       IntraArea  3      110
  10.1.12.0/24   10.1.104.2  104   ospf       IntraArea  3      110
  10.1.100.0/24  10.1.101.1  101   ospf       IntraArea  2      110
  10.1.100.0/24  10.1.104.2  104   ospf       IntraArea  2      110
  10.1.101.0/24  VLAN101     101   connected             1      0
  10.1.102.0/24  10.1.104.2  104   ospf       IntraArea  2      110
  10.1.103.0/24  10.1.101.1  101   ospf       IntraArea  2      110
  10.1.104.0/24  VLAN104     104   connected             1      0
  127.0.0.0/8    reject            static                0      0
  127.0.0.1/32   lo0               connected             1      0
```

2. Access-1 still has its static default route. When will Access-1 use the OSPF routes and when will it use the static default route?

   Access-1 will use the OSPF routes for routing traffic to all the subnets in your network because these routes are more specific. Access-1 will only use the static default route when it needs to route traffic to a destination for which it does not have a specific route.

Access-2

3. View IP routes on Access-2 and see that it has learned routes to all subnets. You can filter the display for just OSPF routes.

```
Access-2(config)# show ip route ospf

                          IP Route Entries

  Destination     Gateway       VLAN  Type    Sub-Type    Metric  Dist.
  -------------   ----------    ----  -----   ---------   ------  -----
  10.1.11.0/24    10.1.102.2    102   ospf    IntraArea   3       110
  10.1.11.0/24    10.1.103.1    103   ospf    IntraArea   3       110
  10.1.100.0/24   10.1.102.2    102   ospf    IntraArea   2       110
  10.1.100.0/24   10.1.103.1    103   ospf    IntraArea   2       110
  10.1.101.0/24   10.1.103.1    103   ospf    IntraArea   2       110
  10.1.104.0/24   10.1.102.2    102   ospf    IntraArea   2       110
```

Core-1

4. View the IP routes on Core-1.

```
Core-1(config)# show ip route

                          IP Route Entries

  Destination     Gateway       VLAN  Type          Sub-Type    Metric  Dist.
  -----------     ----------    ----  -----         ---------   ------  -----
  10.1.11.0/24    10.1.101.3    101   static                    1       1
  10.1.12.0/24    10.1.103.4    103   static                    1       1
  10.1.100.0/24   VLAN100       100   connected                 1       0
  10.1.101.0/24   VLAN101       101   connected                 1       0
```

```
Destination     Gateway      VLAN Type       Sub-Type   Metric Dist.
-----------     ----------   ---- -----      ---------  ------ -----
10.1.102.0/24   10.1.100.2   100  ospf       IntraArea  2      110
10.1.102.0/24   10.1.103.4   103  ospf       IntraArea  2      110
10.1.103.0/24   VLAN103      103  connected             1      0
10.1.104.0/24   10.1.100.2   100  ospf       IntraArea  2      110
10.1.104.0/24   10.1.101.3   101  ospf       IntraArea  2      110
127.0.0.0/8     reject            static                0      0
127.0.0.1/32    lo0               connected             1      0
```

5. Why don't OSPF routes to 10.1.11.0/24 and 10.1.12.0/24 appear in the table?
6. Remove the static routes.

```
Core-1(config)# no ip route 10.1.11.0/24 10.1.101.3
Core-1(config)# no ip route 10.1.11.0/24 10.1.100.2
Core-1(config)# no ip route 10.1.12.0/24 10.1.103.4
```

7. View the IP routing table and confirm that the OSPF routes to 10.1.11.0/24 and 10.1.12.0/24 now appear.

```
Core-1(config)# show ip route

                        IP Route Entries

Destination     Gateway      VLAN Type       Sub-Type   Metric Dist.
-----------     ----------   ---- -----      ---------  ------ -----
10.1.11.0/24    10.1.101.3   101  ospf       IntraArea  2      110
10.1.12.0/24    10.1.103.4   103  ospf       IntraArea  2      110
10.1.100.0/24   VLAN100      100  connected             1      0
10.1.101.0/24   VLAN101      101  connected             1      0
10.1.102.0/24   10.1.100.2   100  ospf       IntraArea  2      110
10.1.102.0/24   10.1.103.4   103  ospf       IntraArea  2      110
10.1.103.0/24   VLAN103      103  connected             1      0
10.1.104.0/24   10.1.100.2   100  ospf       IntraArea  2      110
```

```
Destination     Gateway     VLAN  Type       Sub-Type    Metric  Dist.
-----------     ----------  ----  -----      ---------   ------  -----
10.1.104.0/24   10.1.101.3  101   ospf       IntraArea   2       110
127.0.0.0/8     reject            static                 0       0
127.0.0.1/32    lo0               connected              1       0
```

8. In the real world, you should remove the static routes from Core-2 as well, but in this scenario the next hops for the static routes are the same for those of the OSPF routes. You don't need to take the time to remove the routes for the purposes of this task.

Windows server

9. Access the Windows server desktop and open a command prompt.

10. Execute a trace route to the Windows PC (10.1.12.12).

tracert 10.1.12.12

```
tracing a route to client.hpnu01.edu [10.1.12.12] over a maximum of 30
hops:

  1   <1 ms <1 ms 1 ms  vlan11_defaultrouter.hpnu01.edu [10.1.11.1]
  2   <1 ms <1 ms 1 ms  core-2_access-1_link.hpnu01.edu [10.1.104.2]
  3   <1 ms <1 ms 1 ms  access-2_core-2_link.hpnu01.edu [10.1.102.4]
  4   <1 ms <1 ms 1 ms  client.hpnu01.edu [10.1.12.12]
```

11. Execute a trace route to 10.1.12.1.

tracert 10.1.12.1

```
Tracing route to vlan12_defaultrouter.hpnu01.edu [10.1.12.1] over a
maximum of 30 hops

  1   <1 ms <1 ms 1 ms  vlan11_defaultrouter.hpnu01.edu [10.1.11.1]
  2   <1 ms <1 ms 1 ms  core-1_access-1_link.hpnu01.edu [10.1.101.1]
  3   <1 ms <1 ms 1 ms  vlan12_defaultrouter.hpnu01.edu [10.1.12.1]
```

Task 6: Save

Save the current configuration on each of the switches using the **write memory** command.

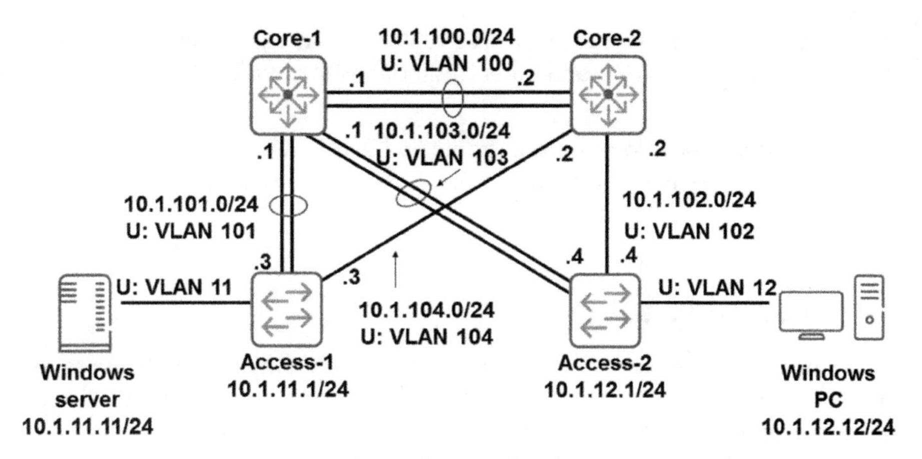

Figure 8-41: Test network topology with a basic OSPF configuration

Learning check

To solidify what you have learned, answer these questions.

1. After you completed the tasks and saved the configuration, which switches were neighbors with each other? You can confirm by using the show and display commands that you used. Consider why some switches become neighbors and others do not.

2. What neighbor state should you look for to ensure that the neighbors have connected successfully and exchanged their topologies?

3. How did you enable OSPF on VLAN interfaces on the ArubaOS switches?

4. Access-1 does not have any neighbors on 10.1.11.0/24 (VLAN 11). Similarly, Access-2 does not have any neighbors on 10.1.12.0/24 (VLAN 12).

   - Why did you still enable OSPF on these VLAN interfaces?

   - Are there any drawbacks to enabling OSPF on these VLAN interfaces?

5. After you finished configuring OSPF, for which networks did Access-1 have routes? What about the other switches? You do not need to list all of the networks, but note what you observed.

6. What did you need to do to make the OSPF routes to 10.1.11.0/24 and 10.1.12.0/24 appear in the Core-1 routing table?

7. You did not delete the static routes on Access-1. Why is Access-1 using OSPF routes to forward traffic to 10.1.12.0/24 and other destinations in your local topology?

8. You noticed that Access-1 is load-sharing traffic to 10.1.12.0/24 over routes through Core-1 and Core-2. After you removed the static route, Core-1, on the other hand, is using one route for this network.

Answers to Learning check

1. After you completed the tasks and saved the configuration, which switches were neighbors with each other? You can confirm by using the show and display commands that you used. Consider why some switches become neighbors and others do not.

   Access-1 is neighbors with Core-1 and Core-2.

   Access-2 is neighbors with Core-1 and Core-2.

   Core-1 is neighbors with Core-1, Access-1, and Access-2.

   Core-2 is neighbors with Core-2, Access-1, and Access-2.

   OSPF routers become neighbors with the other routers on the same subnet. In this topology, the neighbors have direct Ethernet links because you dedicate a subnet to each of those links. But in other topologies, OSPF routers could become neighbors with non-directly connected routers as long as they are on the same subnet.

2. What neighbor state should you look for to ensure that the neighbors have connected successfully and exchanged their topologies?

   Full

3. How did you enable OSPF on VLAN interfaces on the ArubaOS switches?

   You enabled OSPF from the VLAN context. You set the area in the same command.

4. Access-1 does not have any neighbors on 10.1.11.0/24 (VLAN 11). Similarly, Access-2 does not have any neighbors on 10.1.12.0/24 (VLAN 12).

   - Why did you still enable OSPF on these VLAN interfaces?
   - Are there any drawbacks to enabling OSPF on these VLAN interfaces?

   You wanted the switches to advertise the subnets in OSPF so that other routing switches could learn routes to them.

   If you enable OSPF on the interface, it is possible that you could leak routing information to endpoints or establish an OSPF neighbor relationship with an unauthorized device. That unauthorized device could then disrupt your routing topology.

You can set the VLAN interface as a passive interface to prevent adjacencies from forming. Alternatively, you could leave the interface out of the OSPF configuration and have the routing switch redistribute connected (direct) routes. The switch would then advertise the subnet like an external route. You can attend other training to learn more about these options.

5. After you finished configuring OSPF, for which networks did Access-1 have routes? What about the other switches? You do not need to list all of the networks, but note what you observed.

   Every switch has a route to every network in your topology. If the switch supports the network itself, it has a direct route for it. For all other networks, it has learned an OSPF route.

6. What did you need to do to make the OSPF routes to 10.1.11.0/24 and 10.1.12.0/24 appear in the Core-1 routing table?

   You had to delete the static routes to those destinations. The static routes had a lower administrative distance than the OSPF routes, so they were taking precedence.

7. You did not delete the static routes on Access-1. Why is Access-1 using OSPF routes to forward traffic to 10.1.12.0/24 and other destinations in your local topology?

   When you use the default administrative distances, ArubaOS switches prefer static routes (administrative distance 1) to OSPF routes (administrative distance 110). However, Access-1 does not have any specific static routes, only default routes. Default routes only apply if the routing table does not have a specific route to the destination.

8. You noticed that Access-1 is load-sharing traffic to 10.1.12.0/24 over routes through Core-1 and Core-2. After you removed the static route, Core-1, on the other hand, is using one route for this network.

   - Why has Access-1 added two routes to 10.1.12.0/24 to its routing table?
   - Why does Core-1 have only one route to 10.1.12.0/24 in its table?
   - How could you make Access-1 prefer its 10.1.12.0/24 route through Core-1?

   As you learned, OSPF chooses between routes based on metric. Both Access-1 routes to 10.1.12.0/24 have the same metric, so both are added to the routing table. Access-1 actually has a higher bandwidth connection to Core-1, but OSPF is running on the Layer 3 interface (the VLAN), and the metric for every VLAN is the same by default. The paths through Core-1 and Core-2 have the same number of Layer 3 hops (each hop is a VLAN interface), so they have the same metric. OSPF uses path cost, not hops, to choose routes, but because every interface has the same cost in this network, hop count dictates the cost.

   The situation is similar on Core-1. Core-1, as well as all other routing switches, is using the same cost on every VLAN interface. However, the route to 10.1.11.0/24 through Access-1 has fewer hops than the route through Core-2 or Access-2, so its metric is lower. Core-1 prefers that route.

If you want to make Access-1 prefer its 10.1.12.0/24 route through Core-1, you should set the metric (OSPF cost) for the VLAN that connects to Core-1 lower. Set the OSPF cost for the VLAN that connects to Core-2 higher.

Summary

This chapter taught you how to create a variety of static routes. You also learned to use metrics and administrative distance to affect which routes the switch adds to its routing table.

You considered the complexity of using static routes within a complicated, redundant topology, and you learned about the advantages of a dynamic routing protocol for such environments. You then set up a basic OSPF solution.

You also learned how to interpret an IP routing table and trace routing paths.

– Configuring static IP routes
– Choosing between routes to the same destinatio
– Configuring a basic OSPF solution

Figure 8-42: Summary

Learning check

1. What ArubaOS switch command defines a static route to 10.1.1.0/24 with a next-hop address of 10.2.2.2?

2. Which of these is the preferred route?

    a. 10.1.10.0/24 10.1.1.2 metric 10 administrative distance 10

    b. 10.1.10.0/24 10.1.2.2 metric 1 administrative distance 10

    c. 10.1.10.0/24 10.1.3.2 metric 10 administrative distance 1

    d. 10.1.10.0/24 10.1.4.2 metric 5 administrative distance 1

Answers to Learning check

1. What ArubaOS switch command defines a static route to 10.1.1.0/24 with a next-hop address of 10.2.2.2?

   ip route 10.1.1.0/24 10.2.2.2

2. Which of these is the preferred route?

   a. 10.1.10.0/24 10.1.1.2 metric 10 administrative distance 10

   b. 10.1.10.0/24 10.1.2.2 metric 1 administrative distance 10

   c. 10.1.10.0/24 10.1.3.2 metric 10 administrative distance 1

   d. 10.1.10.0/24 10.1.3.2 metric 5 administrative distance 1

9 Virtual Switching Framework (VSF)

EXAM OBJECTIVES

✓ Understand how VSF works and the advantages that it provides

✓ Configure and verify a simple VSF fabric

ASSUMED KNOWLEDGE

Before reading this chapter, you should have a basic understanding of:

- Link aggregation
- VLANs

INTRODUCTION

This chapter introduces Virtual Switching Framework (VSF), an ArubaOS technology that allows multiple Aruba switches of the same model to act as a single virtual device. VSF helps to provide fast failover, scalability, manageability, and high availability.

VSF introduction

The first section of this chapter introduces you to when you would use VSF and how the technology works.

VSF use case

VSF lets two Aruba 5400R Switch Series switches function as a single virtual switch called a VSF fabric. In your test network topology, you will combine Core-1 and Core-2 into a single VSF fabric. In effect, you will have a single modular switch at the core/aggregation layer of your network; that "switch" is distributed physically in two chassis.

In your test network topology, in effect, you now have a single modular switch at the core/distribution layer of your network; that "switch" is distributed physically in two chassis.

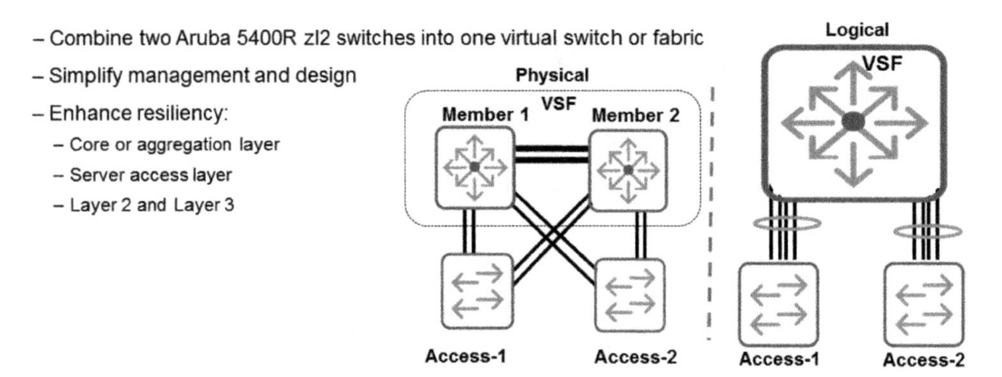

Figure 9-1: VSF use case

You can then manage the two switches as a single entity. You can also treat the switches as a single entity from the point of designing Layer 2 and 3 protocols. VSF, therefore, introduces many benefits in management simplicity, redundancy at both Layers 2 'and 3, operational efficiency, and design simplicity, particularly at the aggregation layer or core of the network. The VSF fabric can also provide high availability and fast failover at both Layer 2 and 3 for servers connected with multiple NICs.

Consider these benefits in more depth:

- **Management simplicity**—You no longer need to laboriously connect to, configure, and manage switches individually. You perform a configuration on the primary switch, and that configuration is distributed to all associated switches automatically, considerably simplifying network setup, operation, and maintenance.
- **Better resiliency and simpler design at Layer 2**—Chapter 6 introduced you to RSTP and MSTP, demonstrating how important these protocols are for eliminating loops in redundant Layer 2 networks. But you also learned about some of the drawbacks:
  - **Relatively slow network convergence**—RSTP or MSTP convergence can take up to one second, which can be too long for modern applications.
  - **Management and troubleshooting complexity**— MSTP and RSTP take time to configure properly, especially in a large network. You must manage the switches individually, and you need to set up spanning-tree instances on each switch in turn, making sure that the parameters for one switch match those of the rest of the region. The real problem emerges in maintenance, however. Troubleshooting spanning-tree-related issues is no easy task, usually requiring a great deal of time to locate the root cause of the failure.

- **Blocked links**—RSTP blocks all but one set of redundant parallel paths. MSTP permits the use of some redundant links but does not load balance very granularly. Half (or more) of the available system bandwidth can be squandered in a backup role, off-limits to data traffic—not a very good use of the network infrastructure investment.

In Chapter 7, you saw how link aggregations compare favorably to RSTP and MSTP. However, traditionally, you can only create link aggregations between the same two devices. In a truly redundant Layer 2 network, a switch needs links to more than one other switch, so the network must use RSTP or MSTP in conjunction with link aggregations.

But, as you saw, VSF lets you create link aggregations that span both switches in the fabric. Spanning tree is no longer necessary to eliminate loops on switch-to-switch links because at a logical level no loop exists. You still use spanning tree to eliminate accidental loops. And, as you saw, the VSF fabric can also provide resilient redundant connections to servers.

VSF provides all the advantages of a completely link-aggregation based Layer 2 design:

- Faster failover
- Load-balancing
- Simple set up and design
- Better stability and less need for troubleshooting

- **Better resiliency and simpler design at Layer 3**—In the practice tasks in previous chapters, you have confronted issues with lack of redundancy at Layer 3. First, your subnets only had a single default router. In Chapter 7, you created a highly redundant Layer 2 design with link aggregations and MSTP. But Core-1 was the only router for your servers and clients, and if it failed, clients and servers could no longer reach each other.

  A standard protocol does exist for solving this problem (Virtual Router Redundancy Protocol [VRRP], which is not covered in this guide). But VSF solves the problem quite elegantly. Now the VSF fabric can be the single default router, but it actually consists of two physical devices. If one of those devices fails, the other takes over seamlessly.

  You also saw the complexity of designing a redundant routing topology that uses static routing. A dynamic routing protocol provides better redundancy, but still introduces complexity, particularly as routing devices and redundant links between them multiply. VSF helps to simplify the situation by creating fewer logical routing devices and simpler topologies between them. You do need to configure protocols to reduce the impact of a VSF master failure, including graceful restart for OSPF; however, these are beyond the scope of this guide.

Physical and logical view of VSF

You can think of a VSF fabric as a modular switch that has been extended over two chassis.

All switches have:

- A management plane, which provides administrators and management solutions access to management interfaces such as the CLI
- A control plane, which runs protocols and builds routing and MAC forwarding tables
- A forwarding plane, which processes and forwards packets

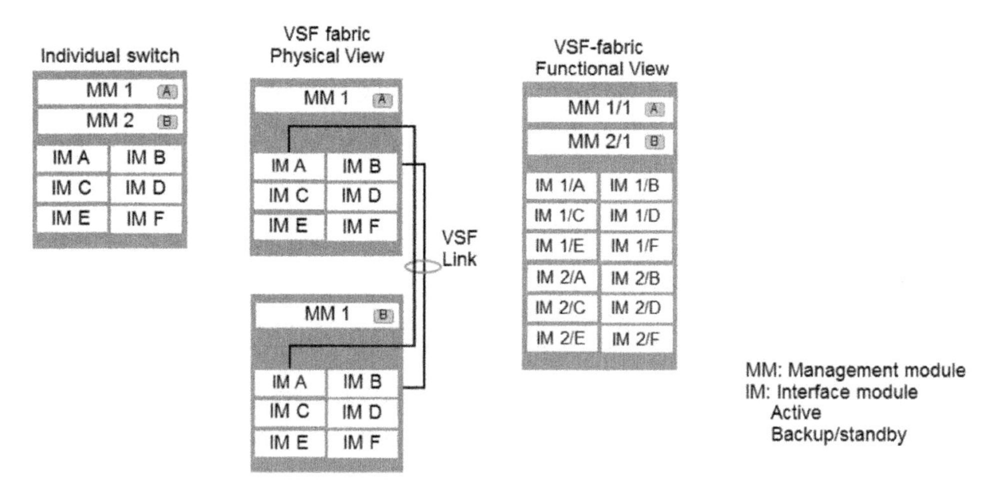

Figure 9-2: Physical and logical view of VSF

A modular switch has a management module that supports the management and control plane while its interface modules provide the forwarding plane. These interface modules also proxy portions of the control plane. If the modular switch has two management modules, as shown in the figure, one management module is active while the other is on standby.

A VSF fabric also has only active management plane and one active control plane, which are provided by the management module in the commander switch. The VSF commander proxies the control plane to the standby member. Both members of the fabric participate in the forwarding plane. Interface modules from both members are combined as if over one large switch connected by the VSF link.

VSF requirements

A VSF fabric must consist of two Aruba 5406R or two Aruba 5412R switches. You cannot combine 5406R and 5412R switches in the same VSF fabric. The two switches can support different interface modules although it is recommended that you install the same modules in both switches for consistency.

To implement VSF, you must also verify that your 5400R switches are running software KB16.01.xxx and later.

The VSF fabric must use only v3 modules, including for both VSF links and other links. The switch must operate in v3 only mode. In fact, when you enable VSF on the switch, it automatically changes to v3 only mode, disabling any v2 modules installed in the switch.

When part of a VSF fabric, a 5400R switch can only use one management module, and enabling VSF de-activates a second management module. It is recommended, though, that you install only one management module in the switch. If the management module of the commander fails, the standby member takes over, as described later in this chapter.

- Two 5400R switches of the same model
- Software KB16.01.xxx and later
- v3 modules and v3-only operation mode
- Recommended: One management module only per switch
  - If present, second module disabled

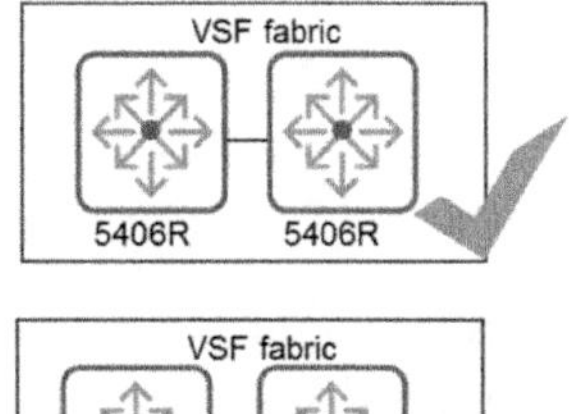

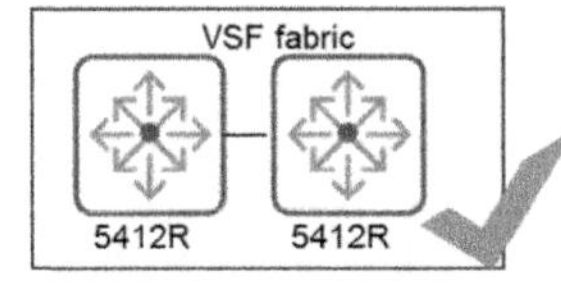

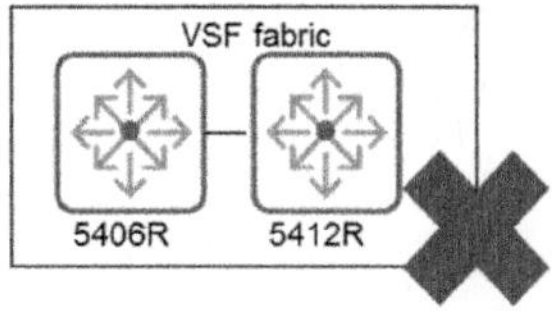

Figure 9-3: VSF requirements

VSF member roles

You will now examine the components of a VSF fabric in more detail. Each VSF fabric is identified by a domain ID. It is best practice to assign every VSF fabric in your network a unique domain ID. If your network includes Comware switches that use Intelligent Resilient Framework (IRF), you should also make sure that each VSF fabric has a different domain ID from the IRF fabrics' domain IDs.

The VSF fabric consists of exactly two physical devices or members. Each member is identified by a unique ID within the fabric, 1 or 2.

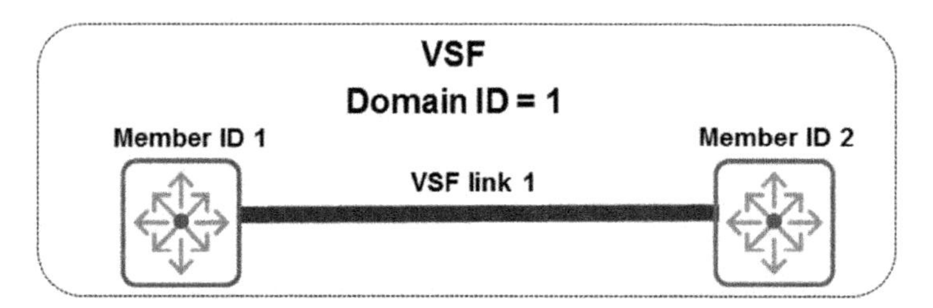

Figure 9-4: VSF member roles

Commander

One member is the commander. In this figure, the member with ID 1 is commander. (Although typical for the commander, member ID 1 is not required; you will learn how a commander is elected a bit later.)

The commander provides the active control plane, which means that it handles protocol functions. Only the commander builds MAC forwarding tables, ARP tables, and routing tables. The commander also runs all IP routing protocols.

The commander provides the active management plane. Users and management solutions can connect through this plane to the CLI, Web user interface, and SNMP agent to manage the VSF fabric configuration. Note that, although the commander provides the only active management plane, you can connect to a console port on the standby switch and receive access to the CLI. The standby member proxies you to the commander management plane.

In addition to providing the device intelligence, the commander also has active physical interfaces, and it receives and forwards traffic.

Standby

The other member is the standby member. You can think of it as interface modules in a modular switch. It does not have its own active control and management functions. Instead it receives the configuration for its interfaces from the commander. It also receives relevant portions of the MAC addressing tables and routing tables from the commander. It can then receive and forward traffic correctly.

Unlike simple interface modules, though, the standby member retains the ability to take over the management and control functions if the commander fails. Therefore, even though the standby member does not have active management and control functions, it permanently receives copies of the current configuration and stores them. It also proxies the control plane so that it is ready to take over as the commander if necessary.

VSF link

VSF members connect to each other over a VSF link in a chain topology. You can think of this link a bit like the backplane paths in a modular switch. It carries:

- **VSF protocol traffic** for managing the fabric
- **Data traffic** that arrives on one member's interface that needs to be forwarded on the other member's interface

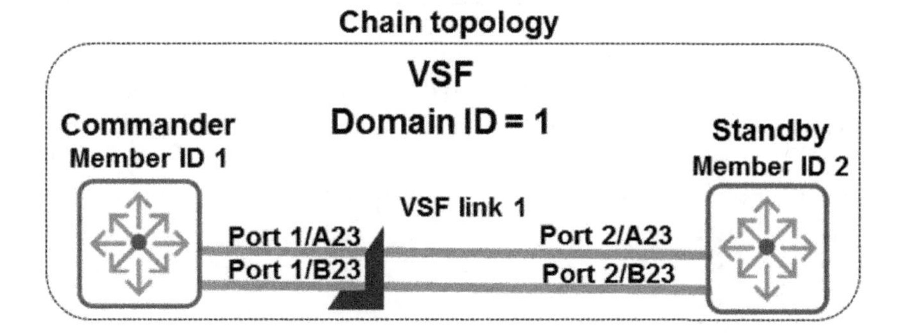

Supports up to eight links:

- Standard 10GbE or 40 GbE
- All same speed (any media)
- Not part of trunk (link aggregation), distributed trunk, mesh, or ISC

Figure 9-5: VSF link

The VSF link consists of a logical VSF link defined on each member and associated with up to eight physical interfaces. You should typically use multiple interfaces to enhance redundancy and to increase bandwidth on the VSF link. To follow best practices, choose at least two links that are on different modules in case one of the modules fails. The figure shows an example of a VSF link with two interfaces on each member. As you see, the member ID is used in all interface IDs, including those for the VSF link and those used to connect to other devices.

The interfaces in a VSF link must be 10GbE or 40GbE, and all interfaces assigned to the link must be the same speed. However, the interfaces can be different media. For example, on link could use fiber, and another link could use copper.

Important

The physical interfaces associated with the VSF link cannot be members of link aggregations (trunks), meshes, distributed trunks, or in an interswitch connect (ISC) associated with distributed trunking (DT). Remove any such configurations from the interface before configuring it as part of the VSF link.
When you add the interface to the VSF link, all other configurations (such as VLAN) are removed automatically.

Note

DT is a technology for enabling two logically separate switches to participate in the same link aggregation. It is covered in other HPE training.

After the physical interface is bound to the VSF link, you cannot configure settings on it. The interface is now dedicated to VSF protocol traffic and to data traffic between members. If you have bound multiple physical interfaces to a VSF link, the interfaces load balance this traffic much like a link aggregation does. (The link supports multiple queues, enforcing any traffic prioritization applied to frames using quality of service mechanisms such as 802.1p or Differentiated Services [DiffServ]. QoS is covered in other HPE training.)

VSF configuration

In the next section of this chapter, you'll look at ways to configure VSF.

VSF configuration process: Plug and play

You'll find it very simple to establish a VSF fabric.

Begin with the 5400R switch that you want to be commander. This switch might already by operating in your network. For example, a customer might have a 5400R switch at the network core and now wants to add redundancy to the core. You'll set up a VSF fabric that uses the current 5400R switch as the commander, preserving all of the current configurations.

It is recommended that you create the VSF fabric during a scheduled network outage because:

- The switch that you're configuring as commander must reboot during the process.
- You should remove any active links during the configuration process to prevent temporary issues (either by disconnecting cables or disabling ports).

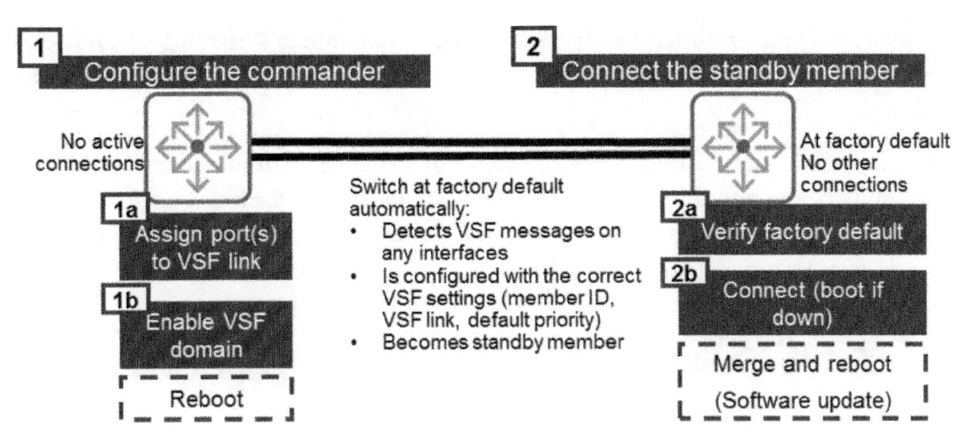

Figure 9-6: VSF configuration process: Plug and play

Follow these simple steps to set up the fabric:

1. Configure the switch that you've selected as commander:

   a. Create a VSF link. As you create the link, you specify:

   – The member ID—Typically, 1 for the commander, although it can be 2

   – The link ID—By default 1 in software KB16.01.xxx

   – The interfaces assigned to the link—Remember the rules and guidelines covered earlier. Also note that the interfaces lose their configuration when you add them to the VSF link. If you've disabled the interfaces, they will become enabled again. Do not cable these interfaces at this point.

   b. Enable VSF, specifying a unique domain ID for this fabric.

   c. Enabling VSF saves the configuration and reboots the switch. The switch automatically adjusts the configuration so that it is valid for VSF mode, adding the VSF member ID to each interface ID. It is now the commander of a VSF fabric that consists of one member. (You'll learn how to confirm this setup later.)

2. You're now ready to connect the standby member.

   a. Begin by making sure the switch has no active connections and is at factory default settings. It might be a new switch or one that you have reset to default settings. It will be taking its configuration from the commander, so it no longer needs the settings.

   b. Connect the standby member to the commander interfaces that you configured as part of the VSF link. Generally, the switch is powered down at this point; boot the switch.

   c. When a 5400R switch is at factory default settings, it detects VSF messages on any of its interfaces and automatically uses those messages to configure itself with the correct VSF domain and link settings. A factory default switch that joins a VSF fabric always defers to the current commander, taking the standby role. The standby switch merges with the fabric and reboots.

If the standby member's software version does not match the commander's version, the commander updates the standby member, and the standby member reboots again. As the standby member reboots and joins the fabric, it takes its configuration from the commander. All of its interfaces are at factory default settings.

Create a Virtual Switching Framework (VSF) Fabric Using Plug-and-Play

Your customer wants to increase redundancy at the core without complex protocols. You will now set up VSF on your Core-1 switch. You will then reset your Core-2 switch to factory default settings and join it to the VSF fabric. The tasks in this section build on the configuration from Chapter 8.

After completing the tasks in this section, you will be able to create a Virtual Switching Framework (VSF) fabric on two Aruba 5400R zl2 Switches using the plug-and-play method.

- Configure Core-1 as VSF commander
- Reset Core-2 to factory defaults and join it to the fabric

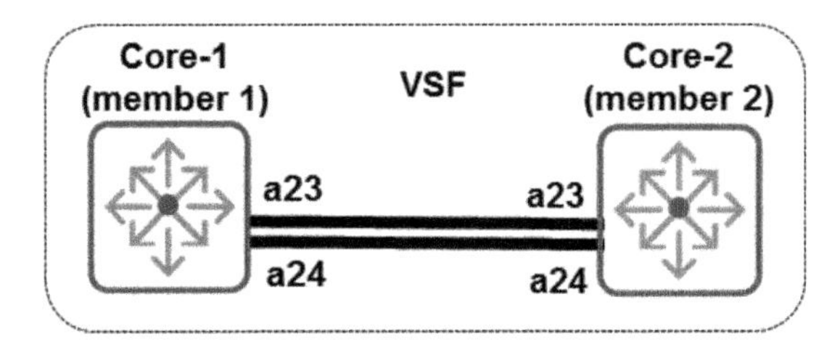

Figure 9-7: Create a Virtual Switching Framework (VSF) Fabric Using Plug-and-Play

Task 1: Configure Core-1 as the commander

You will begin by disabling the interfaces on Core-1, which is the switch that will become commander. You will then set up a VSF link and domain on Core-1.

Core-1

1. Access a terminal session with Core-1. Move to the global configuration mode.
2. Disable all interfaces on Core-1.

```
Core-1(config)# interface a1-a24 disable
```

3. After you join Core-1 and Core-2 in a VSF fabric, they will no longer connect over the link aggregation. Remove the link aggregation.

```
Core-1(config)# no trunk a19,a20
```

4. Remove VLAN 100, the VLAN for that link aggregation.

```
Core-1(config)# no vlan 100
```

5. Specify this switch as VSF member 1 and configure the link. Specify two 10GbE interfaces for the link.

```
Core-1(config)# vsf member 1 link 1 a23,a24

All configuration on this port has been removed and port is placed in VSF mode.
```

6. Enable VSF, specifying 1 for the domain ID.

```
Core-1(config)# vsf enable domain 1
```

7. Confirm that you want to save the configuration and reboot.

```
To enable VSF, the REST interface will be disabled.

This will save the current configuration and reboot the switch.

Continue (y/n)?  y
```

Task 2: View VSF settings on the commander

After the switch reboots, access its CLI again to view the VSF settings.

Core-1

1. The switch has altered its configuration to reflect that it is now VSF member 1. View its interfaces to see one change.

```
Core-1# show interface brief

 Status and Counters - Port Status
```

```
                            | Intrusion                              MDI  Flow Bcast
 Port          Type         | Alert   Enabled Status Mode      Mode Ctrl Limit
 ------------- ------------ + ------- ------- ------ --------- ---- ---- -----
 1/A1          100/1000T    | No      No      Down   1000FDx   Auto off  0
<-output omitted->
 1/A13-Trk2    100/1000T    | No      No      Down   1000FDx   Auto off  0
 1/A14-Trk2    100/1000T    | No      No      Down   1000FDx   Auto off  0
<-output omitted->
 1/A23         SFP+DA3      | No      Yes     Down   10GigFD   NA   off  0
 1/A24                      | No      Yes     Down   .              off  0
```

2. Note the enabled status for interfaces 1/A23 and 1/A24. How did these interfaces become enabled?
3. View the VSF settings on Core-1. Note the domain ID. Also verify that VSF is active, although no VSF stack has formed yet.

```
Core-1# show vsf
 VSF Domain ID    : 1
 MAC Address      : 1c98ec-ab5b3f
 VSF Topology     : No Stack Formed
 VSF Status       : Active
 Uptime           : 0d 0h 6m
 VSF Oobm-MAD     : Disabled
 Software Version : KB.16.01.0006

 Mbr
 ID  Mac Address    Model                          Pri  Status
 --- -------------  ----------------------------   ---  ---------
 1   1c98ec-ab4b00  HP J9850A  Switch  5406Rzl2   128  Commander
```

4. What role does Core-1 have in the VSF fabric at this point? What is its priority?
5. View the running-config. Note that this switch's type and MAC address have been bound to the VSF settings that you configured before enabling VSF.

```
Core-1(config)# show run
; J9850A Configuration Editor; Created on release #KB.16.01.0006
; Ver #0c:11.7c.59.f4.7b.ff.ff.fc.ff.ff.3f.ef:9b
hostname "Core-1"
module 1/A type j9990a
vsf
   enable domain 1
   member 1
      type "J9850A" mac-address 1c98ec-ab4b00
      priority 128
      link 1 1/A23,1/A24
      link 1 name "I-Link1_1"
      exit
<-output omitted->
```

6. View the VSF link. The link is currently down.

```
Core-1# show vsf link
 VSF Member 1
                 Link     Peer   Peer
 Link Link-Name  State    Member Link
 ---- ---------- -------- ------ ----
 1    I-Link1_1  Down     0      0
```

7. You can also view more detailed information about this VSF member.

```
Core-1# show vsf member 1

Member ID         : 1
Mac Address       : 1c98ec-ab4b00
Type              : J9850A
Model             : HP J9850A Switch 5406Rzl2
Priority          : 128
```

```
Status              : Commander
ROM Version         : KB.16.01.0006
Serial Number       : SG62G49470
Uptime              : 0d 0h 37m
CPU Utilization     : 5%
Memory - Total      : 709,308,416 bytes
Free                : 535,109,120 bytes
VSF Links -
#1 : Inactive
```

Task 3: Provision the standby member on Core-1

Next you will provision member 2 on Core-1. This is an optional step, but it lets you configure the member 2 interfaces in advance. Your test network environment might have permanent connections on Core-2, so you want to make sure that these interfaces are disabled on the switch as soon as it joins the fabric.

Core-1

1. Access the terminal session for Core-1 and move to global configuration mode.
2. Provision member 2 on Core-1. When you provision a new member after VSF is enabled, you must specify its type.

```
Core-1(config)# vsf member 2 type j9850a
This will save the current configuration. Continue (y/n)?  y
```

3. Now that you've provisioned member 2, you can pre-configure the member 2 interface settings. First add the member interface module, which is in slot a.

```
Core-1(config)# module 2/A type j9990a
```

Important

If your switch is using a different type of module, you must specify that module's model number instead. Enter **show run** on Core-2 and look for the module line to verify.

4. Disable all the interfaces on Core-2 except the interfaces that will be part of the VSF link. Now when Core-2 joins the VSF fabric, these interfaces will be shut down.

```
Core-1(config)# interface 2/a1-2/a22 disable
```

Important

If Member 2 (Core-2) has permanent connections to other devices, it is very important to disable all of the ports to prevent loops.

Task 4: Join Core-2 to the VSF fabric

You are now ready to join Core-2 to the VSF fabric.

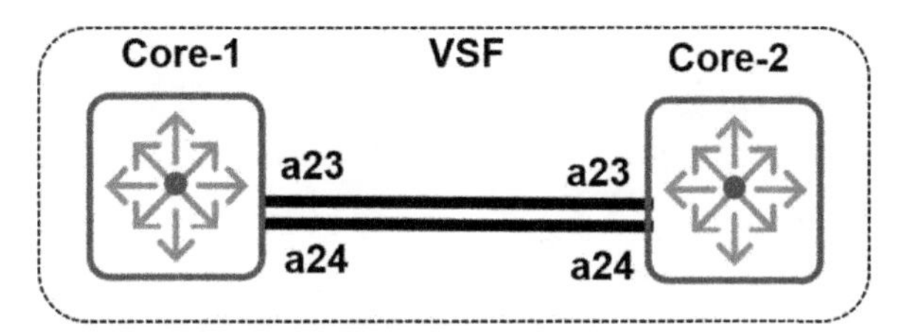

Figure 9-8: Task 4: Join Core-2 to the VSF fabric

Core-2

1. Establish a terminal session with Core-2 and move to global configuration mode.
2. As an experiment, activate the Core-2 interfaces that connects to the interfaces in the Core-1 VSF link.

```
Core-2(config)# interface a23,a24 enable
```

3. Establish the physical connections.
4. Wait a few moments. As you see, nothing happens. A 5400R switch only automatically joins the VSF fabric when it is at factory defaults. Reset Core-2 to factory defaults now, confirming the reset.

```
Core-2(config)# erase startup-config
The current configuration will be deleted and the device rebooted.
Continue (y/n)? y
```

5. Monitor the boot process from the terminal session and wait a minute or two. You should then see this message:

```
Rebooting to join VSF fabric with domain ID 1
```

   Core-2 will then reboot again.

Core-1

While Core-2 is rebooting to join the fabric, view the state of the VSF fabric on Core-1.

6. Return to the terminal session with Core-1.
7. View the VSF link. At first, it will be down.
8. Wait a minute or so. The link should come up, and you should see both sides of the link.

```
Core-1# show vsf link

 VSF Member 1

                     Link      Peer   Peer
 Link Link-Name      State     Member Link
 ---- ---------- -------- ------ ----
 1    I-Link1_1  Up        2      1

 VSF Member 2

                     Link      Peer   Peer
 Link Link-Name      State     Member Link
 ---- ---------- -------- ------ ----
 1    I-Link2_1  Up        1      1
```

9. Also view the state of the VSF fabric. You should see that VRF is active and in a chain topology. The standby member (Core-2) is still booting.

```
Core-1# show vsf

 VSF Domain ID    : 1
 MAC Address      : 1c98ec-ab5b3f
 VSF Topology     : Chain
 VSF Status       : Active
 Uptime           : 0d 1h 44m
 VSF Oobm-MAD     : Disabled
 Software Version : KB.16.01.0006
```

```
Mbr
ID   Mac Address    Model                      Pri  Status
---  -------------  -------------------------  ---  ---------------
1    1c98ec-ab4b00  HP J9850A Switch 5406Rzl2  128  Commander
2    70106f-0d2100  HP J9850A Switch 5406Rzl2  128  Standby Booting
```

Core-2

You will explore how the session with Core-2 behaves now that this switch has joined the VSF fabric.

10. Return to the terminal session with Core-2. You should see this message.

```
USB automounter: Successfully mounted device 1 lun 0 partition 0 at ufa0
initialization done.
Press any key to connect to the commander.
```

11. Press any key and then press any key again.
12. As you see, the hostname is Core-1. You've connected to the management plane for the VSF fabric, which is using the commander (Core-1) configuration.
13. View the status of interfaces of member 2 (Core-2). As you see, they are disabled because you provisioned them in advance.

```
Core-1# show interface brief
 Status and Counters - Port Status

                         | Intrusion                          MDI   Flow Bcast
  Port    Type           | Alert    Enabled Status Mode       Mode  Ctrl Limit
  ------  ---------  + -------  ------- -----  ------     ----  ---- -----
<-output omitted->
  2/A1    100/1000T  | No       No      Down   1000FDx    Auto  off  0
  2/A2    100/1000T  | No       No      Down   1000FDx    Auto  off  0
  2/A3    100/1000T  | No       No      Down   1000FDx    Auto  off  0
  2/A4    100/1000T  | No       No      Down   1000FDx    Auto  off  0
  <-output omitted->
```

Task 5: Save

Save the configuration on the VSF fabric with the **write memory** command.

Learning check

You will now answer several questions to help solidify the concepts that you have learned.

1. What steps did you take to configure VSF on Core-1? Why did this switch become commander?

2. What steps did you take to join Core-2 to the VSF fabric? Why didn't you need to configure any VSF settings on this switch?

3. What changes to the physical interface IDs did you observe?

4. After the VSF fabric formed, what configuration was it running?

5. After the VSF fabric formed, what did you observe in the terminal session with Core-2?

6. Why was it important for member 2 to have its interfaces disabled as soon as it joined the VSF fabric and rebooted? How did you make sure that the interfaces were disabled?

Answers to Learning check

1. What steps did you take to configure VSF on Core-1? Why did this switch become commander?

   You configured the VSF link (member 1 and link ID 1, using interfaces A23 and A24). (It is best practice to use interfaces on different modules, but your switches in the test network environment have only one interface module.)

   You also enabled VSF, setting the domain ID to 1.

   Core-1 first became the commander because it was the only member of the fabric. It remained the commander when Core-2 joined the fabric because Core-2 was at factory defaults and joined the fabric by detecting Core-1's messages. When a factory default switch joins the VSF fabric, the current commander always remains commander.

2. What steps did you take to join Core-2 to the VSF fabric? Why didn't you need to configure any VSF settings on this switch?

   You simply erased Core-2's configuration and connected it to the Core-1 VSF link. 5400R switches with the proper software to support VSF automatically join a VSF fabric when they are at factory default settings and connected to a commander VSF interface.

3. What changes to the physical interface IDs did you observe?

   The interface IDs now have a member ID such as 1/A1 or 2/A23.

4. After the VSF fabric formed, what configuration was it running?

   It was running the Core-1 configuration because Core-1 was commander.

5. After the VSF fabric formed, what did you observe in the terminal session with Core-2?

   You saw the Core-1 hostname because Core-2 was part of the VSF fabric. You can connect to the console port on either VSF member, but the management plane is always provided by the commander.

6. Why was it important for member 2 to have its interfaces disabled as soon as it joined the VSF fabric and rebooted? How did you make sure that the interfaces were disabled?

   After the standby member reboots, it is part of the VSF fabric and takes its configuration from the commander. All of its interfaces are at factory default settings. In the test network environment, your switches are permanently connected together, and the access layer switches enable some of the connected interfaces. To prevent loops, you needed to shut down the interfaces until you were ready to configure them.

   You made sure that the interfaces were disabled by adding member 2 to the commander configuration, specifying the member 2 module, and configuring the member 2 interfaces in advance. This is called provisioning, and you will learn more about it in a moment.

VSF provisioning use cases

Earlier, you practiced provisioning information about the standby member on the commander before you established the fabric. While not required, this step can give you greater control over the fabric in advance.

Provision information about standby member on commander before establishing fabric:

– Control which device can join the fabric (strict provisioning)

– Pre-configure standby member VSF priority and link name (strict or loose provisioning)

– Pre-configure standby member modules and interfaces (strict or loose provisioning)

Provisioning options	Required information
Strict	Type + MAC
Loose	Type

Figure 9-9: VSF provisioning use cases

You used loose provisioning, in which you just specify the member type. Type refers to model number, J9850A for a 5406R or J9851A for a 5412R. You can also use strict provisioning—in which you specify the standby member's type and MAC address (the switch's base MAC address). Strict provisioning controls which device is allowed to join the fabric; only the switch with the specified MAC address can join.

Provisioning also lets you configure the standby member's VSF priority and link name yourself rather have VSF do it automatically when the fabric forms. For example, you might want to pre-configure the standby member with VSF priority 1 so that you can be sure that the commander remains commander if both switches ever reboot at the same time.

You might want also to pre-configure the standby member interfaces as you did earlier. Then when the standby member joins the fabric, it is immediately ready to support traffic. After provisioning the member on the commander, follow these steps:

1. Add interface modules on the standby member to the commander's configuration with this command:

```
Switch(config)# module <standby member ID>/<module ID> type <module type>
```

2. Configure the interface settings just as you would for a module that was already installed in this switch.

If your only purpose in provisioning is to pre-configure the standby member settings, use loose provisioning, in which you only specify the device type. This provisioning method does not offer any extra control over the device that joins the fabric because VSF already prohibits switches of different models from merging in a fabric. But it does provide a place for pre-configuring the member settings.

You can combine the two use cases to both control which device can join the fabric and pre-configure that device's settings. In this case, use strict provisioning.

VSF provisioning process

You must set up provisioning in between enabling VSF on the commander and connecting the standby member.

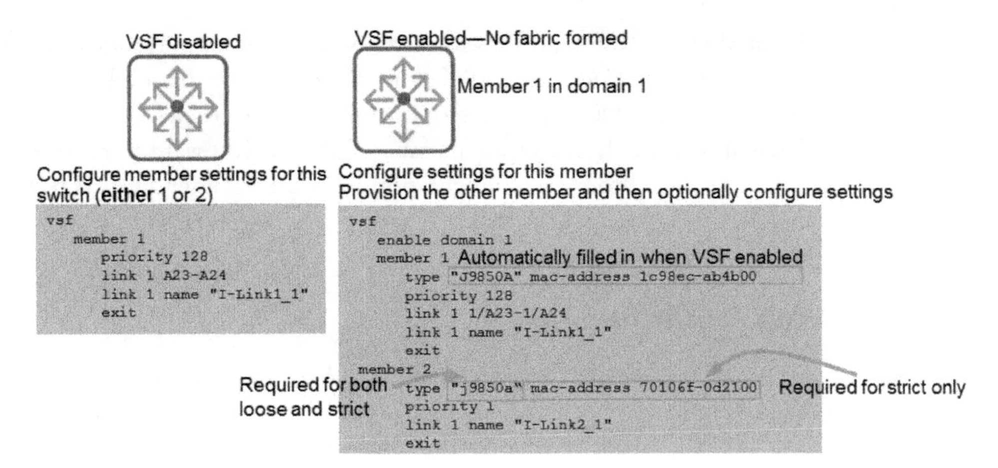

Figure 9-10: VSF provisioning process

When you begin configuring VSF, while VSF is still disabled, you can configure the settings for only *one* member, either member 1 or member 2. The switch becomes this member when you enable VSF, and the switch reboots. In other words, configuring a VSF link for member 1 is configuring this switch as member 1 and setting up its link. If you try to configure a setting for the other member at this point, you will see an error indicating that you've already set a different member ID for this switch. If you've accidently used the wrong ID and want to configure VSF settings with the other member ID, you would need to first remove the current member (with the **vsf member <ID> remove** command).

After you enable VSF and the switch reboots, the member settings that you configured automatically become bound to the switch's type and MAC address, as shown in the configuration in the figure. Note that if you enable VSF before configuring any member settings, the switch automatically becomes member 1 and starts using the default VSF priority and link name (the link interfaces are not defined). If you want to choose the member ID, set up the VSF link before enabling VSF.

After VSF is enabled, you can continue configuring this member's settings. At this point, you can also enter the **member <ID> type** command to provision the other member. Specify the same type as this switch (otherwise, you will receive an error). If you want to use strict provisioning, also add the other member's MAC address. You can then use the **member <ID> link** and **member <ID> priority** commands to configure settings for the member that has not yet joined the fabric.

The member will accept these settings when it joins. If you used strict provisioning, only a switch with the correct MAC address is allowed to join. A switch with a different MAC address will reboot as if it is joining the fabric but fail to actually do so. You will see a message that the switch cannot join the VSF fabric because the fabric already has too many members (the pre-provisioned member is counting as a member). With loose provisioning, any member of the correct type can join, and VSF automatically fills in its MAC address.

You should understand that after the VSF fabric has formed, the configuration includes settings for two members, each with the type and MAC address filled in. The end configuration looks the same whether you pre-provisioned it or let VSF fill in the settings automatically. Therefore, the configuration is bound to the specific hardware. If one of the members fails and you want to replace it, you must first remove the failed member from the configuration with the **vsf member <ID> remove** command. You can then join the replacement switch.

Manual VSF fabric configuration

The plug-and-play setup provides the easiest way to establish a VSF fabric, but you can take other approaches as well. A manual configuration allows you to join two switches, neither of which is at the factory defaults. However, after the fabric forms, only the commander configuration remains.

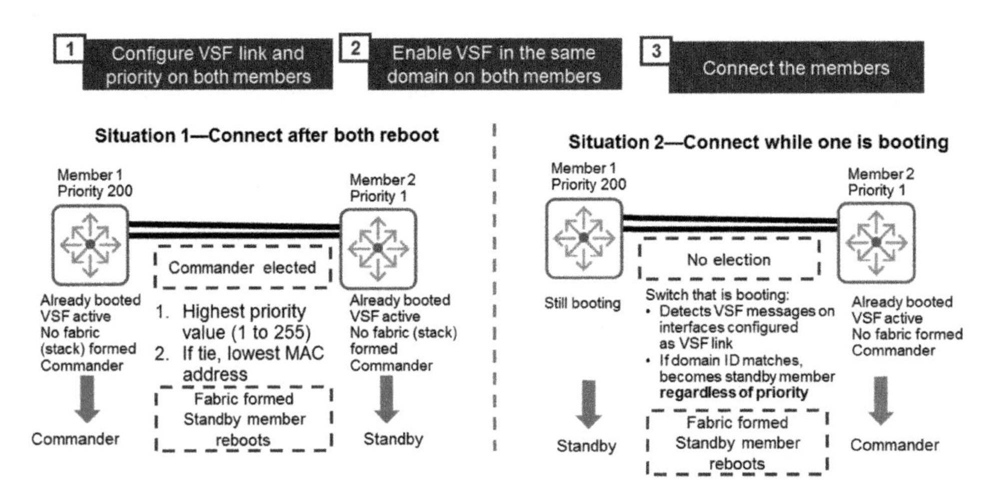

Figure 9-11: Manual VSF fabric configuration

You should begin by disconnecting both switches from all other devices *including from each other*. You then configure both members in a similar manner to configuring the commander for a plug-and-play setup. You assign ports to a VSF link, configuring one switch with member 1 link 1 and the other switch with member 2 link 1.

Before you enable the VSF domain, though, you should set a priority. The VSF priority (values between 1 and 255, default 128) determines which switch will be elected commander; the switch with higher value wins.

Make sure to enable the VSF domain with the same ID on both switches.

After you configure these settings and enable the VSF domain on each switch, it reboots.

If you want to use the VSF priority to determine the commander, allow *both* switches to reboot while still disconnected. Verify that both switches have booted and are running VSF. Each switch should consider itself the commander of a VSF fabric.

Then connect the switches on the interfaces that you added to the VSF link. When the link comes up, the switches start to exchange VSF protocol messages. They verify that they are in the same VSF domain and hold an election. The switch with the higher priority becomes commander. If both switches have the same priority, the switch with the lower MAC address becomes commander.

The VSF fabric forms and begins using the commander configuration. The standby member reboots. When it comes back up, its interfaces are added to the fabric. You would need to configure those interfaces just as with the plug-and-play method.

Note that an election also occurs if both members boot at about the same time. For this reason, even if you use the plug-and-play configuration method, you might want to assign the switch that you want to be commander a higher priority than the other switch. Then, if both switches happen to reboot at the same time, the desired switch remains commander.

It is important for you to understand that an election for the master *only* occurs when two switches that are configured for VSF connect when both are fully booted or are booting at the same time. If a switch that is configured for VSF detects VSF messages on its VSF interfaces while it is booting, it checks the domain. As long as the domain matches its domain, it acts much like a switch at factory defaults. It defers to the current commander on the other side of the link *even if the current commander has a lower priority*, and the booting switch joins the VSF fabric as a standby member.

Finally, note that you still provision the standby member settings on the commander when you are performing a manual VSF configuration. You should always provision only the switch that you intend to assign the higher priority and make commander. That is, you specify the type and optionally MAC address of the future standby member on the future commander.

Although you can technically provision the other member on the switch that will become the standby member, these settings will have no effect. When the standby member merges with the commander's fabric, it takes the commander's settings in preference to its own. Even if you used strict provisioning on the future standby member, it does *not* check the commander MAC address against the provisioned MAC address.

VSF fabric operation within the network

You now know how to set up the VSF fabric. Next, you'll explore connecting the fabric to other devices. You'll also examine how the fabric forwards traffic and fails over in various situations.

Connecting the VSF fabric to other devices

You've established the VSF fabric, and you're now ready to connect it to other devices. Physically, you can create a highly redundant design between the VSF fabric at the core or aggregation layer and the connecting tiers. The figure shows a two-tier topology with an access layer and a VSF fabric at the core. The tiers connect with a highly resilient design in which each access layer switch has one or more links to each core switch. However, because the VSF fabric operates as a single logical switch, you can combine all of those links in a single link aggregation.

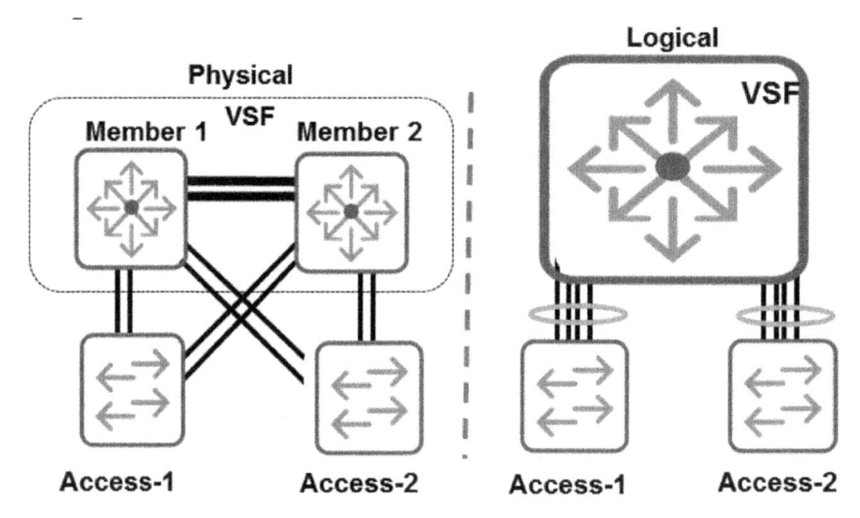

Figure 9-12: Connecting the VSF fabric to other devices

Similarly, a VSF fabric can provide resilient connectivity for servers that support link aggregation (sometimes called NIC teaming or NIC bonding). The server and VSF fabric treat the link like one logical link, but the server has the high availability that comes from connecting to multiple switches. The VSF fabric provides flexibility for the links, accommodating any of the methods for using multiple NICs used by the server OS. These methods include LACP and manual (or static) link aggregation, in which case the VSF fabric also establishes a link aggregation on its side. (Generally, you should match the type, but ArubaOS-Switch software allows an LACP link aggregation that receives no LACP messages—as opposed to inconsistent messages—to come up.)

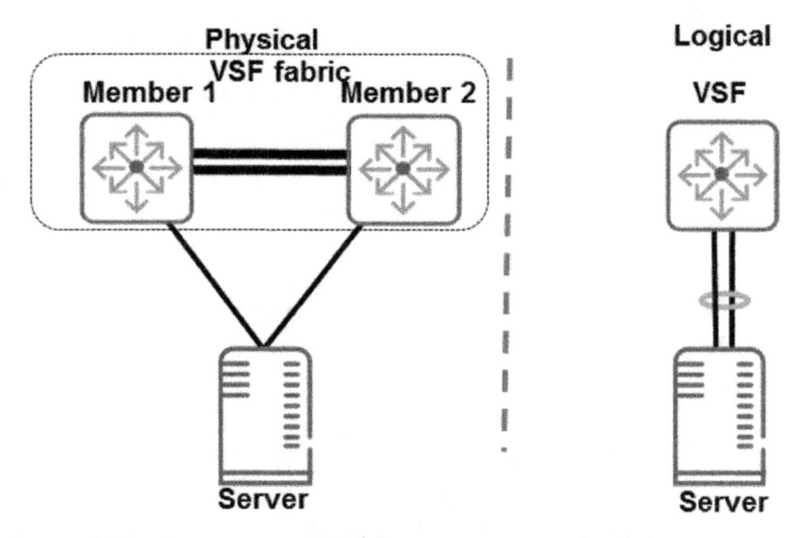

Figure 9-13: Connecting a VSF fabric to a server with a link aggregation

A server might also use a form of NIC bonding that doesn't require awareness on the switch side. For example, a VMware ESXi virtual switch can use source port forwarding for the NIC bonding mode. In this case, the ESXi host virtual switch assigns different VMs to different links. From the switch point of view, each interface connected to the bonded NICs receives traffic from different MAC addresses, just as if the interfaces were connected to different physical servers. Therefore, the VSF fabric should treat the two links as if they are connected to different devices, and you do *not* configure link aggregation.

Configure and Provision a VSF Fabric Manually

You will now configure a link aggregation between the VSF fabric and Access-1, as well as an aggregation between the fabric and Access-2.

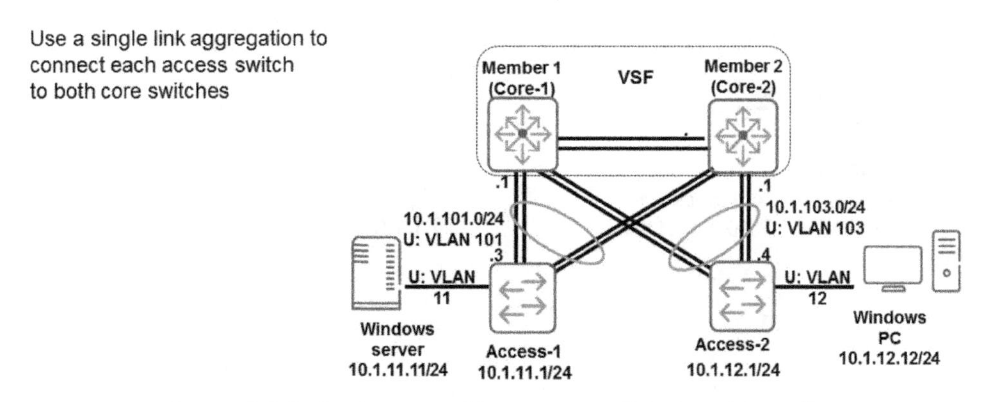

Figure 9-14: Configure and Provision a VSF Fabric Manually

This configuration builds on the tasks you completed in the previous section.

After completing the tasks in this section, you will be able to configure distributed link aggregations on a VSF fabric.

The figures show the final physical and logical topology for your test environment.

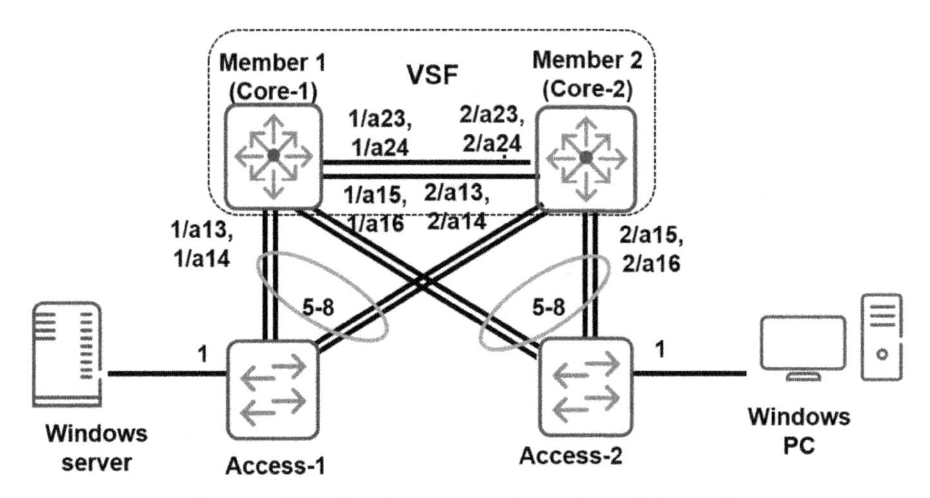

Figure 9-15: Final test network topology (physical)

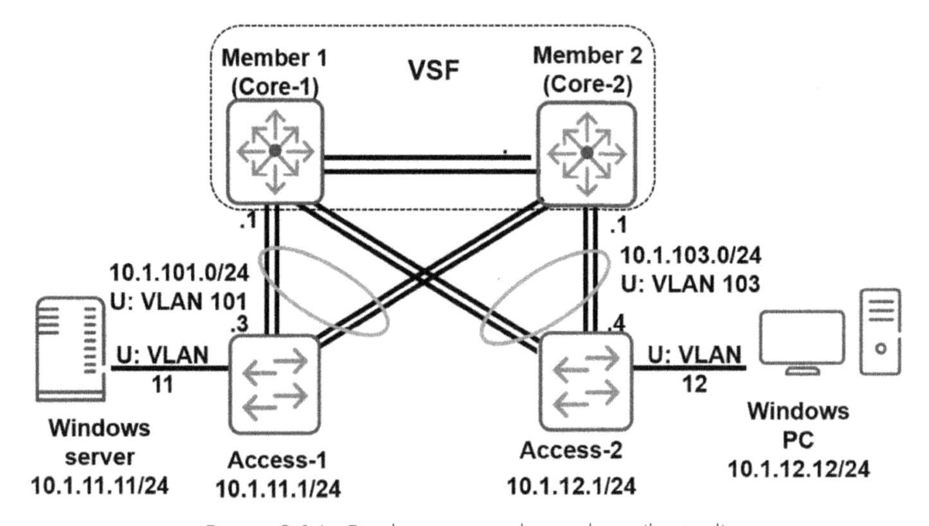

Figure 9-16: Final test network topology (logical)

Task 1: Configure a VSF priority on member 1

Core-1 (member 1) became commander of the VSF fabric because you joined Core-2 (member-2) using the plug and play method. For this scenario, you want Core-1 to remain commander in situations in which an election occurs (such as if both switches boot at the same time). Therefore, you will now configure a higher VSF priority on member 1.

Core-1

1. Access the terminal session for the VSF fabric. You can use the Core-1 or Core-2 console.
2. Move to global configuration mode.
3. Configure a priority of 255 on member 1.

```
Core-1(config)# vsf member 1 priority 255
```

Task 2: Connect the VSF fabric to the access layer switches

You are now ready to connect the VSF fabric to the access layer switches. You will combine all four links between Access-1 and both core switches into a single link aggregation. You will do the same for the four links between Access-2 and the core.

The figures show the current topology and the topology after you have completed this task.

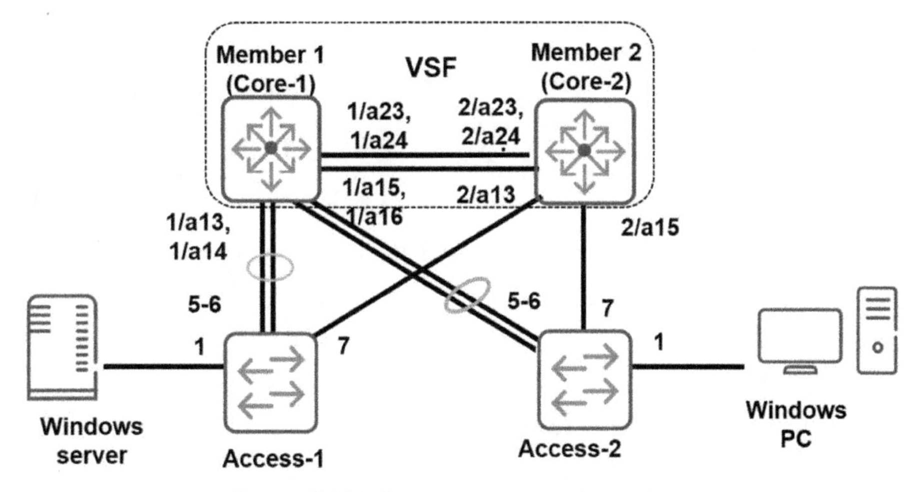

Figure 9-17: Current test network topology

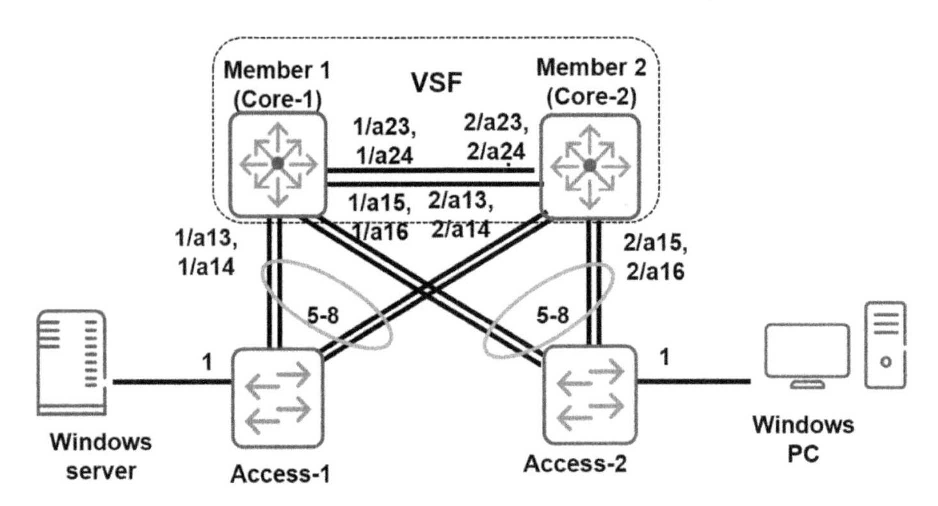

Figure 9-18: Task 2: Connect the VSF fabric to the access layer switches

1. The Core-1 VSF fabric already has the link aggregations that you configured in previous chapters. Review these now.

```
Core-1(config)# show trunks

 Load Balancing Method: L3-based (default)

  Port    | Name                              Type        | Group  Type
  ------ + ------------------------------ --------- + ------ --------
  1/A13  | Access-1                          100/1000T  | Trk2   LACP
  1/A14  |                                   100/1000T  | Trk2   LACP
  1/A15  | Access-2                          100/1000T  | Trk3   LACP
  1/A16  |                                   100/1000T  | Trk3   LACP
```

2. As you see, Trk2 to Access-1 and Trk3 to Access-2 are using only interfaces on member 1.
3. You will now add interfaces on member 2 (Core-2) to the link aggregations. Begin with the link aggregation to Access-1. You do not have to specify the LACP type because the link aggregation already uses that type.

```
Core-1(config)# trunk 2/a13,2/a14 trk2
```

4. Now add the interfaces to the link aggregation to Access-2.

```
Core-1(config)# trunk 2/a15,2/a16 trk3
```

5. Enable the interfaces in the link aggregations.

```
Core-1(config)# interface 1/a13-1/a16,2/a13-2/a16 enable
```

Important

You are enabling the interfaces now so that you can explore LACP behavior. In the real world, follow the best practice of establishing the links **after** you have set up the link aggregation on **both** sides. Otherwise, you can create loops.

6. View the LACP status on the link aggregations.

```
Core-1(config)# show lacp

                                LACP

        LACP    Trunk   Port            LACP     Admin  Oper
 Port   Enabled Group   Status  Partner Status   Key    Key
 -----  ------- ------- ------- ------- ------   -----  ------
 1/A13  Active  Trk2    Up      Yes     Success  0      963
 1/A14  Active  Trk2    Up      Yes     Success  0      963
 1/A15  Active  Trk3    Up      Yes     Success  0      964
 1/A16  Active  Trk3    Up      Yes     Success  0      964
 2/A13  Active  Trk2    Blocked No      Failure  0      963
 2/A14  Active  Trk2    Down    No      Success  0      963
 2/A15  Active  Trk3    Blocked No      Failure  0      964
 2/A16  Active  Trk3    Down    No      Success  0      964
```

7. As you see, one member 2 interface in each link aggregation is blocked, and one interface is down. Think about the current topology and what you know about LACP. Be ready to explain the behavior in the learning check.

Access-1

Now set up the link aggregation on Access-1.

8. Access the terminal session with Access-1 and move to the global configuration context.

9. View link aggregations and verify that trk1 is already operating in LACP mode and includes the two interfaces that connect to member 1.

```
Access-1(config)# show trunk

Load Balancing Method:  L3-based (default)
```

```
Port     | Name                           Type       | Group Type
------   + ------------------------------ ---------- + ---------
5        | Core-1                         100/1000T  | Trk1  LACP
6        |                                100/1000T  | Trk1  LACP
```

10. Add the interfaces that connect to member 2.

```
Access-1(config)# trunk 7,8 trk1
```

11. One of the interfaces that connects to member 2 is already enabled. Enable the other interface.

```
Access-1(config)# interface 8 enable
```

12. VLAN 104 used to be assigned to interface 7. Now interface 7 is part of the link aggregation assigned VLAN 101. Clean up the configuration by removing VLAN 104 and the default route through this interface.

```
Access-1(config)# no vlan 104
Access-1(config)# no ip route 0.0.0.0/0 10.1.104.2
```

13. If necessary in your test network environment, establish the physical connection between Access-1 and the second member 2 interface.

14. Verify that the link aggregation has successfully brought up all links.

```
Access-1(config)# show lacp

                                  LACP

       LACP     Trunk   Port              LACP     Admin  Oper
Port   Enabled  Group   Status  Partner   Status   Key    Key
-----  -------  ------- ------- -------   ------   -----  ------
5      Active   Trk1    Up      Yes       Success  0      562
6      Active   Trk1    Up      Yes       Success  0      562
7      Active   Trk1    Up      Yes       Success  0      562
8      Active   Trk1    Up      Yes       Success  0      562
```

15. View information about the LACP peer. Think about how the VSF fabric appears to Access-2—check the system ID.

```
Access-1(config)# show lacp peer
LACP Peer Information.
```

```
System ID: 6c3be5-6208c0

  Local  Local                      Port               Oper     LACP      Tx
  Port   Trunk  System ID           Port  Priority   Key      Mode      Timer
  ------ ------ --------------     ----- ---------  -------  --------  -----
  5      Trk1   1c98ec-ab5b3f      13    0          963      Active    Slow
  6      Trk1   1c98ec-ab5b3f      14    0          963      Active    Slow
  7      Trk1   1c98ec-ab5b3f      205   0          963      Active    Slow
  8      Trk1   1c98ec-ab5b3f      206   0          963      Active    Slow
```

Access-2

16. Access the terminal session with Access-2 and move to global configuration mode.
17. Repeat the same steps as on Access-1 to add the interfaces that connect to VSF member 2 to trk1. If you need help, the commands are on the next page.

```
Access-2(config)# trunk 7,8 trk1
Access-2(config)# interface 8 enable
```

18. Clean up the configuration by removing VLAN 102.

```
Access-1(config)# no vlan 102
```

19. If necessary in your test network environment, establish the physical connection between Access-2 and the second member 2 interface.
20. Verify that the links have been successfully added.

```
Access-2(config)# show lacp

                                  LACP

         LACP      Trunk   Port               LACP      Admin   Oper
   Port  Enabled   Group   Status  Partner    Status    Key     Key
   ----- -------   ------- ------- -------    -------   ------  ------
   5     Active    Trk1    Up      Yes        Success   0       562
   6     Active    Trk1    Up      Yes        Success   0       562
   7     Active    Trk1    Up      Yes        Success   0       562
   8     Active    Trk1    Up      Yes        Success   0       562
```

Task 3: Verify connectivity and view VSF ports

You will now test connectivity in your environment.

Access-1

1. Access the terminal session with Access-1.
2. View routes on Access-1. You should see *one* route to VLAN 12.

```
Access-1(config)# show ip route

                        IP Route Entries

  Destination     Gateway     VLAN  Type         Sub-Type   Metric Dist.
  ------------- ----------- ----  ---------    ---------  ------  -----
  0.0.0.0/0       10.1.101.1  101   static                  1       1
  10.1.11.0/24    VLAN11      11    connected               1       0
  10.1.12.0/24    10.1.101.1  101   ospf         IntraArea  3       110
  10.1.101.0/24   VLAN101     101   connected               1       0
  10.1.103.0/24   10.1.101.1  101   ospf         IntraArea  2       110
  127.0.0.0/8     reject            static                  0       0
  127.0.0.1/32    lo0               connected               1       0
```

Windows server

3. Access the Windows server and open a command prompt.
4. Ping the Windows PC (10.1.12.12) and verify success.

```
ping 10.1.12.12
```

Note

If the ping fails, check that the PC is set up to receive a DHCP address. Use the **ipconfig/release** and **ipconfig/renew** commands to renew the address.

Also, the PC will typically receive address 10.1.12.12, but might sometimes receive a different IP address. Use the **ipconfig** command in the PC command prompt to check. If the client has received a different address in the 10.1.12.0/24 subnet, use that address in the **ping** and **tracert** commands.

5. Trace the route.

```
tracert 10.1.12.12

Tracing route to pc.hpnu01.edu [10.1.12.12] over a maximum of 30 hops

  1 <1 ms <1 ms  <1 ms  vlan11_defaultrouter.hpnu01.edu [10.1.11.1]

  2  1 ms  1 ms   1 ms  core-1_access-1_link.hpnu01.edu [10.1.101.1]

  3 <1 ms <1 ms  <1 ms  access-2_core-1_link.hpnu01.edu [10.1.103.4]

  4 <1 ms <1 ms  <1 ms  pc.hpnu01.edu [10.1.12.12]

Trace complete.
```

Note

The traceroute will include the intuitive hostnames for the devices if your testing environment is set up with DNS services, which map the IP addresses to the hostnames. You can use the -d option (traceroute -d 10.1.12.12) to run the traceroute without looking up hostnames if these aren't available in your test environment.

Core-1

6. Return to the terminal session with Core-1. View a member 1 port in the VSF link. As you can see, the interface is labeled as a VSF port. The interface is transmitting and receiving some traffic to maintain the VSF fabric.

```
Core-1(config)# show interface 1/a24

 Status and Counters - Port Counters for VSF port 1/A24

  Name  :

  MAC Address       : 1c98ec-ab5be9

  Link Status       : Up

  Totals (Since boot or last clear) :

   Bytes Rx         : 108,336,080          Bytes Tx         : 108,639,600

   Packets Rx       : 189,261,432          Packets Tx       : 135,879,247
```

```
Errors (Since boot or last clear) :
 FCS Rx          : 0                 Drops Tx        : 0
 Alignment Rx    : 0                 Collisions Tx   : 0
 Runts Rx        : 0                 Late Colln Tx   : 0
 Giants Rx       : 0                 Excessive Colln : 0
 Total Rx Errors : 0                 Deferred Tx     : 0

Others (Since boot or last clear) :
 Discard Rx      : 0                 Out Queue Len   : 0
 Unknown Protos  : 0
Rates (5 minute weighted average) :
 Total Rx(Kbps) : 124,656            Total Tx(Kbps) : 124,656
 Packets Rx (Pkts/sec) : 112,514     Packets Tx (Pkts/sec) : 112,515
 Utilization Rx  : 01.24 %           Utilization Tx  : 01.24 %
```

Task 4: Save

Save the configuration on the VSF fabric and both access switches using **write memory**.

Learning check

You will now answer several questions to help solidify the concepts that you have learned.

1. Why didn't member 2 (Core-2) become commander after you raised its priority?

2. After you added the member 2 interfaces to the link aggregations (trunks) to Access-1 and Access-2, one link in each aggregation was blocked and one was down. Why did this happen?

3. After you finished configuring the link aggregation on both sides, describe the link aggregation to the VSF fabric from the point of view of Access-2.

4. Describe what you observed when you traced a route from the server to the client. How many hops is the VSF fabric?

5. How many default routes is Access-1 using? Does it still have redundancy in case Core-1 (the physical switch) fails?

6. How did you observe traffic on the VSF link? What did you observe?

Answers to Learning check

1. Why didn't member 2 (Core-2) become commander after you raised its priority?

   The VSF fabric already had an active commander. In this case, an election isn't held, so member 1 (Core-1) continues to act as commander even though Core-2 has a higher priority now. If both members rebooted at the same time, though, Core-2 would become the commander.

2. After you added the member 2 interfaces to the link aggregations (trunks) to Access-1 and Access-2, one link in each aggregation was blocked and one was down. Why did this happen?

   Each access switch has an interface connected to Core-2 as an individual switch. This link is not yet part of the link aggregation, so it isn't sending the proper LACP messages. LACP on the Core-1 VSF fabric has blocked it. The other interface is down because you haven't yet enabled it on the access switch side.

3. After you finished configuring the link aggregation on both sides, describe the link aggregation to the VSF fabric from the point of view of Access-2.

   To Access-2, the link aggregation appears exactly like a link aggregation to a single device. LACP detects that all the links connect to the same partner whether a link is physically connected to Core-1 or Core-2. Core-1 and Core-2 are a single logical device running just one LACP protocol.

4. Describe what you observed when you traced a route from the server to the client. How many hops is the VSF fabric?

   The packet was routed to Access-1, to the VSF fabric, to Access-2, to the client. The VSF fabric is one hop.

5. How many default routes is Access-1 using? Does it still have redundancy in case Core-1 (the physical switch) fails?

   Access-1 is using just one default route, through 10.1.101.1 (the VSF fabric's IP address on the VLAN between Access-1 and the fabric). Similarly, Access-2 is using just one default route.

   But Access-1 still has redundancy in case member 1 (Core-1) fails. Member 2 (Core-2) would take over as the commander of the VSF fabric and continue routing traffic. You should enable OSPF graceful restart (non-stop OSPF routing, though) to ensure seamless failover.

6. How did you observe traffic on the VSF link? What did you observe?

   You can use the same show commands that you use on other interfaces. You should have seen some traffic crossing the link. This was mainly traffic for maintaining the fabric and synchronizing the control plane. As you will learn over the next pages, the amount of traffic that the VSF link must carry depends on the topology and other factors. (Using link aggregations to connect the VSF fabric to other devices and balancing links over both members tends to decrease the amount of traffic on the VSF link.)

Tracing Layer 2 traffic: Broadcasts, multicasts, and unknown traffic

You should understand how VSF fabrics handle traffic forwarding. Although the commander runs protocols and builds forwarding tables, all traffic does *not* need to pass through it. Instead traffic is forwarded much more efficiently.

VSF uses a distributed forwarding and routing model. The members in the VSF virtual device synchronize forwarding and routing tables, but each member forwards frames or routes packets independently.

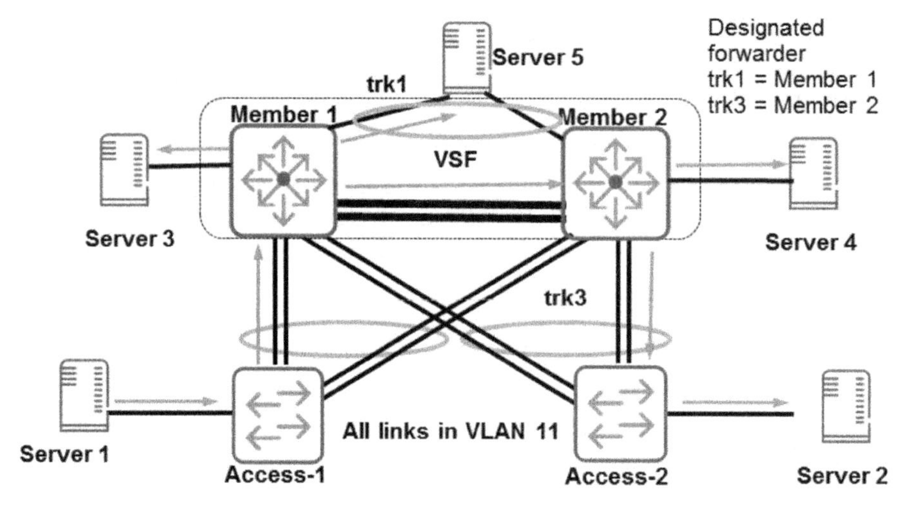

Figure 9-19: Tracing Layer 2 traffic: Broadcasts, multicasts, and unknown traffic

Over the next several pages, you will look at the traffic forwarding in detail, beginning with flooding of Layer 2 traffic. When the VSF fabric needs to flood a broadcast, a unicast with an unknown destination MAC address, or an unknown multicast (sometimes called BUM traffic), it floods the frame on all interfaces on both members except the interface that received the traffic. If necessary, the frame crosses the VSF link.

Remember that a link aggregation is a logical interface, so the VSF fabric doesn't flood the frame on every link in every link aggregation. Instead, VSF designates one of the members as the forwarder for each link aggregation. In other words, when a member receives a BUM frame, it floods the frame locally on all link aggregations for which it is the designated forwarder, as well as any local interfaces that aren't part of a link aggregation. If the frame needs to be flooded on any link aggregations for which the other member is the designated forwarder, the member sends the BUM frame over the VSF link. The other member then forwards it.

You cannot control which member is the designated forwarder for each link aggregation. VSF automatically assigns this role. The more link aggregations the fabric has, the more load balanced the role will be. You can view the designated forwarders using the **show vsf trunk-designated-forwarder** command.

Note that you can connect servers like Server 3 and Server 4 to both members of the VSF fabric with link aggregations (like Server 5) and should do so for a higher availability design. The figure shows a design with one link on these servers so that you can see the VSF member behavior when a member has the sole connection to a device.

Tracing traffic flow: Layer 2 and Layer 3 unicasts

The VSF fabric, like a typical Ethernet switch, uses the source MAC addresses in traffic to create a MAC forwarding table so that it can stop flooding unicasts. Both the commander and standby member can learn MAC addresses, which they share to minimize traffic flooding.

When a member device receives a frame to be forwarded at Layer 2, it finds the outbound interface of the frame by searching its Layer 2 forwarding table. It then forwards the packet out the outbound interface, which might be local or on the other member. If the outbound interface is on the other member, the source member forwards the packet on the VSF link, first encapsulating the packet with a VSF header that indicates how the packet should be forwarded.

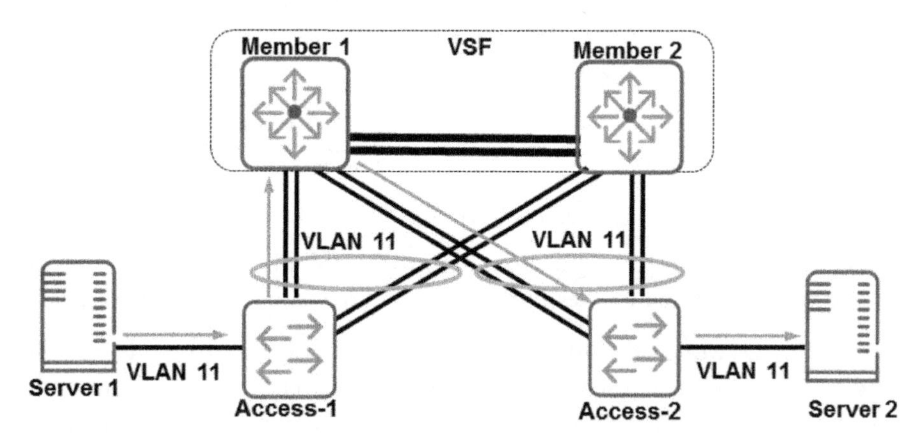

Figure 9-20: Tracing traffic flow: Layer 2 and Layer 3 unicasts

A VSF fabric, just like any switch implementing link aggregation, learns MAC addresses on a link aggregation as a whole, and it selects one link in the aggregation for forwarding each conversation. However, instead of using a hash to choose a link from *any* of the links, the VSF member prefers its own local links. (If the member has multiple local links in the aggregation, then it uses the typical hashing mechanism to choose between those.) In a design like this one—in which both VSF members have local links in all link aggregations—members can forward traffic very efficiently with a minimum of traffic that must travel over VSF links.

For example, consider a topology like the one in your test network but using an IP topology in which VLANs extended across the network. Assume that a server connected to Access-1 attempts to reach a server connected to Access-2. For this example, the network has been active for some time, and the switches have learned the proper forwarding ports for both devices' MAC addresses.

1. The server sends the traffic on its link to Access-1.
2. Access-1 knows that the forwarding port for server 2 is trk1. It chooses one of its four links for the traffic; for the sake of the example, it chooses one of the links to Core-1.
3. The VSF fabric has learned trk3 as the forwarding port for server 2. So Core-1 chooses one of its local links in this aggregation and forwards the traffic to Access-2.
4. Access-2 forwards the traffic to server 2.

A similar traffic flow holds for unicasts that need to be routed at Layer 3, as illustrated in the figure below.

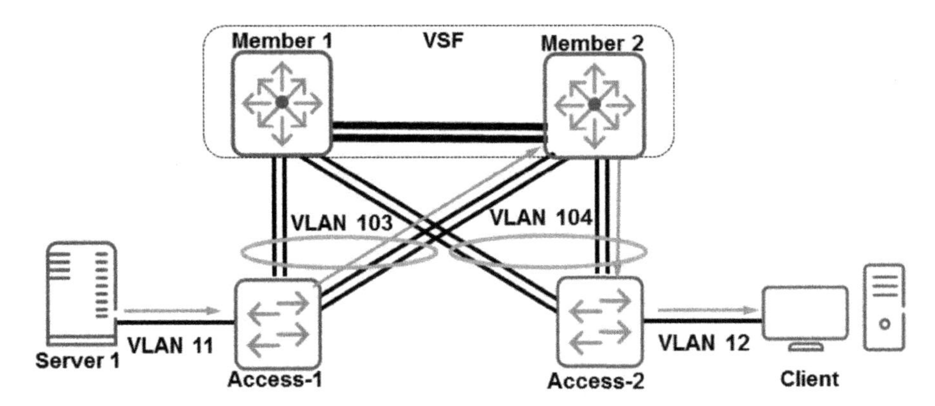

Figure 9-21: Tracing traffic flow: Layer 2 and Layer 3 unicasts

The commander builds IP routing tables. It programs the tables into the forwarding information bases (FIBs) on its own interface modules and on the standby member interface modules.

When either member receives packets to route at Layer 3, it scans this FIB to identify the next hop and the forwarding egress port. It assembles the correct new Layer 2 header, as routers do, and forwards the packet out the correct egress port, which could be local or on another member. Just as with Layer 2 forwarding, a packet that needs to egress on another member travels across the VSF link.

However, often the packet is being forwarded on a link aggregation, so the member simply uses a local link in that aggregation.

In this way, all members of the device can actively route traffic. However, to the outside network, the packets have traveled just one hop—as if one device routed the packets.

Now consider a topology exactly like the one in your test environment. The server attempts to send traffic to the client.

1. The server sends the packet to Access-1, its default router.
2. Access-1 routes the traffic to 10.1.101.1 in VLAN 101. This VLAN is on the link aggregation to the VSF fabric, so Access-1 chooses one of the links in this aggregation for this traffic. For this example, Access-1 selects a link to physical Core-2 switch.
3. The VSF fabric has a route to 10.1.12.0/24 through 10.1.104.4 in VLAN 104. Core-2 routes the traffic to VLAN 104 and selects trk3, which is the only interface in VLAN 104, as the forwarding interface. Core-2 can determine the correct next hop and forwarding interface even though it is not the commander because the commander has programmed it to do so. Core-2 selects one of its local links in trk3 for forwarding the traffic.
4. Access-2 receives the packet and routes it to the client.

You've examined the traffic flow over link aggregations. If a member receives a unicast that needs to be forwarded on a non-aggregated interface on the other member, of course, the member forwards the traffic on the VSF link.

VSF failover

VSF provides very fast failover for a number of situations. If a single link in a link aggregation fails, traffic fails over to other links in the same aggregation. If an entire device fails, traffic in link aggregations fails over to the links on the other member.

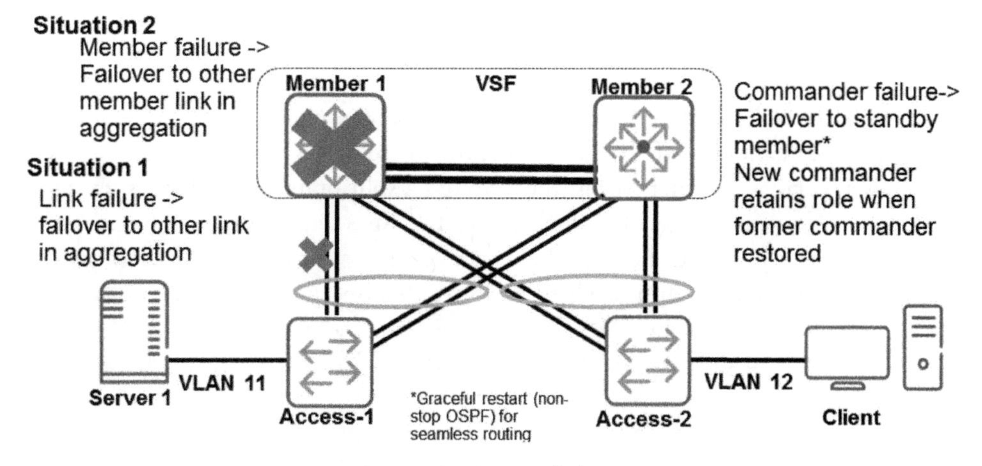

Figure 9-22: VSF failover

The failed device could be the commander, in which case the standby member takes over as the new commander using the control plane information that it has been mirroring. The new commander can seamlessly continue forwarding and routing traffic. If it is running a routing protocol such as OSPF, however, the new commander needs to re-establish adjacencies and rebuild routing tables. Graceful restart, configured by enabling non-stop OSPF routing on the VSF fabric, enables the new commander to complete these functions without disrupting connectivity. For graceful restart to work, OSPF neighbors must act as graceful restart helpers (which Aruba switches do by default).

When a former commander is restored, the new commander continues acting as commander. VSF always favors the current commander. If you want the former commander to take over as commander again, you should reboot the current commander.

VSF link failure without MAD

As you've learned, you should attempt to create a resilient VSF link with multiple physical links that are connected on interfaces on different modules. However, some situations can still cause all of the physical links in the VSF link to fail, leading to a split fabric.

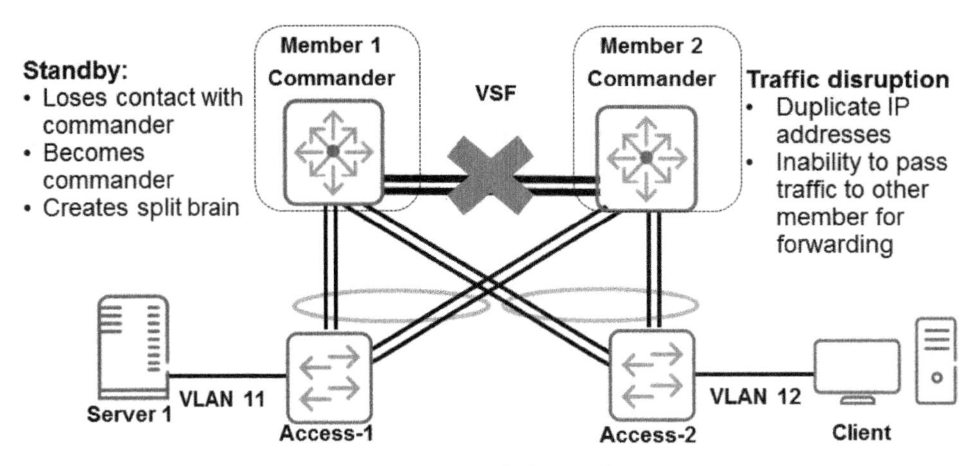

Figure 9-23: VSF link failure without MAD

If a split occurs, the standby member loses contact with the commander. From the standby member point of view, the commander has failed, so the standby member becomes commander as well. Now multiple devices are active; in other words, the VSF has a split brain. Both devices use the same IP addressing and run the same protocols, which can cause issues with routing, for example. In addition, connectivity issues can occur because the devices cannot send data traffic between each other. If traffic arrives on one member and needs to be forwarded on another member, the traffic is now dropped.

MAD

The final section of this chapter explains how to enable Multi-Active Detection (MAD) on the VSF fabric to prevent undesirable behavior if the VSF link fails.

MAD to protect against split brain

When MAD is operating, the VSF link can fail without disrupting traffic. The standby still loses contact with the commander on the VSF link. However, the standby member can use MAD to detect that the commander is still up (you'll see exactly how in just a moment). The standby member then becomes an inactive fragment and shuts down all of its interfaces. The commander remains up and an active fragment. The link aggregations all fail over gracefully to the links connected to the active fragment.

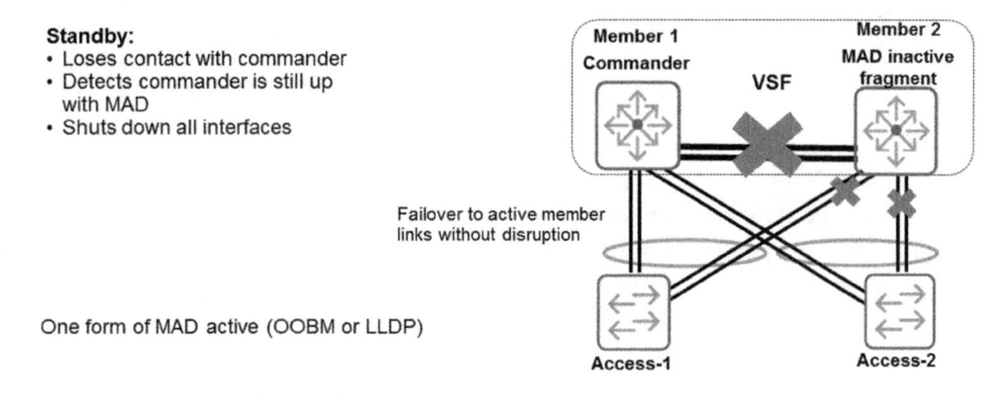

Figure 9-24: MAD to protect against split brain

When the VSF link is restored, the standby member rejoins the fabric. Until that point, you have limited management control over the standby member.

VSF supports two types of MAD: OOBM and LLDP. You can configure *one* method on the VSF fabric.

OOBM MAD

The OOBM MAD technology uses the commander and standby member's OOBM ports to detect if the VSF link fails. For OOBM MAD to work, the commander and standby member each require their own OOBM IP address, which is different from the global OOBM IP address for the VSF fabric as a whole. Management stations contact the VSF fabric OOBM IP address to manage the switch. The individual member OOBM addresses, on the other hand, are dedicated to MAD.

The OOBM ports can connect to a switch that connects to other devices such as management stations. Then you can use the OOBM ports for connecting the switches to an out-of-band management network, as well as for MAD. However, if you choose you can alternatively connect the OOBM ports directly together. In this latter case, the OOBM ports are dedicated to MAD (you cannot also manage the VSF fabric through them). After you have determined that the ports are set up correctly and the switches have IP addresses, you can enable OOBM MAD.

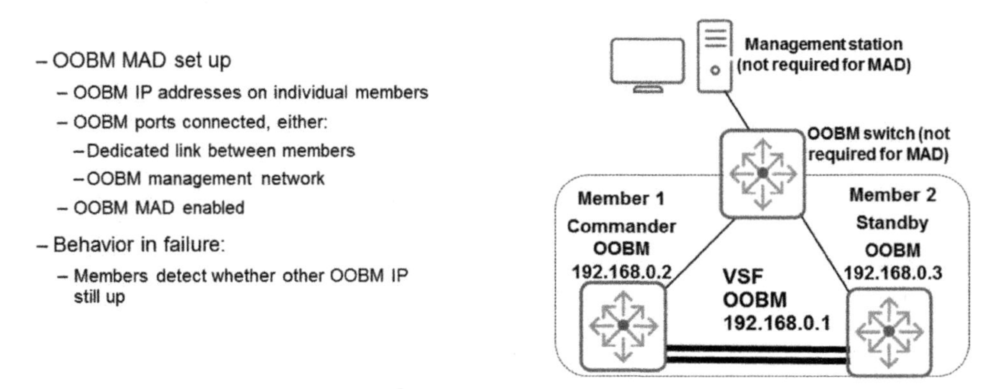

Figure 9-25: OOBM MAD

If a VSF link fails, the standby member can now attempt to contact the commander OOBM IP address. If successful, the standby member knows that the VSF link failed but that the commander is still up. The standby member can then become an inactive fragment and shut down all of its interfaces. (After the VSF link is restored, the member will automatically rejoin the fabric.

LLDP MAD setup

As an alternative to OOBM MAD, you can enable LLDP MAD. This form of MAD leverages an existing connection for detecting the other member's status if the VSF link fails. The connection must meet these criteria:

- It is a link aggregation that uses LACP.
- The link aggregation includes links on both members of the VSF fabric.
- LLDP is enabled on all of the links.

The device on the other side of the link is the MAD assistant. In addition to supporting the LACP link aggregation, the assistant must meet these requirements:

- It must have an IP address that is reachable by the VSF fabric.
- It must have an SNMPv2c community with read access (either read-only or read-write). SNMPv2c must also be enabled.

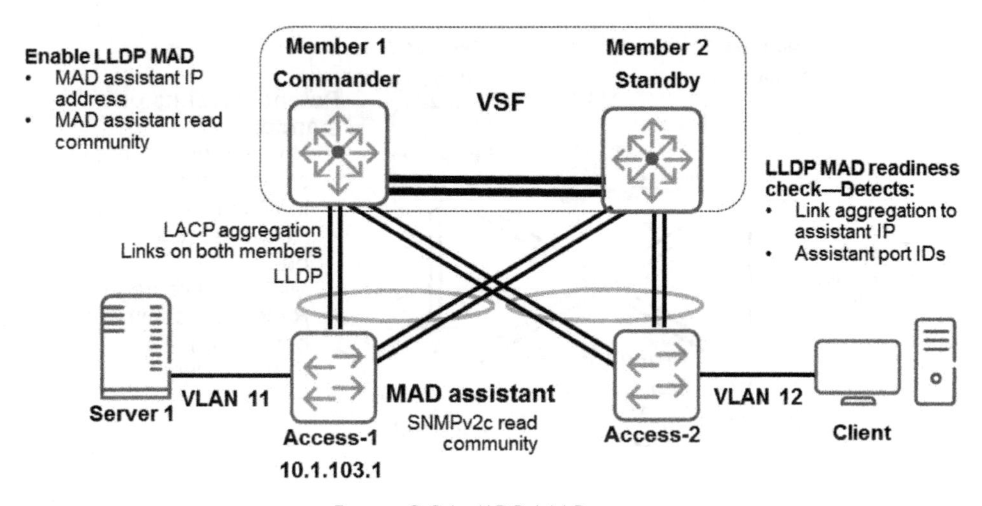

Figure 9-26: LLDP MAD setup

After you check these criteria, you can enable LLDP MAD, specifying the MAD assistant IP address and the SNMPv2 read community.

The VSF fabric then verifies that LLDP MAD is ready to operate. It detects the link aggregation on which it reaches the IP address specified for the MAD assistant and verifies that it has at least one link on each member in this aggregation. The VSF fabric then uses LACP to determine the port IDs that the assistant assigns to its side of the links and LLDP to verify the information. The fabric needs to know the port IDs, as the ports are identified in SNMP so that it can query their status about them in a VSF link failure situation.

When the fabric has collected all of this information, LLDP MAD is active and ready to detect a VSF link failure. You'll learn later how to verify that the readiness check proceeded correctly.

LLDP MAD behavior

If the VSF link fails, the standby member sends a probe to the MAD assistant (an SNMP GET message). This probe queries the assistant for the status of all of its ports in the link aggregation using the port IDs detected earlier. The assistant selects an active link in the link aggregation and sends a probe response.

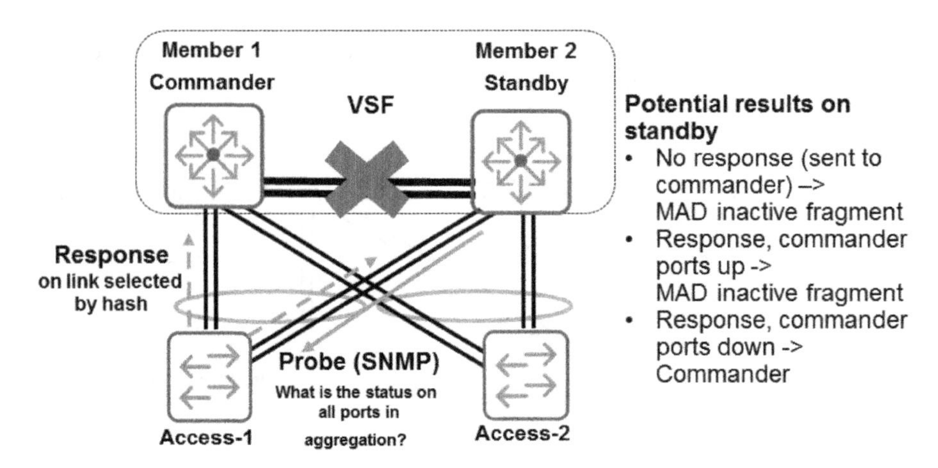

Figure 9-27: LLDP MAD behavior

One of three results occurs on the standby member:

- The standby member does not receive a probe response because the commander is still up, and the assistant sent the response on one of the links to the commander. When the standby member does not receive a timely response, it knows that the commander is up, and so it becomes a MAD inactive fragment and shuts down all interfaces.
- The standby member receives a probe response, and the response indicates that the links to the commander are still up. Again, the standby member becomes a MAD inactive fragment and shuts down all of its interfaces.
- The standby member receives a probe response, and the response indicates that the links to the commander are down. In this case, the standby member determines that the VSF link is down because the commander is down. The standby member becomes the new commander.

In all of these cases, MAD enables the active VSF fragment to continue forwarding traffic while the inactive one doesn't interfere.

Maintain the VSF Fabric and Set Up LLDP MAD

Next, you will explore several VSF maintenance tasks. You will then disconnect the VSF fabric.

– Explore maintenance tasks.
– Set up and test LLDP MAD.

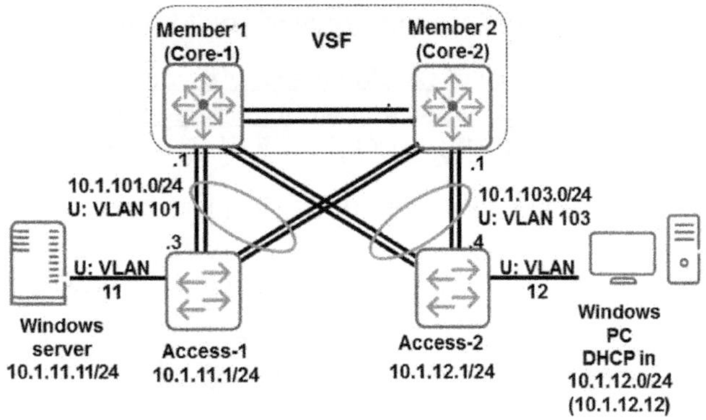

Figure 9-28: Maintain the VSF Fabric and Set Up LLDP MAD

In the real world, you would continue to use the VSF fabric to provide redundancy at the core as you add wireless and management services. However, you want to make sure that both Core-1 and Core-2 are available as separate switches for other students sharing your equipment.

While disconnecting the fabric, you will also take the opportunity to review rules and procedures for managing VSF links and members.

Learning check

You will now answer several questions to help solidify the concepts that you have learned.

1. Which VSF maintenance tasks can you complete when both members are active?

2. Which VSF maintenance tasks require you to shut down the standby member?

3. A member has failed, and you want to add a new switch in its place. What must you do first?

4. What did you see when you checked the LLDP MAD parameters and status before the VSF link failed and after it failed?

Answers to Learning check

1. Which VSF maintenance tasks can you complete when both members are active?

   You can add an interface to a VSF link. You can remove an interface from a VSF link, as long as the link still has an interface. You can change the domain ID.

2. Which VSF maintenance tasks require you to shut down the standby member?

   You must shut down the standby member if you want to remove a VSF link entirely. You need to remove a VSF link when you want to change from using links of a different speed such as migrating from 10GbE to 40GbE.

   You must also shut down the standby member if you want to remove it from the VSF fabric

3. A member has failed, and you want to add a new switch in its place. What must you do first?

   Remove the failed member ID from the VSF fabric.

4. What did you see when you checked the LLDP MAD parameters and status before the VSF link failed and after it failed?

   Checking the parameters indicates the MAD readiness state. You can see the link aggregation through which the switch reaches the MAD assistant and the probe set, which are the interfaces in that link aggregation. The LAG connectivity is full if the VSF fabric has interfaces on both members in this LAG.

   The status shows whether the VSF fabric is split and whether this fragment is active or inactive. The fragment is always active if the fabric is not split. In normal operation, you do not see any probe requests. The standby member sends this only when the link fails.

   After the VSF link failed, you saw that the commander was the active fragment and that the standby was the inactive fragment. You saw that the standby member had sent a probe request.

Summary

By enabling two 5400R switches to operate as a single logical switch, VSF enhances resiliency at the server access, distribution, or core layers. At the same time, VSF delivers the enhanced resiliency in a simpler way. Simple link aggregations replace complex spanning tree or routing topologies.

- VSF use cases and benefits
- VSF plug-and-play configuration, manual configuration, and provisioning
- VSF traffic forwarding and failover

Figure 9-29: Summary

You learned several strategies for establishing a VSF fabric so that you can choose the method that works best for you—whether the simple plug-and-play setup or a manual configuration, both with or without provisioning. You also understand when the VSF priority controls which switch becomes controller and when the existing controller maintains the role regardless of priority.

You also examined the components and operation of a VSF fabric. The controller provides the active management and control planes, but both members contribute forwarding planes and can actively forward traffic on the most efficient path. If a link or even member fails, VSF permits seamless failover helping customers maintain business continuity for network services.

Learning check

1. A standby member of a VSF fabric receives a unicast frame. The MAC forwarding table indicates trk1 as the forwarding interface for the destination MAC address. The commander and standby members both have one interface in trk1. When does the standby member forward the frame on its local interface in trk1?

    a. only when it is the designated forwarder for that link aggregation

    b. only when the hash algorithm selects its local interface

    c. only when the commander's link has failed

    d. always

2. An administrator wants to use the plug-and-play method to establish a VSF fabric. What should the administrator check for the standby member before connecting it to the commander VSF ports?

    a. It has a VSF link configured on the proper ports.

    b. It has a VSF priority lower than the commander's.

    c. It is at factory default settings.

    d. It is configured with the same VSF domain ID as the commander.

Answers to Learning check

1. A standby member of a VSF fabric receives a unicast frame. The MAC forwarding table indicates trk1 as the forwarding interface for the destination MAC address. The commander and standby members both have one interface in trk1. When does the standby member forward the frame on its local interface in trk1?

   a. only when it is the designated forwarder for that link aggregation

   b. only when the hash algorithm selects its local interface

   c. only when the commander's link has failed

   d. always

2. An administrator wants to use the plug-and-play method to establish a VSF fabric. What should the administrator check for the standby member before connecting it to the commander VSF ports?

   a. It has a VSF link configured on the proper ports.

   b. It has a VSF priority lower than the commander's.

   c. It is at factory default settings.

   d. It is configured with the same VSF domain ID as the commander.

10 Wireless for Small-to-Medium Businesses (SMBs)

EXAM OBJECTIVES

✓ Understand the basics of wireless communications and 802.11 standards

✓ Define a WLAN and differentiate between wireless security options

✓ Configure basic settings on Aruba IAPs

✓ Form an Aruba Instant cluster

ASSUMED KNOWLEDGE

Before reading this chapter, you should have a basic understanding of:

- IP addressing
- Virtual LAN (VLAN) assignments

INTRODUCTION

This chapter introduces you to wireless networks, focusing on setting up a wireless network for a small-to-medium business (SMB) using Aruba Instant Access Points APs (IAPs).

Introduction to wireless technologies

While this book is not designed to make you an expert in wireless communications, the first section of this chapter does teach you enough about wireless technologies that you will be able to understand basically how devices use wireless technology to communicate. Specific implementation examples discussed later in this chapter, use Aruba IAP solutions.

Wireless communications with the 802.11 standard

IEEE 802.11 defines the standard for wireless communications. At the physical layer, communications use radio frequency (RF) technology, encoding information through modulations of radio waves. The radio waves propagate through the air allowing any receiver tuned to the proper frequency to receive the signal. Each 802.11 standard uses a specific frequency band or bands. A frequency band is further divided into narrower frequencies called channels. A wireless device can only communicate with other wireless devices tuned to the same channel.

- Ethernet frames encapsulated by an 802.11 header
- Wireless signals transmitted on a channel (a specific frequency either in the 2.4 or 5 GHz band)
- Shared medium (one device transmits while others listen)

Standard	Frequency
802.11b	2.4GHz
801.11g	2.4GHz
801.11a	5GHz
802.11n	2.4GHz or 5GHz
802.11ac	5GHz

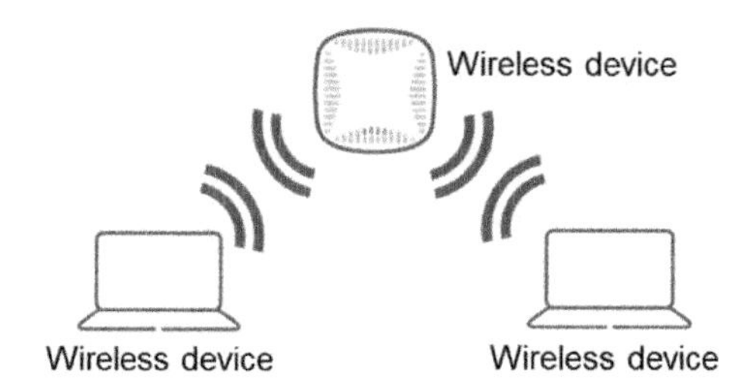

Figure 10-1: Wireless communications with the 802.11 standard

As wireless technologies have evolved, new standards have emerged. 802.11b, standardized in 1999, was the first standard to be widely adopted, and 802.11a and 802.11g were widely used for many years; however, these standards are now largely obsolete. Now most wireless equipment supports 802.11n and 802.11ac.

These standards differ primarily at the physical layer in how they modulate the signal; the latest standards support much higher data rates. The table in Figure 10-1 indicates some of the high level differences. You will look at data rates for each standard and special 802.11n and 802.11ac features in a bit more detail later in this chapter.

At the Data Link layer, all of the 802.11 standards are quite similar. Devices use Ethernet frames for data. They encapsulate these frames with 802.11 headers for delivery in the wireless network. 802.11 also defines various management and control frames for maintaining the network and controlling data transmissions, but these subjects are beyond the scope of this book.

The wireless medium is shared, much like Ethernet was historically shared when devices connected through hubs. Only one device can transmit on a channel at once. Other devices must listen and wait their turn. 802.11 devices use Carrier Sense Multi-Access Collision Avoidance (CSMA/CA) to listen for other devices that are transmitting, to back off if necessary, and to take a turn to transmit themselves.

Infrastructure mode communications

Wireless devices such as laptops and smart phones can communicate directly together in an ad hoc network. This is the type of wireless network that you can create yourself on your laptop.

However, enterprise environments usually feature infrastructure mode networks, which are the focus for this guide. An infrastructure mode network uses the same basic communication rules that you already learned, but in this mode, an access point (AP) establishes the wireless network. It identifies the network with a wireless network name or Service Set Identifier (SSID), which it advertises in periodic beacons. When you see a list of wireless networks on your wireless devices, you are seeing SSIDs beaconed by APs. See Figure 10-2.

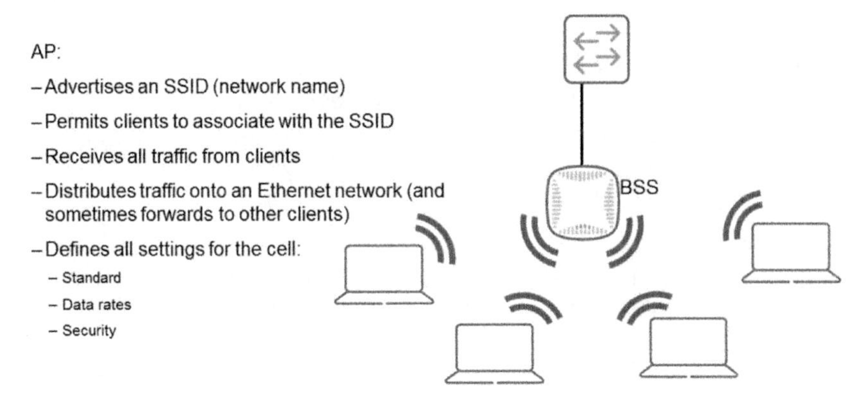

Figure 10-2: Infrastructure mode communications

A wireless station or client (a device such as a laptop, smartphone, tablet, or similar wireless device) joins the wireless network by associating with the SSID on the AP. (The AP can use security to control which clients are allowed to associate; you'll learn about these options later in this chapter.)

Note

APs can hide the SSID rather than beacon it. In this case, the client must know the SSID to associate with the AP. However, most clients can detect hidden SSIDs (from wireless frames sent by other clients already associated with the SSID). Therefore, this option doesn't provide security and isn't often used in modern deployments.

An AP and all of the stations associated with it compose a basic service set (BSS). You probably will not hear the BSS term very often. An AP and its stations are also often called a wireless cell.

Each station sends all of its traffic to the AP, and the AP is responsible for sending the traffic toward its destination. Similarly, the AP receives traffic destined to the station and forwards this traffic on to the station. In addition to connecting to wireless stations, the AP is connected to a wired network. As the interface between the wired and the wireless networks, the AP receives wireless traffic from stations, decapsulates the 802.11 header, and forwards an Ethernet frame on to the wired network. Likewise, the AP receives and forwards traffic from the wired network to the correct wireless stations. In this way, wireless devices can communicate on an Ethernet LAN. Most client traffic is usually destined to corporate resources or the Internet, so the AP bridges wireless client traffic on its uplink, typically to an Ethernet LAN. The AP also forwards traffic from one wireless client to another; you can also filter inter-client communications such as these for higher security, if you desire.

The AP manages the settings for the wireless network, including among others:

- The channel used

  If you have multiple APs in the same general area, each AP should operate on a non-overlapping channel so that the cells' communications do not interfere with each other. For the 2.4 GHz range, non-overlapping channels are at least five channels apart (for example, channel 1 and channel 6 are non-overlapping). Hewlett Packard Enterprise (HPE) training provides more details about channel planning. Aruba APs support Adaptive Radio Management (ARM), which lets the APs automatically adjust their channel and transmit power to avoid interfering with each other.

- The wireless standard or standards that can be used

  A wireless cell uses just one standard, but some standards provide backward compatibility with others. For example, 802.11g can be backward compatible with 802.11b, and 802.11n can be backward compatible with 802.11b, 802.11g, and 802.11a. 802.11ac can be backward compatible with 802.11n and 802.11a.

- The data rates that devices must support to connect to the network (basic set)

- The data rates that devices can use in the network (supported set)

  You'll learn more about data rates on the next pages.

- Security settings, discussed later

The AP advertises many of these settings in a beacon so that stations can agree to them when they associate.

Data rates and throughput

As mentioned in the previous section, an AP defines the data rates that a client must support to associate to the SSID and the data rates that the client can use. The data rate defines the type of modulation applied to the signal and determines the rate of transmission for a frame. Each 802.11 standard specifies multiple data rates.

– Each standard supports multiple data rates:

Standard	Frequency	Maximum data rate (Mbps)
802.11b	2.4GHz	11
801.11g	2.4GHz	54
801.11a	5GHz	54
802.11n	2.4GHz or 5GHz	600
802.11ac	5GHz	Currently 1733 in Aruba AP 330 Series*

*Standard 6933 Mbps, not yet supported by any vendors

– Connection's data rate depends on:
  – Signal level between devices
  – Use of special 802.11n and 802.11ac features
– Throughput depends on many factors such as:
  – Data rate
  – Number of clients in cell
  – Overhead
  – Retransmissions

Figure 10-3: Data rates and throughput

A job aid at the end of this chapter shows you the precise data rates. Highlights include:

- 802.11b, 802.11g, and 802.11a are legacy standards that support data rates up to 11 Mbps (b) or 54 Mbps (g and a).
- 802.11n supports theoretical rates up to 600 Mbps.
- 802.11ac supports theoretical rates up to 6.933 Gbps. However, no equipment supports these rates yet. As of 2016, the Aruba AP 330 Series supports among the highest data rates with a maximum of 1.733 Gbps.

At various times when you've connected to a wireless network, you might have noticed that your connection has different data rates. If a standard supports multiple rates, what makes a client and AP use one rate rather than another?

For 802.11a, 802.11b, and 802.11g, signal strength is the primary factor for the data rate. Different forms of signal modulation can transmit more information within the signal, but the signal must be stronger for these forms of modulation to work.

For 802.11n and 802.11ac, signal strength is the primary factor for determining the modulation and coding scheme (MCS). Different MCSs are associated with different sets of data rates. With these standards, the data rate is also affected by the use of special 802.11n and 802.11ac features, which are described in the next section.

You should understand that the connection's data rate doesn't translate directly to the throughput possible on the connection. Because the wireless medium is shared, each device only receives a fraction of the opportunities to transmit or receive data; these opportunities are sometimes called "airtime." For example, consider a scenario in which 10 devices are sharing the wireless medium, all are using 135 Mbps for their data rate, and all are trying to use the network equally. Each device will receive at most 13.5 Mbps throughput. In fact, due to factors such as control frame overhead and retransmissions, each device will receive less than that. (Multi-user multiple input-multiple output

MIMO [MU-MIMO] makes an exception to the rule that only one device can receive data at a time; you'll learn about MU-MIMO in a moment.)

802.11n and 802.11ac features

802.11n and 802.11ac introduce several features to enhance the data rate, some of the most important of which are illustrated in Figure 10-4. Note that *both* the AP and the client must support a feature for the feature to work.

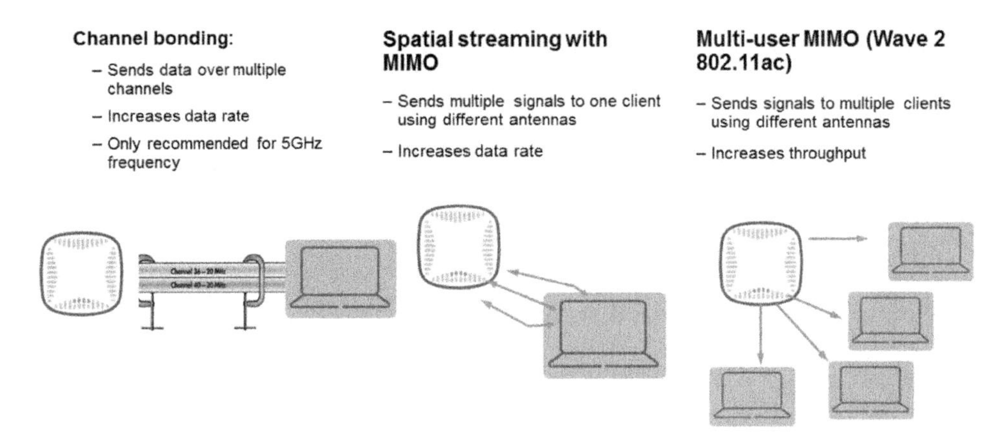

Figure 10-4: 802.11n and 802.11ac features

Channel bonding

Wireless devices send wireless signals over a frequency of a certain width, called a channel. A normal 802.11 channel has a width of 20 MHz. Channel bonding allows the wireless device to send signals over a wider frequency, leading to higher bandwidth for the communications. 802.11n channel bonding permits devices to use two 802.11n channels for their communications, which doubles the data rate (or actually slightly more than doubles the rate because devices can use a small area between channels as well). 802.11ac permits even wider channels (20 MHz, 40 MHz, 80 MHz, and 160 MHz).

Channel bonding increases data rate, but does come at the cost of fewer channels available for neighboring APs. Typically, you should only implement channel bonding on 802.11ac radios and 802.11n radios that use the 5GHz frequency band. Due to the narrow width of the 2.4GHz band, the wide channels of channel bonding can lead to neighboring cells using the same frequencies and interfering with each other.

Spatial streaming

802.11n and 802.11ac use Multi-Input Multi-Output (MIMO) technology. MIMO allows a wireless device to send and receive different data streams over different antennas. Each of these data streams is called a spatial stream because it uses a different antenna and follows a slightly different path to the receiver. Each spatial stream encodes different data, so using two spatial streams doubles the data rate. Using three spatial streams triples the data rate, and so on.

Multi User MIMO (MU-MIMO)

802.11ac Wave 2 introduces MU-MIMO. MU-MIMO, like spatial streaming MIMO, allows a wireless device to use different antennas to send distinct signals over different paths. However, instead of sending the data signals to the same receiver, the device sends each signal to a *different* receiver. MU-MIMO does not increase the data rate for the AP-to-client connection. However, because multiple clients can receive data at the same time, their throughput is increased. 802.11ac supports downlink MU-MIMO exclusively; that is, an AP can transmit signals to multiple clients at once, but only one client can transmit to the AP at a time. Because most traffic is destined to clients in a typical wireless deployment, MU-MIMO can still have a significant effect on throughput.

The AP must receive a great deal of information from MU-MIMO clients so that it can locate the clients and modulate the simultaneous signals such that each signal is maximized on the correct client and minimized on other clients. This is why clients must support MU-MIMO as well as the AP for the AP to use this technology. For example, a cell has six clients, two MU-MIMO-capable clients and three non-MU-MIMO-capable clients. The AP can send data to the two MU-MIMO capable clients simultaneously. It can then send data to one of the non-MU-MIMO capable clients. It can't send data to a MU-MIMO-capable and a non-MU-MIMO-capable client at the same time, even if it has the capability to support more spatial streams.

Summary

Refer to Job Aid 10-1 at the end of this chapter for a list of the 802.11 standards, their frequencies, and their data rates when using various features.

WLAN

You will now learn about defining a wireless network. A wireless network is often called a wireless LAN (WLAN). A WLAN is defined by the SSID.

As well as the SSID, the WLAN defines various other settings, including security settings and the VLAN on which user traffic is forwarded. You will look at these settings in a moment.

- WLAN
  - Multiple APs with the same SSID
  - Defines settings such as security and VLAN
- Clients can roam between the APs
- APs distribute their traffic into the same Ethernet network

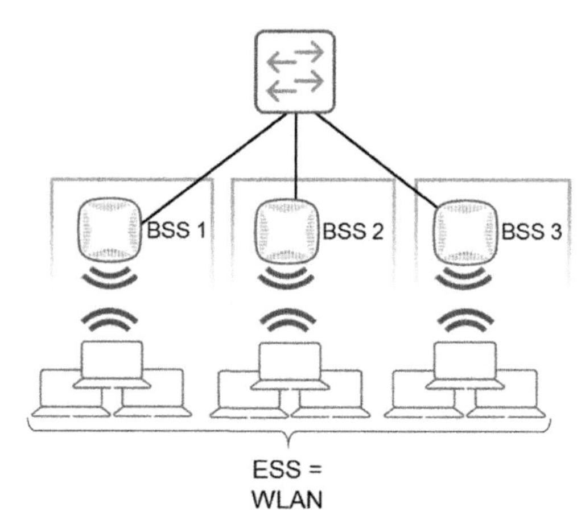

Figure 10-5: WLAN

Multiple APs can support the same SSID. Earlier you looked at an infrastructure mode cell, or BSS. When multiple APs support the same SSID, they form an extended service set (ESS), which is equivalent to the WLAN. (See Figure 10-5.) Wireless devices can roam from one AP to another AP within the same WLAN; that is, as a client moves, it can disassociate from one AP and re-associate to a new AP with a better signal without disrupting the connection or the user noticing. The APs should be forwarding the clients' traffic into the same Ethernet network, so the devices can continue to receive traffic without disruption.

WLAN security: Encryption and authentication

The shared wireless medium has a serious security drawback. Any device can eavesdrop on any other device's communications. For this reason, most wireless networks use a combination of:

- **Authentication**—Ensures that only authorized users can connect to the WLAN.
- **Encryption**—Encrypts traffic so that only the correct wireless station and the AP can read it. Encryption (data privacy) is typically joined with a data integrity algorithm, which ensures that data is not intercepted and changed in transit.

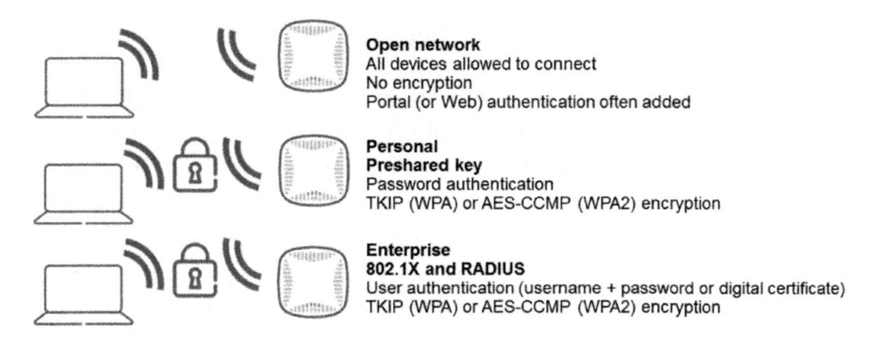

Figure 10-6: WLAN security: Encryption and authentication

When you don't implement authentication or encryption, the WLAN is open. See Figure 10-6. The AP allows any device to associate with the SSID and establish a connection. In enterprise environments, you typically only use open WLANs for guest access, and you set up a portal, also called Web-based authentication, which occurs after the device connects to the open WLAN. You've probably experienced this type of authentication at a hotel or coffee shop. You connect to the WLAN without entering any password. Then when you open a Web browser, you're redirected to a login page. Aruba IAPs support portal authentication services, but this book focuses on setting up a WLAN for employee access.

For an employee WLAN, it is recommended that you use WPA2 with Advanced Encryption Standard (AES) for encryption. Wi-Fi Protected Access (WPA) with Temporal Key Integrity Protocol (TKIP) for encryption is a less secure alternative. The 802.11n and 802.11ac standards require the use of WPA2. If you select a different form of encryption, devices will *not* be able to use the 802.11n or 802.11ac data rates.

Both WPA and WPA2 let you choose between two forms of authentication.

WPA2 with preshared keys (PSK), sometimes called Personal mode, provides basic authentication for smaller networks without backend authentication servers. The PSK is essentially a password. You configure the password in the WLAN (or wireless network) settings, and users must enter the password to connect to the network.

In a corporate or government environment, you should usually combine WPA2 with 802.1X RADIUS authentication instead—sometimes called Enterprise mode. 802.1X authentication requires users to log in to a RADIUS server with individual credentials such as a username and password or a digital certificate. Aruba IAPs provide a local user database and RADIUS services, making it very easy for you to set up a small-scale Enterprise solution. For companies with more employees, you can also integrate IAPs with a network RADIUS server or with an LDAP directory such as OpenLDAP or Windows Active Directory (AD).

Enterprise mode WPA/WPA2 requires the wireless client to have an 802.1X client. Most OSs support such clients natively.

WLAN security: MAC authentication

Some customers want to control not only which users connect to the WLAN, but also which devices they use. MAC authentication (MAC-Auth) denies connections to all devices except those with MAC addresses that match accounts on a RADIUS server. See Figure 10-7. The RADIUS server can be the IAP internal RADIUS server or an external server. The IAP supports sending the MAC address to the server with a variety of formats, such as colon or hyphen delimiters and lower or upper case letters; the submitted MAC address must match the username in the account exactly. The IAP also sends the MAC address for the password, so the account on the RADIUS server should specify the MAC address in the same address for the password.

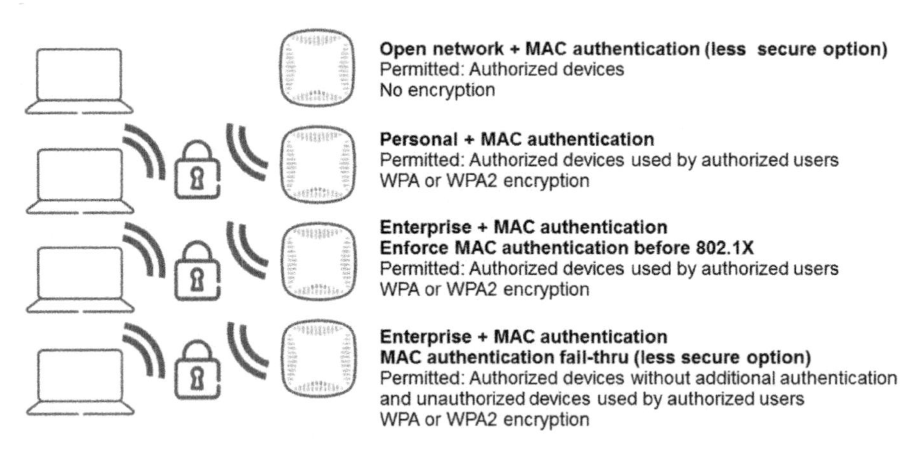

Figure 10-7: WLAN security: MAC authentication

When you add MAC authentication to an open WLAN, the IAP uses the MAC address as the only criteria for determining whether a device can connect. It is recommended that you combine MAC authentication with another form of security because MAC authentication on its own does not provide much security. Unauthorized users can easily sniff wireless traffic from authorized devices, discover authorized MAC addresses, spoof one of them, and connect.

Adding MAC authentication to a WPA/WPA2 Personal (PSK) mode WLAN causes the IAP to check *both* the device MAC address *and* the password submitted by the user.

For WPA/WPA2 Enterprise (802.1X) mode WLANs, you have a choice:

- The IAP can enforce both MAC authentication and 802.1X for all clients—in other words, only authorized users on authorized devices can connect to the network. For this option, choose **Perform MAC authentication before 802.1X** in the Instant User Interface (UI).
- The IAP can enforce MAC authentication *or* 802.1X authentication with MAC authentication performed first and 802.1X only performed if the device fails MAC authentication. That is, authorized devices can always connect without the user having to submit credentials. Unauthorized

devices might be able to connect if the user can submit valid credentials. This option is called **MAC authentication fail-thru** in the Instant UI. (Again, this option provides less security because unauthorized users could spoof an authorized MAC address and obtain access without performing 802.1X.)

In both cases, the IAP uses the same RADIUS server, which can be the internal server, for 802.1X and MAC authentication.

VLAN for the WLAN

When you define a WLAN on an AP, you also define the VLAN in which the AP forwards wireless traffic on the Ethernet LAN. Aruba IAPs support three types of VLAN assignment when you are using a network Dynamic Host Configuration Protocol (DHCP) server to assign wireless clients their IP addresses.

Note

An IAP can also act as a DHCP server and assign clients IP addresses from a DHCP scope defined for the WLAN. The IAP then uses Network Address Translation (NAT) to translate the client IP addresses to its own address on the LAN. However, this feature is beyond the scope of this book.

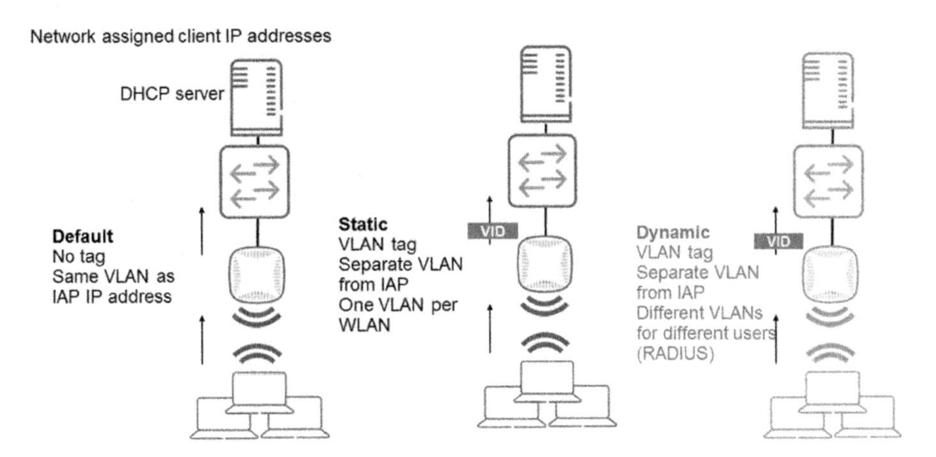

Figure 10-8: VLAN for the WLAN

Default

When you choose a default for the WLAN's VLAN assignment, the IAP forwards the wireless client traffic without a tag, placing clients in the untagged VLAN on the switch port. See Figure 10-8. The IAP also receives its IP address on the untagged VLAN so the wireless clients are in the same VLAN as the IAPs' management traffic. (You can set up a tagged VLAN for the IAPs, but this is a more advanced configuration.)

Static

Choosing static for the WLAN's VLAN assignment allows you to specify a VLAN ID. The IAP then forwards the client traffic with a VLAN tag that has this ID. Now the clients and the IAPs are in different VLANs. It is very important that you remember to tag the connected switch port for the selected VLAN so that the switch can receive the traffic. The VLAN must also exist in the Ethernet LAN, have a default router, DHCP services, and so on—just like a VLAN that you would add for wired clients. You can choose to place wireless clients in an existing VLAN or create a new VLAN dedicated to them.

Dynamic

Dynamic VLAN assignments work with 802.1X authentication. The IAP receives the VLAN assignment for each individual client from the RADIUS server during the authentication process. If the IAP is using its internal server, you would need to set up the assignment for the user account. Different clients can be assigned to different VLANs. You must tag the switch port connected to the IAP for all of the VLANs to which clients might be assigned. This book focuses less on this option as it requires more advanced setup.

Provisioning an Aruba IAP

In the next section, you will learn about how to connect to a factory default Aruba IAP and configure a WLAN on it.

Provisioning an IAP through the Instant SSID

Aruba has designed the IAP for simplicity in provisioning. The IAP can use a power adapter, but most administrators choose simply to connect the AP to a Power over Ethernet (PoE) capable switch, which then provides the AP with both power and Ethernet connectivity.

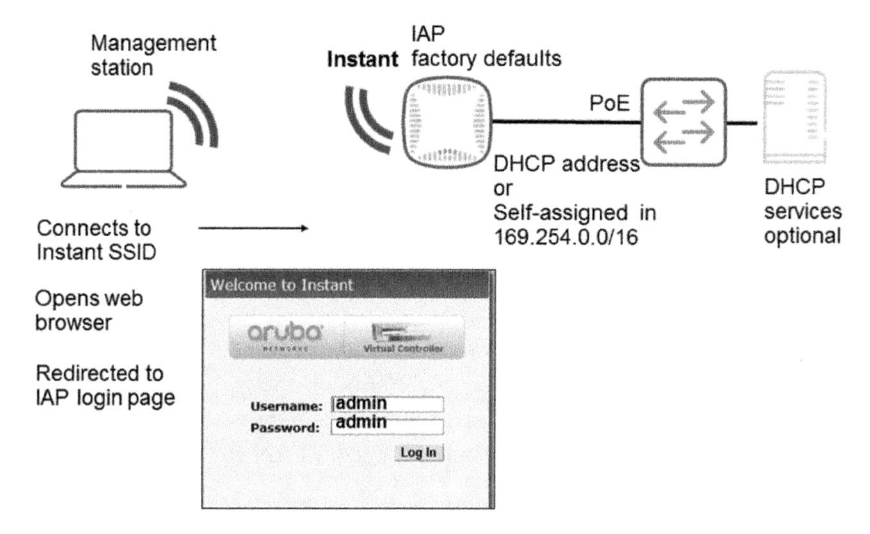

Figure 10-9: Provisioning an IAP through the Instant SSID

To reach the IAP Web browser interface, you don't need to set anything up through the CLI in advance. If you connect the IAP to a VLAN with DHCP services, the IAP will receive a DHCP address on its Ethernet interface. Otherwise, the IAP will configure a self-assigned default IP address in 169.254.0.0/16 range on this interface. You can then set a static IP address after you connect.

In either case, you do not need to know the address to reach the IAP UI. The IAP provides the Instant SSID for this purpose. By default, the IAP broadcasts this SSID on its 2.4GHz radio until you have accessed the IAP and set up another WLAN on it. To provision the IAP, simply connect your management station to the Instant SSID. Open a Web browser, and the IAP automatically redirects you to its UI login page. See Figure 10-9.

You then enter the default credentials (**admin** for both the username and password), and you're ready to configure the IAP.

Configure Wireless Services on an Aruba Instant Access Point

The tasks in section were designed for an environment in which multiple people are completing the same tasks. In such an environment, it is difficult to use the Instant SSID to provision the IAP. Everyone's APs broadcast the same Instant SSID, so you cannot be sure that you are connecting to your AP when you connect to that SSID. Therefore, you will take a slightly different approach to provisioning the AP as you complete the tasks.

You will now connect one of your IAPs to Core-2 and set up a WLAN on the IAP.

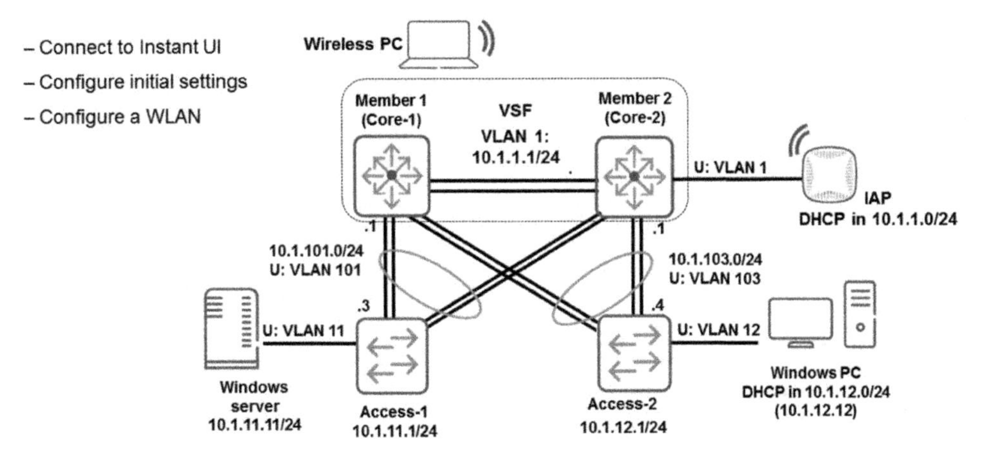

Figure 10-10: Configure Wireless Services on an Aruba Instant Access Point

For the reasons mentioned earlier, rather than connect to the Instant SSID to reach the IAP Web browser interface, you will connect to the IAP through its Ethernet connection. You will then configure some initial system settings and a WLAN for employees. This WLAN will enforce the recommended security for an enterprise: WPA2 (AES-CCMP) with 802.1X authentication.

Task 1: Connect the IAP to the network

Your IAP will receive a DHCP address, and you will connect it in VLAN 1. You will first set up IP settings in VLAN 1 on the VSF fabric. You will then check the VLAN settings on the member 2 switch port that connects to your IAP, after which you will enable the port.

Important

Your switches should be running the configuration established at the end of the final task in Chapter 9.

Core-1

1. Access the VSF fabric CLI through Core-1 (or Core-2).
2. Move to global configuration mode.
3. Core-1 should already have VLAN 1, 10.1.1.1/24, configured on it. (You can check with the *show ip* command if you are not sure).
4. Enable OSPF on the VLAN 1 so that the access layer switches can learn a route to it. You can configure VLAN 1 as a passive OSPF interface because Core-1 is the only router in its subnet.

```
Core-1(config)# vlan 1 ip ospf area 0

Core-1(config)# vlan 1 ip ospf passive
```

Note

This step is not required because access layer switches could also use their default routes to reach 10.1.1.0/24. However, you will often use OSPF in more complex topologies, and it is good to get in the habit of enabling OSPF on interfaces that you want to advertise.

5. Set the server's IP address as the helper address for VLAN 1.

```
Core-1(config)# vlan 1 ip helper-address 10.1.11.11
```

6. The PoE+ ports connected to the IAPs power the IAPs even when the ports are disabled. Remove power from the ports so that you can bring the IAPs up after you enable the ports.

```
Core-1(config)# no interface 2/a3,2/a4 power-over-ethernet
```

7. Verify the VLAN assignment for the port connected to the IAP that you are using in your test environment (see the figure above). The port should be assigned to VLAN 1 as an untagged port.

```
Core-1(config)# show vlan port 2/a4 detail
```

```
Status and Counters - VLAN Information - for ports 2/A4

 VLAN ID Name                 | Status      Voice Jumbo Mode
 ------- -------------------- + ---------- ----- ----- --------
 1       DEFAULT_VLAN         | Port-based No    No    Untagged
```

8. Enable this port.

```
Core-1(config)# interface 2/a4 enable
```

9. Power the port.

```
Core-1(config)# interface 2/a4 power-over-ethernet
```

10. If you are working with local equipment, connect the AP to the port physically. See Figure 10-11.

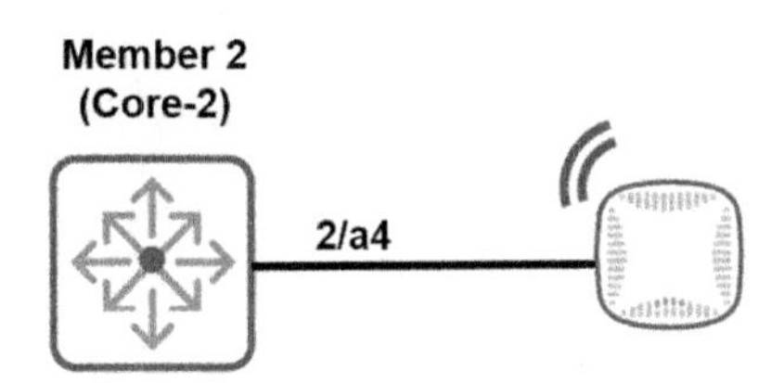

Figure 10-11: Connect the AP to the port

11. View PoE+ information on the port connected to the IAP. The Delivering status indicates that the port is sending power to a connected device.

    Also note how the port is allocating power to the device, the device's power class, and port's priority.

```
Core-1(config)# show power-over-ethernet 2/a4

 Status and Counters - Port Power Status for port 2/A4

  Power Enable       : Yes
                                       LLDP Detect         : enabled
  Priority           : low             Configured Type     :
  AllocateBy         : usage           Value               : 17 W
  Power Class        : 4               Detection  Status : Delivering

  Over Current Cnt   : 0               MPS Absent Cnt      : 0
  Power Denied Cnt   : 0               Short Cnt           : 0

  Voltage            : 54.6 V          Current             : 121 mA
  Power              : 25.0 W          Pre-std Detect      : off
```

12. View LLDP information on the port. Find the management IP address. Also examine the PoE+ requests that the IAP used LLDP to send.

```
Core-1(config)# show lldp info remote-device 2/a4

 LLDP Remote Device Information Detail

  Local Port    : 2/A4
  ChassisType   : mac-address
  ChassisId     : 40 e3 d6 c1 5a 3a
  PortType      : mac-address
  PortId        : 40 e3 d6 c1 5a 3a
  SysName       : 40:e3:d6:c1:5a:3a
  System Descr  : ArubaOS (MODEL: 325), Version Aruba IAP
  PortDescr     : eth0
  Pvid          :

  System Capabilities Supported  : bridge, wlan-access-point
  System Capabilities Enabled    :

  Remote Management Address
     Type    : ipv4
     Address : 10.1.1.12
Poe Plus Information Detail

    Poe Device Type             : Type2 PD
    Power Source                : Only PSE
    Power Priority              : Unknown
    Requested Power Value       : 25 Watts
    Actual Power Value          : 25 Watts
```

13. Record the management IP address:

Note

If you do not see a remote management address, wait a minute and then enter the command again.

If you then see a self-assigned address (begins with 169 or an IPv6 address), you need to troubleshoot VLAN 1 connectivity to the server. On Core-1, enter **ping 10.1.11.11** and **ping source 1 10.1.11.11**. Based on which pings fail, you can determine whether Core-1 cannot reach to 10.1.11.0/24 at all or whether the server lacks connectivity to 10.1.1.0/24. Examine routing tables. Also check the server IP address and gateway.

If both pings succeed, check DHCP services in the Server Manager on the server. Make sure that the DHCP server is authorized (right-click the server name and click Authorize if necessary). Check the VLAN 1 scope.

Task 2: Access the Instant UI

Typically, to provision an Aruba IAP, you simply need to connect to the Instant SSID and open a Web browser. You can also access the Instant UI through the IAP's Ethernet port.

Windows PC

1. Access the desktop for the Window's PC.
2. Open a Web browser such as Windows IE.
3. Navigate to the management IP address for the IAP, which you located in the previous task.
4. You will see a certificate error because the IAP redirects you to an HTTPS session, and it uses a self-signed certificate by default. See Figure 10-12. Choose to continue to the website.

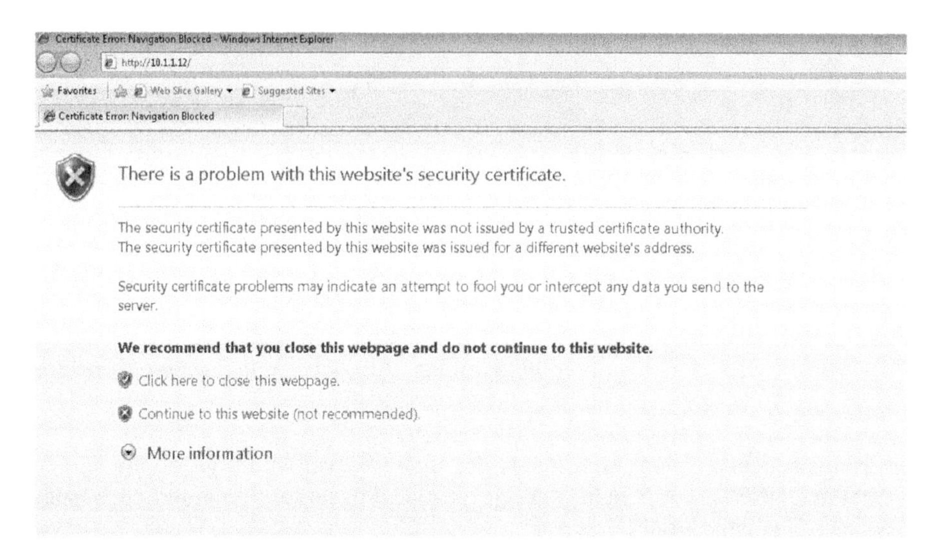

Figure 10-12: Certificate error

5. Log in to the Instant UI with the default credentials: admin for both the username and password. See Figure 10-13.

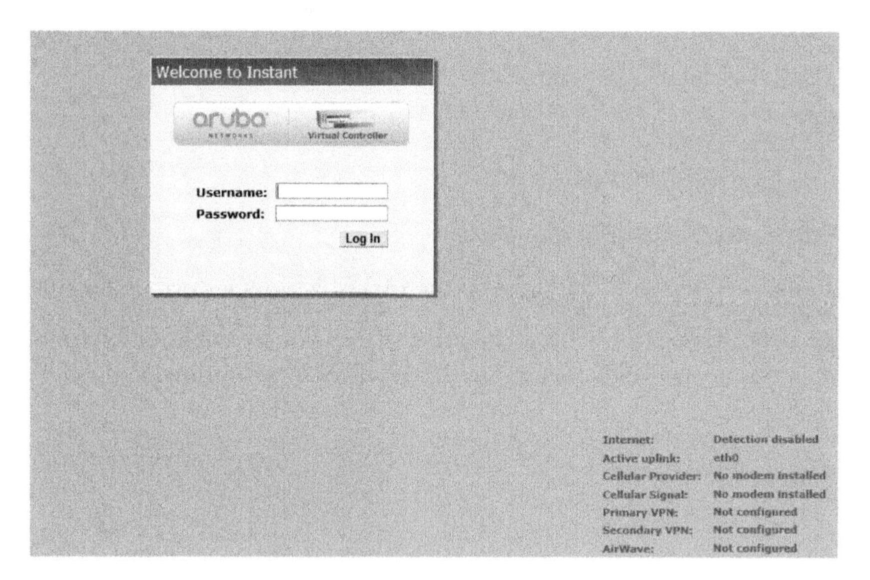

Figure 10-13: Log in to the Instant UI

6. On your first login to the IAP, you will see a message about trying Aruba Central, which is a cloud-based management solution for IAPs. See Figure 10-14. Click Close.

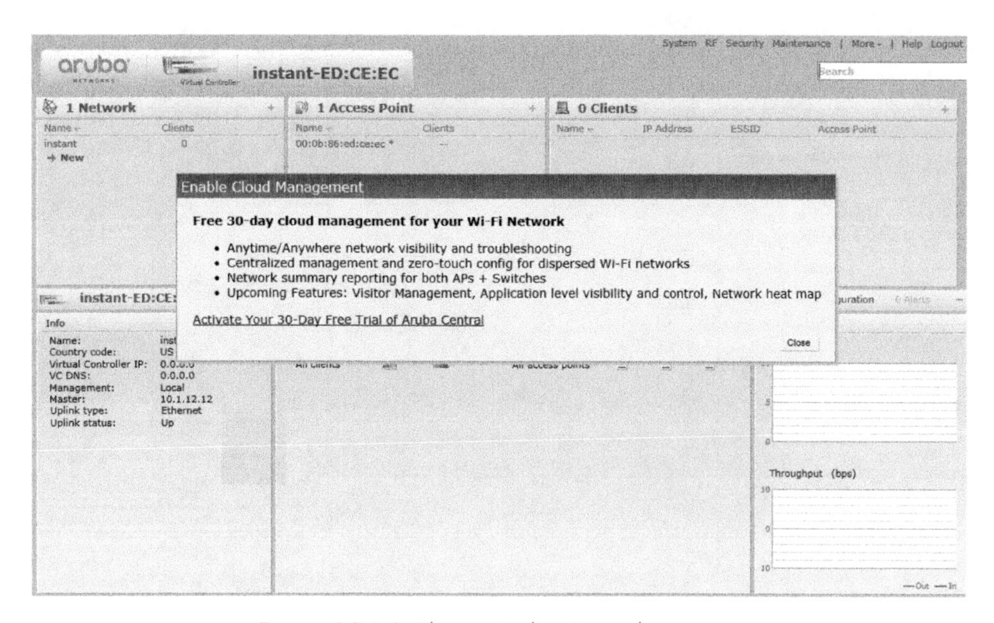

Figure 10-14: The try-Aruba-Central message

7. You should now see the Dashboard. The top of the window and the middle bar show the system name for this IAP. Currently, the name is derived from "instant" and the final three octets of the IAP's MAC address.

8. The top portion of the UI has a Network pane, in which you manage WLANs. Note that this pane currently shows the Instant WLAN.

Important

If your IAP has a non-default network instead of the Instant one, you must reset the configuration. Select **Maintenance**, click the **Configuration** tab, and click **Clear Configuration**. Then click **Clear All Configurations and Restart**.

Wait for the IAP to restart and log in again. (Note that you might see a log in error when you first try to log in. This occurs because the IAP takes a couple of minutes to determine that it's master and supports the UI.)

---

9. The top portion of the UI also includes panes for managing the Access Point settings and for monitoring clients. See Figure 10-15.

10. The middle bar lets you choose the view for the bottom portion of the interface. Currently, Monitoring is selected, allowing you to view system information, as well as RF and usage monitoring information.

11. Find the links at the top of the interface, which you can use to access additional configuration and management settings.

12. Find the drop-down menu at the bottom of the interface. Change the UI language to your native language, if you prefer.

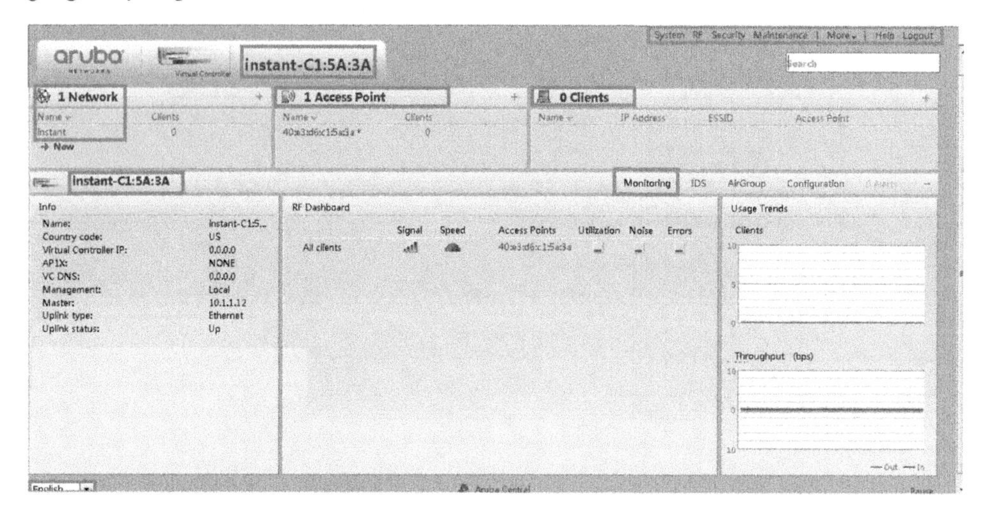

Figure 10-15: The Aruba IAP UI

Task 3: Set system settings

In this task, you will define a new name for your IAP.

1. You should be in the Instant UI. Click the **System** link at the top of the UI. See Figure 10-16.

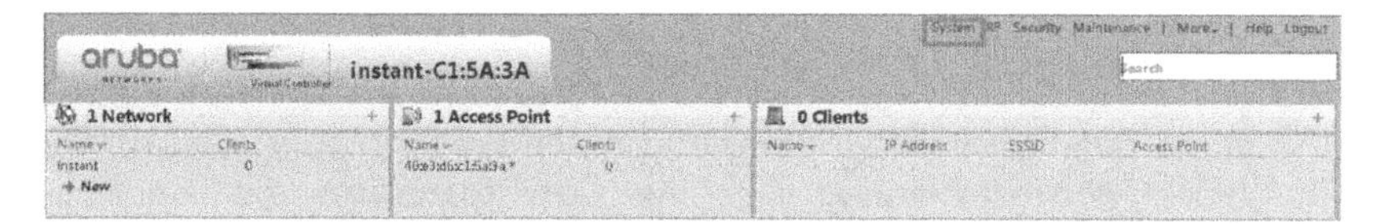

Figure 10-16: Click the **System** link at the top of the UI

2. Change the Name to **ATPN*k*-S*s*-IAPs** in which k is the number of the kit that you're using and s is a random number. For example, you might enter: **ATPN1-S8-IAPs**. See Figure 10-17.
3. Enter an intuitive description of the system location.
4. In the real world, you would click the Admin tab to change the login credentials to secure credentials that are unique to your company.
5. Click OK.
6. Note that the system name changes in the UI. See Figure 10-17.

Figure 10-17: The system name changes in the UI

Task 4: Create a WLAN

You will now create a WLAN for employees. The WLAN will use Enterprise (802.1X) security with WPA2 encryption.

1. You should be in the Instant UI. Click **New** in the Network pane. See Figure 10-18.

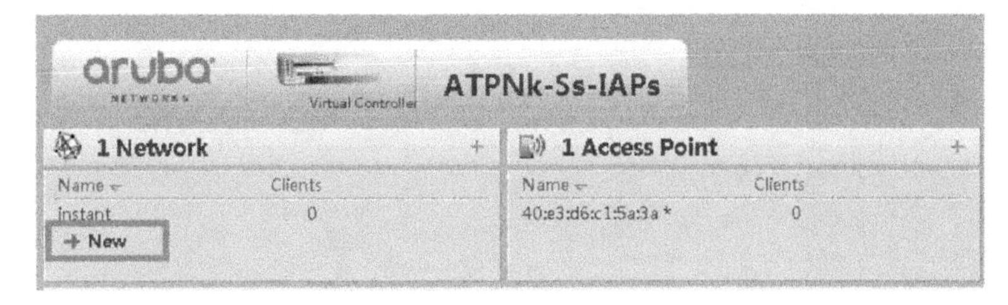

Figure 10-18: Click **New** in the Network pane

2. A wizard displays, which helps you set up a WLAN.
3. The Name setting defines the SSID. Enter ATPNk-Ss, replacing k with your kit number and s with your random number.

Important

You must fill in the correct variables for your kit and number. Otherwise, you will not be able to identify your SSID and connect to it with your wireless client. For example, if you are using kit 1 and number 8, enter **ATPN1-S8**.

4. For Primary usage, select Employee. The wizard will then display the options typical for a WLAN used by employees. See Figure 10-19.
5. Click Next.

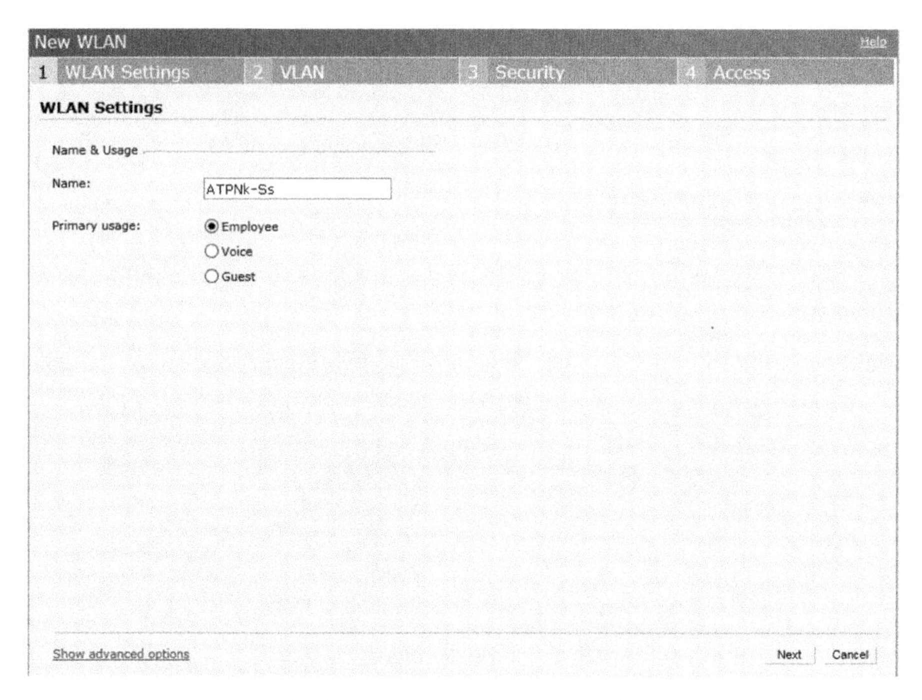

Figure 10-19: Select Employee and click Next

6. For Client IP assignment, choose Network assigned, which configures the IAP to simply pass DHCP requests onto the LAN. Your Windows server will provide DHCP addresses to the employee wireless clients. See Figure 10-20.
7. For Client VLAN assignment, select Default, which forwards the client's traffic as untagged.
8. Click Next.

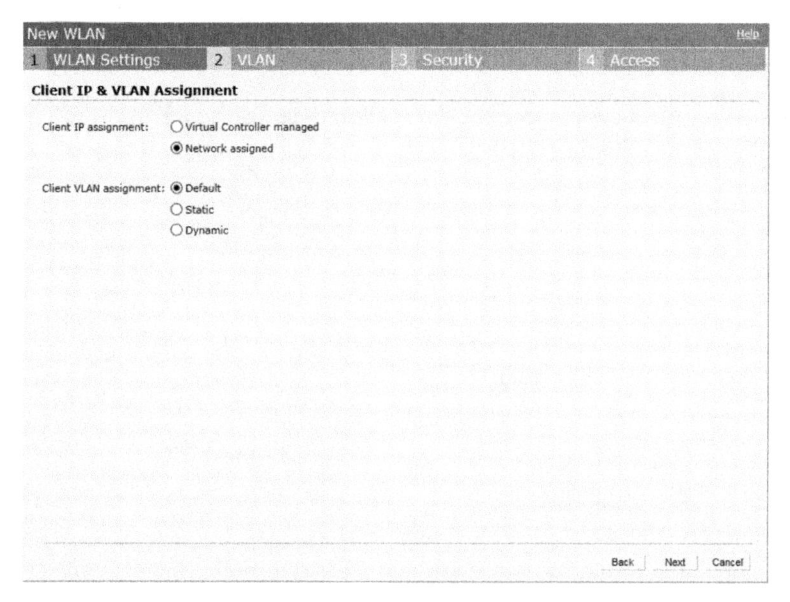

Figure 10-20: Select Network assigned and Default, then click Next

9. In the left bar, choose Enterprise for the Security Level.

10. For Key management, choose WPA-2-Enterprise.

    This option sets AES-CCMP for the wireless encryption and data integrity. These are the most secure options and are required for clients to use 802.11n and 802.11ac data rates over 54Mbps. If you have some legacy devices, you can select both WPA and WPA2.

11. Make sure that Authentication server 1, lists InternalServer. This option lets you set up local user credentials on the IAP for use with 802.1X authentication.

12. Click the Users link next to Internal server. See Figure 10-21.

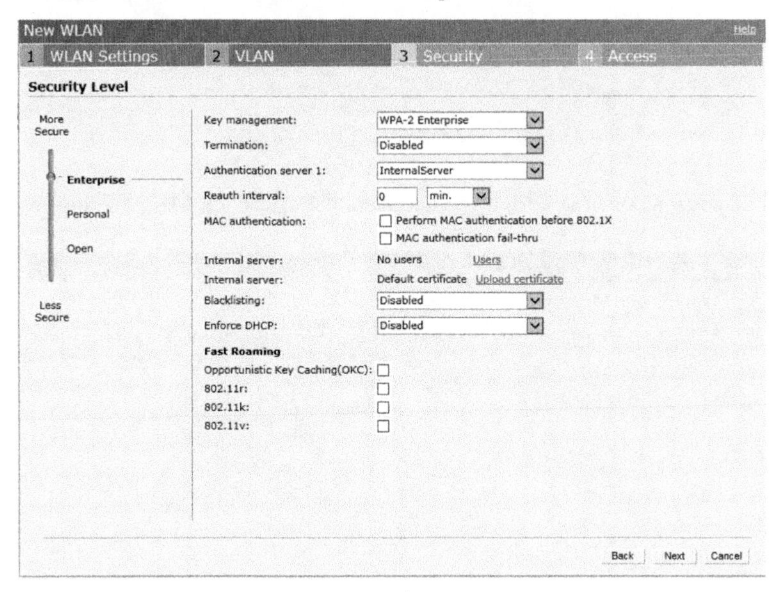

Figure 10-21: Click the Users link

13. Add a user account, which you will use for testing the WLAN. See Figure 10-22.

    a. Set the Username to employee.

    b. Enter password for Password and Retype.

    c. From the Type drop-down menu, choose Employee.

    d. Click Add.

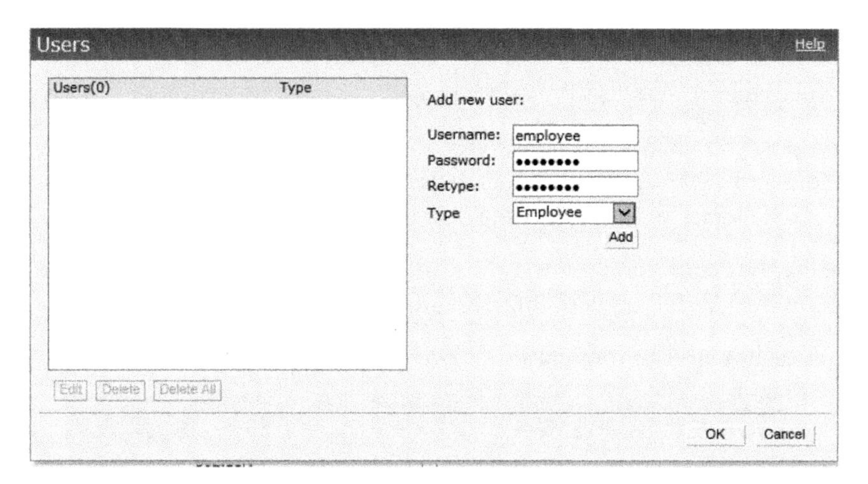

Figure 10-22: Add a user account

e. Verify that your account is added to the Users list and click **OK**. See Figure 10-23.

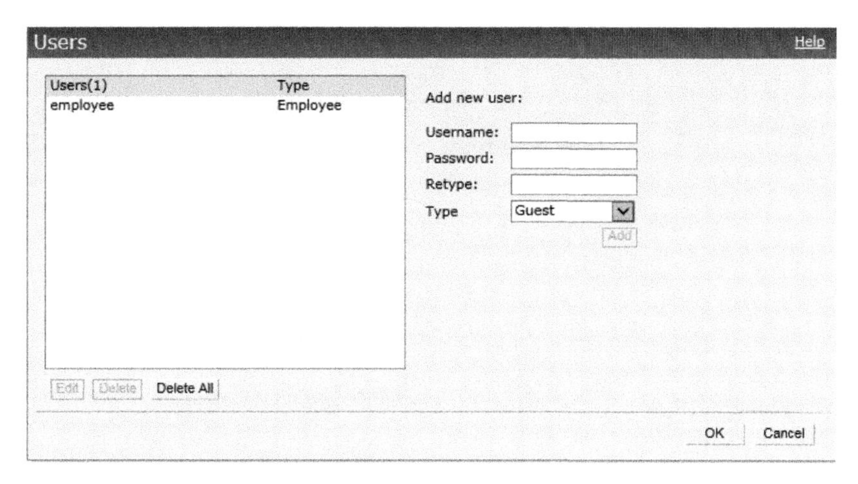

Figure 10-23: Verify that your account is added to the Users list

14. You should now see one user in the Internal server.

15. A RADIUS server requires a certificate to perform 802.1X authentication. The IAP uses a certificate by GeoTrust Global, by default. When you connect to the SSID on your wireless client, you will see that you have to choose to trust this certificate.

16. Leave the other settings at their defaults and click Next. See Figure 10-24.

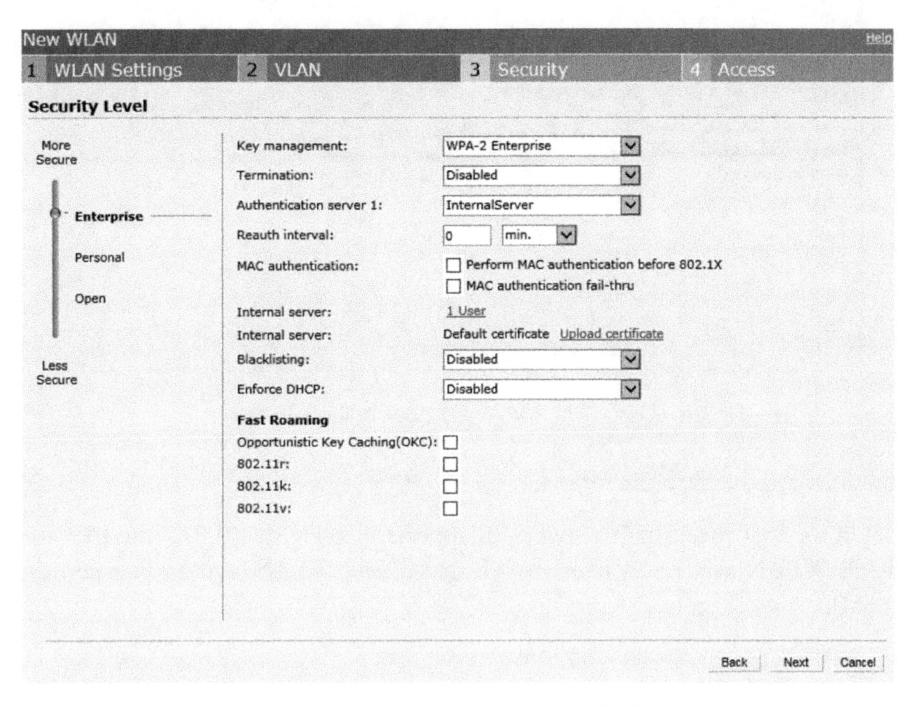

Figure 10-24: Leave other settings at their default and click Next

17. You can set up access rules to control the wireless users' traffic. However, these rules are beyond the scope of this book. Accept the default setting, which is unrestricted access. See Figure 10-25.

    Click Finish.

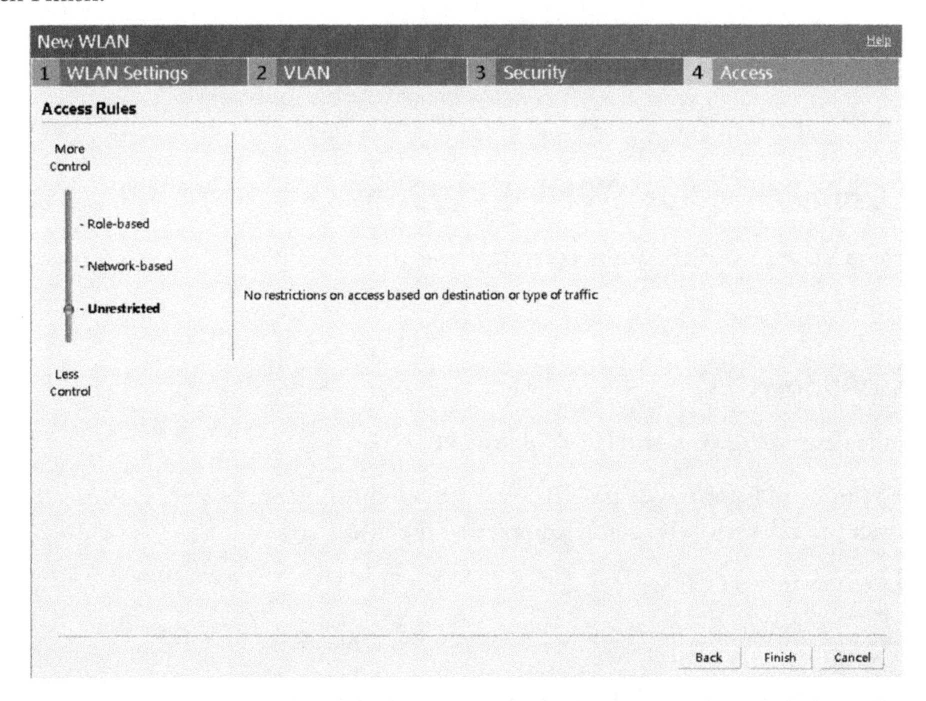

Figure 10-25: Accept the default setting, which is Unrestricted, and click Finish

18. You should now see your WLAN in the Networks list. See Figure 10-26.

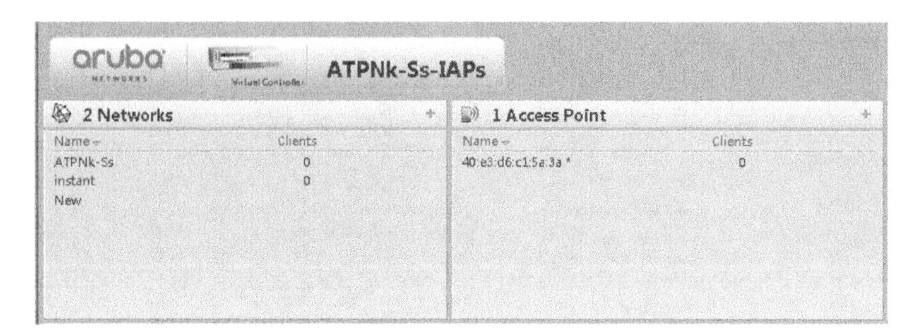

Figure 10-26: The WLAN is now in the Networks list

19. Wait a moment and refresh the interface.

20. You should see that the Instant network disappears. See Figure 10-27. The IAP automatically deletes this WLAN as soon as you have created a new WLAN and are not connected to the Instant SSID.

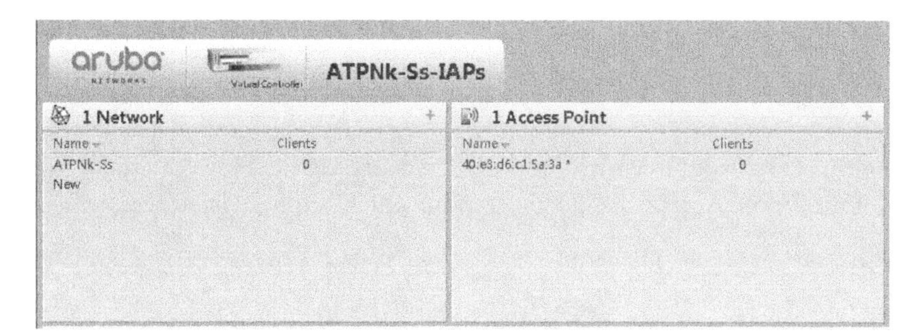

Figure 10-27: The Instant network disappears

Task 5: Connect to the ATPNk-Ss WLAN

You will now test your configuration by connecting to the ATPN*k*-S*s* WLAN on your wireless client.

Wireless Windows PC

1. Access the desktop for your wireless Windows PC.
2. If your Windows PC has a wireless NIC, you can use this PC for this step. However, take care to disconnect the Ethernet NIC before completing the next steps.
3. Log in to the wireless PC.
4. Click the network connection icon in the system tray to see a list of SSIDs.

Important

If you don't see the icon in the system tray, restart the client. (You can run **shutdown /r**).

5. Select your own SSID, using the kit number and chosen random number that identifies it and click Connect. See Figure 10-28.

Figure 10-28: Select your SSID and click Connect

6. After a moment, you will be prompted to enter credentials. Enter employee and password. See Figure 10-29.

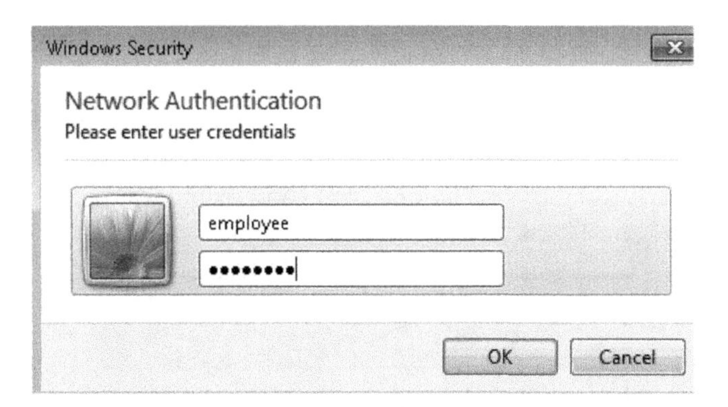

Figure 10-29: Enter your credentials and click OK

7. After a moment, you will see a message that the server (the IAP) has an untrusted certificate. See Figure 10-30.
8. Click the arrow next to Details to see more about the certificate.

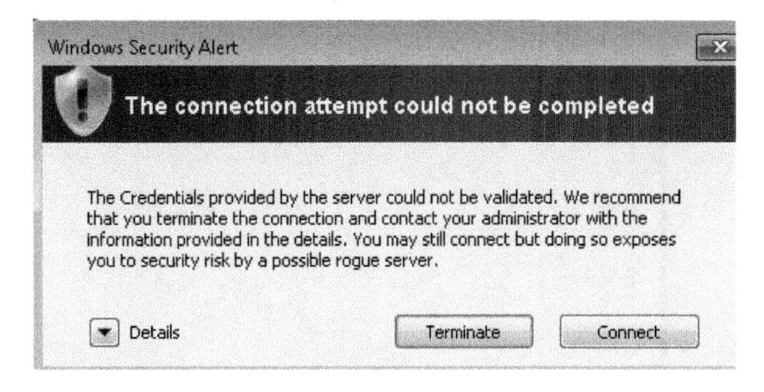

Figure 10-30: Security alert for untrusted certificate

9. Click Connect to accept the certificate and server as trusted. See Figure 10-31.

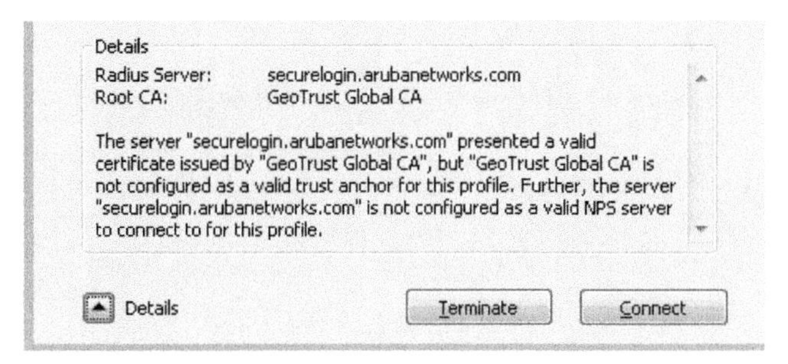

Figure 10-31: Click Connect to accept the certificate

10. The connection should now establish. You might see an alert on the connection icon because the connection doesn't have Internet access in the test environment. You can ignore that alert.

Windows PC

You will now check the client status in the Instant UI.

11. Access the desktop for your Windows PC. Return to the Instant UI.
12. You should see the client in the Client list. Note the client's IP address. Think about why the client has an IP address in 10.1.1.0/24.

Task 6: Save

1. Save the configuration on the VSF fabric.
2. The configuration on the IAP is automatically saved. Backing up the configuration is optional. If you want to do so, click the Maintenance link at the top of the Instant UI. Then select the Configuration tab and click **Backup Configuration**. See Figure 10-32.

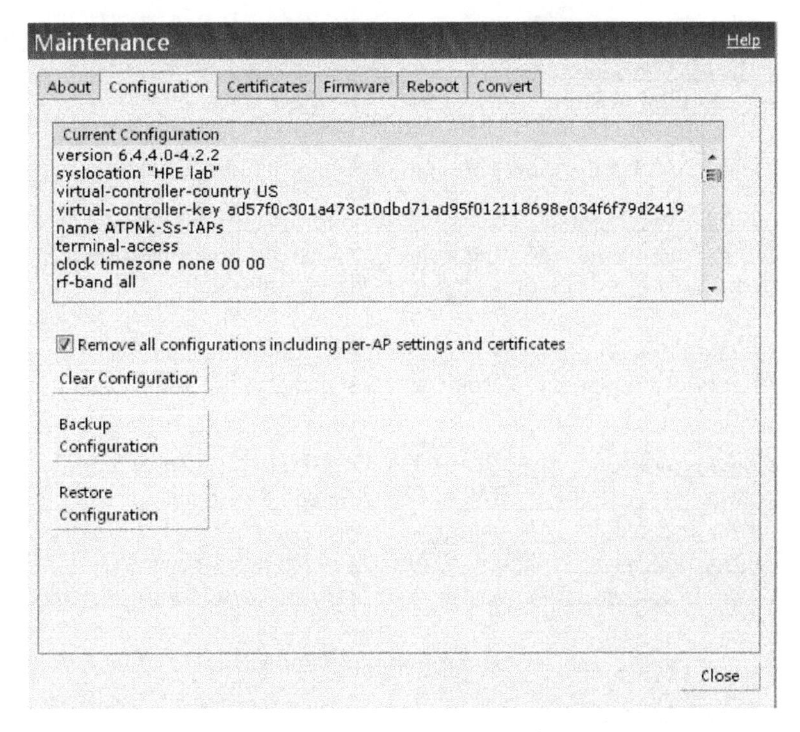

Figure 10-32: Click Back up Configuration

Learning check

You will now answer questions that will help you retain what you have learned.

1. What information were you able to obtain about the AP by looking at LLDP information on Core-2?

2. When did the Instant SSID disappear from the configuration?

3. What IP address did the wireless client receive after you connected it to the wireless network? How is the AP forwarding the client traffic to the switch? What could be a drawback to this behavior?

4. What PoE information did you see on the port that connects to the IAP?

Answers to Learning check

1. What information were you able to obtain about the AP by looking at LLDP information on Core-2?

   You can collect a great deal of information. You can see the AP's management IP address, its system name, its MAC address, its firmware version, and so on. You also can see information about the requested PoE settings.

2. When did the Instant SSID disappear from the configuration?

   It disappeared after you created another WLAN and refreshed the browser. If you were connected through the Instant SSID, the Instant SSID would have disappeared after you created another WLAN and disconnected from the Instant SSID.

3. What IP address did the wireless client receive after you connected it to the wireless network? How is the AP forwarding the client traffic to the switch? What could be a drawback to this behavior?

   The client received an IP address in VLAN 1, subnet 10.1.1.0/24. When you set up the WLAN, you chose the default for the VLAN assignment. This configures the IAP to send the wireless client's traffic into the LAN untagged. Therefore, the client is assigned to the untagged VLAN on the AP's port (here, VLAN 1), which is the same VLAN that the AP uses for its IP address. This could make your network infrastructure management VLAN less secure.

4. What PoE information did you see on the port that connects to the IAP?

   You saw that the port was powering the connected device. The device had requested power using LLDP. You might have seen that it requested 18W or more, depending on the IAP model.

PoE

In the preceding set of tasks, you saw how the switch powered the IAP using PoE, a popular option because you can deploy the AP almost anywhere and connect it with a single cable. Implementing PoE is often as simple as connecting the AP to a port on a switch that supports PoE. Sometimes, though, you will need to plan a power budget to ensure that the switch can supply enough power to all the PoE devices connected to it. Or you will need to troubleshoot a switch that is not supplying power as you expect. A bit more knowledge about PoE will help you in these situations.

A device that receives PoE power is called a powered device (PD). The switch (or other device) that provides the power is the power sourcing equipment (PSE). See Figure 10-33. When a device connects to a PSE, the PSE detects whether the device is a PD so that the PSE knows whether to supply power (specifically, the PSE checks whether the device has applied the correct resistance and capacitance between the powered pairs of wires in the cable).

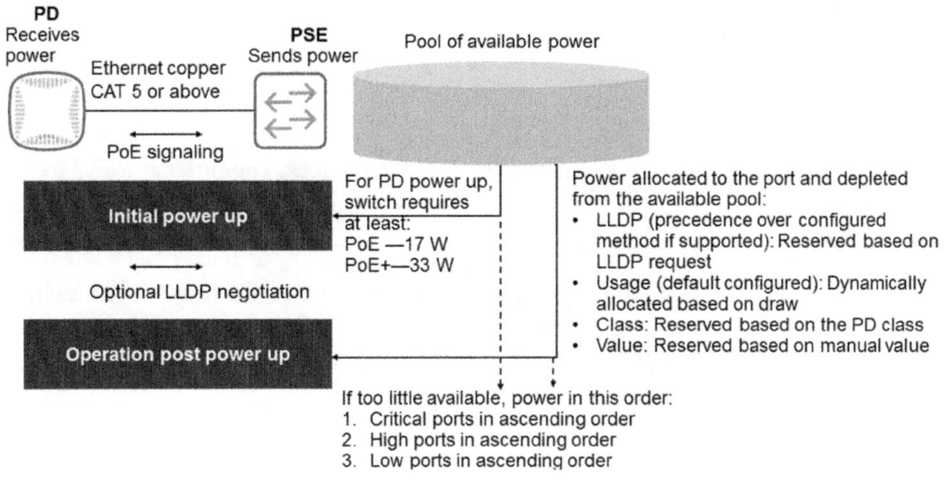

Figure 10-33: PoE

PoE/PoE+ classes

The PD might also indicate its class using a signature resistance. Table 10-1 shows the PoE and PoE+ power classes. As you see, a class indicates the range of power that the PD requires. Only PoE+ devices can use class 4. A PSE classifies a PD that does not indicate a class as class 0, which can receive power up to the PoE maximum of 12.94 W (perhaps slightly higher if the cable loss is low). The third column in the table indicates that the PSE supplies more power than the PD actually draws due to power lost over the length of the cable.

Table 10-1: PoE and PoE+ power classes

Class	PD power drawn (W)	Minimum Power required at PSE (PoE switch) (W)
0	0.44-12.94	15.4
1	0.44-3.84	4
2	3.84-6.49	7
3	6.49-12.95	15.4
4 (802.3at or PoE+ only)	12.95-25.5	30

Initial power up

When a PD connects to a PoE port, the switch checks the amount of power that it has available (that is, power that is not currently allocated to other PDs). To power up a PD, regardless of the PD's class or power requirements, an ArubaOS switch must have at least:

- 17W to supply power to a PoE PD
- 33W to supply power to a PoE+ PD

If the switch doesn't have enough power available, the switch checks the port's priority to determine whether it should stop powering another port and power this one instead. You'll learn how priority works in just a moment.

The 17W or 33W requirement is simply an initial check of the available power. The switch doesn't necessarily allocate all of this power to the port. How the switch allocates power depends on two factors: whether the PD and port use LLDP to communicate about the PoE or PoE+ requirements and the type of allocation configured on the port.

LLDP allocation

The PSE and PD can negotiate the power required by the device precisely using LLDP-MED type length value (TLV) extensions. PoE+ includes power management through LLDP TLVs as part of the standard. The PoE and PoE+ LLDP TLVs are different, and ArubaOS switch ports support only the PoE+ TLV by default (**lldp config <int_ID_list> dot3TlvEnable poeplus_config**). If the PD is a PoE device that supports LLDP-MED instead, you can enable the port to accept the LLDP-MED messages with the **interface <int_ID_list> poe-lldp-detect enable** command. (If both types of TLVs are enabled and received, the PoE+ TLVs take precedence.)

When a PD sends a power request using LLDP TLVs *and* the port supports that type of TLV, the allocation method for the port becomes "lldp" *regardless of the PoE allocation method configured on the port.* The port reserves the number of Watts requested by the PD, subtracting that amount from the available power. The switch keeps the requested power in reserve for the port even if the PD isn't drawing all of the requested power at the moment. This behavior can help to prevent the switch from beginning to power too many devices and running out of power when the PD needs more power.

The PD can also use LLDP to dynamically request less power (for example, if it enters sleep mode) or more power at any point during its operation.

Usage allocation

If the PD doesn't use LLDP or the port doesn't support the TLVs, the configured PoE allocation method takes effect.

By default, ArubaOS switches supply power to each port according to its usage. Only the power that a port is actively supplying is subtracted from the switch's pool of available power. Therefore, if a port is drawing less power than the peak requirements, the switch has more power available for other ports. However, if too many ports then start to draw more power, the switch might not be able to meet the needs. As mentioned above, it supplies power to the ports according to their configurable priority (first critical, then high, then low). If multiple ports have the same priority, the switch prioritizes the ports according to port ID.

Class allocation

Alternatively, you can allocate power according to class. When a port uses this option, it reserves the maximum power required by the PD's class when the device first connects. It subtracts the full value from the pool of available power. Even if the PD is not drawing power at that level, the switch cannot use the reserved power on another port—unless a port with a higher priority connects. Because allocating power by class reserves as much power as the port ever needs, the class option helps to prevent situations in which the switch begins to power more devices than it can power at peak operation.

Value allocation

You can also manually specify the power allocated to a specific port, which you might do when you want to guarantee a specific power to a device that does not have a PoE class. When the port starts to power a PD, it reserves the full manually specified value regardless of whether the PD is drawing that much power. As always, the allocation isn't necessarily a guarantee that the switch will continue to allocate that power to the port. If the switch runs out of power, it uses priority to choose which ports to power.

Priority

Whenever the switch needs to allocate more power than it has available, whether because a new device connects or because ports using the usage allocation method begin to supply more power, the switch uses the PoE priority to decide which PDs to power.

If the port has a higher priority than any of the current ports receiving power, the switch removes power from the lowest priority port and gives it to this port instead. Priority is determined first by an administratively set value. The switch provides power preferentially to all critical priority ports, then to high priority ports, and then to low priority ports (the default priority). When the switch can power some, but not all, ports of a particular priority, it powers the ports in ascending order (such as 1 through 48, or A1 through A24, then B1 through B24, and so on). For example, a switch might have four critical priority ports, four high priority ports, and 40 low priority ports. It has enough power to power the four critical ports and the four high priority ports, but only 20 other ports. It would power the 20 low priority ports with the lowest IDs.

To power a port, the switch must be able to give the port its full allocation as dictated by LLDP or the allocation method. For example, if a switch has 20W left available, and the port with the next highest priority connects to a PD that has requested 22W with LLDP, the switch does not power that port.

You might want to set high or critical priority on the ports that connect to IAPs. Then if a new PD connects to a low priority port, the new PD does not cause an AP port to power down, simply because the port has a lower ID.

Device profiles

ArubaOS switches running software K/KA/KB/WB16.01 or above support device profiles to make it even simpler to deploy Aruba APs such as the IAPs. Typically, you have a standard configuration that you want to apply to all ports that connect to APs. This configuration would include the untagged VLAN in which the IAPs have their IP addresses, the tagged VLAN assignments for static (and dynamic) VLANs assigned to WLANs, the PoE settings such as a critical PoE priority, and so on.

A device profile lets you define these settings once on a switch in advance of connecting the APs. You associate the device profile with the aruba-ap type. You must also enable the aruba-ap type for the device profile to take effect. Installers can then connect the IAPs to any switch port without coordinating in advance. The switch detects the Aruba AP type in an AP's LLDP messages and automatically applies the settings in the device profile.

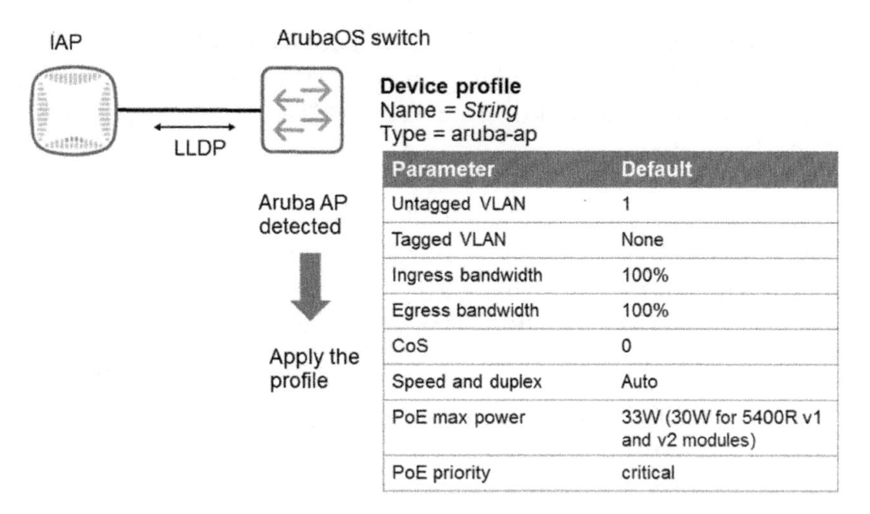

Parameter	Default
Untagged VLAN	1
Tagged VLAN	None
Ingress bandwidth	100%
Egress bandwidth	100%
CoS	0
Speed and duplex	Auto
PoE max power	33W (30W for 5400R v1 and v2 modules)
PoE priority	critical

Figure 10-34: Device profiles

Figure 10-34 lists the settings defined in the default device profile, which is associated with the aruba-ap by default. You can alter any of these settings. You can also create a new device profile, which is also populated with these settings by default. The ingress bandwidth and egress bandwidth settings are rate limits, and the class of service (CoS) settings define a quality of service (QoS) level for traffic received from the AP. Not all features are supported on all ArubaOS switches. You should check your switch documentation.

Clustering Aruba IAPs

You will now learn how to expand your wireless services seamlessly using an IAP cluster.

Autonomous versus controlled APs

Aruba Instant makes it simple to add more APs to the solution and expand your wireless network. To understand how, you should learn a little more about the two basic ways that an AP can operate.

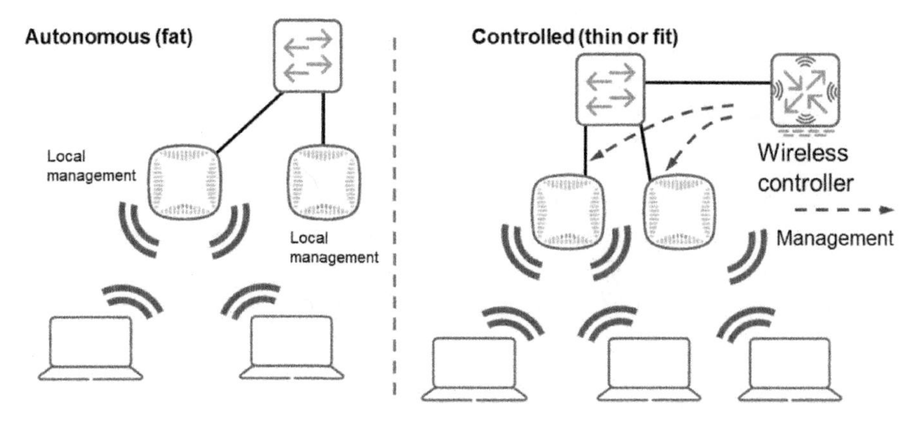

Figure 10-35: Autonomous versus controlled APs

APs can operate on their own, or autonomously. Each autonomous AP is configured and functions on its own. These APs are more common in smaller networks with few APs that need to be managed. Autonomous APs are sometimes also called "stand-alone" or "fat" APs. See Figure 10-35.

Every Aruba IAP has the ability to act as an autonomous AP.

However, most corporate environments require multiple APs to meet their coverage needs. As you deploy more APs, it becomes more difficult to manage the APs and coordinate their settings. For larger networks, customers often deploy controlled APs. Controlled APs establish the wireless network like autonomous APs. However, all of their settings are managed by a centralized wireless controller.

Some controlled APs, sometimes called "thin" APs, encapsulate all traffic from wireless users and tunnel it to the wireless controller. The controller can then apply certain security and QoS settings and forward the traffic on the Ethernet LAN. Other controlled APs, sometimes called "fit" APs, receive their settings from a controller but forward wireless traffic directly onto the Ethernet LAN themselves. Aruba offers controlled APs that operate in either way, depending on how you configure the controller.

You can learn about implementing Aruba controlled AP solutions in Aruba training.

Aruba IAP cluster and virtual controller (VC)

An Aruba Instant cluster gives you the unified management of a controller-based solution without requiring a controller appliance.

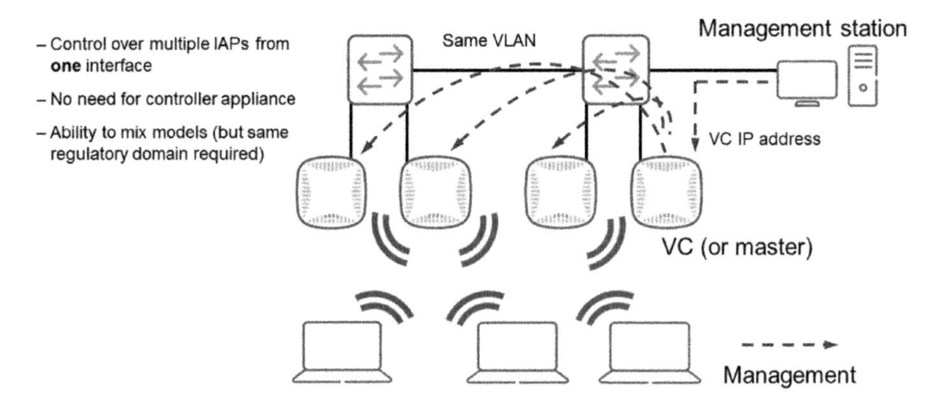

Figure 10-36: Aruba IAP cluster and virtual controller (VC)

An IAP cluster consists of multiple IAPs that are managed by a virtual controller (VC), which is simply one of the IAPs that has been elected to act as the single point of management access. The IAP acting as VC is also called the master IAP. The VC can have its own IP address, separate from the IP address of the master IAP acting as VC, so that no matter which IAP is acting as VC at the moment, you can access the VC IP address. The settings that you configure in the VC Instant UI are propagated across the IAP cluster.

Clusters can consist of IAPs of different models. However, the IAPs must be in the same regulatory domain such as US or rest of world (ROW).

Automatic cluster formation: No master election

Establishing an IAP cluster is as simple as connecting multiple IAPs on the same VLAN. The IAPs can automatically elect a master. However, understanding a bit more about the process can help you avoid unwanted behavior.

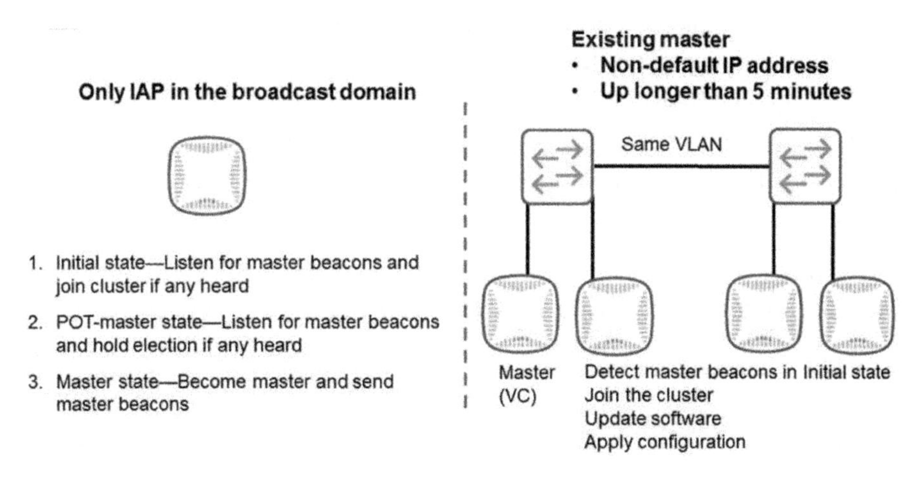

Figure 10-37: Automatic cluster formation: No master election

First, consider a scenario in which you power up and connect a single IAP. After the IAP first boots, it remains in an Initial state for about a minute. In this state, the IAP listens for master beacons. If it detects any beacons, it joins that IAP's cluster. However, in this case, the IAP is alone, so it does not detect any beacons, and it moves to the POT-master state. The IAP is still listening for master beacons, but now if it detects any, it triggers an election. Again, the isolated IAP passes through the state without hearing any beacons. Finally, the IAP becomes master—here, of a cluster of one-and it begins sending master beacons.

Now imagine that this IAP has been up and acting as master for more than five minutes. It has a non-default IP address, whether a DHCP address or a static address that you assigned. Now you connect more IAPs to the VLAN and boot the IAPs up. These IAPs pass into the Initial state, but they detect the existing master's beacons during this state. Therefore, they join the existing cluster without any election occurring. The existing master remains master.

If the master has been up for less than five minutes, an exception can occur. An IAP with a 3G/4G uplink can take over as master even though it detects the master beacons in its Initial state. Also, the existing master must have a non-default IP address. If it is using a self-assigned (169.254.0.0/16) IP address, an IAP with a non-default address will preempt the master role.

When the IAPs join the cluster, they check their software version against the VC's. If the version does not match (both major and minor version), these IAPs attempt to update their software automatically as follows:

- If the VC (master IAP) and the IAP joining the cluster run the same class image, the new IAP downloads the software directly from the VC.

  For example, an IAP 215 and an IAP 225 both run the Centaurus class image.

- If the new IAP runs a different class image from the VC (for example, an IAP 205 and an IAP 215) and the cluster is under AirWave management (discussed in the next chapter), then the new IAP downloads the image from the AirWave server. The AirWave server must have the proper image loaded on it in advance and must be reachable by the new IAP.

- If the new IAP runs a different class image from the VC and the cluster is not under AirWave management, the IAP downloads the image from Aruba Activate. The IAP must have Internet connectivity to reach Activate.

When the IAP is on the same software version as the master, it boots with VC configuration. It now supports any WLANs configured on the VC, uses the radio settings configured on the VC, and implements all settings in the configuration.

Automatic cluster formation: Master election

Often you can rely on the IAPs to automatically choose the best master. If you want to follow this approach, boot up and connect the IAPs in the same VLAN at about the same time. The IAPs will pass through the states discussed previously. Because multiple IAPs are passing through the states at about the same time, they will hear each other's master beacons in the POT-master or master state. Therefore, they will hold a master election.

A master election might also occur because the IAPs booted up fully before they were able to reach each other over the broadcast domain. And, after a cluster has formed, if the current master fails, an election is held for the new master.

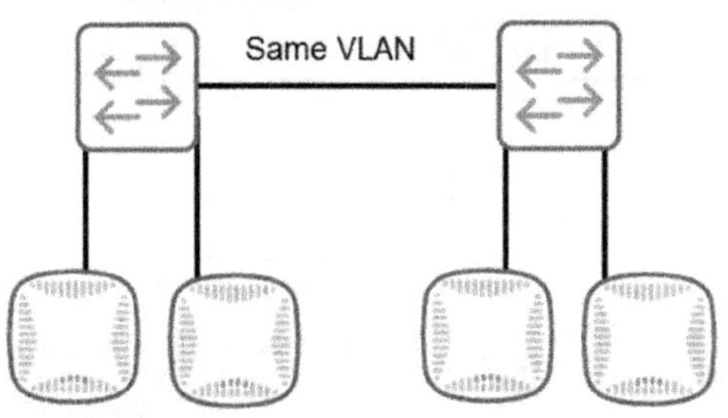

Figure 10-38: Automatic cluster formation: No master election

As you see in Figure 10-38, the election follows these rules:

1. An IAP manually configured as preferred master becomes master.

   Use this setting to ensure that a particular IAP—such as the master of a cluster on which you've already configured settings—remains master as you add IAPs. As you saw previously, an existing master should remain master as you boot up and connect new IAPs. However, errors could occur. Perhaps a connectivity issue such as a port in the wrong VLAN or a disabled port (a disabled port still supplies PoE power) causes an IAP to enter the master state before it hears the master's beacons. In that case, an election would occur, and the new IAP might win if you haven't set the preferred master setting.

2. If no or multiple IAPs are the preferred master, an IAP with non-default (self-assigned) IP address
3. If a tie occurs, the IAP with a 3G/4G uplink is preferred.
4. If a tie occurs, the IAP with the most capable hardware becomes master. For example, an IAP 135 is preferred to an IAP 105 because the IAP 135 has more system resources.
5. If a tie occurs, the IAP with longest uptime is elected.
6. If a tie occurs, the IAP with highest MAC address becomes master.

Cluster distribution of responsibilities

You saw how you could manage the IAP cluster as simply as a single IAP. At the same time, the IAP cluster is designed for an efficient distribution of services. Every IAP in the IAP cluster provides a control plane and data plane. The master simply maintains the management plane. The IAPs receive their radio, WLAN, VLAN and other configurations from the master. However, they largely enforce this configuration independently. Each IAP forwards its connected wireless devices' traffic directly on to the LAN without sending the traffic through the VC. This approach prevents bottlenecks, particularly when the wireless network needs to support a high throughput.

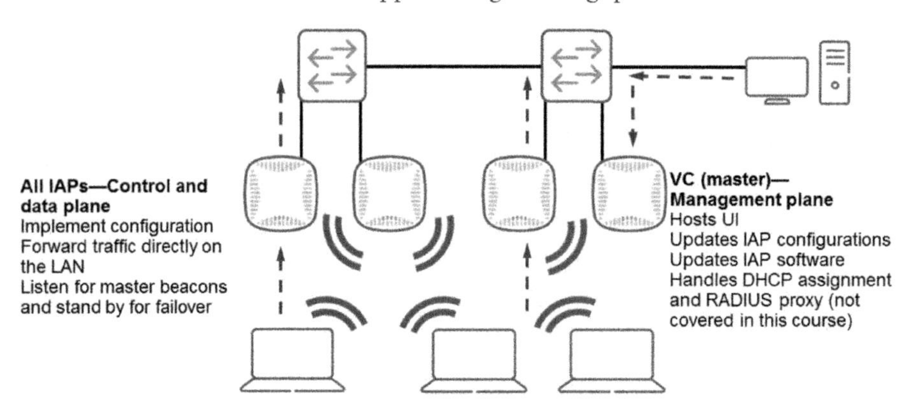

Figure 10-39: Cluster distribution of responsibilities

The master is primarily responsible for hosting the Instant UI. See Figure 10-39. You must always direct the web browser to the VC IP address or the master IP address. If you access another IAP's IP address, you won't be able to log in to the UI, and you will instead be redirected to the VC address. As you make configuration changes in the Instant UI, the master tracks the changes and sends them to the IAPs, which periodically check in for new configurations. The master also handles software updates.

Finally, the master does fulfill a few special roles. For example, it coordinates the assignment of DHCP addresses if you implement the VC assigned option for client IP addresses on a WLAN. The master can also proxy RADIUS requests to an external server on behalf of all IAPs in the cluster. A detailed look at these features is beyond the scope of this book.

Create an Aruba Instant Access Point cluster

You will now configure a few additional settings to ensure that your first IAP will act as VC as you introduce additional IAPs. You will then add a second IAP to your environment and see how it and the first IAP form a cluster.

You will also adjust the WLAN configuration so that wireless users are assigned to a different VLAN from the IAPs' and switches' management VLAN. This approach helps to preserve security for the network infrastructure. In addition, you will assign wireless users to a new VLAN dedicated to them.

Finally, you will create a device profile on the VSF fabric to automatically apply the correct settings for an AP connection to any switch port that connects to an Aruba AP.

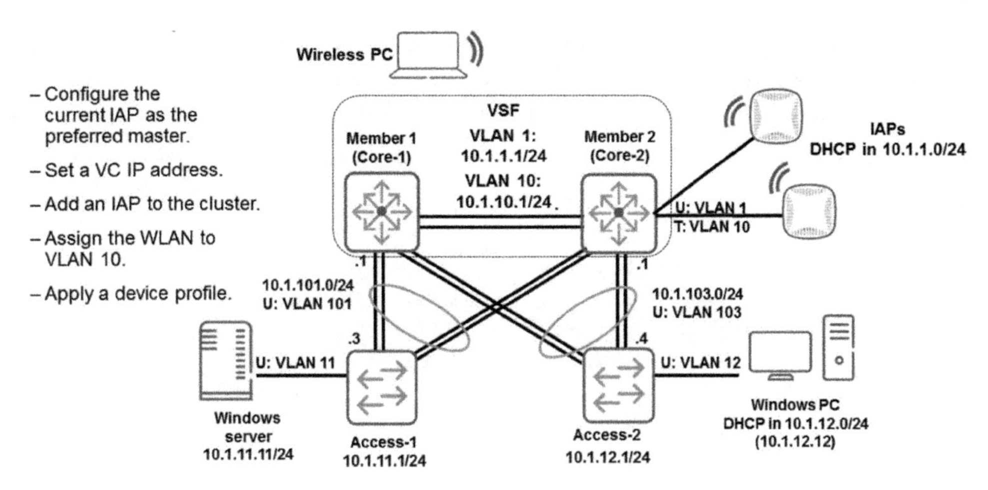

Figure 10-40: Create an Aruba Instant Access Point cluster

Task 1: Ensure that your existing IAP becomes master

Important

Your switches and IAP should be running the configuration established at the end of the preceding set of tasks.

You have already set up a WLAN on one IAP. You now want to expand wireless services by adding an additional IAP. (In the real world, you could just as easily add many more IAPs.)

Before you connect and power up the new IAP, you want to ensure that your existing IAP will be elected master. It should remain master without an election because the new IAP should detect its master beacons as soon as it boots up and while it is still in the Initial state. However, you want to make sure by using the preferred VC option. Otherwise, if a new IAP is somehow elected master, you could lose your configuration.

You will also set a static IP address for the VC that is different from the master IAP IP address.

Windows PC

1. Access the desktop for the Window's PC.
2. Open a Web browser such as Windows IE.
3. Navigate to the management IP address for the IAP, which is often 10.1.1.12.
4. Log in with the default credentials (**admin** and **admin**).

5. Now that you know a bit more about IAP clusters, you can see that this IAP—as the only IAP in your network—has actually elected itself master. The system name that you previously set was for the IAP cluster, and you configured the WLAN on the IAP cluster of one. See Figure 10-41.

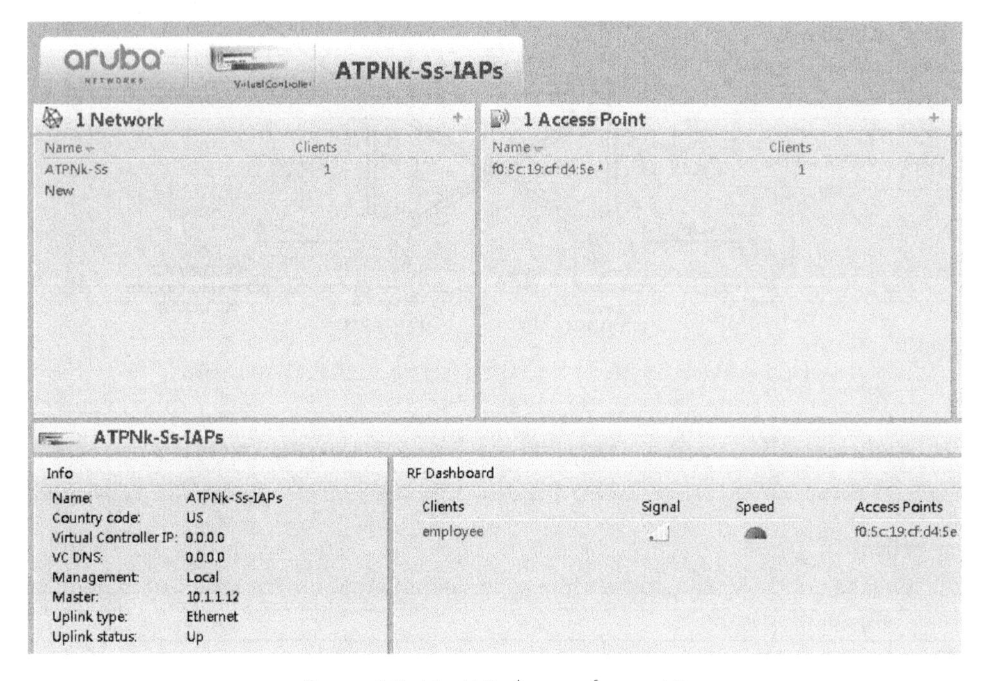

Figure 10-41: IAP cluster of one AP

You want to assign the VC an IP address of its own, separate from the master IAP.

6. Click the **System** link at the top of the UI. (See Figure 10-42.)

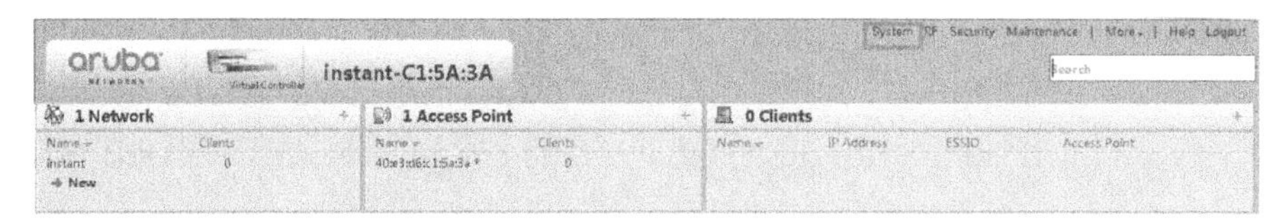

Figure 10-42: Click the System link

7. For the Virtual Controller IP, enter **10.1.1.5**.

   The IP address is within the same subnet as the IAPs' DHCP addresses, but it is excluded from DHCP and reserved for use by the VC.

Note

You can assign the VC an IP address in a different subnet from the IAPs, but you would then need to go to the advanced system settings and select Custom for the Virtual Controller network settings. Then you would need to configure the network mask, default gateway, and VLAN associated with this subnet. You would also have to assign this VLAN as a tagged VLAN on all switch ports connected to IAPs (any of which could become master in a failover situation).

8. Click **Show advanced options.**
9. Note that Auto join is enabled.

   Auto join must be enabled for your other IAP to automatically join the cluster when you power it on and connect it to the network. This feature should be enabled by default. If you disable this feature, you will see your IAP listed in the VC Instant UI interface in red, and you can then manually add the IAP. See Figure 10-43.

10. Click **OK.**

System | Help

General | Admin | Uplink | L3 Mobility | Monitoring | WISPr | Proxy | Time Based Services

Name:	ATPNk-Ss-IAPs
System location:	HPE lab
Virtual Controller IP:	10.1.1.5
Dynamic Proxy:	RADIUS TACACS
MAS integration:	Disabled
NTP server:	
Timezone:	International-Date-Line-
Preferred band:	All
AppRF visibility:	Disabled
Virtual Controller network settings:	Default
Auto join mode:	Enabled
Terminal access:	Enabled
Console access:	Enabled
Telnet server:	Disabled
LED display:	Enabled
Extended SSID:	Enabled
Deny inter user bridging:	Disabled
Deny local routing:	Disabled
Dynamic CPU management:	Automatic

Hide advanced options | OK | Cancel

Figure 10-43: Auto join mode is enabled

Now you will ensure that the current IAP remains master.

11. Select the AP, which is using its MAC address as its default name, in the Access Point list.
12. After you select the AP, you will see an **edit** link. Click the link.

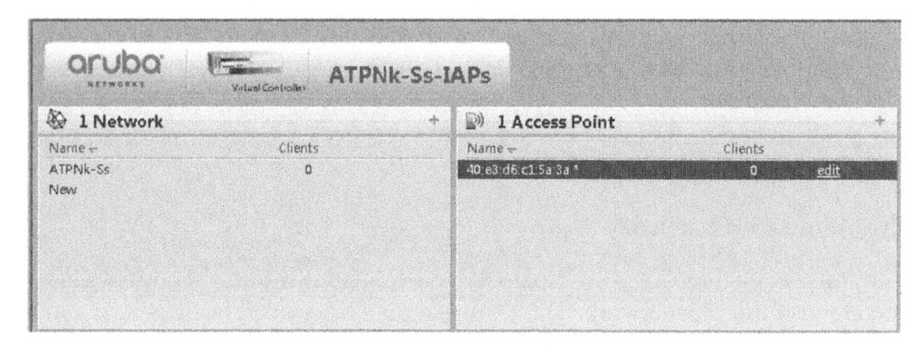

Figure 10-44: Click the edit link

13. Change Name to an intuitive name that indicates the IAP's location. Figure 10-45 shows **IAP-North** as an example.
14. For Preferred master, set Enabled. This setting gives this IAP preference over other IAPs in the master election.
15. As you see, you could give the IAP a static IP address from this window, if you wanted. For this task, though, let the IAP continue to receive its IP address through DHCP.
16. Click **OK**.

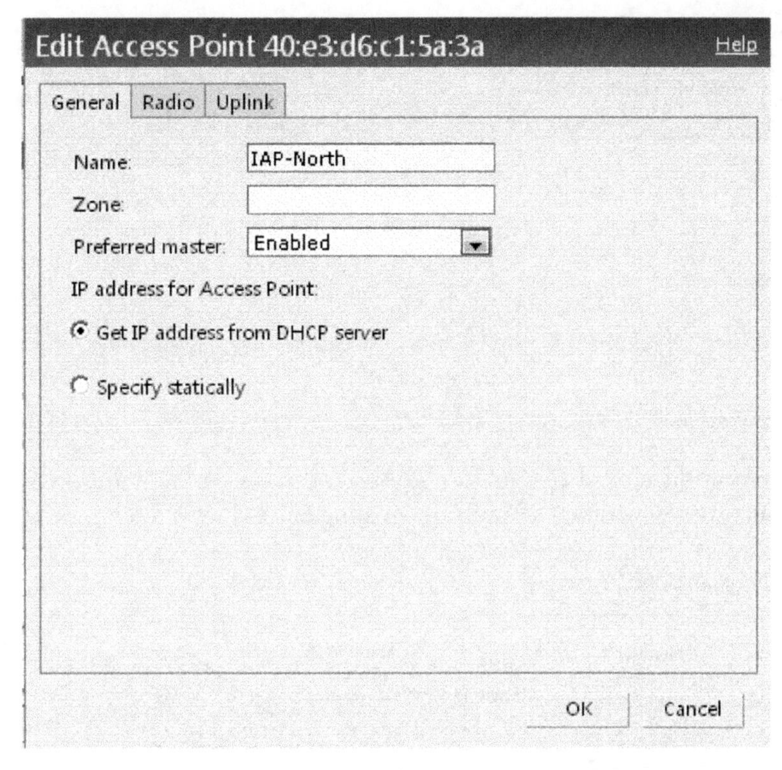

Figure 10-45: Change the AP's name to an intuitive name

17. The new settings require an IAP reboot. See Figure 10-46. (Note that if this were your only IAP, you should only change the master preference after normal work hours or during a scheduled network outage.) Click **OK**.

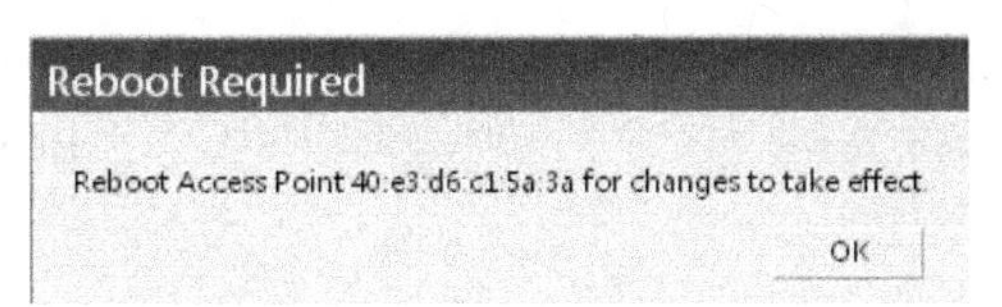

Figure 10-46: New settings require a reboot

18. After a minute or so, you can log back into the UI. Use the VC IP address, 10.1.1.5.
19. You should now see the new name for the AP. See Figure 10-47.

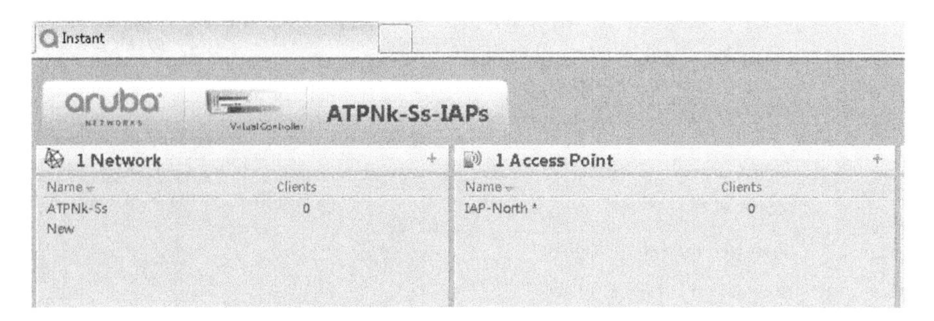

Figure 10-47: The UI now displays the AP's new name

Task 2: Connect the second IAP and establish the cluster

You will now connect the second IAP 325 to VLAN 1, the same VLAN to which the current IAP connects. You will observe that the IAP joins the existing IAP's cluster. See Figure 10-48.

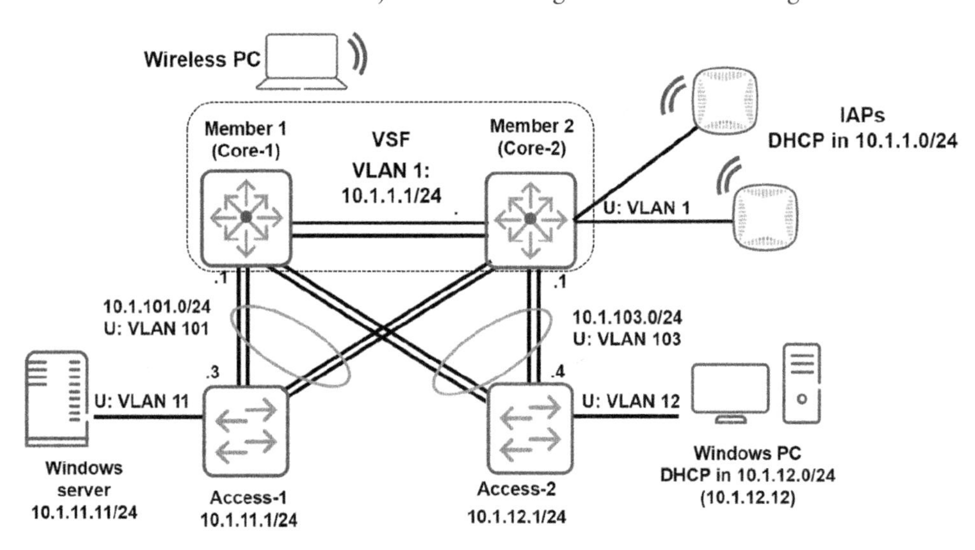

Figure 10-48: New IAP joins the existing IAP's cluster

Core-1

1. Access a terminal session with the VSF fabric through the Core-1 console connection (or Core-2).
2. Verify that the interface that connects to the second IAP is untagged in VLAN 1.

```
Core-1(config)# show vlan port 2/a3 detail
 Status and Counters - VLAN Information - for ports 2/A3
```

```
VLAN ID Name                  | Status     Voice Jumbo Mode
------- -------------------- + ---------- ----- ----- --------
1       DEFAULT_VLAN          | Port-based No    No    Untagged
```

3. Enable this port.

```
Core-1(config)# interface 2/a3 enable
```

4. Enable PoE on this port.

```
Core-1(config)# interface 2/a3 power-over-ethernet
```

5. If necessary, establish the physical connection between Core-2 and the second IAP. See Figure 10-49.

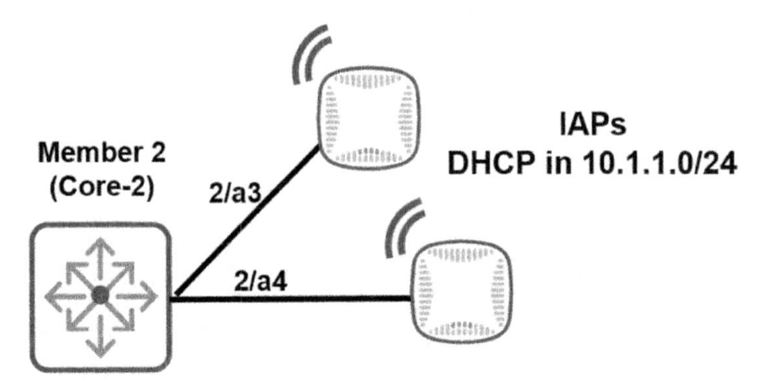

Figure 10-49: If necessary, establish a physical connection between Core-2 and the second IAP

6. Use show commands to verify that the interface comes, the IAP powers up, and the IAP receives an IP address. Note the commands that you used.

Windows PC

7. Return to Instant UI that is open on the Windows PC desktop. If necessary, log in again.
8. Wait about five minutes. You should see a second AP in the Access Points list. See Figure 10-50.

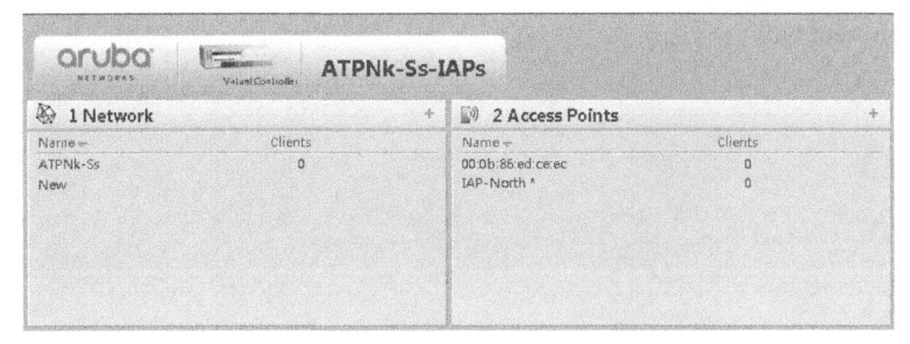

Figure 10-50: The UI now displays the second AP

9. Select the new IAP and click the **edit** link.
10. Assign the IAP a name such as **IAP-South**. See Figure 10-51.
11. Click **OK**.

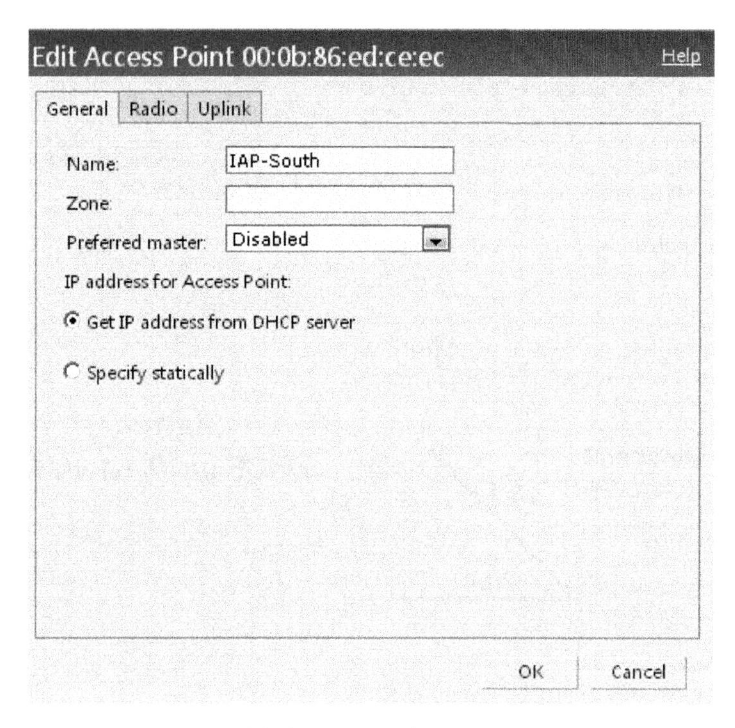

Figure 10-51: Assign the IAP a name

Task 3: Assign the WLAN to a VLAN

You will now edit the WLAN settings, and the changes will apply to both APs. You will assign the ATPN*k*-S*s* WLAN to a new VLAN, VLAN 10. You could then set up access control lists (ACLs) to protect the VLAN on which managers access the switches and IAP cluster, if you wanted.

1. Click ATPNk-Ss in the Networks list.
2. Click the **edit** link. See Figure 10-52.

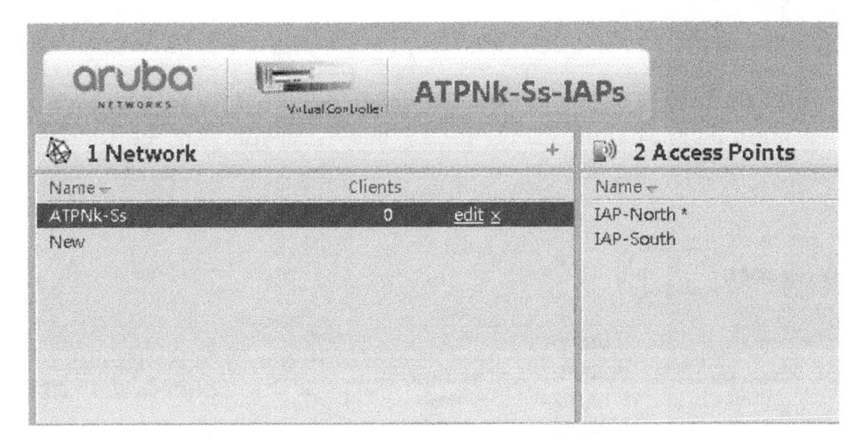

Figure 10-52: The edit link

3. Click **2 VLAN** at the top of the page.
4. For Client VLAN assignment, select Static.
5. For VLAN ID, enter 10. See Figure 10-53.

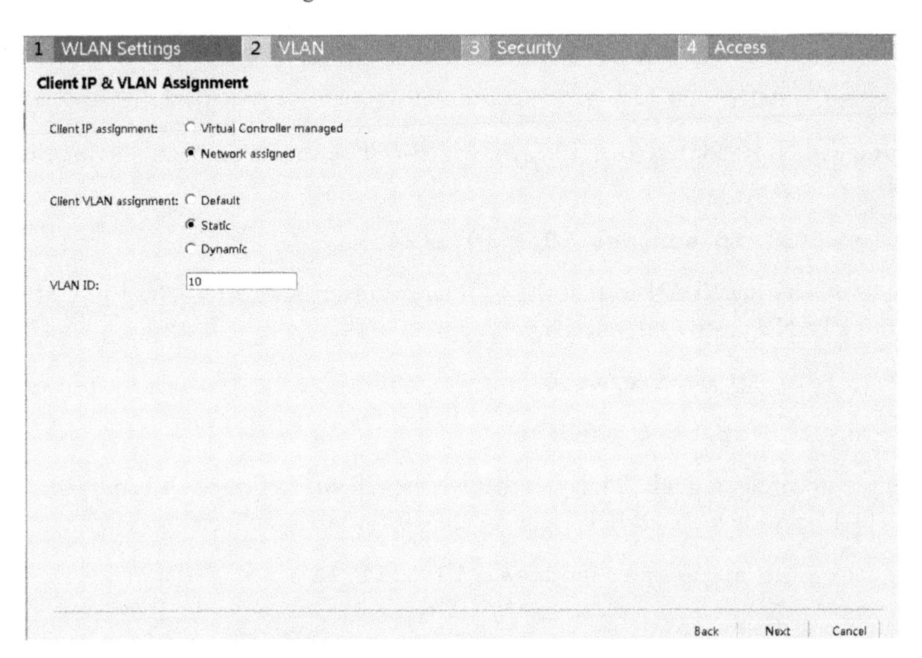

Figure 10-53: Type 10 in the VLAN ID field

6. Click **4 Access** at the top of the page. Accept the current settings and click **Finish**.

Task 4: Create the VLAN for wireless users in the wired network

You have configured the Instant cluster to forward wireless users' traffic in a VLAN reserved for those users. You now must establish this VLAN in the wired network. In the topology you see in Figure 10-54, IAPs connect to member 2 in the core VSF fabric, so you just need to set up the VLAN on the VSF fabric.

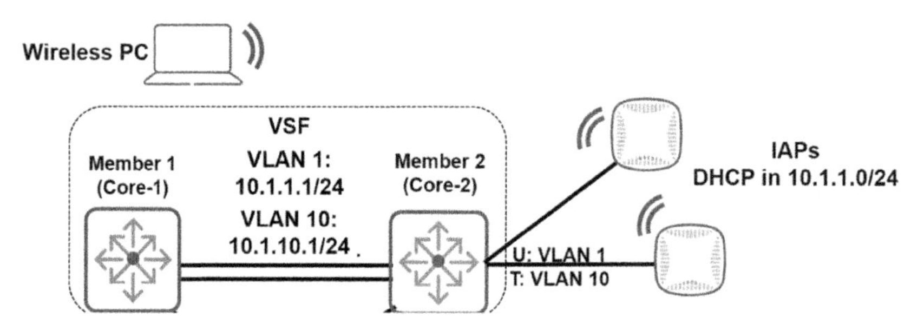

Figure 10-54: IAPs connect to member 2 in the core VSF fabric

1. Access the terminal session with the VSF fabric. Move to the global configuration mode.
2. Create VLAN 10.

```
Core-1(config)# vlan 10
```

3. Assign the switch IP address 10.1.10.1/24, which will be the default gateway for clients in the VLAN.

```
Core-1(vlan-10)# ip address 10.1.10.1/24
```

4. Enable OSPF on the VLAN so that the VSF fabric advertises 10.1.10.0/24 in OSPF. You can make this VLAN a passive interface, if you want.

```
Core-(vlan-10)# ip ospf area 0
Core-1(vlan-10)# ip ospf passive
```

5. Remember to configure DHCP relay so that wireless clients can receive IP addresses. The server at 10.1.11.11 already has a DHCP scope for the 10.1.10.0/24 subnet.

```
Core-1(vlan-10)# ip helper-address 10.1.11.11
Core-1(vlan-10)# exit
```

Task 5: Create a device profile to automatically tag AP ports for the wireless user VLAN

IAPs will forward traffic for clients connected to ATPNk-Ss with a VLAN 10 tag. The final step in establishing the VLAN for wireless users is to tag *both* IAPs' ports for that VLAN. Otherwise, the switch will drop the wireless users' traffic.

Rather than configure this VLAN assignment directly on the port, you will configure a device profile with the assignment. The switch will then automatically apply the VLAN assignment, as well as other settings in the device profile, on whichever port an IAP connects.

1. Make sure that you are in global configuration mode on the VSF fabric (Core-1 hostname).
2. Adjust the "default-ap-profile" to include VLAN 10 as a tagged VLAN.

```
Core-1(config)# device-profile name default-ap-profile tagged-vlan 10
```

Important

Type out the full name of the profile "default-ap-profile." The name is case-sensitive.

3. View your device profile. Note that your tagged VLAN settings have been added. Also review the default settings for the other settings.

```
Core-1(config)# show device-profile config

 Device Profile Configuration

 Configuration for device-profile : default-ap-profile

  untagged-vlan      : 1

  tagged-vlan        : 10

  ingress-bandwidth  : 100%

  egress-bandwidth   : 100%

  cos                : 0

  speed-duplex       : auto

  poe-max-power      : 30W

  poe-priority       : critical

 Device Profile Association
```

```
Device Type    : aruba-ap
Profile Name   : default-ap-profile
Device Status : Disabled
```

4. The "default-ap-profile" is already bound to the aruba-ap type, as you can see in the output above. However, the profile is not yet enabled. Enable the profile for Aruba APs now.

```
Core-2(config)# device-profile type aruba-ap enable
```

5. The APs are sending LLDP messages, so the switch should apply the device profile. Verify that the interfaces connected to the IAPs have applied the profile. You might need to wait about a minute for the LLDP messages to arrive.

```
Core-1(config)# show device-profile status
 Device Profile Status
 Port     Device-type Applied device profile
 -------- ----------- ----------------------
 2/A3     aruba-ap    default-ap-profile
 2/A4     aruba-ap    default-ap-profile
```

6. Also verify that the VLAN assignment has indeed been applied.

```
Core-1(config)# show vlan port 2/a3,2/a4 detail
 Status and Counters - VLAN Information - for ports 2/A3
  VLAN ID Name                 | Status     Voice Jumbo Mode
  ------- -------------------- + ---------- ----- ----- --------
  1       DEFAULT_VLAN         | Port-based No    No    Untagged
  10      VLAN10               | Port-based No    No    Tagged
 Status and Counters - VLAN Information - for ports 2/A4
  VLAN ID Name                 | Status     Voice Jumbo Mode
  ------- -------------------- + ---------- ----- ----- --------
  1       DEFAULT_VLAN         | Port-based No    No    Untagged
  10      VLAN10               | Port-based No    No    Tagged
```

7. The switch is now ready to receive traffic from wireless users in the ATPN*k*-S*s* WLAN. Verify connectivity between the VLAN 10 and the server at 10.1.11.11.

```
Core-1(config)# ping source 10 10.1.11.11
10.1.11.11 is alive, time = 1 ms
```

Task 6: Connect the client to the WLAN and monitor the connection

You will now connect the wireless client to the WLAN and verify that it receives an IP address in VLAN 10, subnet 10.1.10.0/24. You will also view information about the connection in the Instant UI.

Wireless Windows PC

1. Access the wireless Windows PC desktop.
2. The wireless PC might have already connected to your SSID automatically. The PC might also have connected and failed to receive an IP address while you were configuring the new settings. Double-check the connection.
3. If necessary, disconnect the PC from the current SSID and connect to your ATPN*k*-S*s* SSID. Enter the credentials if prompted (the credentials might be stored).
4. Verify that the connection establishes. See Figure 10-55.

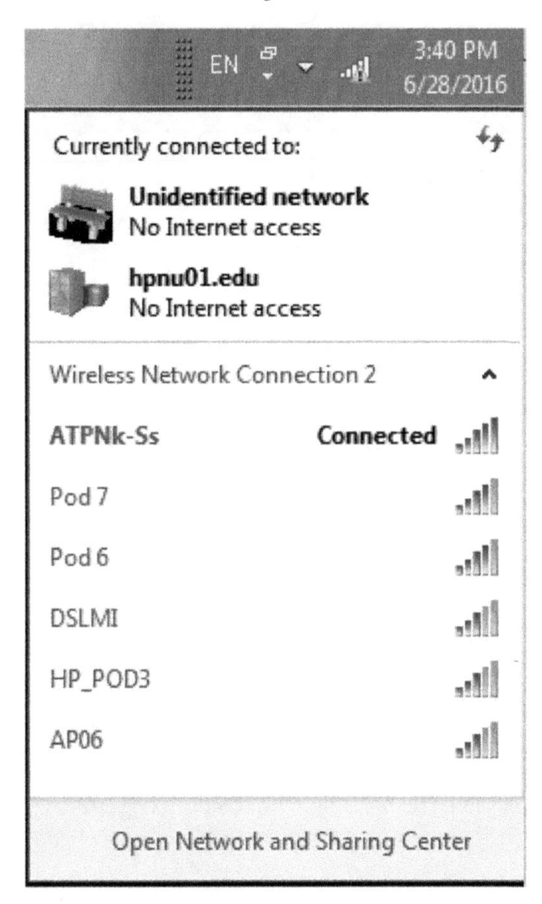

Figure 10-55: Verify that the connection is established

Windows PC

5. Return to Instant UI that is open on the Windows PC desktop. If necessary, log in again or refresh the interface.
6. You should see the client. Note that the client now has an IP address in VLAN 10.
7. Click the client name. What information can you see about the client connection in bottom pane?

Task 7: Save

1. Save the configuration on the VSF fabric.
2. The configuration on the IAP cluster is automatically saved. Backing up the configuration is optional. If you want to do so, click the Maintenance link at the top of the Instant UI. Then select the Configuration tab and click **Backup Configuration**. See Figure 10-56.

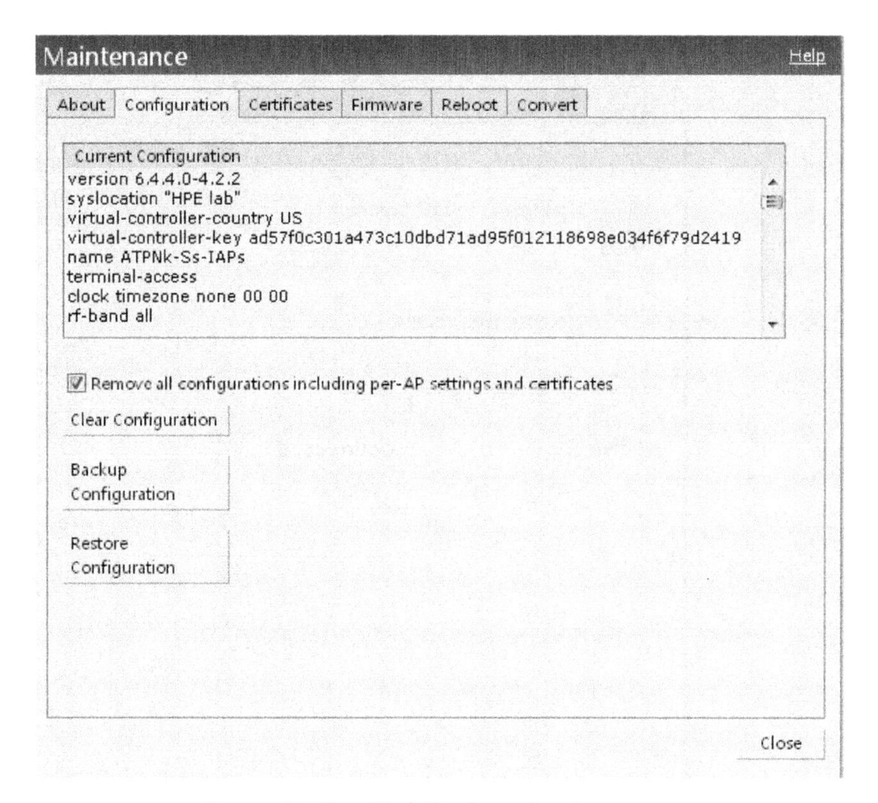

Figure 10-56: Click Back up Configuration

Learning check

1. Did you need to configure any settings on the second IAP to join it to the cluster? What did you need to do?

2. What IP address did the wireless client receive after you connected it to the wireless network? How is the AP forwarding the client traffic to the switch? Why did you need to change the VLAN assignment on both IAPs' switch ports?

Answers to Learning check

1. Did you need to configure any settings on the second IAP to join it to the cluster? What did you need to do?

   You didn't need to configure any settings. You just verified that you were connecting the IAPs in the same VLAN. (You also configured the existing IAP as the preferred master.)

2. What IP address did the wireless client receive after you connected it to the wireless network? How is the AP forwarding the client traffic to the switch? Why did you need to change the VLAN assignment on both IAPs' switch ports?

   The client received an IP address in 10.1.12.0/24 because the IAP is sending the traffic with a VLAN 12 tag. You had to tag the switch ports for both IAPs because both IAPs forward traffic locally. Traffic does not have to pass through the VC.

   You can use the Instant UI to collect additional information about client connections on your IAP cluster. Figure 10-57 and Figure 10-58 are provided below for your reference.

   You can click the client in the Clients list and make sure that Monitoring is selected in the middle bar. Then you can see information about the signal, the connection data rate, throughput, and so on. See Figure 10-57.

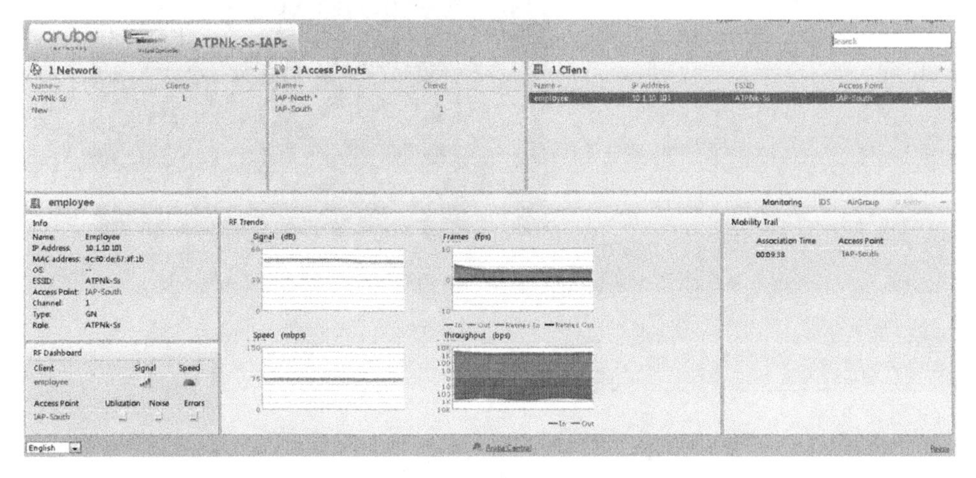

Figure 10-57: Client monitoring

When you need to troubleshoot or obtain more detailed information, click **More** at the top of the UI and select **Support**. You can then choose the statistics that you want to view from the Command menu. Choose the AP on which to collect the statistics or all APs. Then click **Run**. In Figure 10-58, you are viewing 802.1X statistics. This client passed authentication. You can see an EAPOL success and a RADIUS Accept message.

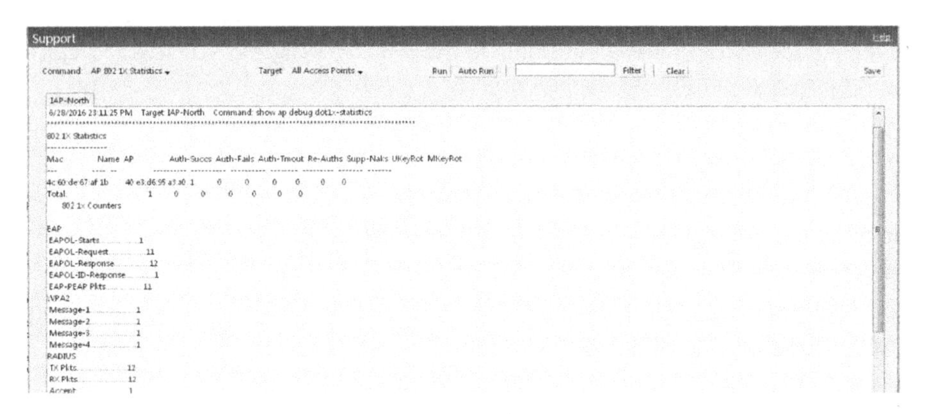

Figure 10-58: Tech support

Summary

This chapter has introduced you to basic 802.11 concepts that help you to set up an Aruba IAP correctly. You can connect to an IAP and provision it with its initial settings. You know how to configure enterprise-level security on a WLAN and how to assign that WLAN to a VLAN. You can also form an IAP cluster. See Figure 10-59.

- Basics of wireless communications
- Provisioning an Aruba IAP
- Configuring a WLAN with Enterprise (802.1X) WPA2 security
- Establishing an IAP cluster

Figure 10-59: Summary

Learning check

This final learning check will help you review the information this book provides.

1. Match the following IEEE standards with the appropriate description. Each option can only be used once.
   a. 802.11a
   b. 802.11b
   c. 802.11g
   d. 802.11n
   e. 802.11ac

   ___ Supports speeds up to 54 Mbps and uses the 5 GHz frequency

   ___ Supports MIMO for theoretical data rates up to 600 Mbps

   ___ Supports speeds up to 54 Mbps and uses the 2.4 GHz frequency

   ___ Supports MIMO for theoretical data rates in the Gbps range

   ___ Was the first IEEE wireless standard to achieve widespread adoption and uses the 2.4 GHz frequency

2. A WLAN defines all the following settings for an IAP, except which?
   a. SSID
   b. Power settings for the AP radios
   c. Wireless protection such as WPA2
   d. MAC authentication

3. What is a requirement for an IAP cluster to form?
   a. IAPs support same WLAN.
   b. IAPs are the same model.
   c. IAPs are in the same VLAN.
   d. IAPs have the same self-assigned IP address.

Answers to Learning check

1. Match the following IEEE standards with the appropriate description. Each option can only be used once.
   a. 802.11a
   b. 802.11b
   c. 802.11g
   d. 802.11n
   e. 802.11ac

   a Supports speeds up to 54 Mbps and uses the 5 GHz frequency

   d Supports MIMO for theoretical data rates up to 600 Mbps

   c Supports speeds up to 54 Mbps and uses the 2.4 GHz frequency

   e Supports MIMO for theoretical data rates in the Gbps range

   b Was the first IEEE wireless standard to achieve widespread adoption and uses the 2.4 GHz frequency

2. A WLAN defines all the following settings for an IAP, except which?
   a. SSID
   b. Power settings for the AP radios
   c. Wireless protection such as WPA2
   d. MAC authentication

3. What is a requirement for an IAP cluster to form?
   a. IAPs support same WLAN.
   b. IAPs are the same model.
   c. IAPs are in the same VLAN.
   d. IAPs have the same self-assigned IP address.

Job aid

This job aid summarizes 802.11 standards, their frequencies, and their data rates.

Not all data rates for 802.11n and 802.11ac are shown for simplicity. Instead, for each set of features, the range of available data rates are shown from the lowest data rate (the lowest MCS using the long guard interval [GI]) to the highest data rate (the highest available MCS using the short GI). Remember that the MCS depends on the signal strength between the client and AP. So, for example, a client and AP are using four spatial streams and four bonded channels, but have a poor signal. The data rate will be at the lower end of 117 – 1733.3 Mbps range. But if the client has a good signal, the data rate will be at the higher end of this range.

Also note that some 802.11ac features will not be available in the first products that come to market; for example, APs and clients will initially support a maximum of four spatial streams. In addition, enterprises will typically only use up to four bonded channels. Aruba 330 Series APs support these maximums of four spatial streams and four bonded channels (or two spatial streams and eight bonded channels), delivering up to 1733 Mbps on the 802.11ac radio.

Job aid—Table 10-2: 802.11 standards

Standard	RF frequency band	Number of spatial streams	Use of channel bonding	Data rates (Mbps)
802.11b	2.4 GHz	N/A	N/A	1, 2, 5.5, 11
802.11g	2.4 GHz	N/A	N/A	6,9,12,18,24,36,48,54
802.11a	5 GHz	N/A	N/A	6,9,12,18,24,36,48,54
802.11n	2.4 GHz or 5 GHz	One spatial stream	MSC 0 – MSC 7	
			No channel bonding	6.5 – 72.2
			Channel bonding (2)	13.5 – 150
		Two spatial streams	MSC 8 – MSC 15	
			No channel bonding	13 – 144.4
			Channel bonding (2)	27 – 300
		Three spatial streams	MSC 16 – MSC 23	
			No channel bonding	19.5 – 216.7
			Channel bonding (2)	40.5 – 450
		Four spatial streams	MSC 24 – MSC 31	
			No channel bonding	26 – 288.9
			Channel bonding (2)	54 – 600
802.11ac	5 GHz	One spatial stream	MSC 0 – MSC 9	
			No channel bonding*	6.5 – 86.7
			Channel bonding (2)	13.5 – 200
			Channel bonding (4)	29.3 – 433.3
			Channel bonding (4+4)	58.5 – 866.7
		Two spatial streams	MSC 0 – MSC 9	
			No channel bonding*	13 – 173.3
			Channel bonding (2)	27 – 400
			Channel bonding (4)	58.5 – 866.7
			Channel bonding (4+4)	117 – 1733.3
		Three spatial streams	MSC 0 – MSC 9	
			No channel bonding	19.5 – 288.9
			Channel bonding (2)	40.5 – 600
			Channel bonding (4)**	87.8 – 1300
			Channel bonding (4+4)*	175.5 – 2340

Job aid—Table 10-2: Continued.

Standard	RF frequency band	Number of spatial streams	Use of channel bonding	Data rates (Mbps)
		Four spatial streams	MSC 0 – MSC 9	
			No channel bonding*	26 – 346.7
			Channel bonding (2)	54 – 800
			Channel bonding (4)	117 – 1733.3
			Channel bonding (4+4)	234 – 3466.7
		Five spatial streams	MSC 0 – MSC 9	
			No channel bonding*	32.5 – 433.3
			Channel bonding (2)	67.5 – 1000
			Channel bonding (4)	146.3 – 2166.7
			Channel bonding (4+4)	292.5 – 4333.3
		Six spatial streams	MSC 0 – MSC 9	
			No channel bonding	39 – 577.8
			Channel bonding (2)	81 – 1200
			Channel bonding (4)*	175.5 – 2340
			Channel bonding (4+4)	351 – 5200
		Seven spatial streams	MSC 0 – MSC 9	
			No channel bonding*	45.5 – 606.7
			Channel bonding (2)	94.5 – 1400
			Channel bonding (4)**	204.8 – 3033.3
			Channel bonding (4+4)	409.5 – 6066.7
		Eight spatial streams	MSC 0 – MSC 9	
			No channel bonding*	52 – 693.3
			Channel bonding (2)	108 – 1600
			Channel bonding (4)	234 – 3466.7
			Channel bonding (4+4)	468 – 6933.3
*MSC9 is not available.				
**MCS6 is not available				

11 Aruba AirWave

EXAM OBJECTIVES

✓ Configure AirWave management settings on an IAP cluster

✓ Configure SNMP v2c settings on ArubaOS switches

✓ Discover ArubaOS switches in AirWave and bring switches and IAPs under monitoring and management

✓ Implement Zero Touch Provisioning (ZTP) for Aruba IAPs and ArubaOS switches

ASSUMED KNOWLEDGE

Before reading this chapter, you should have a basic understanding of:

- Simple Network Management Protocol (SNMP)
- Switches
- Access Points (AP)

INTRODUCTION

In this guide, you have focused on configuring and managing ArubaOS switches from the command line interface (CLI). In larger networks, though, you will find it helpful to have a centralized tool for monitoring, managing, configuring, and troubleshooting all of your network infrastructure devices. Similarly, although Instant clustering dramatically simplifies wireless network management, some companies need to manage many IAP clusters in different locations.

This chapter introduces you to Aruba AirWave, a powerful unified wired and wireless network management tool. You will learn about AirWave capabilities and how AirWave uses Simple Network Management Protocol (SNMP), Telnet or SSH, and HTTPS to monitor and manage devices. You will also learn how to discover network infrastructure devices with AirWave so that AirWave can manage them and also how to use AirWave to provision new switches and APs automatically and instantly.

Note
The latest version of AirWave as of the publication of this guide is 8.2.

Introduction to Aruba AirWave

The only management platform designed to prioritize mobility the way users increasingly do, Aruba AirWave provides unified management of wired and wireless users and devices. It grants network administrators deep insight into traffic at the application and network service level.

Figure 11-1: Introduction to Aruba AirWave

AirWave helps administrators to move beyond simple management and monitoring to network optimization. They can quickly troubleshoot and resolve issues, helping the business to maintain continuity. AirWave also helps administrators to implement best practices for network security. Customizable reports help to support companies' efforts to prove compliance with regulations.

Key AirWave capabilities

AirWave offers a simple discovery process for Aruba IAPs that you will practice in the tasks in the "Manage Devices in Aruba AirWave and Use ZTP" section of this chapter.

You can also quickly discover all the network infrastructure devices on a network, including HPE switches, as well as ArubaOS switches, and third-party devices from vendors such as Cisco, Juniper, and Motorola. AirWave automatically classifies the devices by type, and you can configure it to automatically place devices in the most appropriate device group.

Comprehensive visibility
Device discovery and classification
Reporting, analysis, and diagnosis
Views of the entire access network, wireless and wired
Aruba Clarity for at-a-glance troubleshooting

Centralized management
Firmware updates
Configurations pushed from templates
Automated compliance enforced across entire distributed organization

Easy, intuitive interface
Role-based administrative access
Organized and managed by business or organizational unit

Figure 11-2: Key AirWave capabilities

After you add discovered devices to AirWave, AirWave monitors the devices, providing comprehensive views of the entire access network, both wireless and wired. As you will learn later, you can view details about clients, network usage, device status, and other crucial information at a glance. AirWave also detects and reveals network usage deviations, helping you to diagnose and resolve issues as quickly as possible. Clarity, an engine within AirWave, offers you at-a-glance views of issues that might cause disruptions in client connectivity such as failed authentications and DHCP requests, enabling you to quickly resolve user issues.

AirWave also centralizes device management, including firmware updates and configurations. You can specify a firmware version and configuration template for a device group, and AirWave audits devices for compliance, automatically updating devices to the desired firmware and configuration, if you choose.

IT staff administer AirWave from a console, which supports a variety of methods for authenticating managers, including integration with directory services, making it easy to grant specific rights to specific business units. For example, role-based management can grant different users rights to view and manage different devices.

AirWave deployment options

Customers can purchase AirWave as an appliance, which includes the software and hardware, or as software only. The software is distributed as an .iso image, which includes AirWave bundled with Linux CentOS, and can be installed on a bare metal server or a Red Hat Enterprise Linux (RHEL) Basic Server without any customization or applications. You can also install the image as a virtual machine (VM) on Microsoft Windows Hyper-V. Customers with VMware ESXi environments can alternatively obtain an ova image for a VM.

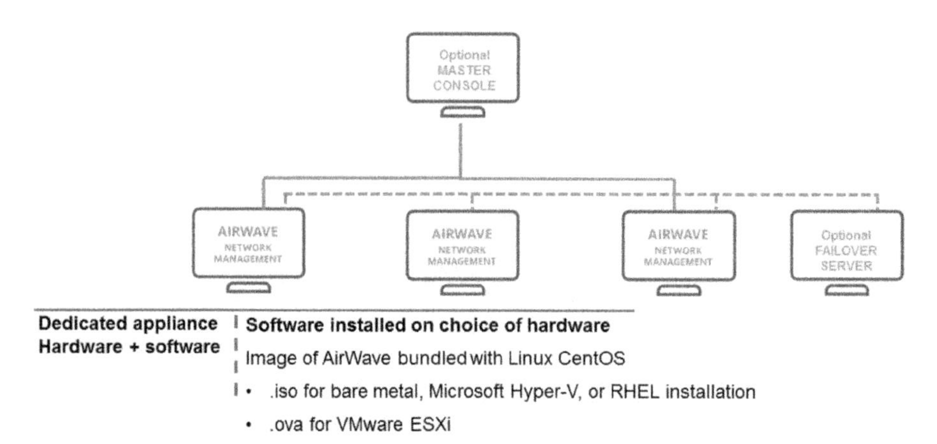

Figure 11-3: AirWave deployment options

It is important to plan the server to meet processor, RAM, disk, and network requirements, which vary based on the number of devices that AirWave will manage, as well as the types of features planned. Refer to the *Airwave 8.2 Server Sizing Guide* for detailed guidelines.

This guide outlines hardware requirements for a server that can support up to 1500 devices and for a server that can support up to 4000 devices. If the customer has more devices, you can deploy multiple AirWave servers. In this case, you should recommend the optional AirWave Master Console, which acts as the single management interface for multiple AirWave servers, providing centralized management for a scalable management solution.

Customers who require high availability can deploy AirWave failover servers. A failover server monitors the status of one or more active AirWave servers and takes over a server's functions if it fails. The failover server should have the same hardware as the server that it monitors (if it monitors more than one servers, it should meet the highest requirements).

Discovering and managing an existing Aruba Instant cluster

A company that has an expanding number of Instant clusters might decide that it is time to manage those clusters centrally. To discover each already established Instant cluster in AirWave, you simply configure a few settings in the **System** > **Admin** page of the Instant UI.

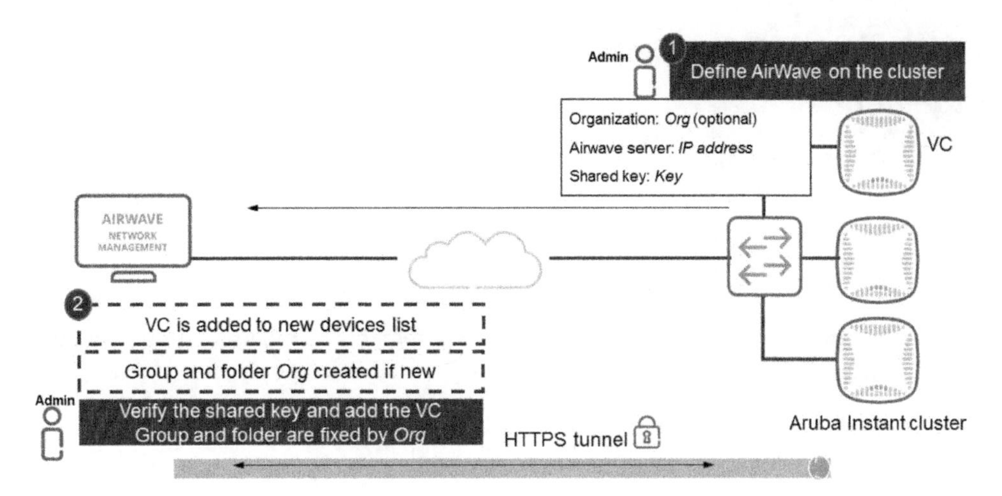

Figure 11-4: Discovering and managing an existing Aruba Instant cluster

You must set the AirWave IP address and a shared key. The Virtual Controller (VC) for the cluster then contacts the AirWave server, which adds the VC as a new device. Devices in the new devices list, however, are not yet monitored or managed by AirWave. You must authorize them. The shared key, which only authorized network administrators should know, protects AirWave from managing rogue devices. You do not set the shared key anywhere on the AirWave Management Platform (AMP). Instead, you view the shared key when checking the VC in the list of new devices. If the key is correct, you can then add the VC to AirWave. At that point, AirWave also discovers and adds all IAPs in the cluster.

You can optionally set an Organization string in the **System** > **Admin** page. This string defines the AirWave group and folder to which the VC and its IAPs are assigned. You'll learn more about AirWave groups and folders later in this chapter. If the group and folder do not yet exist, AirWave creates them. Be careful to avoid typos, which would lead to the creation of a new group and folder and your IAPs placed in these instead of the correct ones. You *cannot* override the group and folder set by the organization string when you add the VC.

If you don't set an Organization string, AirWave places the IAP in the default Access Points group and the Top folder. You cannot override this behavior by selecting a different group and folder when you authorize the IAP. You can, however, move the IAPs to a different group and folder after you add them, if you so choose.

After you add the device, AirWave uses HTTPS to establish a secure, encrypted management tunnel with an autonomous IAP or the VC (master) of an Instant cluster. The IAP or VC pushes information such as client connections, RF status, and so on over this tunnel. AirWave can also send messages to alter the Instant configuration.

Rather than use the shared key method, you can set up certificate authentication of Aruba Instant devices. For more information on this option, refer to the AirWave documentation.

Preparing ArubaOS switches for AirWave monitoring and management

You initiate the discovery of other types of network devices, including ArubaOS switches, through the AirWave console, AMP. However, you might need to prepare the devices in advance by configuring them with the correct credentials for the management protocols that AirWave will use:

- SNMPv1 and SNMPv2c = a read only or read write community
- SNMPv3 = a username and, depending on the requirements of the user's group, an authentication password, a privacy password, or both

You should also prepare the devices to support Telnet or SSH logins, which AirWave also uses to monitor and manage devices. The sections that follow explain more.

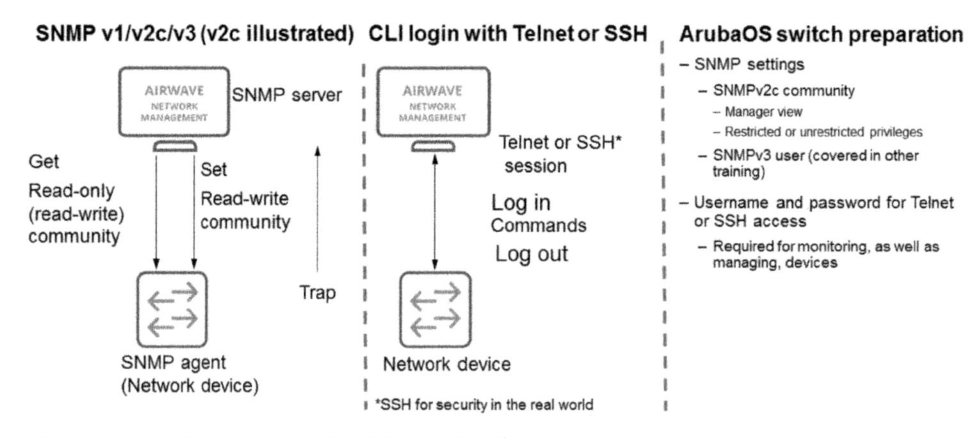

Figure 11-5: Preparing ArubaOS switches for AirWave monitoring and management

SNMP

SNMP is an industry-standard protocol that allows you to manage and monitor a variety of network devices from a central location. AirWave uses SNMP to monitor ArubaOS switches and a variety of other network infrastructure devices.

SNMP works in a client-server relationship. SNMP agents, which function as clients, run on managed devices such as switches. The SNMP server is a management application that requests, handles, and analyzes the information from the managed devices. As an SNMP server, AirWave sends Get messages to the agents to query them about their settings, their logs, their status, and so on. As of the publication of this guide, AirWave does not use SNMP to configure settings on Aruba devices.

Agents mostly only reply to server messages, but they can send SNMP traps to their trap server to update the server when an alarm or event, such as a failed link, occurs. You should ensure that AirWave is configured as a trap server on your devices.

When your solution uses SNMP v2c, the SNMP servers and agents include a community string in the messages. SNMP agents will only accept GET messages if the community matches their read-only or read-write community string. They will only accept SET messages if the community matches their read-write community string.

SNMPv2 is easy to set up but is not secure because the community is included in messages in plaintext. Therefore, the community string cannot function as a true password. You will use SNMPv2 in the practice tasks for simplicity, but you should use SNMPv3 in the real world for security.

Note

SNMPv3 provides for data privacy (encryption) and data integrity. Instead of community strings, SNMPv3 uses user credentials. The server credentials (not transmitted in plaintext) must match user credentials on the managed device, and those credentials are associated with various read and write privileges.

The ArubaOS switches have SNMP enabled and running in v2c mode by default. ArubaOS switches support the "public" community, by default, and they assign unrestricted (write) privileges to it. You should typically delete this community for security purposes. Then you should create your own community. When you define the community, you can set the view—manager or operator—and privilege level—restricted (read) or unrestricted (write). You can create either a manager restricted or manager unrestricted community for AirWave to use (AirWave only requires SNMP read access because it uses SSH to configure the switches). You will match the community string on AirWave, as you will learn how to do later in this chapter.

Logging into the CLI

To complete some functions, AirWave logs into a network device's CLI and sends commands over a terminal session, just like you do as an admin. The process happens in the background when you use a service applying a configuration template just as SNMP GET and SET messages do. You simply see that you make a change through the AirWave management interface, and the change is applied.

AirWave supports Telnet or SSH for logging into switches. You should configure the ArubaOS switch with a username and password for Telnet or SSH access even if you plan only to monitor the device. It is recommended that you use SSH in a real world environment for security. AirWave logs into the ArubaOS switch CLI to check the running-config in order to audit the configurations of monitored devices, and you will see errors if AirWave has no credentials for doing so. Later in this chapter, you will learn how to set up AirWave with the correct login credentials to match the credentials defined on the ArubaOS switch.

Discovering ArubaOS switches in AirWave

After you have set up SNMP on the switches, you can discover them in AirWave.

AirWave uses SNMP and HTTP scanning to discover many devices at once. It must use SNMP, rather than HTTP for ArubaOS switches.

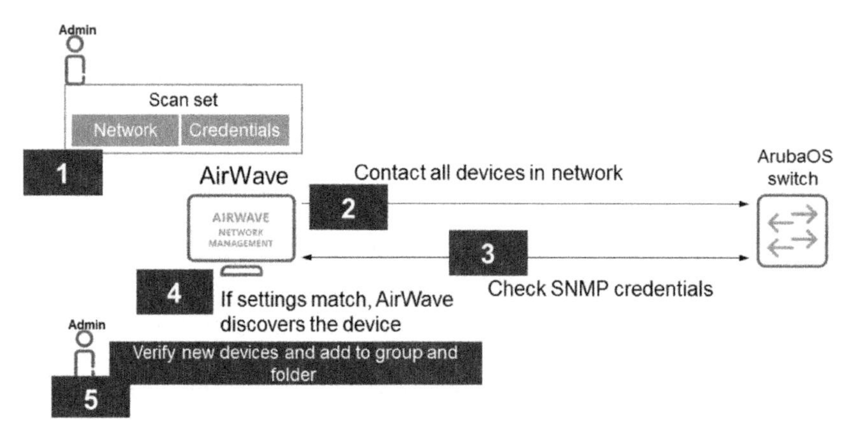

Figure 11-6: Discovering ArubaOS switches in AirWave

To set up the scan, you begin by defining networks, which are the subnets on which network infrastructure devices have their IP addresses. Next you define scan credentials for the protocol that you plan to use to discover the devices. The credentials that you define here must match the credentials that you prepared on the device. You then combine the networks and credentials in a scan set. Note that AirWave only uses these credentials for discovering devices, not for monitoring and managing them. You will learn how to set up those credentials a bit later.

When you initiate a scan with this scan set, AirWave contacts every IP address in the networks and attempts to retrieve information from that device using SNMP or HTTP. This information helps AirWave to determine what type of device it has reached and begin to monitor the device.

Any network infrastructure devices that AirWave succeeds in discovering, AirWave adds to the new device list. You must authorize the device, placing it in a group and folder and specifying a management level. You will learn more about groups, folders, and management levels a bit later in this chapter.

Note

You can configure AirWave to automatically authorize discovered devices and classify them in groups and folders.

You can manually add a device by specifying its IP address, SNMP credentials, and Telnet/SSH credentials. AirWave then discovers and immediately authorizes and adds the device.

For more information on these options, refer to the AirWave documentation.

Discovery and communication settings

It is important for you to understand that AirWave only uses the SNMP credentials that you specify for the scan set during the discovery process. Communication settings, which define settings such as SNMP communities and login credentials for Telnet or SSH, allow AirWave to monitor and manage the devices after they are authorized and added.

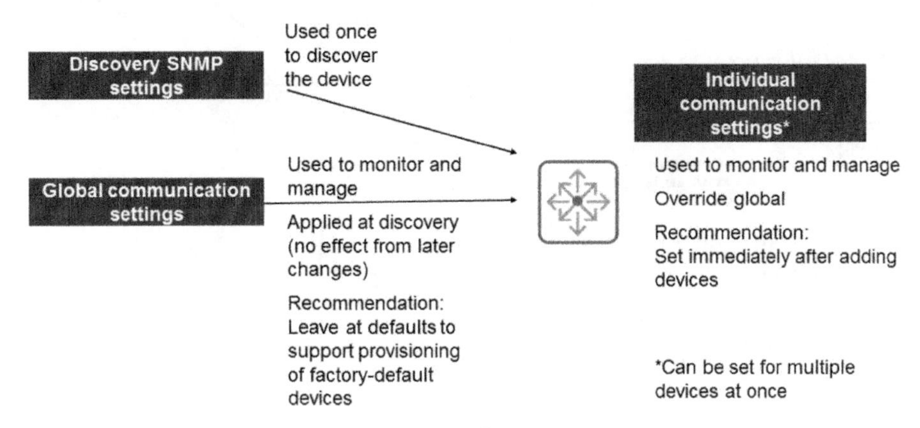

Figure 11-7: Discovery and communication settings

AirWave has global communication settings for many different device types. When you add a device to AirWave, AirWave applies the global configuration settings for that device type to the device. This application is a one-time process; if you later make changes to the global configuration settings, those changes do not apply to devices that are already added to AirWave.

Typically, you should leave the global configuration settings at their default settings so that AirWave can communicate with devices operating at factory defaults. For example, this is required for Zero Touch Provisioning (ZTP), about which you will learn more later, to function correctly. Adjust the individual communication settings for devices as soon as you add them, specifying the correct SNMP community and login credentials for your devices. If you fail to adjust the settings, AirWave will consider the devices down because SNMP polls fail, and AirWave will not be able to monitor configurations or complete other functions.

As you'll see in the later, AirWave allows you to make the same changes to communication settings on many devices at once to streamline the process.

Discover and Monitor Devices in Aruba AirWave

You will now configure SNMP settings on your test network ArubaOS switches and AirWave settings on your Instant cluster. You will then discover the network devices in AirWave, begin monitoring them, and explore the information that you can view.

Note that you will not add Access-2 to AirWave at this time. Instead you will use ZTP to add Access-2 and deploy a configuration to it later in this chapter.

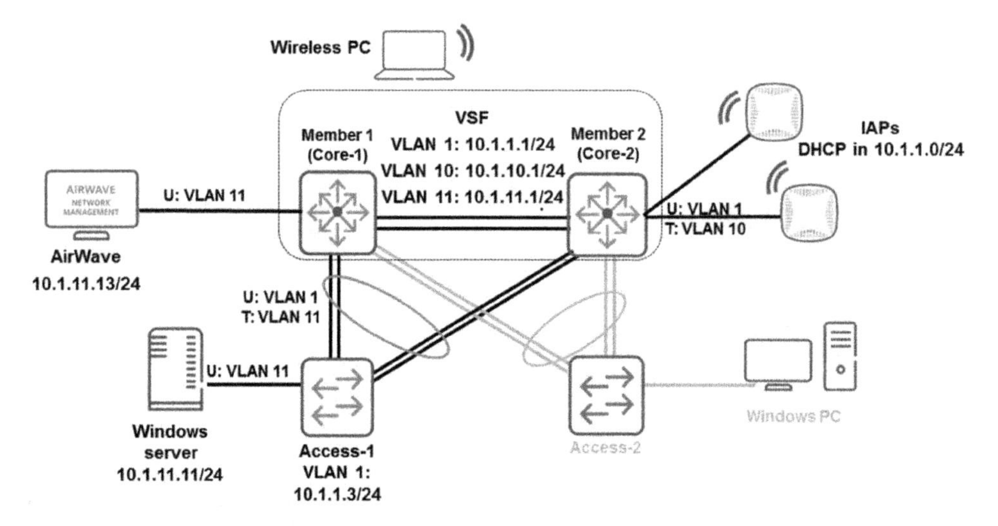

Figure 11-8: Discover and Monitor Devices in Aruba AirWave

After completing the tasks in this section, you will be able to:

- Configure SNMP communities on ArubaOS switches
- Configure an IAP cluster for discovery in Aruba AirWave
- Discover ArubaOS switches and Aruba IAPs in AirWave
- Assign switches and APs to device groups

The figures show the final topology for your test network. In the first task, you will adjust the topology so that Access-1 no longer routes traffic. Instead, it extends VLAN 11 to Core-1, which provides routing. The Access-2 switch and connections are grayed out because you will not use that switch in completing the tasks.

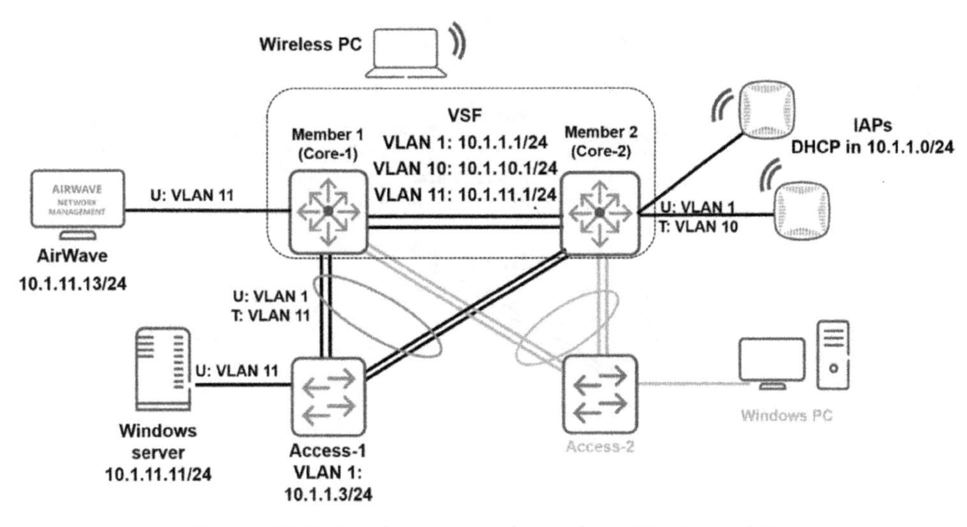

Figure 11-9: Final test network topology (VLAN and IP)

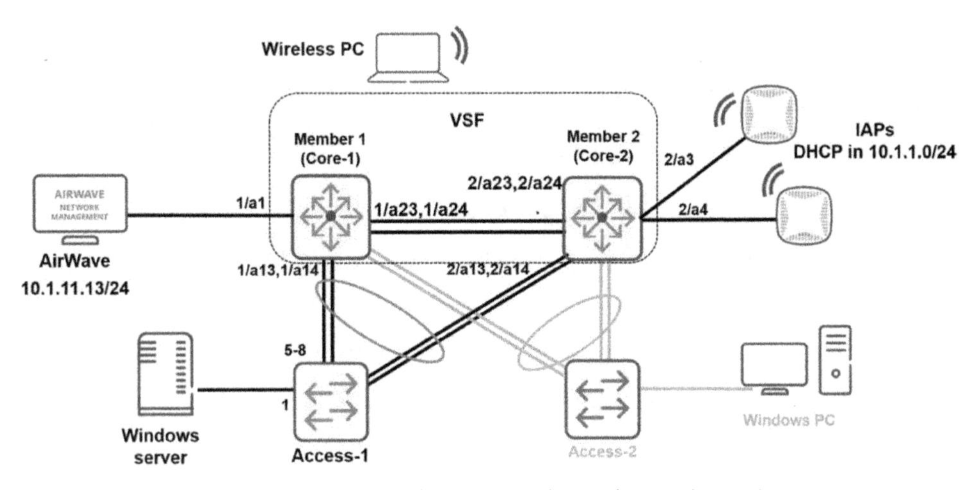

Figure 11-10: Final test network topology (physical)

Task 1: Connect AirWave to the network

AirWave is already installed and assigned an IP address: 10.1.11.13/24. However, AirWave is operating at the factory default settings and has not yet discovered any devices.

You will now configure your VSF fabric and Access-1 switch to connect to AirWave. The figure below shows the new topology, in which Access-1 is once again an edge device that does not perform any routing. The VSF fabric-will be the default router for the 10.1.11.0/24 subnet.

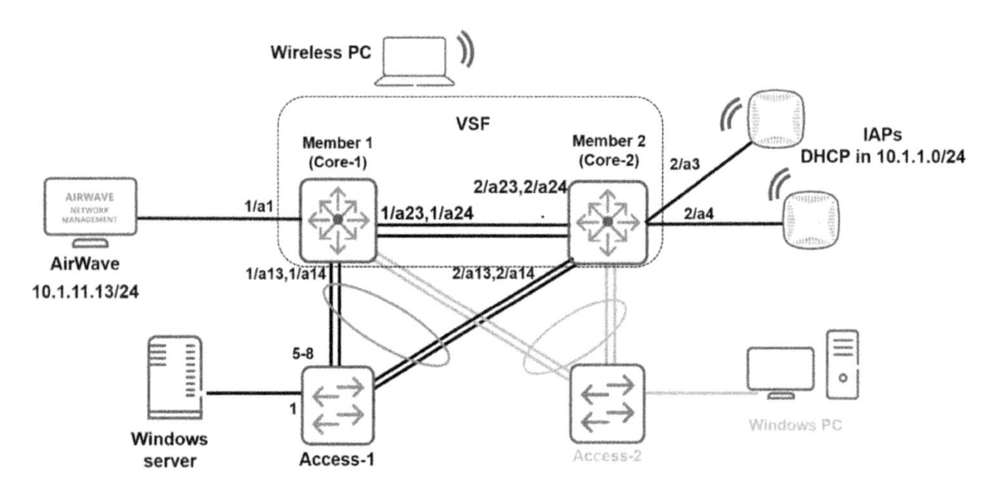

Figure 11-11: Task 1: Connect AirWave to the network

Access-1

1. Access a terminal session with Access-1.
2. Move to global configuration mode.
3. Remove the current VLAN 11 IP address on Access-1.

```
Access-1(config)# no vlan 11 ip address
```

4. Because the trk1 interface will no longer have a VLAN dedicated to it alone, remove the BPDU filter.

```
Access-1(config)# no spanning-tree trk1 bpdu-filter
```

5. Assign VLAN 1 as an untagged VLAN on the trk1 link aggregation.

```
Access-1(config)# vlan 1 untagged trk1
```

6. Assign VLAN 11 as a tagged VLAN on the trk1 link aggregation.

```
Access-1(config)# vlan 11 tagged trk1
```

7. Make sure that Access-1 still has an IP address on VLAN 1, 10.1.1.3 on subnet 10.1.1.0/24, and its default route through VLAN 1.

```
Access-1(config)# show ip route
```

```
                    IP Route Entries

Destination    Gateway        VLAN Type       Sub-Type Metric Dist.
-----------    ------------   ---- -----      -------- ------ -----
0.0.0.0/0      10.1.1.1       1    static              1      1
10.1.1.0/24    DEFAULT_VLAN   1    connected           1      0
127.0.0.0/8    reject              static              0      0
127.0.0.1/32   lo0                 connected           1      0
```

Important

If this IP address and route are not on the switch, enter **vlan 1 ip address 10.1.1.3/24** and **ip route 0.0.0.0/0 10.1.1.1**.

8. Access-1 has some leftover settings that will not affect this task. But you can clean them up with these commands.

```
Access-1(config)# no vlan 101
Access-1(config)# no ip route 0.0.0.0/0 10.1.101.1
Access-1(config)# no router ospf
All OSPF configuration will be deleted. Continue [y/n]? y
Access-1(config)# no ip routing
```

Core-1

You will now set up Core-1 as the VLAN 11 default router and connect AirWave to VLAN 11.

9. Access a terminal session with the VSF fabric through Core-1 and move to the global configuration mode.

10. Because the link aggregation to Access-1 will no longer have a dedicated VLAN, remove the BPDU filter. Also remove the filter from the link aggregation to Access-2, which you will use later in this chapter.

```
Core-1(config)# no spanning-tree trk2,trk3 bpdu-filter
```

11. Assign VLAN 1 as the untagged VLAN on trk2, which connects to Access-1.

```
Core-1(config)# vlan 1 untagged trk2
```

12. Ping 10.1.1.3. (If the ping fails, double-check the IP address on Access-1.)

```
Core-1(config)# ping 10.1.1.3
10.1.1.3 is alive, time = 1 ms
```

13. Configure the 10.1.11.1/24 IP address on VLAN 11.

```
Core-1(config)# vlan 11
Core-1(vlan-11)# ip address 10.1.11.1/24
```

14. Assign VLAN 11 as a tagged VLAN on trk2.

```
Core-1(vlan-11)# tagged trk2
```

15. Ping the server.

```
Core-1(vlan-11)# ping 10.1.11.11
```

16. Assign VLAN 11 as an untagged VLAN on 1/a1, which connects to AirWave.

```
Core-1(vlan-11)# untagged 1/a1
Core-1(vlan-11)# exit
```

17. Enable the interface that connects to AirWave.

```
Core-1(oonfig)# interface 1/a1 enable
```

18. If necessary in your environment, establish the physical connection.

19. Ping AirWave to make sure that it is connected correctly.

```
Core-1(config)# ping 10.1.11.13
```

20. Also ping AirWave using the VLAN 1 IP address as the source address to confirm that AirWave has the correct default gateway and can send routed traffic.

```
Core-1(config)# ping source 1 10.1.11.13
```

 **Note**

The Core-1 configuration will differ slightly from the configuration shown in the config files on the Windows PC. That configuration has the default VLAN assigned to the Trk3 link aggregation because Access-2 is not being used in this set of tasks. It also does not have OSPF settings. Your configuration retains the VLAN 103 assignment on Trk3 and the OSPF settings in case you want to reach the rest of the network from Access-2 and the Windows PC.

If you want to clean up the configuration to match the 11.2 config file, though, you can assign Trk3 to VLAN 1, remove the Trk3 BPDU filter, and remove OSPF. You can also disable the 1/a15,1/a16,2/a15, and 2/16 interfaces that connect to Access-2.

Task 2: Access the AirWave Management Platform

You will now access the AirWave Management Platform (AMP) and start exploring AirWave.

Windows server

1. Open a browser window on the server.

   You might want to choose Firefox because IE sometimes prevents you from accessing some content on the AirWave server, depending on your security settings.

2. Browse to https://10.1.11.13.
3. You might see a warning that the certificate is untrusted. Choose to bypass the warning and go to the site.
4. Log in to the AMP with the default credentials, **admin** for both the username and password.

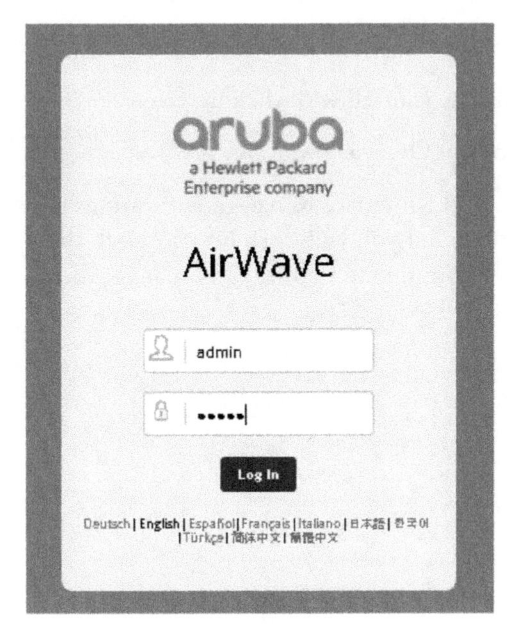

Figure 11-12: Log in to the AMP

5. You are using a trial version of AirWave. The first time that you log in, you will see a message about sending information to Aruba. Click **Cancel**.

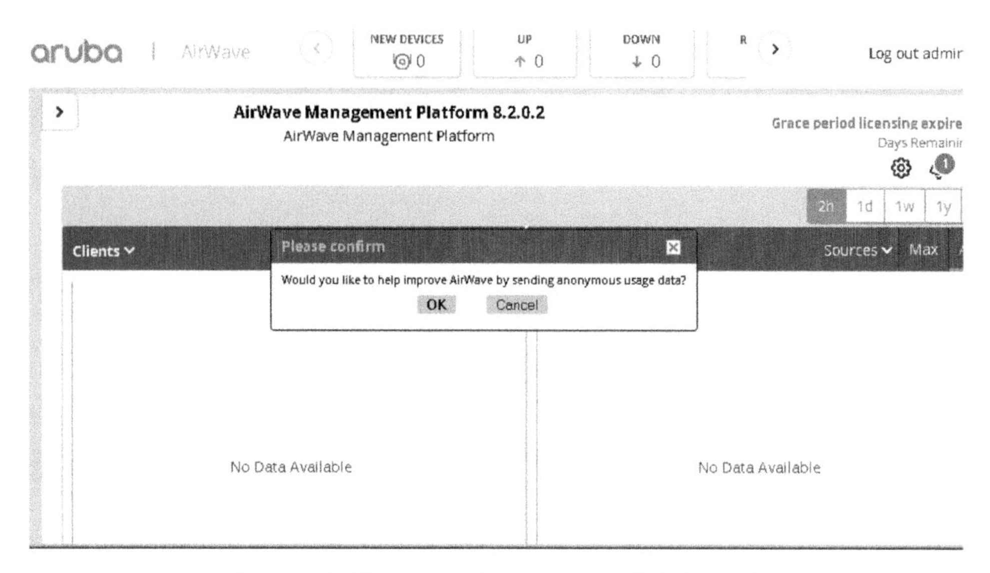

Figure 11-13: Usage data request - click Cancel

6. Take a moment to familiarize yourself with the interface.

   a. You begin at the Home > Overview window.

   b. You use the left side of the interface to navigate to various pages. You will not explore all of these pages at this time, but you will learn how to reach the most important pages for discovering devices and beginning to monitor and manage them.

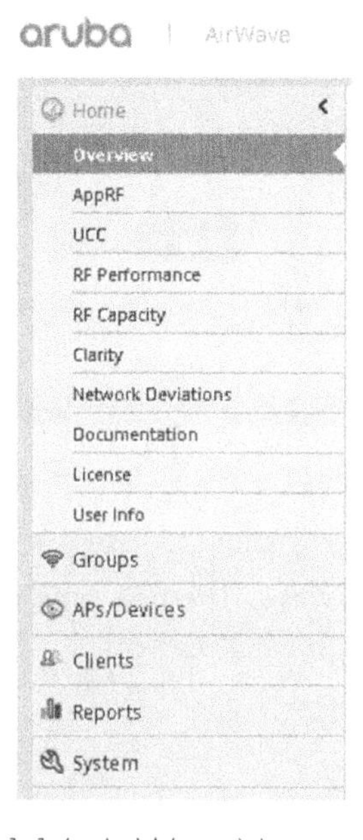

Figure 11-14: AirWave Navigation pane

c. The top of the interface gives you an at-a-glance view of device status and the number of connected clients, as well as detected rogue APs and alerts.

7. In the left navigation bar, click Groups. Note that AirWave currently has one group: Access Points.

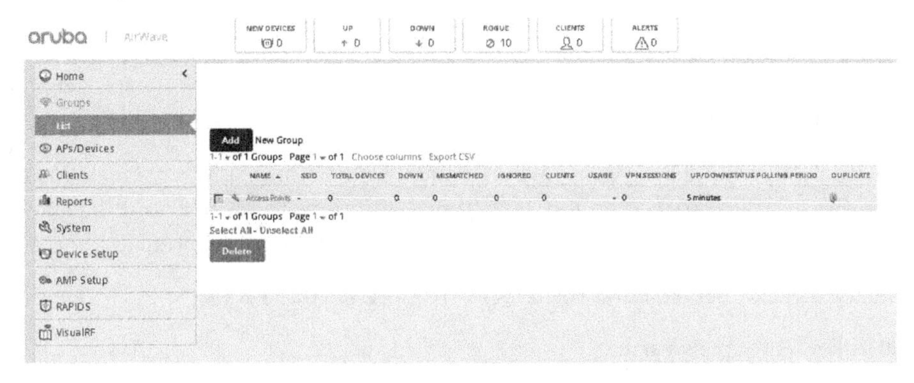

Figure 11-15: Click Groups

8. In the left navigation bar, click APs/Devices. The top of the page has a Folder drop-down menu that you can use to change the view. See that the menu currently includes one folder: Top.

Figure 11-16: Click APs/Devices

Task 3: Configure AirWave settings on the Instant cluster

You will now configure the Instant cluster VC to contact AirWave.

Windows server

1. Still on the Windows server, open another Web browser and access the VC at 10.1.1.5. Log in with the default credentials (**admin** and **admin**).

2. Click **System** in the links at the top of the UI.

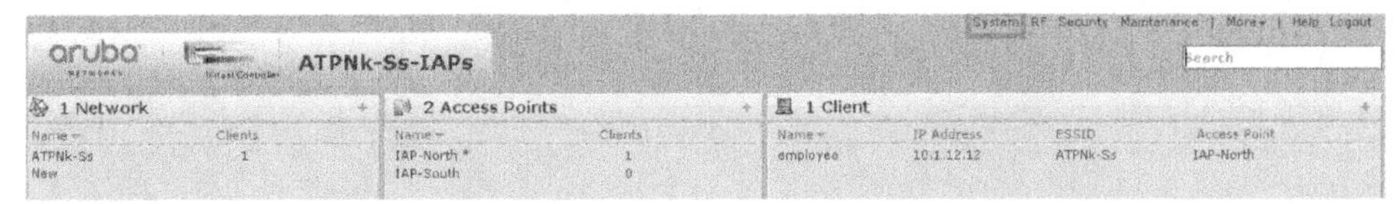

Figure 11-17: Click System

Note

If you used the initialization steps to load the configuration on the Instant cluster, your APs won't be using the names shown in the figure (because your IAPs have different MAC addresses than the IAPs used to create the configuration). You can ignore this discrepancy.

3. Select the **Admin** tab.
4. For Organization, enter **Lab APs**.

   This organization string helps you to organize the device within AirWave. Here, you're organizing the device according to its type and its (fictional) location. You will see precisely how AirWave uses the string to assign the Instant cluster to a group and folder in a moment.
5. For Primary AirWave server, enter the server's IP address, **10.1.11.13**.
6. For shared key, enter **password**.

Important

In the real world, you should follow typical complexity requirements for choosing a shared key. You are using **password** for simplicity in this task.

7. Click **OK**.

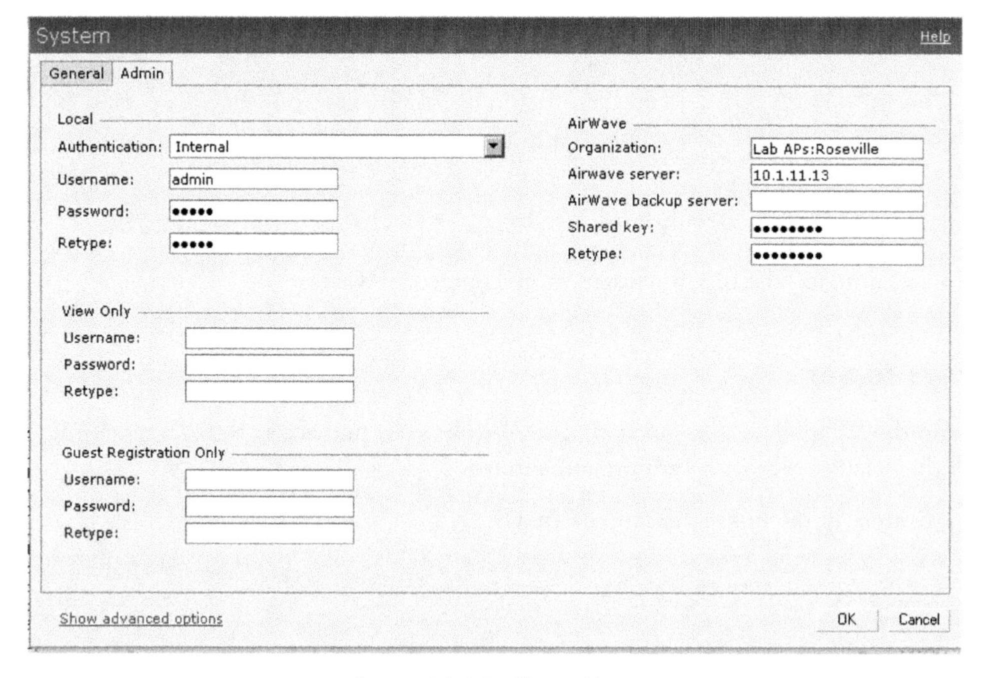

Figure 11-18: Shared key

Task 4: Authorize the Instant cluster in AirWave

The Instant VC now contacts AirWave. You will log into the AirWave Management Platform (AMP), find the VC in the new devices list, and authorize it.

Windows server

1. Return to the browser logged into AMP (if necessary, log in again).
2. Look at the bar at the top of the interface. You should see a new device.

   You might need to wait a minute or two and refresh the interface.

Figure 11-19: New device indicator

3. When you see one new device, click the new devices icon, which takes you to the APs/Devices > New page. Alternatively, you can use the bar on the left to navigate to this page.
4. Here you see your VC, identified by system name. Hover your mouse over the text in the Type column. You will see **password**, the shared key submitted by the Instant VC. Because only authorized network administrators should know the shared key, AirWave administrators can, in this way, verify that an authorized device has contacted AirWave.
5. You will now authorize the device and add it to AirWave. Select the check box next to the VC.
6. Make sure that the Device Action is set to Add Selected Devices and the Management Level is Monitor Only + Firmware Upgrades.

Important

When you are adding an already configured Instant cluster to AirWave, always begin by monitoring the device only, *not* managing it.

You will learn when you can safely begin managing the cluster later in this chapter.

7. The Group and Folder drop-down menus show default options. You don't need to change these. You will see how the organization string submitted by the VC override any differing options.
8. Click **Add**.

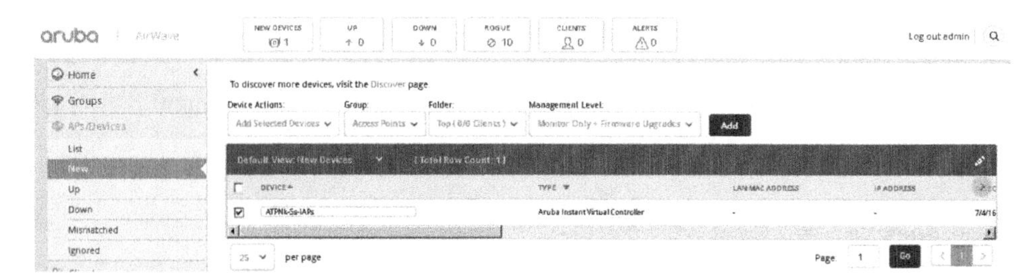

Figure 11-20: Add the selected device

9. The interface informs you that the device will be added to the subfolder specified by the organization string. Click **Apply Changes Now**.

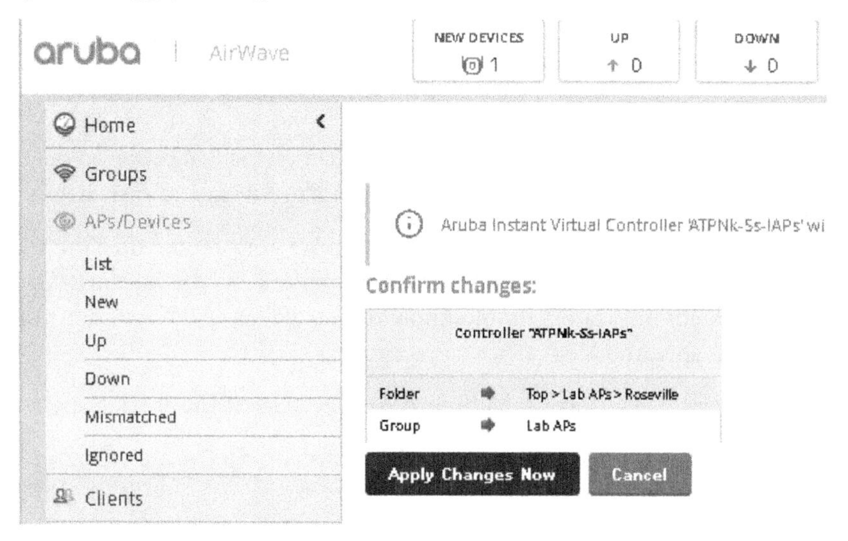

Figure 11-21: Apply changes

10. Use the left bar to navigate to the APs/Devices > List page.

11. The top of this page has a drop-down menu from which you can choose the folder. Use this menu to choose the folder in which AirWave placed the Instant VC. As you see, AirWave automatically created the folder and subfolder based on the Organization string.

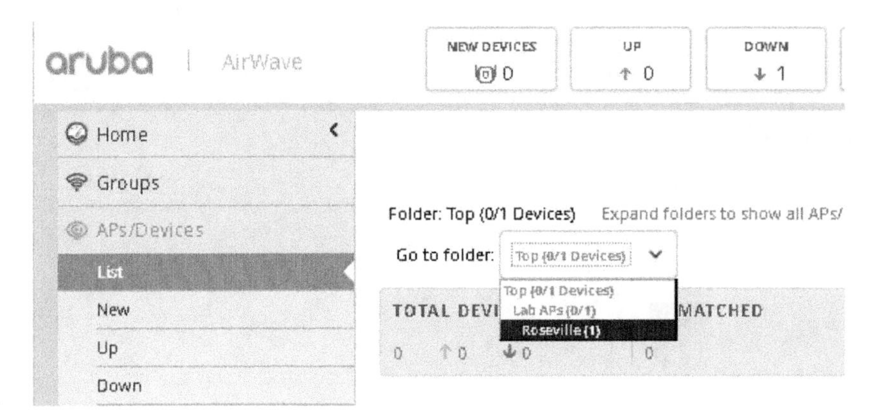

Figure 11-22: Choose the folder

12. The page changes to show the devices in this folder—currently the Instant VC—and graphs and lists created from information collected from those devices.

13. Wait a moment for AirWave to discover the two IAPs in the cluster. You will then see three devices on the page.

Note

The Configuration column for your devices might initially show Error. You can ignore this for now.

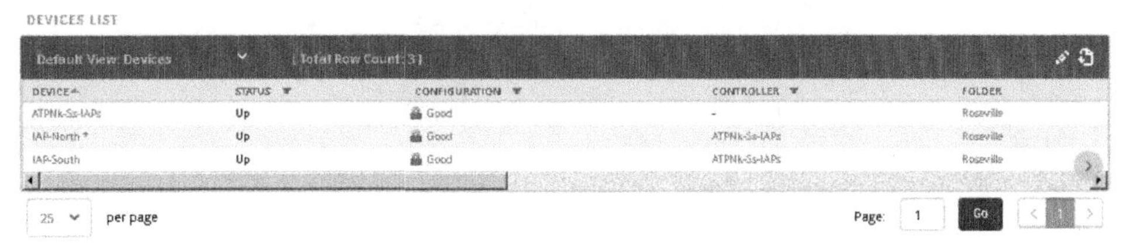

Figure 11-23: Discovered devices

14. Click **Groups** in the left navigation bar.

15. As you see, AirWave automatically created a group from the first part of the Instant VC's organization string.

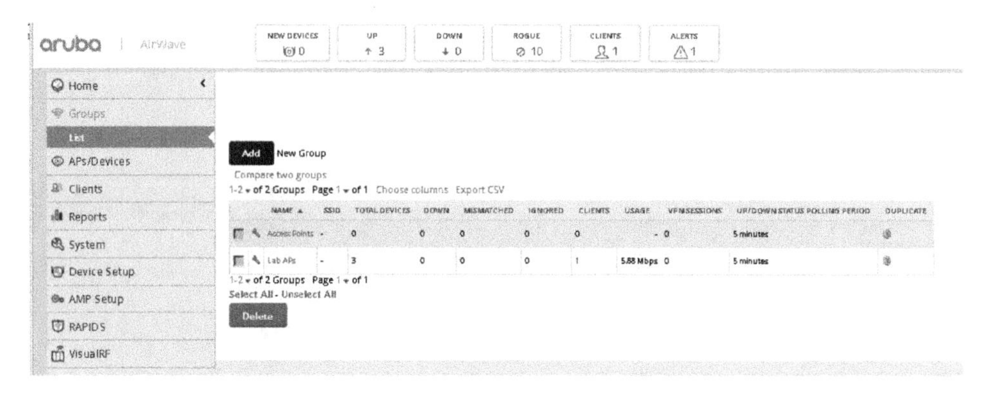

Figure 11-24: List of groups

16. Click Lab APs in the group list.
17. Note the pages that you can use for this group.

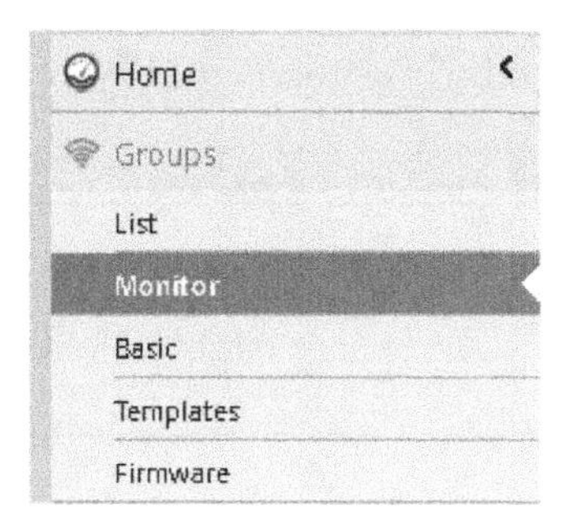

Figure 11-25: Pages for the group

18. Click **Basic.**
19. You will set a lower than default polling period for client statistics so that you can see information more quickly.
    a. Select **Yes** for Override Polling Period for Other Services.
    b. For Client Data Polling Period, select **5 minutes**.

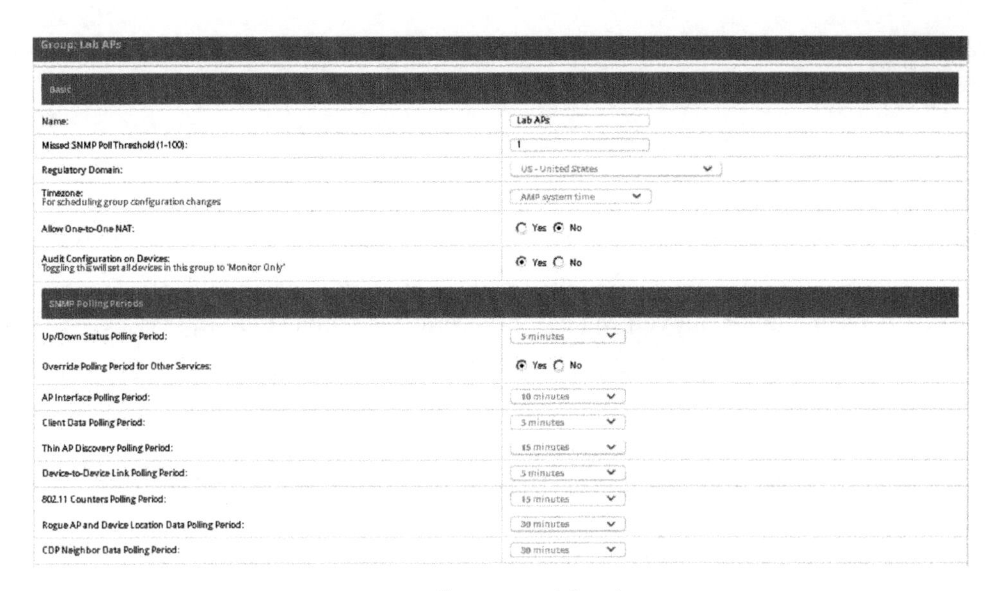

Figure 11-26: Polling period for client statistics

20. Scroll to the bottom of the page and click **Save and Apply**.
21. Click **Apply Changes Now**.
22. Return to the Instant UI interface. You will probably need to log in again.
23. See that the Instant VC now lists AirWave as a manager. However, you can still manage this Instant cluster through the local Instant UI because AirWave is only monitoring the cluster.

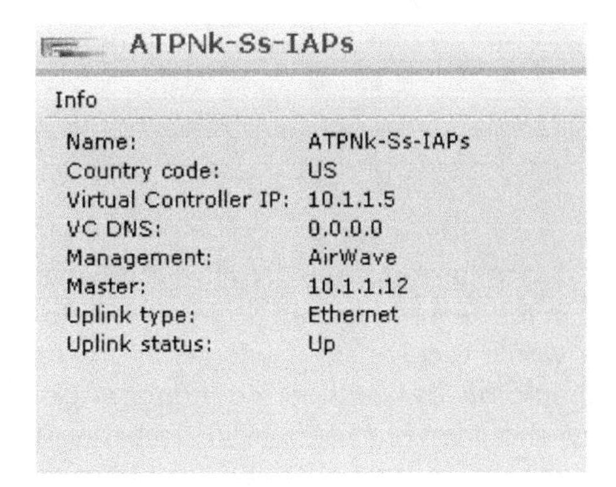

Figure 11-27: Instant UI interface

Task 5: Configure SNMP settings on ArubaOS switches

Next you will set up your ArubaOS switches for discovery in AirWave by configuring them with a new name for the SNMPv2 read-write community. You will also create a user with Level 15 privileges (full read-write privileges) for Telnet access.

You will configure the VSF fabric and Access-1. You will not configure Access-2 for discovery in AirWave. Instead you will add Access-2 later in this chapter.

Core-1

1. Access a terminal session with the VSF fabric and move to global configuration mode.
2. Define the SNMP read-write community. In the real world, you should use a more complex string for the community, but you will use "password" for simplicity in this task.

```
Core-1(config)# snmp-server community password manager unrestricted
```

3. Configure AirWave as the trap server.

```
Core-1(config)# snmp-server host 10.1.11.13 community password
```

4. Remove the default community for security.

```
Core-1(config)# no snmp-server community public
```

5. Verify the settings.

```
Core-1(config)# show snmp-server
 SNMP Communities

  Community Name                   MIB View Write Access
  -------------------------------- -------- ------------
  password                         Manager  Unrestricted
<-output omitted->
```

Important

In this task, you will use a local user account to allow AirWave to log into the switch. You will place the local user in the Level-15 group, but you will not enable command authorization, so the local user would receive full privileges in any case.

In the real world, you should *always* configure a manager password in addition to local user accounts to prevent unauthorized users from accessing the switch. You could then have AirWave use the manager account or a local user account with the proper privileges as you choose.

6. Add a local user with Level 15 privileges to secure Telnet access.

```
Core-1(config)# aaa authentication local-user manager group Level-15
password plaintext
```

 **Important**

The group name is case sensitive. Enter **Level-15** exactly as indicated.

7. When prompted to enter and confirm the password, enter **hpe**.

```
New password for manager: hpe

Please retype new password for manager: hpe
```

8. Verify that you have added the manager account to the group.

```
Core-1(config)# show authorization group Level-15

 Local Management Groups - Authorization Information

  Group Name            : Level-15

  Group Privilege Level : 35

  Users

  ----------------

  manager

  Seq. Num.  | Permission  Rule Expression                      Log

  ---------- + ----------- ------------------------------------ -------

  999        | Permit      configure .*                         Disable

  1000       | Permit      .*                                   Disable
```

Access-1

9. Repeat exactly the same steps on Access-1.

   Try to remember the commands and use the ? key to help.

   Alternatively, you can use the **show history** command on Core-1 to copy and paste commands. Remember to copy the commands to Notepad and remove the numbers. Also remember that the final command will prompt you to enter a password twice. Add **hpe** to the bottom of commands on its own line. Then add another line with **hpe**.

10. When you are finished, validate that the switch has only the “password” community, which is set to manager unrestricted access. Also validate that you have added the manager user to the Level-15 authorization group.

Core-1 (Optional)

LLDP MAD on the VSF fabric requires an SNMP community with read access to the MAD device.

11. You are not using LLDP MAD for this set of tasks, but the best practice would be to return to Core-1 and change the LLDP MAD setting. You should also change the MAD assistant IP address because Access-1 is using a different address.

```
Core-1(config)# vsf lldp-mad ipv4 10.1.1.3 v2c password
```

Task 6: Discover the ArubaOS switches in AirWave

You will now discover the ArubaOS switches in AirWave.

Windows server

1. Access the Windows server desktop and return to the browser window with the AMP. You might need to log in again.
2. From the left navigation bar, select **Device Setup** > **Discover**.
3. Define the network on which your switches have their IP addresses, 10.1.1.0/24.
   a. Click **Add** under Networks.

Figure 11-28: Define the network on which your switches have their IP addresses

   b. Name the network VLAN 1.
   c. Specify the network address and mask: 10.1.1.0 255.255.255.0.
   d. Click **Add**.

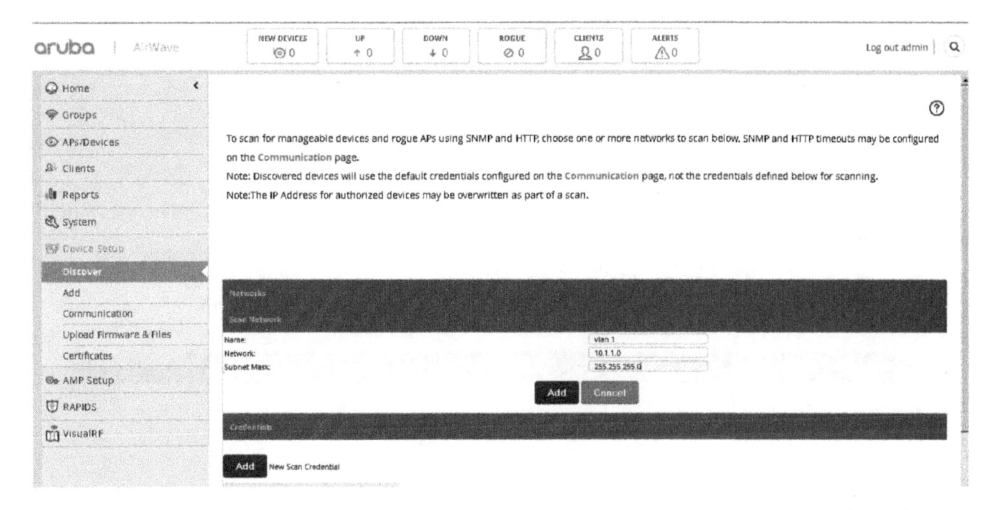

Figure 11-29: Name the network and specify the network address and mask

4. Define scan credentials that match the switches' SNMPv2 community.

   a. Click **Add** under Credentials.

Figure 11-30: Define scan credentials

   b. Give the credentials an intuitive name such as **My ArubaOS switches**.

   c. For the credential type, select **SNMPv2c**.

   d. For the Community String and Confirm community String, specify **password**.

   e. Click **Add**.

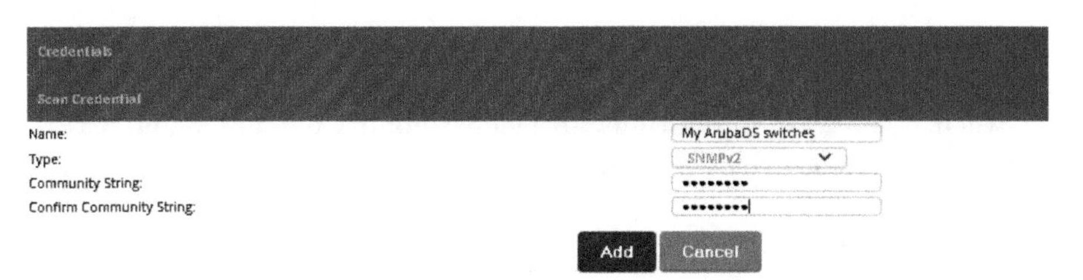

Figure 11-31: Specify Community String and Confirm community String

5. Define a scan set that associates your network with your credentials.

   a. At the top of the page, click **Add** next to New Scan Set.

Figure 11-32: Define a scan set

b. Select the check boxes for VLAN 1 and My ArubaOS switches.

c. You will authorize the devices manually, so use the system default for the new device location, which is the new device list.

d. Click **Add**.

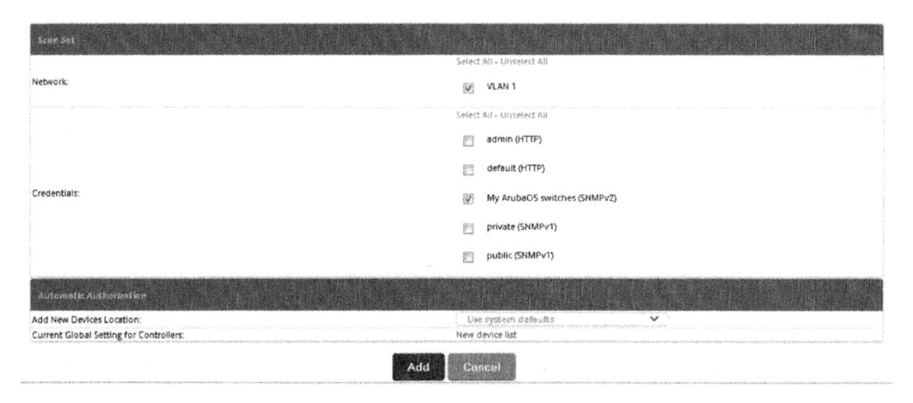

Figure 11-33: Use the system defaults for the New Device Location

6. Begin the scan by selecting the check box for the scan set and clicking **Scan**.

7. You should see that scan is in progress. While AirWave completes the scan, move on to the next task.

Task 7: Create AirWave groups and folders

The ArubaOS switches do not submit an organization string when you discover them in this way, so you will create the groups and folders for them manually on AirWave. You will create a global group named "Lab switches" and two groups that use that global group: Core-5400R and Access-3800. You will also create a "Lab switches" folder and subfolder for a location, such as Roseville, in that folder.

Windows server

1. You should still be logged into the AMP from Windows server.

   First add the groups.

2. Select **Groups** in the left navigation bar.

3. First add the global group.

    a. Click **Add** next to New Group.

Figure 11-34: Add the global group

    b. Name the group **Lab switches** and click **Add**.

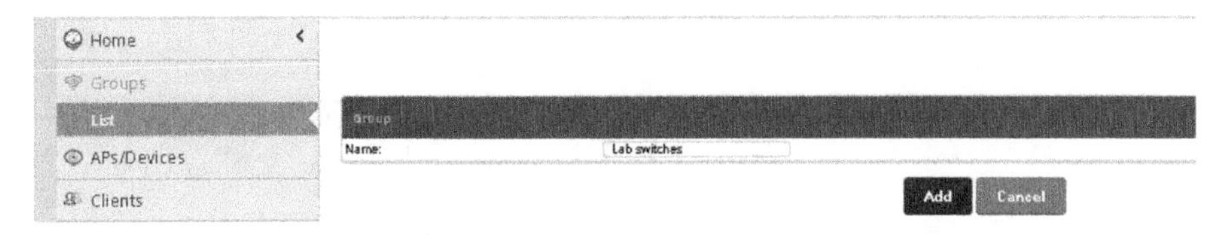

Figure 11-35: Name the group

    c. You are automatically moved to the group's Basic page.

    d. Select **Yes** for the Is Global Group option.

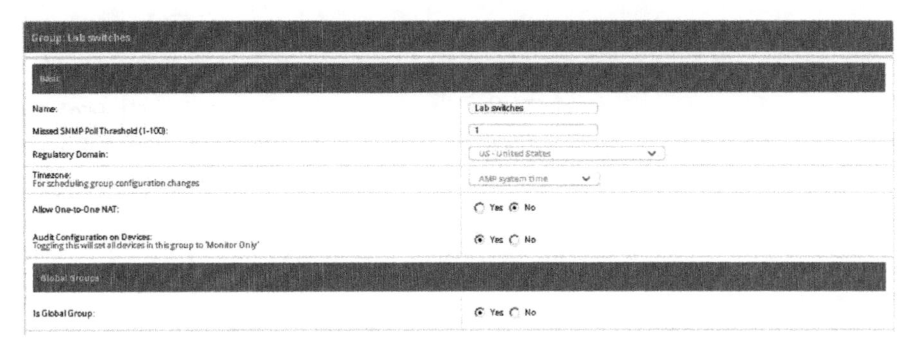

Figure 11-36: Set Is Global Group to Yes

    e. Select **Yes** for Override Polling Period for Other Services.

    f. Change the Device-to-Device Link Polling Period statistics to 2 minutes so that you can see changes more rapidly.

Figure 11-37: Set Device-to-Device Link Polling Period to 2 minutes

g. Choose to show device settings only for selected device types.

Figure 11-38: Show settings for selected device types

h. Select the check box for HPE ProVision Switch.

Figure 11-39: Select HPE ProVision Switch

i. Scroll to the bottom of the page and click **Save and Apply**.

j. Verify that the changes shown below are listed and click **Apply Changes Now**.

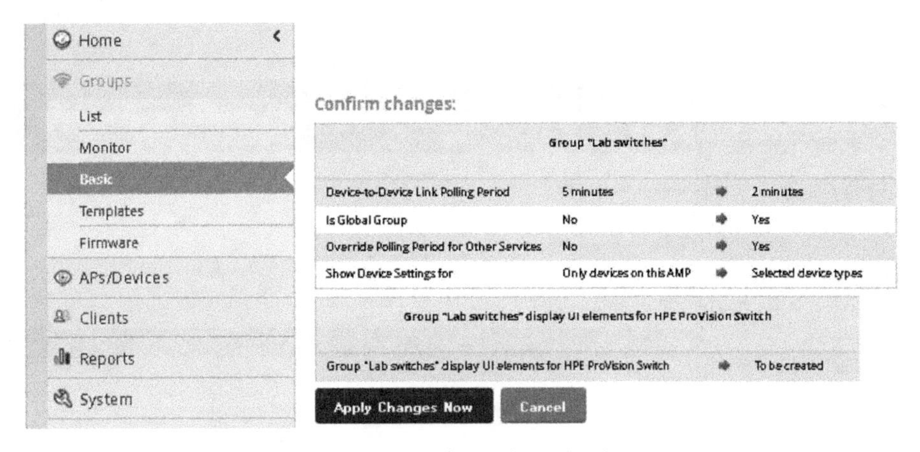

Figure 11-40: Verify and apply changes

4. Now add the Core-5400R group.

   a. Click **Add**.

Figure 11-41: Add new group

   b. Name the group **Core-5400R** and click **Add**.

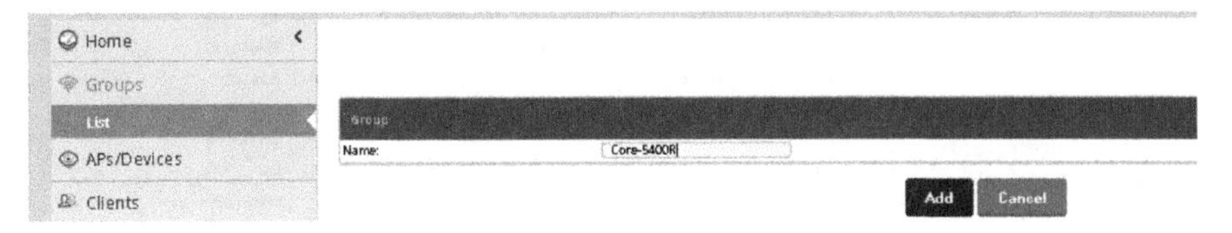

Figure 11-42: Name the group Core-5400R and add it

   c. You are automatically moved to the group's Basic page.

   d. Select **Yes** for the Use Global Group option.

   e. For the Global Group, choose **Lab switches**.

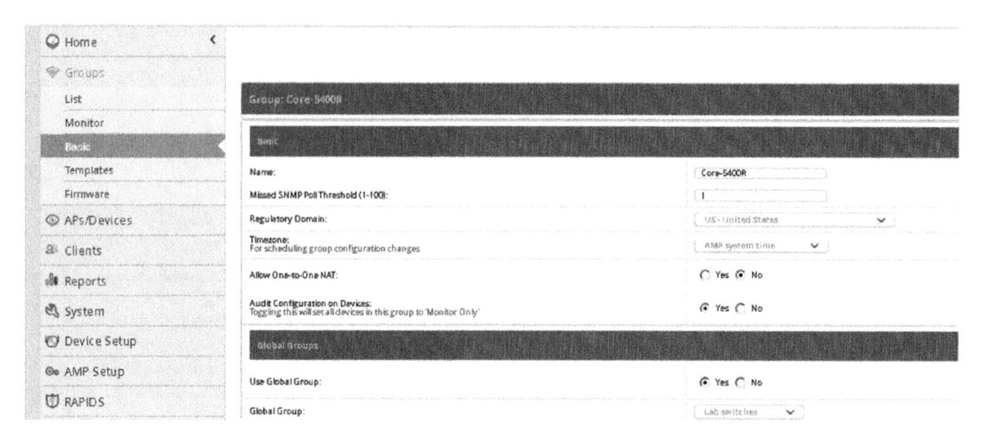

Figure 11-43: Make selections on the group's Basic page

f. Leave the other settings at their default settings and click **Save and Apply** at the bottom of the page.

g. The group has inherited settings from the global "Lab switches" group. Click **Apply Changes Now**.

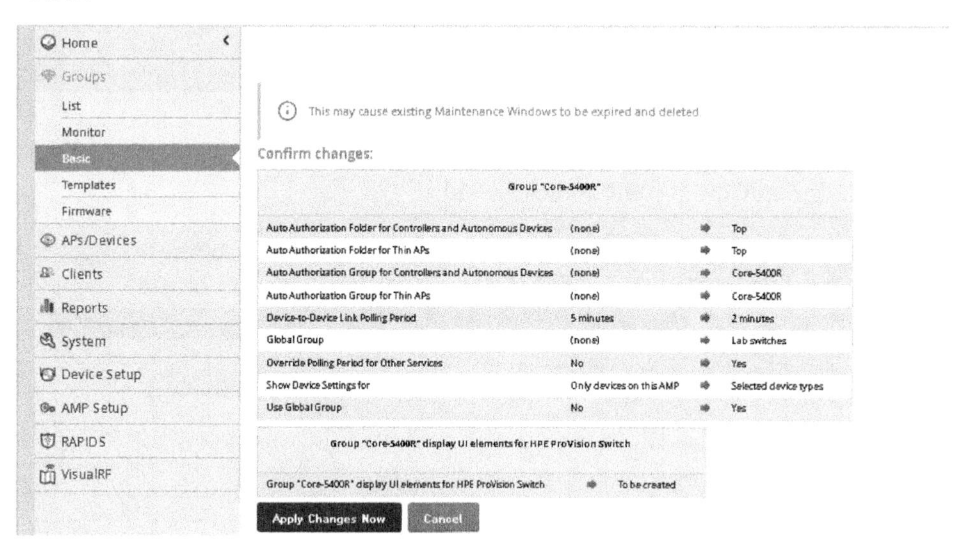

Figure 11-44: Confirm and apply changes for the Core-5400R group

5. Click **List** to return to the list of groups.

6. Repeat Step 4 to add the Access-3800 group. Make sure to include the hyphen (-).

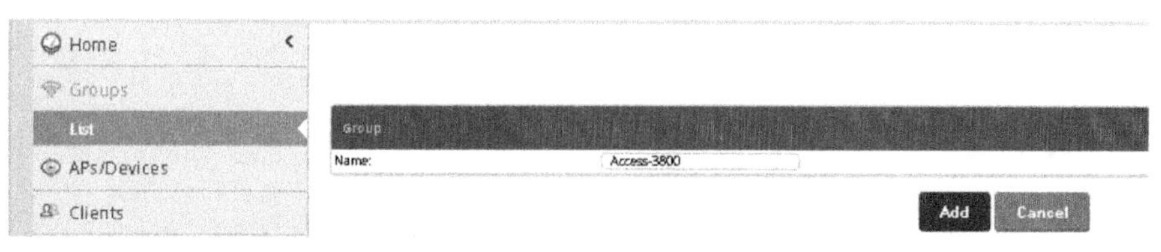

Figure 11-45: Add the Access-3800 group

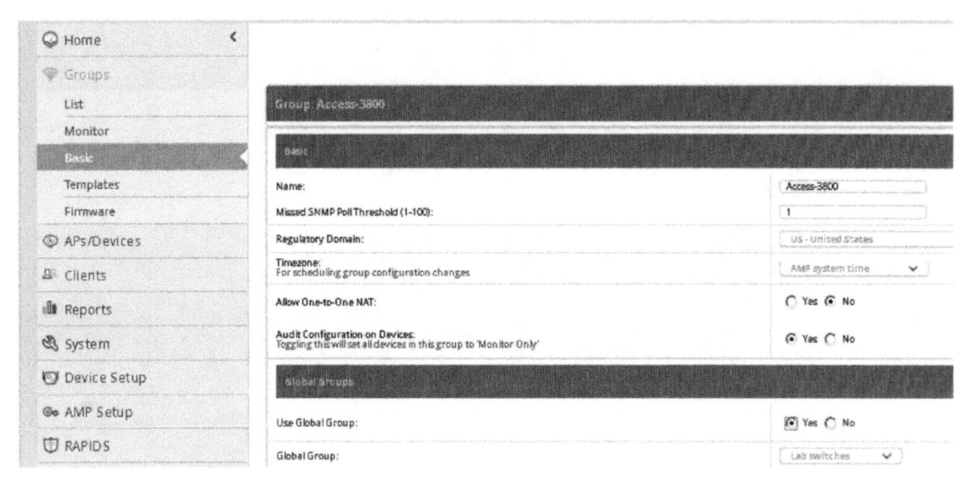

Figure 11-46: Make selections on the group's Basic page

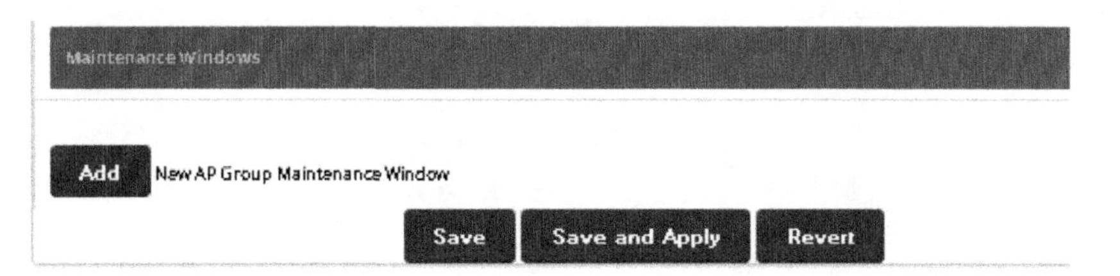

Figure 11-47: Add New AP Group Maintenance Window

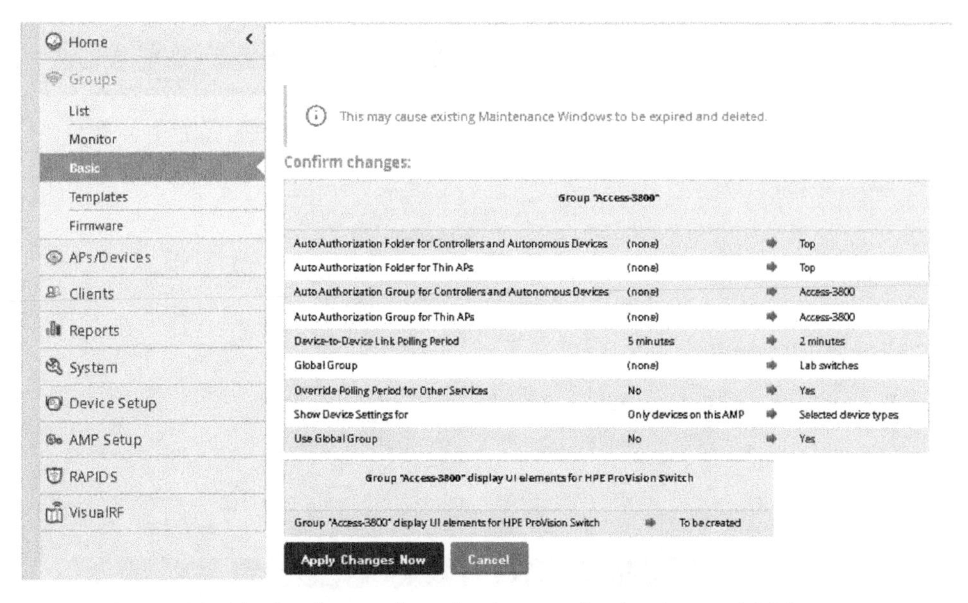

Figure 11-48: Confirm and apply changes for the Access-3800 group

Next add the folders.

7. Move to the APs/Devices > List page.
8. Scroll to the bottom of the page.
9. Click **Add New Folder**.

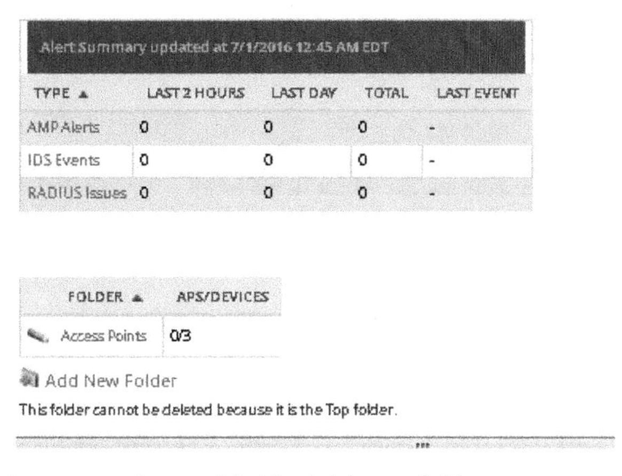

Figure 11-49: Add new folders

10. Name the folder **Lab switches**.
11. Change Parent Folder to Top and click **Add**.

Figure 11-50: Change Parent Folder to Top

12. Now you will add the subfolder. Click Add New Folder again
13. Name this folder **Roseville**.
14. Change Parent Folder to **Lab switches**.
15. Click **Add**.

Figure 11-51: Add subfolder

Task 8: Authorize the ArubaOS switches

You will now authorize the ArubaOS switches, adding them to the groups and folder that you created.

1. You should still be logged into the AMP from Windows server.
2. AirWave should have discovered the switches by now. Move to the APs/Devices > New page.
3. Select the check box for the core VSF fabric.
4. Make sure that the Device Action is set to **Add Selected Devices** and the Management Level is **Monitor Only + Firmware Upgrades**.
5. Choose **Core-5400R** for Group and **Lab switches > Roseville** for the Folder.
6. Click **Add**.

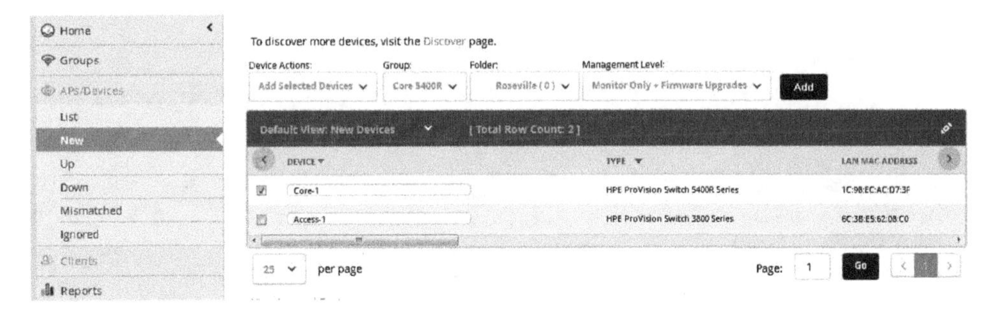

Figure 11-52: Add selected devices: Core 5400R

7. The warning explains that some AMP user roles will not be able to see these devices because they are in a folder that is off-limits to that role. Ignore the warning and click **Apply Changes Now**.

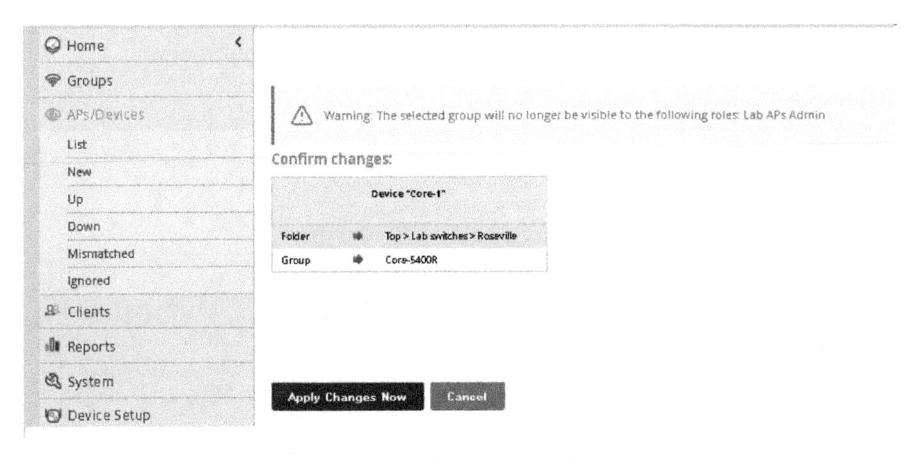

Figure 11-53: Apply changes for device "Core-1"

8. Select the check box for the Access-1 switch.

9. Choose the **Access-3800** group and the **Lab switches** > **Roseville** folder.

10. Click **Add**.

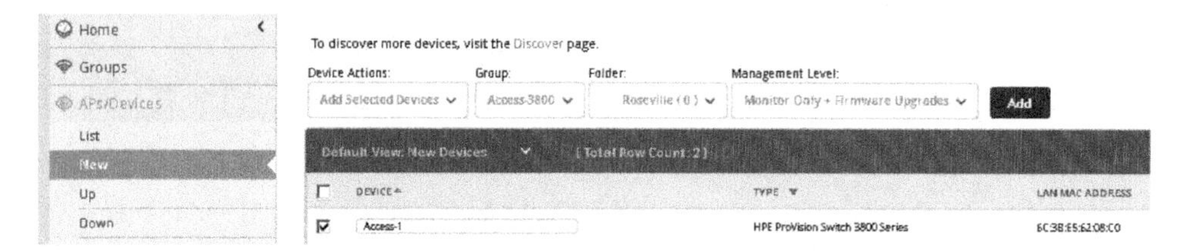

Figure 11-54: Add selected devices: Access-3800

11. Again, ignore the warning and click **Apply Changes Now**.

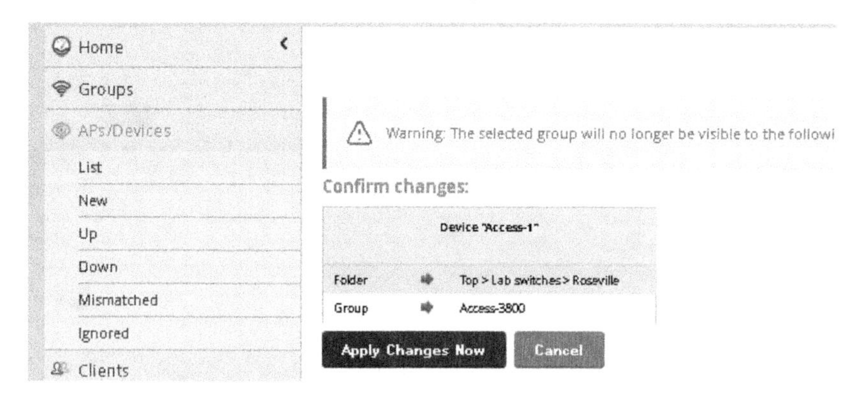

Figure 11-55: Apply changes for device "Access-1"

12. AirWave only uses the SNMP credentials that you specified on the Device Setup > Discovery page for discovering the devices. As the final step in adding the devices, you must configure AirWave to use these credentials to monitor and manage the switches as well.

    Move to the APs/Devices > List page.

13. For Folder, select **Lab switches** > **Roseville**.

Figure 11-56: Select Lab switches > Roseville

14. Scroll to the Devices List.

15. Click the pencil icon next to the list.

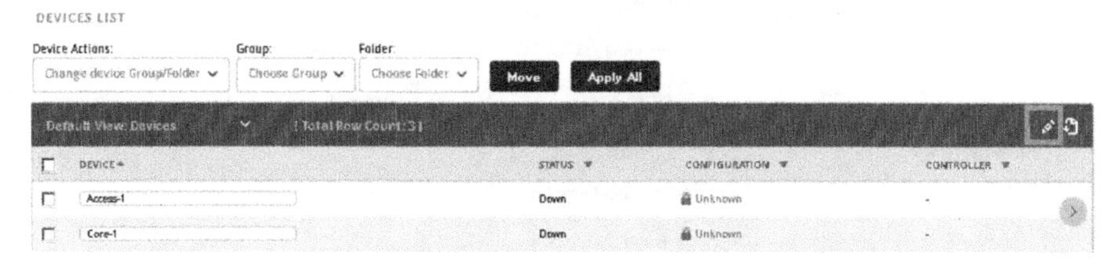

Figure 11-57: Click pencil icon on Devices List

16. You can now apply a variety of actions to the devices. Select the check box at the top of the list to select all devices.

17. For Device Actions, select **Update the credentials used to communicate with these devices**.

18. Click **Update**.

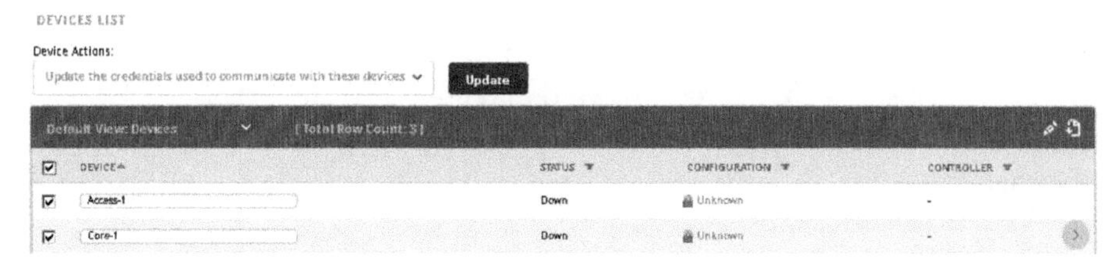

Figure 11-58: Update the credentials used to communicate with these devices

19. Enter **password** for Community String and Confirm Community String.

20. For Telnet/SSH Username, enter **manager**.

21. For Telnet/SSH Password and Confirm Telnet/SSH Password, enter **hpe**.

22. Click **Update**.

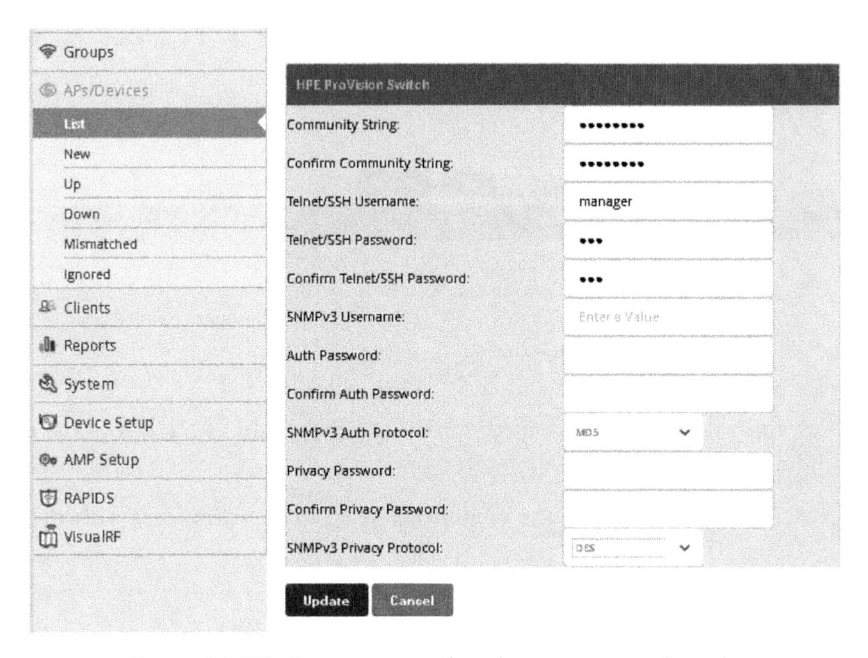

Figure 11-59: Enter password and username credentials

23. Apply the changes.

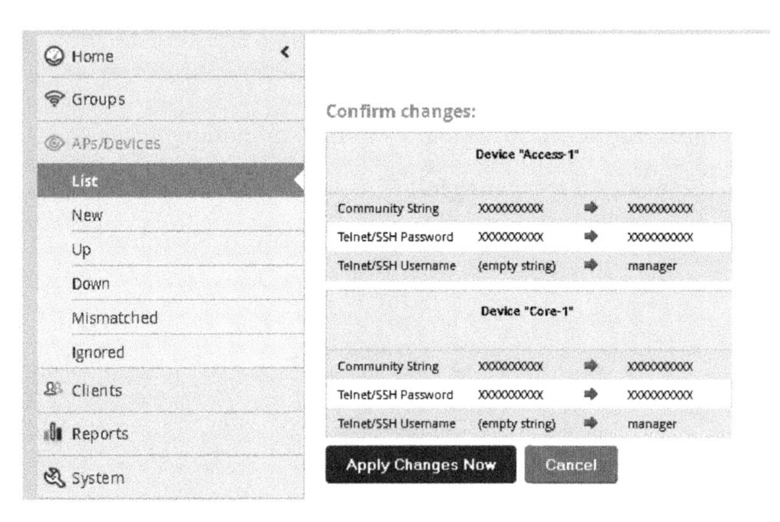

Figure 11-60: Confirm and apply changes

Task 9: Explore monitoring options

You have completed the required configuration. For the final task, you will explore just some of the information that AirWave collects.

1. You should still be logged into the AMP from Windows server.
2. Move to the APs/Devices > List page.
3. Select the Lab APs > Roseville folder.

Figure 11-61: Access the Roseville folder on the List page

4. Scroll through the page and note information that you can see. You can also monitor devices at the group level.
   a. Click **Groups** in the left navigation bar.
   b. Select the Core-5400R group.
   c. Click **Monitor** under Groups.
   d. You can see more useful information for wired devices at the individual device level. Scroll down to the device list.
   e. You will see several devices because AirWave has discovered both members of the VSF fabric. Do not worry if one device is down as long as one Core-1 and the device listed as Core-1-2 are up.

Note

If all devices are still showing as down, AirWave might not have contacted them again after you changed the credentials. You can speed this process up by clicking the pencil icon, selecting all of the switches, selecting **Poll selected devices** from the Device Actions list, and clicking **Poll Now**. Then click Apply Changes Now and return to the device list.

f. Click the Core-1 that is up in the Device List. (You will also see Core-1-2, which is the second member of the VSF fabric. Do not click this device.)

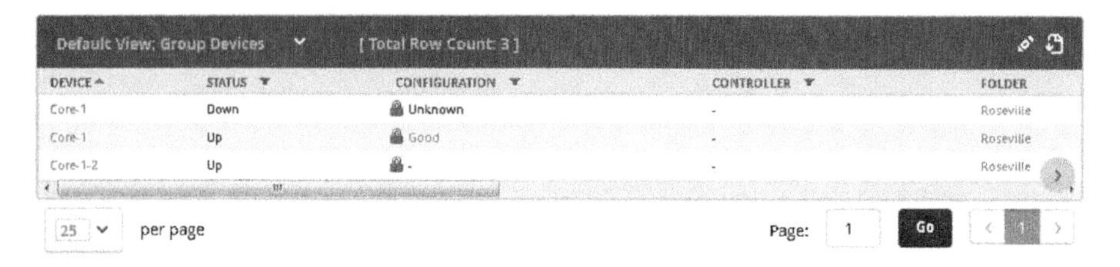

Default View: Group Devices [Total Row Count: 3]

DEVICE ▲	STATUS ▼	CONFIGURATION ▼	CONTROLLER ▼	FOLDER
Core-1	Down	Unknown	-	Roseville
Core-1	Up	Good	-	Roseville
Core-1-2	Up	-	-	Roseville

25 per page Page: 1 Go

Figure 11-62: Information available on the Group Devices page

g. You can now see network usage and CPU utilization.

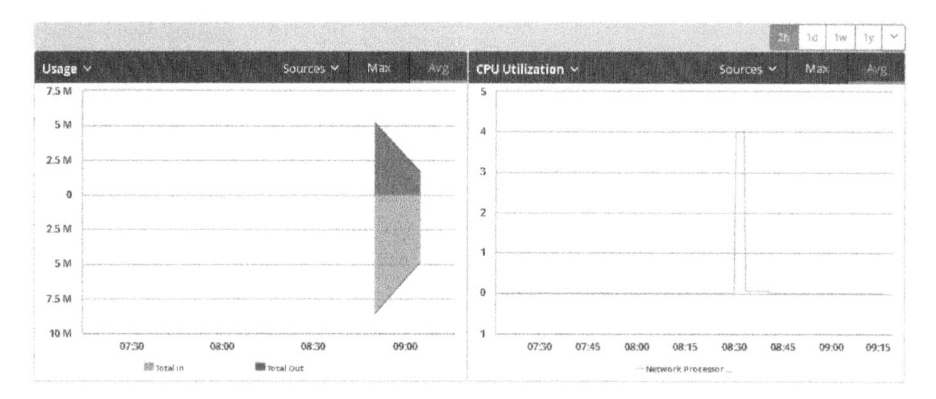

Figure 11-63: Network usage and CPU utilization

h. You can also view the switch's neighbors. Neighbors that are monitored or managed by AirWave are listed as links that you can click to move to that device's page.

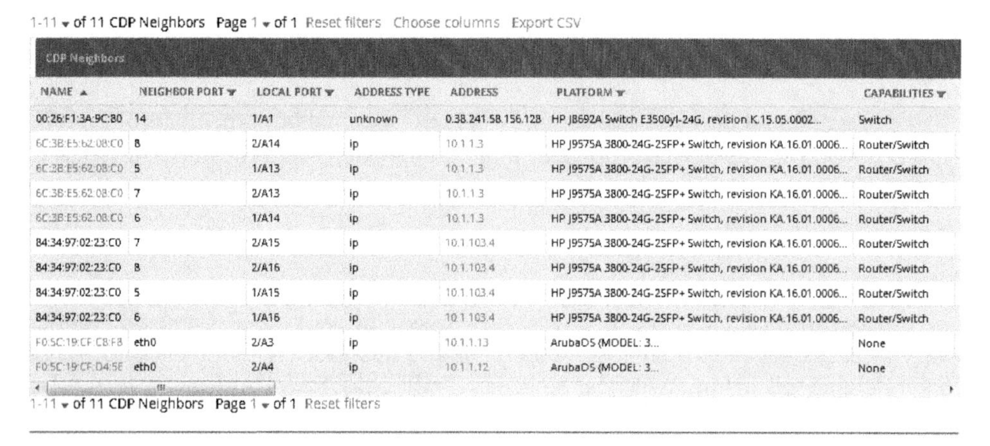

1-11 ▾ of 11 CDP Neighbors Page 1 ▾ of 1 Reset filters Choose columns Export CSV

CDP Neighbors

NAME ▲	NEIGHBOR PORT ▼	LOCAL PORT ▼	ADDRESS TYPE	ADDRESS	PLATFORM ▼	CAPABILITIES ▼
00:26:F1:3A:9C:80	14	1/A1	unknown	0.38.241.58.156.128	HP J8692A Switch E3500yl-24G, revision K.15.05.0002...	Switch
6C:3B:E5:62:08:C0	8	2/A14	ip	10.1.1.3	HP J9575A 3800-24G-2SFP+ Switch, revision KA.16.01.0006...	Router/Switch
6C:2B:E5:62:08:C0	5	1/A13	ip	10.1.1.3	HP J9575A 3800-24G-2SFP+ Switch, revision KA.16.01.0006...	Router/Switch
6C:3B:E5:62:08:C0	7	2/A13	ip	10.1.1.3	HP J9575A 3800-24G-2SFP+ Switch, revision KA.16.01.0006...	Router/Switch
6C:3B:E5:62:08:C0	6	1/A14	ip	10.1.1.3	HP J9575A 3800-24G-2SFP+ Switch, revision KA.16.01.0006...	Router/Switch
84:34:97:02:23:C0	7	2/A15	ip	10.1.103.4	HP J9575A 3800-24G-2SFP+ Switch, revision KA.16.01.0006...	Router/Switch
84:34:97:02:23:C0	8	2/A16	ip	10.1.103.4	HP J9575A 3800-24G-2SFP+ Switch, revision KA.16.01.0006...	Router/Switch
84:34:97:02:23:C0	5	1/A15	ip	10.1.103.4	HP J9575A 3800-24G-2SFP+ Switch, revision KA.16.01.0006...	Router/Switch
84:34:97:02:23:C0	6	1/A16	ip	10.1.103.4	HP J9575A 3800-24G-2SFP+ Switch, revision KA.16.01.0006...	Router/Switch
F0:5C:19:CF:C8:F8	eth0	2/A3	ip	10.1.1.13	ArubaOS (MODEL: 3...	None
F0:5C:19:CF:D4:5E	eth0	2/A4	ip	10.1.1.12	ArubaOS (MODEL: 3...	None

1-11 ▾ of 11 CDP Neighbors Page 1 ▾ of 1 Reset filters

Figure 11-64: View the switch's neighbors

5. Move to the Home > Clarity page.

6. This page shows failed associations, authentications, and DHCP requests to help you rapidly find and troubleshoot issues. (Your test network might not have any failures, so you might see N/A for all sections.)

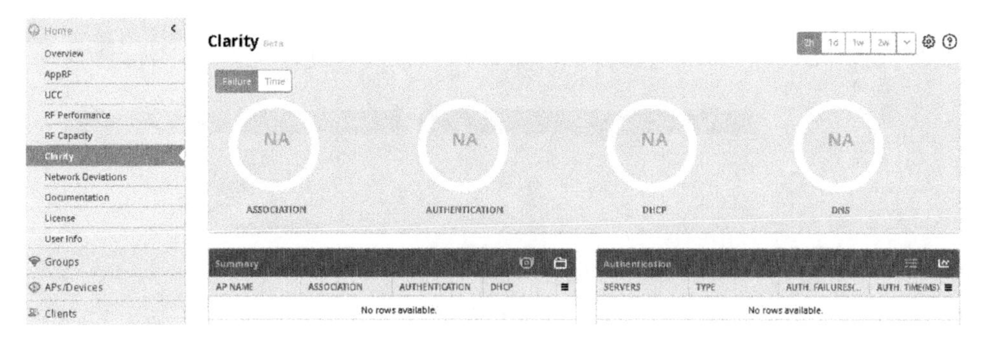

Figure 11-65: Information on the Home > Clarity page

7. Move to the Home > Network Deviations page.

8. AirWave shows how network usage patterns deviate from the norm. Because you just discovered the network infrastructure devices, AirWave hasn't built up a consistent baseline yet, but it will do so over time.

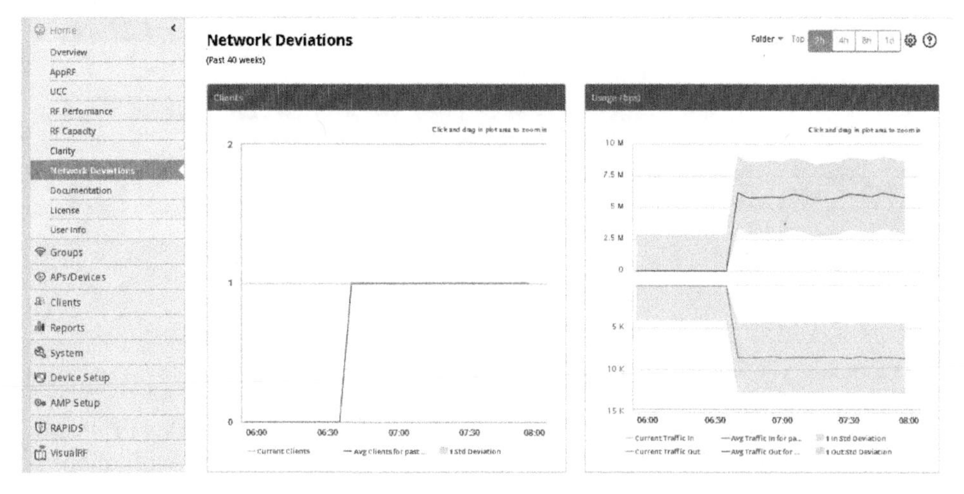

Figure 11-66: Home > Network Deviations page

9. If you have extra time, explore other pages such as the Home > RF Performance and Home > RF Capacity pages. Record the type of information that you can see.

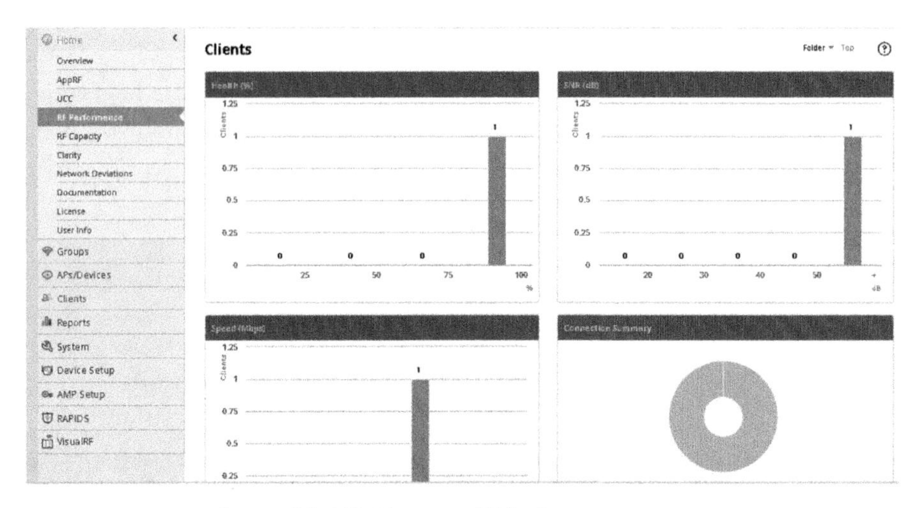

Figure 11-67: Home > RF Performance page

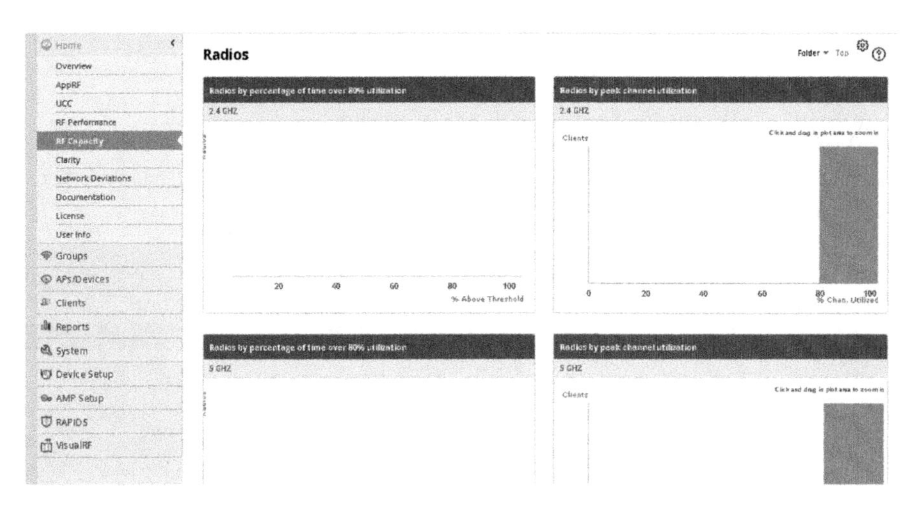

Figure 11-68: Home > RF Capacity page

Caution

If you plan to complete the tasks in the section titled "ZTP with Aruba Activate" later in this chapter, you can keep the passwords on the switches. Also save the configurations on the switches.

If you do not plan to complete those tasks, it is recommended that you remove the passwords now. On **every** switch, enter **no aaa authentication local-user manager**.

Learning check

1. On the ArubaOS switches, you set the community to manager or operator and specified the restricted or unrestricted option.
   - Based on the table, which manager option is associated with read-only access and which with read-write access?
   - When might you want to use the **manager** versus **operator** options for an SNMP community?

Table 11-1: ArubaOS SNMPv2 communities

Community	Read rights	Write rights
manager unrestricted	All managed objects	All managed objects
manager restricted	All managed objects	No objects
operator unrestricted	All managed objects except CONFIG objects	All managed objects except CONFIG objects
operator restricted	All managed objects except CONFIG objects	No objects

2. Begin to think about the role that groups play in AirWave. What options did you see for specifying at the group level?

3. Begin to think about the role that folders play in AirWave. How did you use them in the tasks that you completed?

4. What information did AirWave collect about your network? Note the AirWave capabilities that you explored.

Answers to Learning check

1. On ArubaOS switches, you set the community to manager or operator and specified the restricted or unrestricted option.
   - Based on the table, which manager option is associated with read-only access and which with read-write access?
   - When might you want to use the **manager** versus **operator** options for an SNMP community?

Table 11-2: ArubaOS SNMPv2 communities

Community	Read rights	Write rights
manager unrestricted	All managed objects	All managed objects
manager restricted	All managed objects	No objects
operator unrestricted	All managed objects except CONFIG objects	All managed objects except CONFIG objects
operator restricted	All managed objects except CONFIG objects	No objects

The unrestricted option is associated with write access, and the restricted option is associated with read access. The operator options do not allow access to config options, which AirWave might need to view or write to.

2. Begin to think about the role that groups play in AirWave. What options did you see for specifying at the group level?

   You set configuration templates and firmware auditing settings at the group level. You can also configure monitoring settings such as poll periods at the group level.

3. Begin to think about the role that folders play in AirWave. How did you use them in the tasks that you completed?

   You used the folder for choosing which sets of devices you view in the APs/Devices pages.

4. What information did AirWave collect about your network? Discuss the AirWave capabilities that you explored with your classmates.

   AirWave collected information such as:

   - Device status
   - Interface status
   - Client connections
   - Failed associated and authentication attempts
   - Bandwidth usage over time
   - Deviations in usage
   - RF signal strength
   - Number of devices on each channel
   - Possible rogue APs
   - CDP neighbors on switches
   - Device configurations

Groups and folders

During the tasks that you just completed, you saw how AirWave assigns each network infrastructure device to a group and a folder. You will now explore the effects of these assignments in more detail. What is the difference between a group and a folder? What strategies can you take for grouping devices and assigning them to folders?

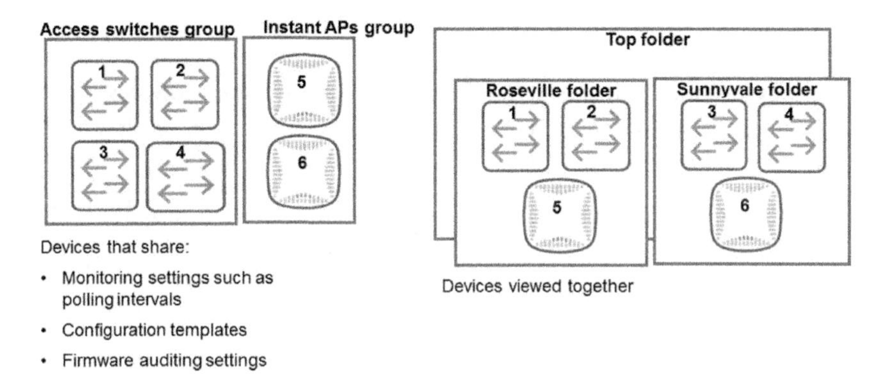

Figure 11-69: Groups and folders

Groups

The group defines settings for the device, including:

- Settings that define how AirWave monitors the device such as how often it polls for particular information
- Firmware version and whether and how AirWave will enforce compliance with this version
- Configuration templates

Because the firmware and configuration are often specific to a type of device, many administrators create groups based on device type. However, a group can have multiple configuration templates that are associated with different device types. This flexibility enables you to combine different types of devices in the same group. For example, you could choose to combine Aruba 5400R and Aruba 3810 switches in the same group, and each device type would have its own template. A group can also have different templates for different firmware; you can choose whether the template only applies to the selected firmware or not. When you configure the firmware version auditing settings, you can also choose the firmware version individually for different types of devices.

Earlier, you created a group for Instant clusters, for the Aruba 5400R zl2 switches at the core, and for the Aruba switches at the access layer. You might also create different groups for different locations such as one group for Site 1 5400R switches and another group for Site 2 5400R switches. However, as you will see, the group configuration templates do include variables to allow for some flexibility so not all environments will require different groups for different sites.

Groups can be hierarchical (maximum, two levels). A global group defines settings for any subscriber groups that use that global group. However, when you change a global group setting you sometimes have the choice whether to implement that change in subscriber groups or not. In this way, you can quickly change settings for multiple groups that have similar devices or similar requirements while maintaining the flexibility to control other settings at the individual group level.

Folders

Folders group devices for viewing in AMP. When you visit the **APs and Devices** page, which shows monitoring information such as connected clients and network usage, you select the folder. You then see the combined information collected from all devices in that folder.

Many administrators choose to create folders based on device location, making it easy for them to see how the network is being used in various buildings or at various sites.

Folders are hierarchical. AirWave begins with a single folder named "Top." You can add subfolders to this folder and then further subfolders within those folders.

Folders and groups do not have to correspond with each other. Devices in different groups can be combined in a single folder. Devices in the same group can be assigned to different folders. For example, you have a device group for all access layer 3810 switches and a device group for all Instant clusters. You might want to assign devices in the switch device group to multiple folders corresponding to the switches' location. You might want to assign the Instant clusters to the same location folders as the switches, as shown in the figure.

Or you might want to assign the devices in different groups to different folders depending on how you are using AirWave and on administrator preference. Some administrators prefer to create some correspondence between groups and folders. In this case, you would create one folder for the switches and a different folder for access points (which are in different groups). You could then create similarly named subfolders for locations within each of these folders. This is the approach that you took in the practice tasks.

Management level

When you authorized your devices in AirWave, you chose Monitoring and Firmware as the management level. You'll now examine the AirWave capabilities at this management level in more detail, as well as at the Managed device level.

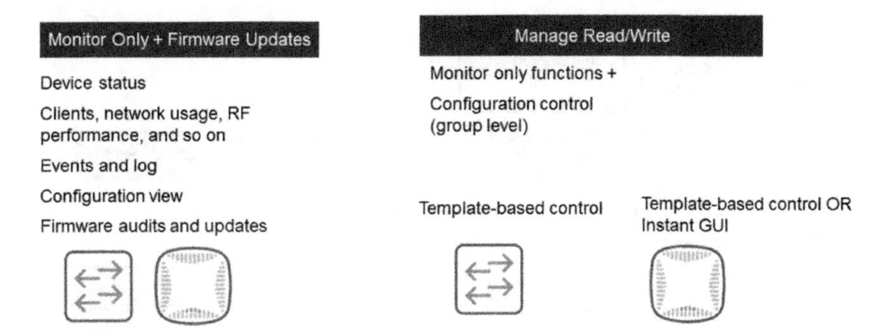

Figure 11-70: Management level

Monitor Only + Firmware Updates

When AirWave controls a device at the Monitor Only + Firmware Updates level, it monitors the device status. It also polls the device for information. You can see the results of these polls in displays of client information, network usage graphs, Clarity information, reports, and so on. AirWave also checks the device configuration and allows you to view the configuration from the AMP and assess the configuration against a template. You can also use AirWave to audit these devices' firmware and automate updates.

Manage Read/Write

AirWave continues to monitor devices at the Manage Read/Write level and to manage the firmware for these devices, but it can then also control other functions. One of the primary ways that AirWave controls a managed device is by enforcing its compliance with the configuration template defined for the device's group. AirWave audits all managed devices against this template, and, if any settings fail to match, AirWave accesses the device and redefines the unaligned settings. AirWave skips any settings with wildcards symbols (%), letting you set those on individual devices.

For managed Aruba Instant APs and clusters, you have a choice. You can control the cluster with a template, as described above, or you can activate the Instant GUI. When you choose the latter option, you can access an Instant GUI page for each AirWave group with Instant devices. This page resembles the Instant UI, but controls settings for all VCs within the group.

Management levels and configuration templates

It is recommended that you initially authorize devices with existing configurations at the Monitoring and Firmware level to prevent unanticipated changes to these configurations. (Factory default devices on which you are implementing ZTP make an exception as you will see.)

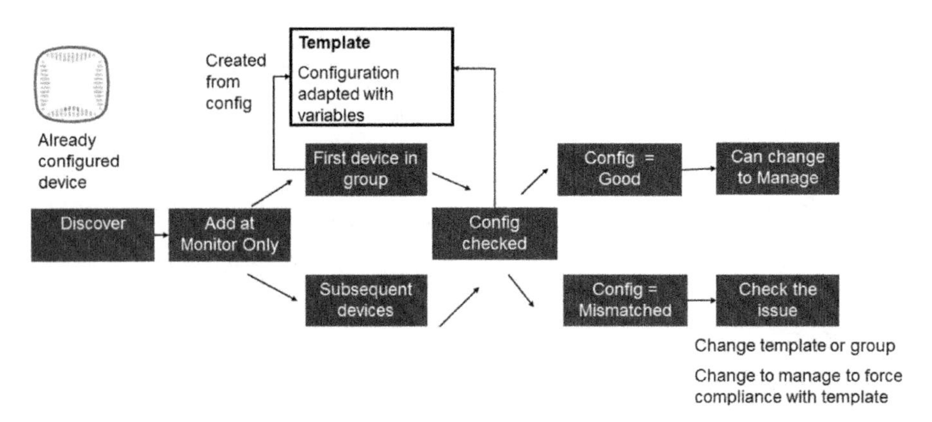

Figure 11-71: Management levels and configuration templates

When you authorize the first device in a group, AirWave automatically imports that device's configuration as the configuration template for the group. It also automatically processes the configuration as a template, replacing some settings that should generally be unique on each device—such as hostname and IP address—with variables.

As you authorize more devices in the group, still at the Monitoring and Firmware level, AirWave assesses the devices' compliance with the configuration template, marking each device as Good, if its configuration matches, or Misconfigured, if the configuration does not. AirWave doesn't take any action to change the misconfigured devices' settings, though, because the devices are only monitored, not managed.

Note

You can manually add templates to the group by importing the configuration from any of these devices. See the AirWave documentation.

Sometimes you will want to leave the devices at the Monitoring and Firmware level permanently—you want to use AirWave's extensive viewing, analysis, and reporting capabilities and to automate firmware updates but to continue using the CLI to make any configuration changes.

If, though, you want to use template-based configuration management and other features that alter device configurations, you must move the devices to Managed device mode. Any device that has Good status for its configuration, you can safely move to Managed device mode. For any device that has the Misconfigured status, you must assess the issue. AirWave has an audit tool that helps you quickly find the misaligned settings. You'll need to decide whether the device is misconfigured and you want AirWave to bring it into alignment by making the device managed, whether the template needs adjustment, whether the group needs a new template (perhaps one specific to this device type or firmware version), or whether the device belongs in a different group with a different configuration template. After you've made the determination and any necessary changes, you can make the device managed.

ZTP with Aruba Activate

Aruba AirWave's template-based management capabilities, when combined with the Aruba Activate service, make it possible for companies to instantly provision new Instant APs and ArubaOS switches. An on-site installer simply connects the device to the network and the device receives its configuration from AirWave without any administrator intervention—true zero touch provisioning (ZTP).

You'll need to complete a little advance setup in AirWave and Aruba Activate. You can then deploy many devices at many locations with very little additional work.

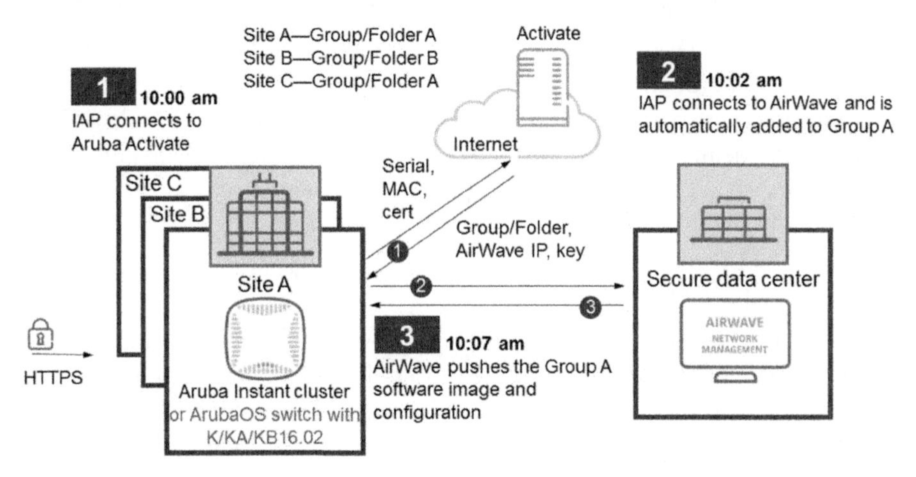

Figure 11-72: ZTP with Aruba Activate

AirWave setup

AirWave will be using group configuration templates to provision devices. To ensure that the correct templates are already in place, you create a device group for each different template that you intend to apply. You then manually configure one IAP cluster or ArubaOS switch, discover that device in AirWave, and authorize the device in the group. This device's configuration will then become the template for other devices deployed with ZTP.

Many users who begin to use ZTP with Aruba Activate are already monitoring or managing their network with AirWave and have the required templates in place.

Activate setup

A company must set up its Activate account, a free service for any Aruba customer, in advance. Aruba adds any supported devices ordered by this customer to the account. Devices are linked with information known by Aruba through the supply chain such as the device's self-signed certificate, ensuring that only this customer's true devices can be provisioned by this customer's AirWave servers.

Administrators set up Activate folders based on their organizational structure and place their new devices in those folders. Activate folders are distinct from AirWave folders although you should typically make the two correspond. You should create an Activate folder for each combination of AirWave group and folder to which you want to assign ZTP devices.

Within each folder, administrators define a provisioning rule, which specifies the Organization string, as well as the AirWave server IP address and the shared secret. The shared secret must be the same for every device that connects to the same AirWave server and must match the secret for the device that you already authorized in AirWave. You'll recognize the information in this rule as similar to the information that you defined manually on an Instant VC to connect the VC to AirWave.

Administrators can also define a move folder rule for the Activate folder. This rule uses information such as serial number and ordering information to automatically move a new device from one folder to another, helping users with many devices quickly sort them into the correct folders. Finally, administrators can optionally define a notification rule for the folder, which defines how Activate should inform administrators when a device is provisioned (such as by email).

ZTP with Activate process

The figure illustrates the process for an IAP to become provisioned with ZTP and Activate. For simplicity, the figure shows a single IAP, but the same process applies for an Instant cluster. The VC (also called master) takes responsibility for obtaining the configuration from AirWave after it is elected. ArubaOS switches with K/KA/KB/WB16.02 software also use the same process.

1. The IAP connects to the network and receives IP settings from a DHCP server. The IAP's network connection must permit it access to the Internet. Once it has become master, the IAP automatically contacts the Activate service using HTTPS.

   Activate recognizes the device by its serial number, MAC address, and self-signed certificate, which the device sends over the secure HTTPS tunnel. Activate uses this secure information to link the IAP to the correct customer account and folder within that account. It then sends the information defined in the provisioning rules to the IAP.

2. In minutes and without any administrative intervention, the IAP has learned everything that it needs to know to contact the AirWave server. The IAP connects to AirWave over HTTPS. As long as the IAP submits the same shared secret as the device that was already authorized in AirWave, AirWave automatically authorizes the IAP as a Managed device and places it in the requested group and folder.

   Administrators also need to activate automatic authorization in advance and set the automatically authorized mode for the device type to Manage Read/Write.

3. Because AirWave enforces firmware and configuration compliance for managed devices, it automatically pushes the correct firmware and configuration for the IAP's group to the IAP. The IAP applies the configuration, and, if it is VC of a cluster, then the configuration applies to the rest of the IAPs in the cluster as well.

Minutes after initial connection, the network infrastructure devices are running the correct configurations.

Important

ArubaOS switches require K/KA/KB/WB16.02 software or higher to support ZTP with Activate.

ZTP with DHCP

Aruba IAPs, Instant clusters, and ArubaOS switches also support a legacy form of ZTP that uses DHCP options, rather the Activate service, to notify the ZTP devices of their AirWave group, folder, server IP address, and shared secret. Typically, you should use the Activate version of ZTP because:

- The complete process uses secure HTTPS communications.
- Less setup is required at the customer site.
- Administrators can assign devices to any Activate folder they want and easily set up different provisioning and move folder rules for devices in different folders.
- Activate has more powerful capabilities such as notifications.

However, you will use the DHCP process in because you don't have an Activate account for your test network environment. You should also understand this process in case you need to use it for users who want ZTP for their ArubaOS switches but are not yet ready to upgrade to 16.02 software.

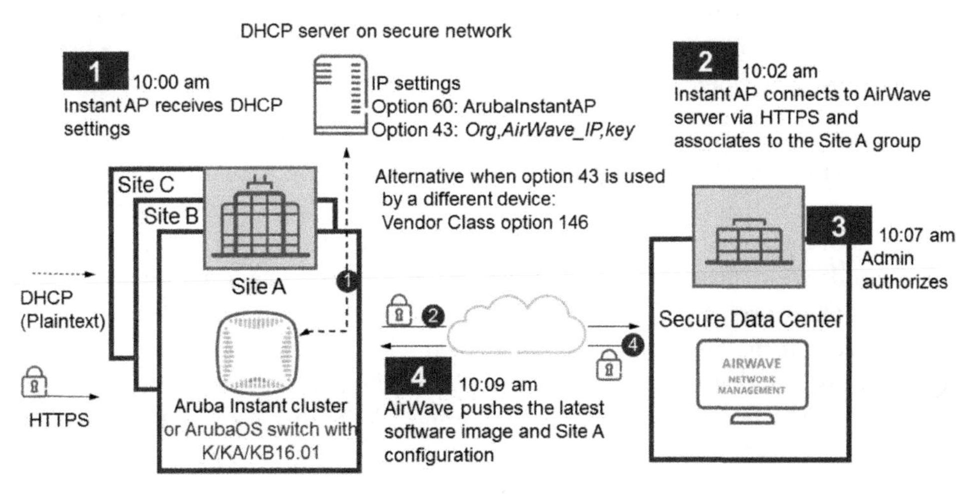

Figure 11-73: ZTP with DHCP

Again, the figure illustrates the process for an IAP, but the same process applies for an ArubaOS switch.

The IAP connects to the network and receives settings from a DHCP server. In addition to the typical IP settings, the server sends two special DHCP options:

- Option 60, which is a custom option that includes the string "ArubaInstantAP." This option signals to the IAP (or ArubaOS switch) that it will receive AirWave configuration settings in a vendor-specific option.
- Option 43, which is the vendor-specific option. This option is a string with the settings that the IAP needs to connect to AirWave, defined with this format:

  Org,AirWave_IP,shared secret

  For an IAP, Org uses the same format as the manual Organization setting. Anything before a colon defines the IAP's device group and highest level folder (with the exception of Top). You can add colons and additional strings to place the IAP in a subfolder. For example, "Access Points:Lab" places the IAP in the Access Points group and the Top > Access Points > Lab folder.

  For an ArubaOS switch, the Org format is slightly different. Anything before a colon defines the ArubaOS switch's group, and anything after the first colon defines the device's folder. For example, "Access-3800:Switches:Lab" places the switch in the Access-3800 group and the Top > Switches > Lab folder.

  You should already be familiar with the role of the AirWave IP and the shared secret.

The IAP now contacts AirWave and the rest of the process continues as with Activate-based ZTP.

It is important that you understand that the DHCP messages between the new device and the DHCP server are in plaintext. (The later communications between the device and AirWave use encrypted HTTPS.) Therefore, you should make sure that these devices communicate over a secure network.

You should typically use option 43 to deliver the string with the AirWave management settings, as described above. However, sometimes your devices are sharing a scope with other devices that are already using option 43 for another purpose. Or you might have two different types of network infrastructure devices in the same subnet. You need to place the devices in different AirWave groups so that each type of device receives the proper configuration. However, you can only specify one string for option 43. In these cases, you should use DHCP vendor classes. A vendor class enables a DHCP client to recognize that the DHCP option applies to it while other devices ignore the option. You create a vendor class that identifies your device type. You then create option 146 for this vendor class and specify the *Org,AirWave_IP,shared secret* in that option. With this second setup, option 60 is still required.

You'll learn how to set up all of these DHCP options.

Manage Devices in Aruba AirWave and Use ZTP

You will now make your Instant cluster and the Access-1 switch managed devices and explore the configuration templates and other management options. You will then erase the configuration on your IAPs and Access-2 switch and practice using ZTP to provision these devices with their configuration. Access-2 will use a configuration that places its ports in VLAN 11, so the Windows PC will receive an IP address in 10.1.11.0/24.

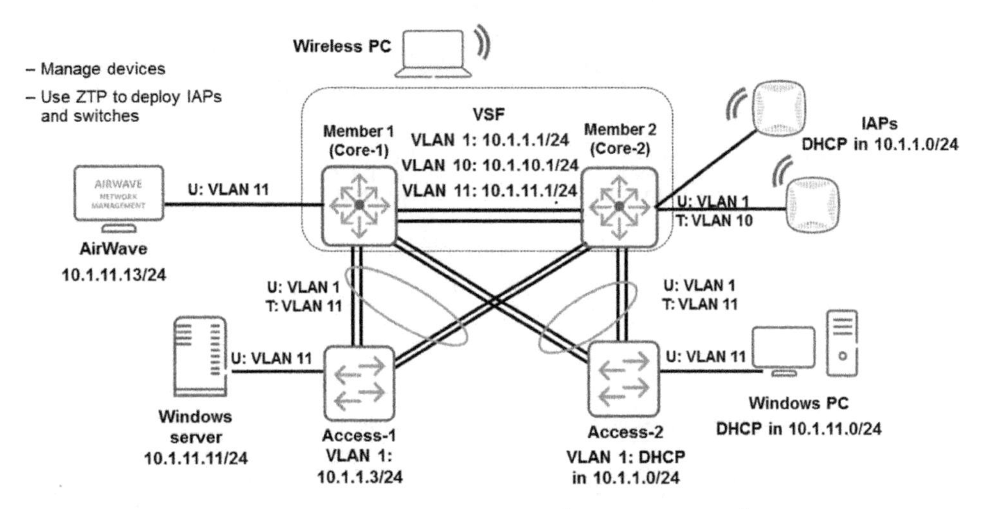

Figure 11-74: Manage Devices in Aruba AirWave and Use ZTP

After completing the tasks in this section, you will be able to:

- Safely change monitored devices in AirWave to managed devices
- Use configuration templates to manage devices at the group level
- Use zero touch provisioning (ZTP) to provision Instant APs and ArubaOS switches

The figure shows the final topology.

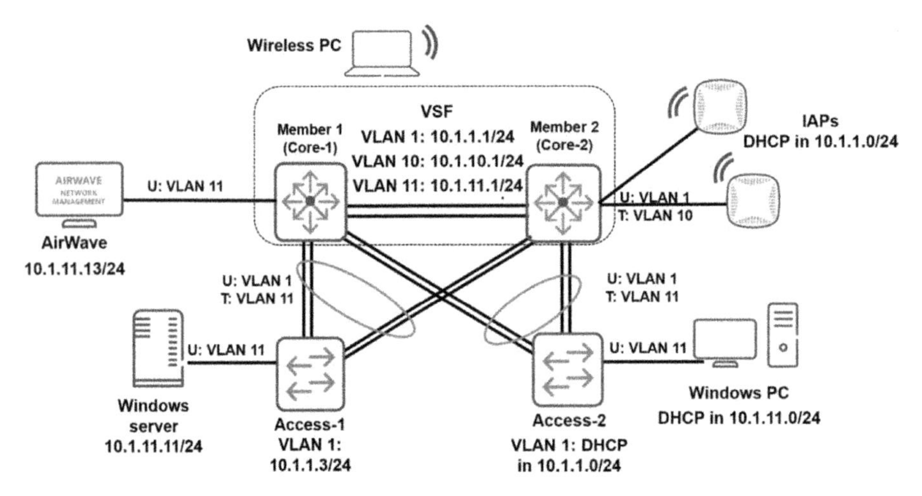

Figure 11-75: Final test network topology (logical)

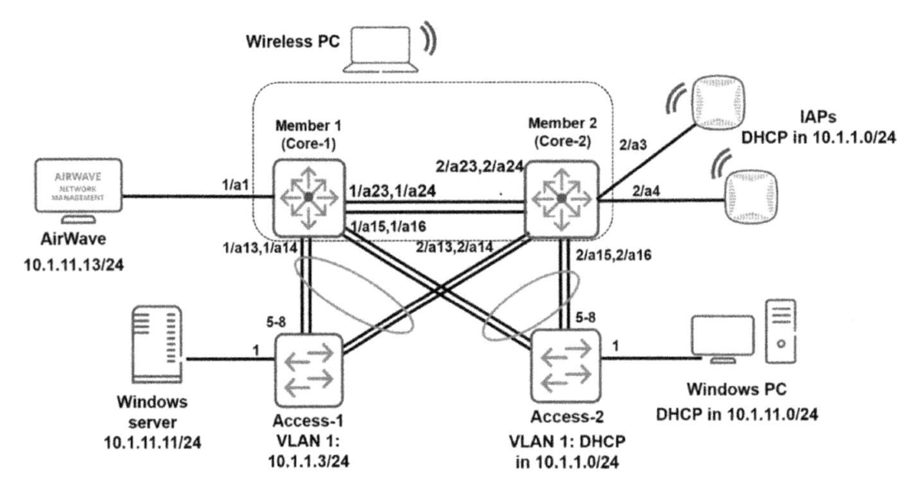

Figure 11-76: Final test network topology (physical)

Task 1: Change devices' management level in AirWave

Important

This task and the tasks that follow build on the topology you created the "Discover and Monitor Devices in Aruba AirWave" section of this chapter. That topology must be in place before you begin.

In this task, you will see that AirWave has automatically created configuration templates for groups. You will check that the Access-1 switch configuration and your Instant APs (IAPs) configuration match these templates. You will then change these devices to managed devices.

Manage an ArubaOS switch

You will begin by managing Access-1.

Windows server

1. Access the Windows server desktop.
2. Open a Web browser and browse to https://10.1.11.13.
3. Log in to the AMP with the default credentials (**admin** and **admin**).
4. Check the configuration template being used by Access-1's group, Access-3800.
    a. Click **Groups** in the left navigation bar.
    b. Click **Access-3800**.

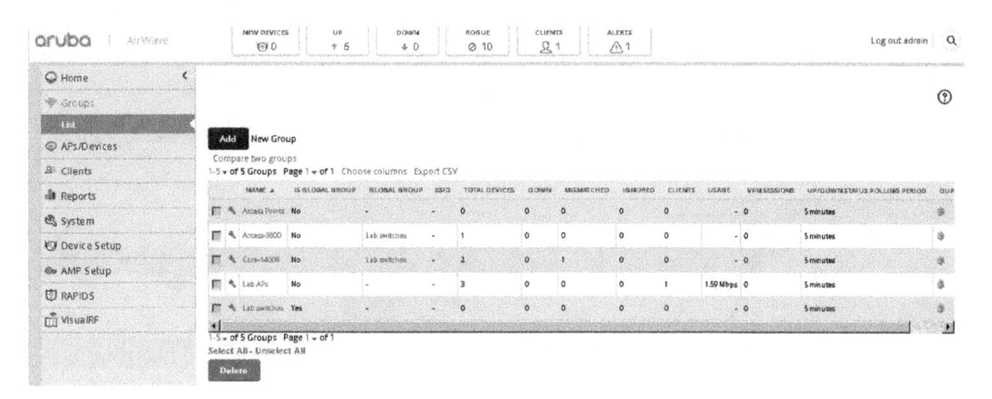

Figure 11-77: Check the configuration template being used

    c. Click **Templates** in the left navigation bar.
    d. You should see that AirWave has automatically created a template based on Access-1's configuration. AirWave creates the template from the first device added to the group.
    e. Click the pencil icon next to the template to view it in more detail.

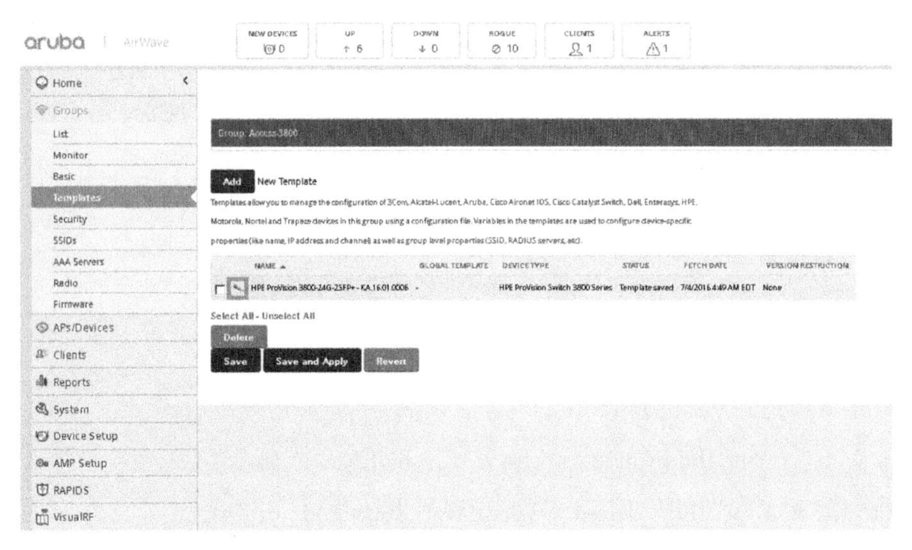

Figure 11-78: Access detailed template information

f. As you see, the template is based on the configuration, but AirWave has automatically changed it in ways to make it generic for all devices in the group. For example, the hostname is a variable. VLAN 1 has script to allow the switch to use a static IP address or DHCP address based on settings that you can configure in AirWave. AirWave detected these settings and automatically configured them for Access-1 when you imported it.

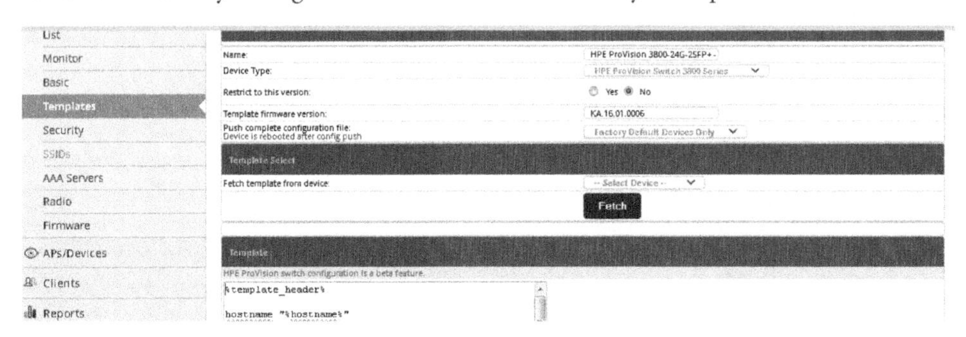

Figure 11-79: View template in more detail

g. Scroll to the bottom of the page and click **Cancel** to stop viewing the template without making any changes.

Important

If the template doesn't match the Access-1 configuration, select the switch from the Fetch template from device list and click Fetch. Then Save the new template and apply the changes.

5. You will now check whether Access-1 conforms with the template and then change it to a managed device.

   a. Move to the APs/Devices > List page. Make sure that the Lab switches > Roseville folder is selected.

   b. Scroll down to the Devices List.

   c. Verify that Access-1's configuration is listed as Good, which means that it matches the template for the group.

   d. Now you can change Access-1 to a managed device without changing its configuration in an unexpected way.

   e. Click the pencil icon at the top right of the list.

   f. Select the Access-1 check box.

   g. For Device Actions, select **Management Level.**

   h. Click **Manage Read/Write**.

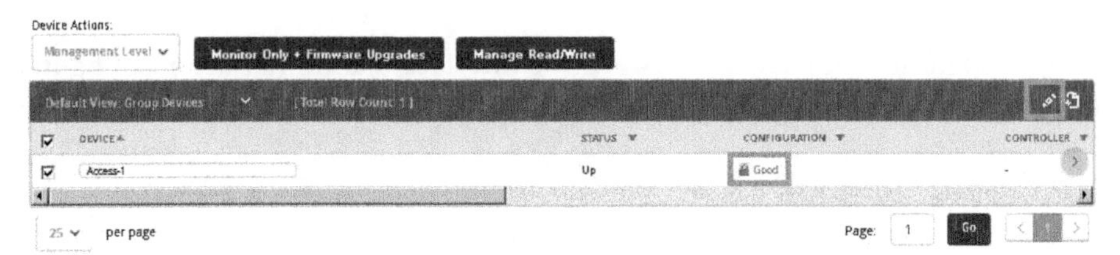

Figure 11-80: Change Access-1 to a managed device

   i. Click **Apply Changes Now**.

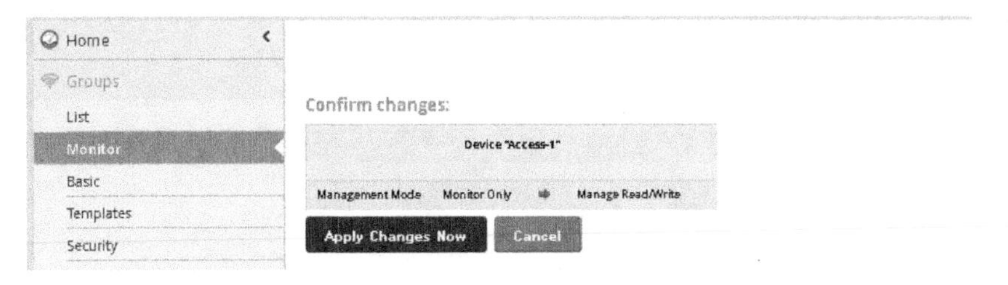

Figure 11-81: Apply changes to Device "Access-1"

6. You will now see how you can use the template to configure your devices centrally. For the purposes of these practice tasks, you want access devices to obtain DHCP addresses on VLAN 11 so that the same template can apply to all devices.

   a. Navigate to the Groups > List window and click the Access-3800 group name.

   b. Click **Templates** under **Group**

   c. Click the pencil icon to edit the template.

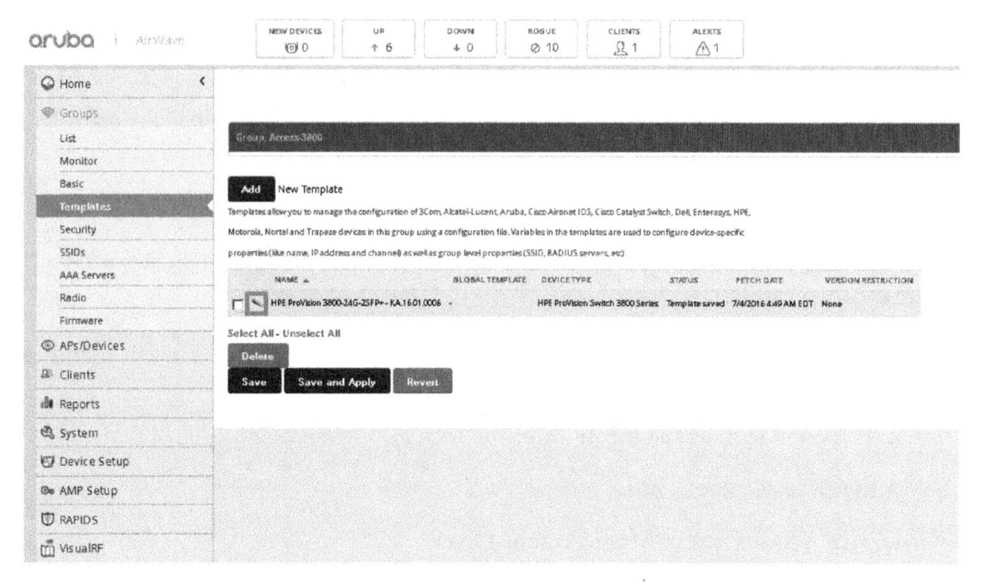

Figure 11-82: Click the pencil icon to edit the template

d. Scroll through the template. In between the interface 5 and interface 7 configurations, add these lines. Make sure to type the lines so that the quotation mark is not a smart quote.

```
interface 6
  name "Core-1 second link"
  exit
```

e. Scroll to the bottom of the page and click **Save**.

f. Click **Save and Apply**.

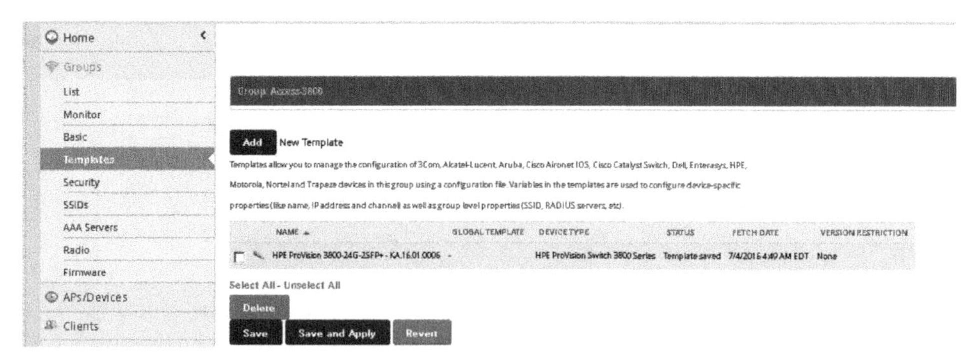

Figure 11-83: Save and apply the changes to the template

g. After you apply the changes, AirWave will update all managed devices in the group to match the template. You could schedule this to occur later during a maintenance window, but for the purposes of this task, click **Apply Changes Now**.

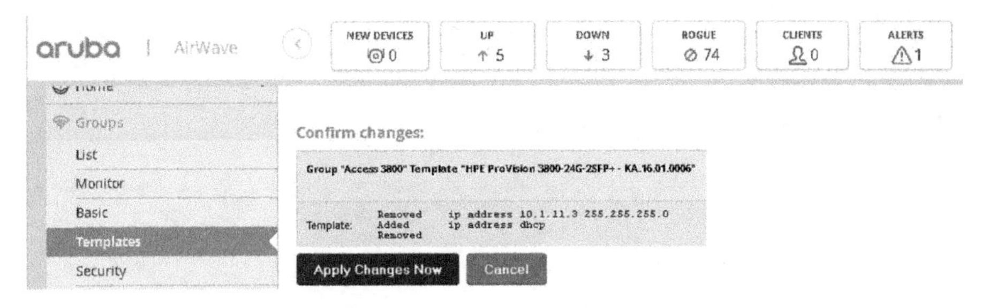

Figure 11-84: Click Apply Changes Now

Access-1

7. Wait a minute or so, and then check the IP address on Access-1. You should see the new interface description.

```
Access-1(config)# show trunk

 Load Balancing Method:  L3-based (default)

  Port  | Name                             Type      | Group Type
  ----- + -------------------------------- --------- + ----- -----
  5     | Core-1                           100/1000T | Trk1  LACP
  6     | Core-1 second link               100/1000T | Trk1  LACP
  7     | Core-2                           100/1000T | Trk1  LACP
  8     |                                  100/1000T | Trk1  LACP
```

Manage an Instant cluster

You will now make your Instant cluster managed devices and manage the cluster from within AirWave.

Windows server

8. Return to the Windows server and the browser logged into AMP.
9. First check the Instant configuration template.
   a. Click **Groups** in the left navigation bar.
   b. Click **Lab APs**.

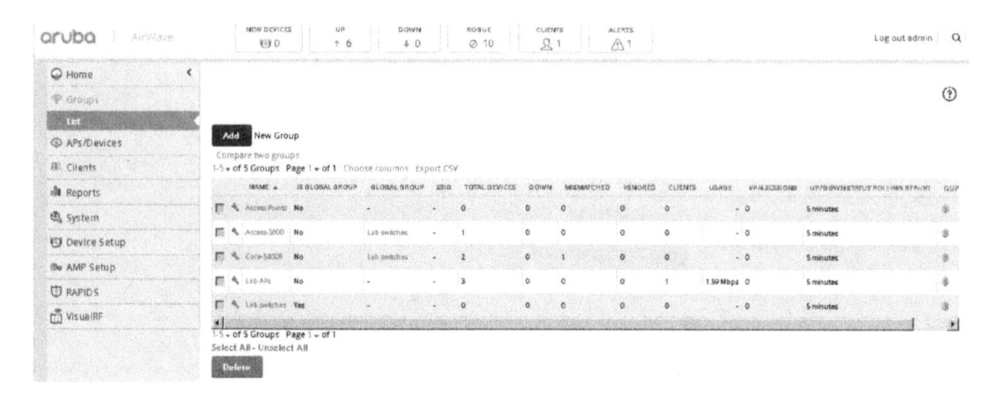

Figure 11-85: Instant configuration template

c. Click **Templates** under **Group**.

d. You should see two templates, one for the VC and one for the IAPs controlled by the VC (thin APs).

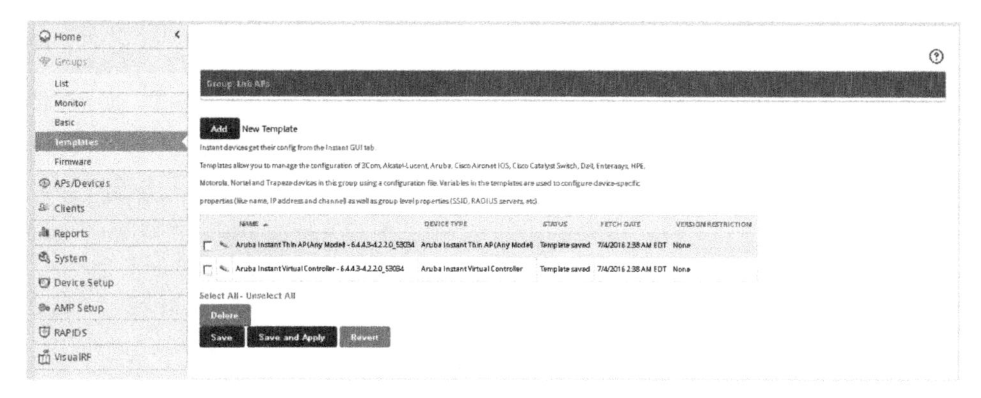

Figure 11-86: Templates for the VC and the IAPs controlled by the VC

10. Now check that the VC and IAPs comply with the configuration and change them into managed devices. The steps are the same as for Access-1. Try to remember the process. Figures are provided on the next page for your reference.

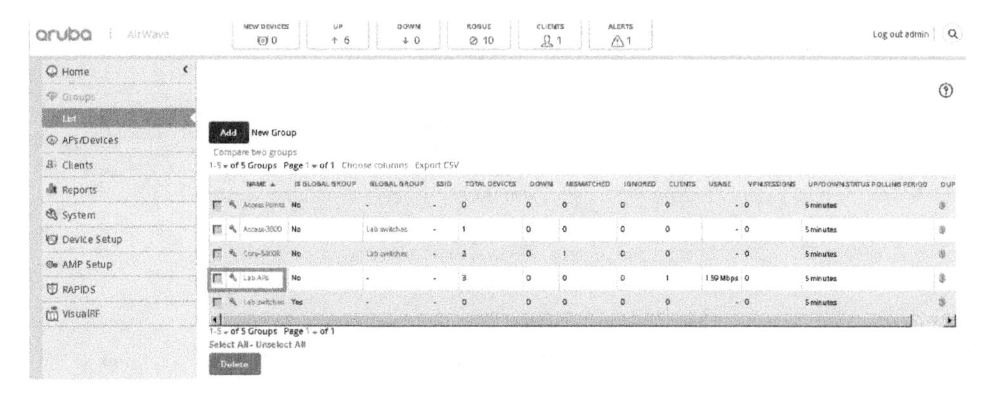

Figure 11-87: Check the configuration template that is being used > Click the pencil icon to view details

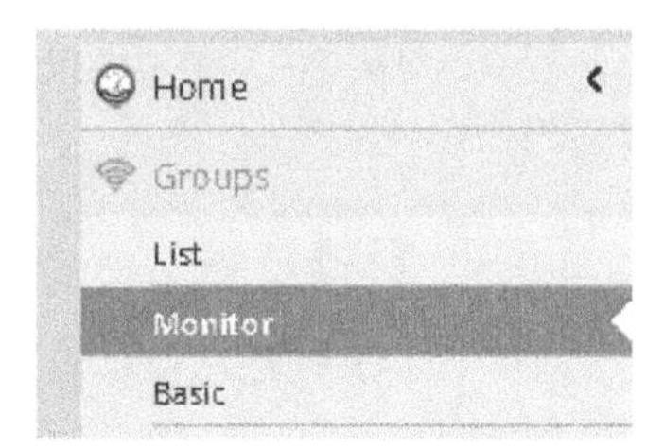

Figure 11-88: Access the Monitor page

Default View: Group Devices [Total Row Count: 3]

DEVICE	STATUS	CONFIGURATION	CONTROLLER	FOLDER
ATPNk-Ss-IAPs	Up	Good	-	Roseville
IAP-North *	Up	Good	ATPNk-Ss-IAPs	Roseville
IAP-South	Up	Good	ATPNk-Ss-IAPs	Roseville

25 per page Page: 1 Go

Figure 11-89: Verify that the configuration for each device is Good > Click the pencil icon

Device Actions:
Management Level | Monitor Only + Firmware Upgrades | Manage Read/Write

Default View: Group Devices [Total Row Count: 3]

DEVICE	STATUS	CONFIGURATION	CONTROLLER
ATPNk-Ss-IAPs	Up	Good	-
IAP-North *	Up	Good	ATPNk-Ss-IAPs
IAP-South	Up	Good	ATPNk-Ss-IAPs

25 per page Page: 1 Go

Figure 11-90: For Device Actions, select Management Level > Click Manage Read/Write

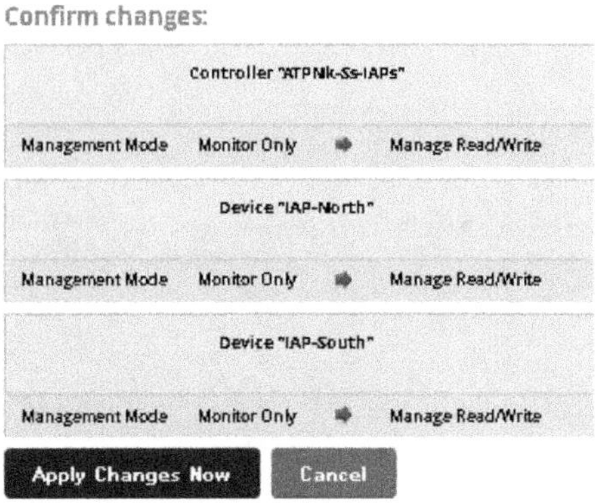

Confirm changes:

Controller "ATPNk-Ss-IAPs"

Management Mode Monitor Only ➡ Manage Read/Write

Device "IAP-North"

Management Mode Monitor Only ➡ Manage Read/Write

Device "IAP-South"

Management Mode Monitor Only ➡ Manage Read/Write

Apply Changes Now Cancel

Figure 11-91: Confirm and apply changes

11. Wait a minute or two.
12. Return to the Instant UI at 10.1.1.5. You will see that the UI now has just a few options because AirWave has taken over the management.

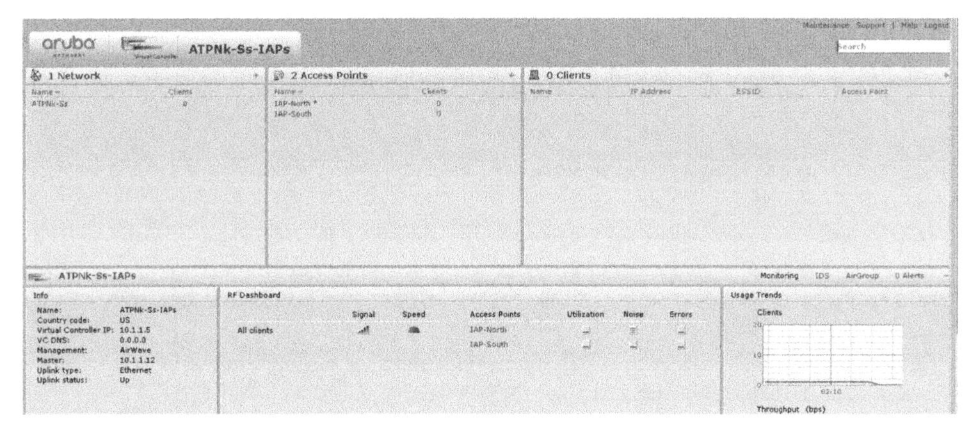

Figure 11-92: UI when AirWave management is in place

13. Return to the AMP.
14. You can use the configuration template to manage the IAPs like you did the ArubaOS switches. You might choose this option if you're familiar with the Instant CLI. But if you prefer the Instant UI which you practiced using in Chapter 10, you can enable that instead.
    a. Move to the Groups > Basic window for the Lab APs group.
    b. Scroll to the Aruba Instant settings.
    c. For Enable Instant GUI Config, click **Yes**.
    d. Scroll to the bottom of the window and click **Save and Apply**.

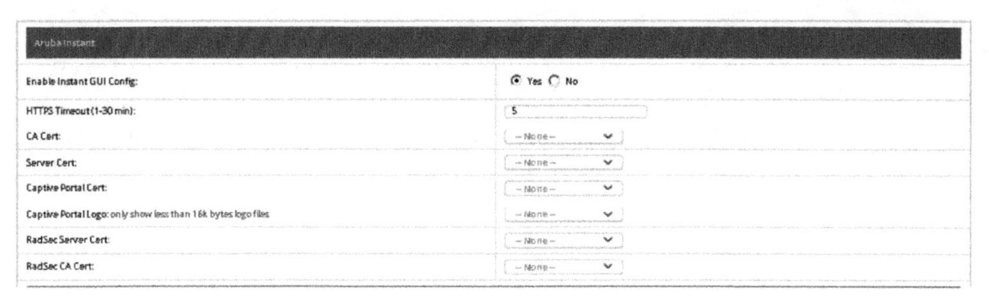

Figure 11-93: Enable Instant GUI Config

e. Click Apply Changes Now.

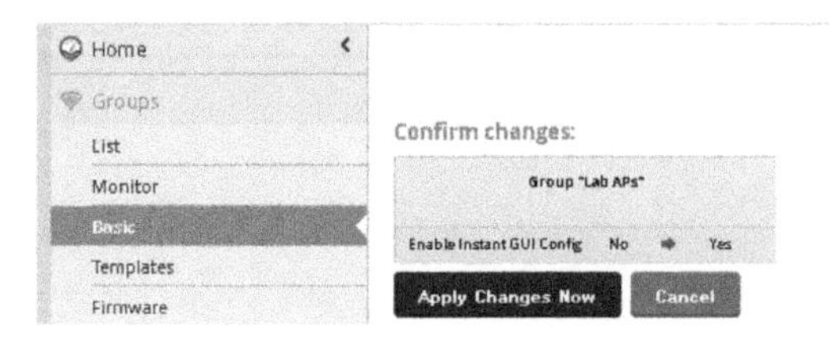

Figure 11-94: Confirm and apply changes

15. Click **Instant Config** under Groups. As you see, you have many of the same options as in the Instant GUI. Click Networks and make sure that you see the ATPNk-Ss network, which was imported from the template.

Important

If you see an error that the IGC engine is not enabled, wait about three minutes and then refresh the window. Do not move on until the Instant GUI is available.

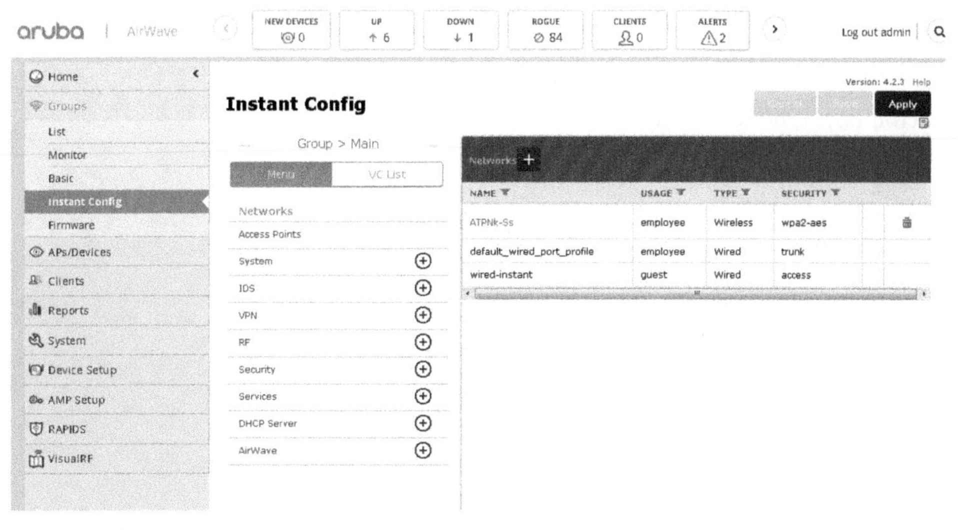

Figure 11-95: Check for the ATPNk-Ss network

Task 2: Set up ZTP for an Instant cluster

You will now simulate a situation in which you are deploying a new Instant cluster. The cluster will be provisioned with the Lab APs configuration. Because you have a limited amount of test equipment, you will delete the existing IAP cluster from AirWave, reset that cluster to factory default settings, and then have these IAPs play the role of the new IAPs.

Note

You will no longer see the template for the Lab APs group because you are using the Instant Config GUI instead. However, AirWave still has a configuration template for this group in the background and will apply it to the new Instant cluster when you add it to the group.

Windows server

1. You should be logged into the AMP.
2. Delete the Instant cluster.
    a. Move to the APs/Devices > List page.
    b. For Go to folder, select **Lab APs** > **Roseville**.

Figure 11-96: Select Lab APs > Roseville folder on the List page

    c. Scroll down to the Devices List.
    d. Click the pencil icon and select all of the devices.
    e. For Device Actions, choose **Delete selected devices**.
    f. Click **Delete**.

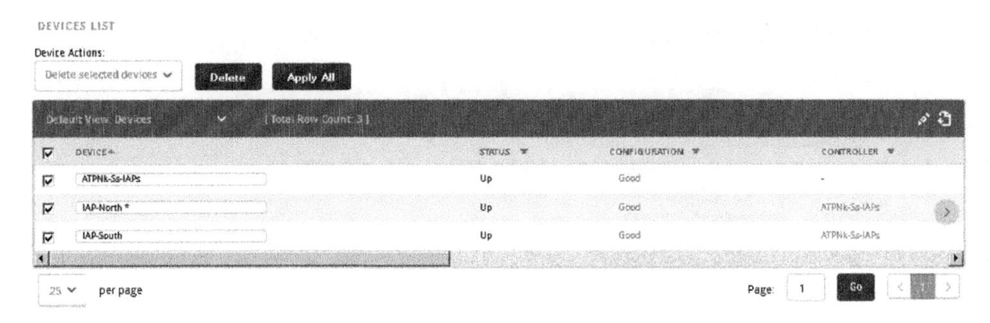

Figure 11-97: Delete selected devices

g. Click **Apply Changes Now**.

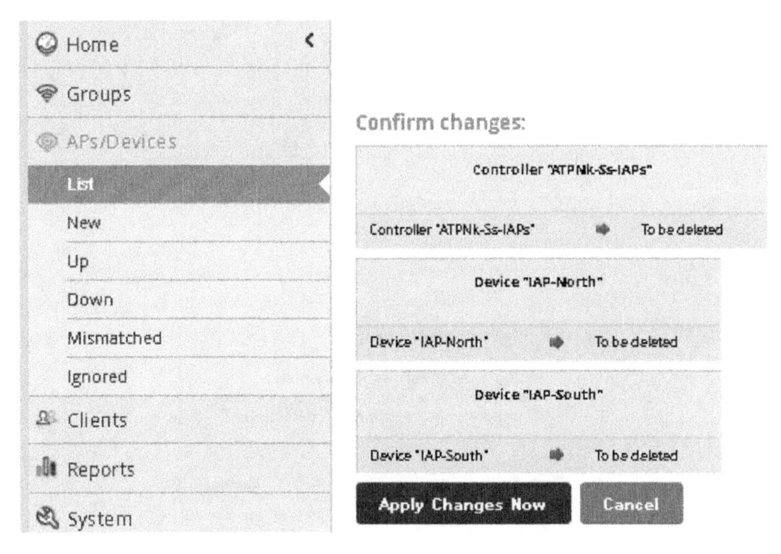

Figure 11-98: Apply Changes Now

3. Reset the Instant cluster to factory default settings.
   a. Access the Instant UI at https://10.1.1.5.
   b. Click the **Maintenance** link.
   c. Click the **Configuration** tab.

Note

If the Configuration tab doesn't display, wait a minute or so for the Instant VC to realize that it is no longer managed by AirWave. Then refresh the interface and click Maintenance again.

d. Click **Clear Configuration**.

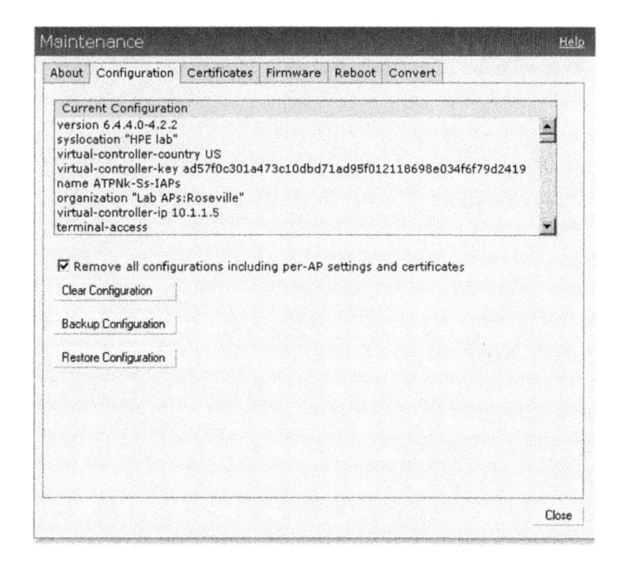

Figure 11-99: Remove all configurations

4. Confirm the clear and restart.

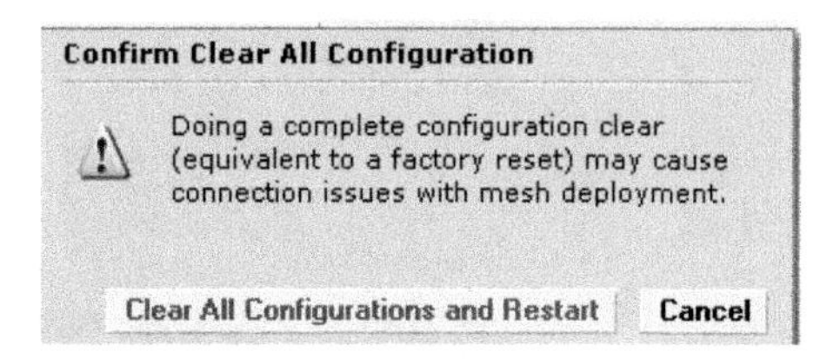

Figure 11-100: Clear All Configurations and Restart

5. Return to the browser with AMP and refresh the page. Verify that you see 0 new devices. If you see one new device, the Instant cluster might have attempted to contact AirWave before you reset the configuration. In that case, click the New Devices icon, select the device, and initiate the Delete Selected Devices action.

Core-1

Disconnect the IAPs from the network until you're ready to perform the ZTP process.

6. Access the terminal session with the VSF fabric.

7. Disable the interfaces that connect to the IAPs.

```
Core-1(config)# interface 2/a3,2/a4 disable
```

Windows server

You are now ready to set up the DHCP options for ZTP:

- Option 60 = ArubaInstantAP
- Option 43 = Lab APs:Sunnyvale,10.1.11.13,password

In option 43, the first part of the Organization must reference an *existing* AirWave group with the proper configuration with this device. This will also be the higher level folder for the IAP. You're specifying Sunnyvale as a subfolder to indicate the location of the new IAPs. The folders don't necessarily have to exist in AirWave already.

8. Begin by creating DHCP option 60.
    a. Return to the Windows server desktop and open the Server Manager.
    b. Expand **Roles** > **DHCP Server** > **server_name**.
    c. Right-click IPv4 and select **Set Predefined options**.

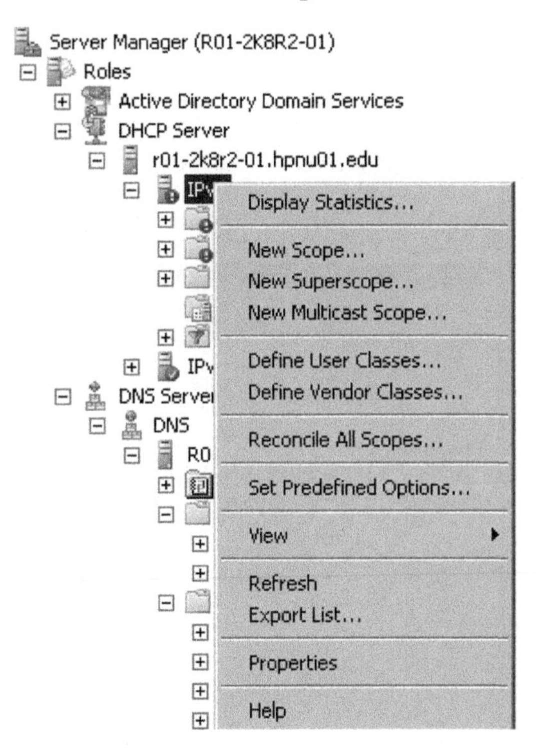

Figure 11-101: Set Predefined options

    d. The Option class should list DHCP Standard Options.

e. Click **Add**.

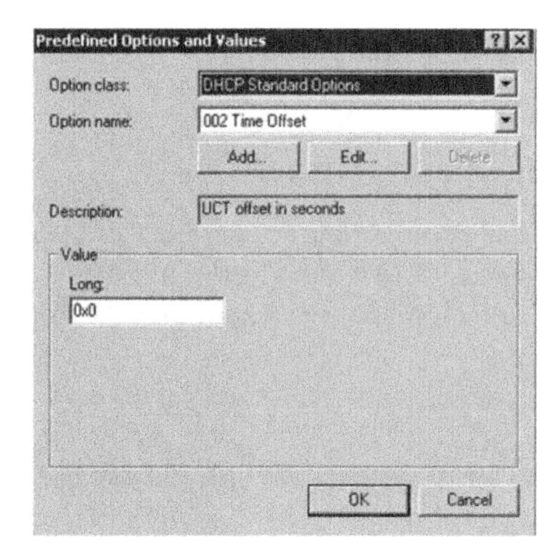

Figure 11-102: Add DHCP Standard Options

f. Set Name to **ZTP** (you can use any name that is intuitive to you.)

g. For Data type, select **String**.

h. For Code, type **60**.

i. The description is optional. Click **OK**.

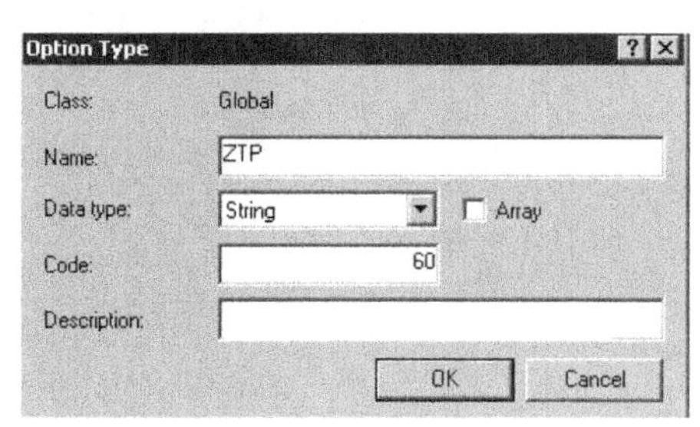

Figure 11-103: Add Option Type information

j. Then click OK again.

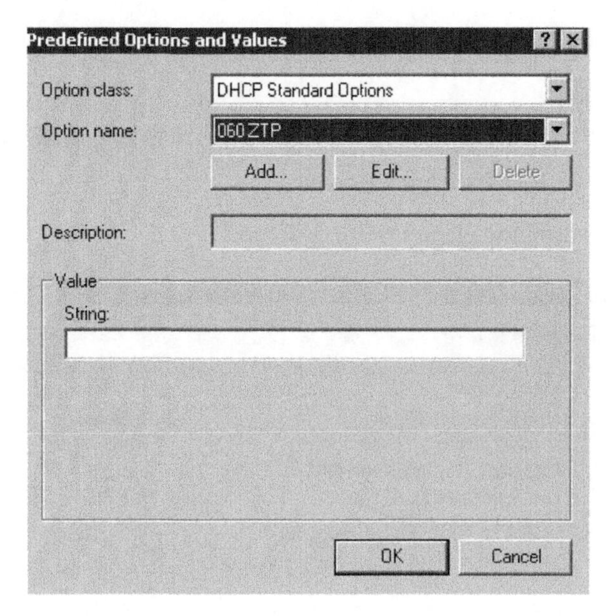

Figure 11-104: Predefined Options and Values

9. Now add option 43, which specifies the AirWave management settings, and DHCP option 60 to the VLAN 1 scope used by the IAPs.

   a. Expand Scope [10.1.1.0] VLAN 1.

   b. Right-click Scope Options and click **Configure Options**.

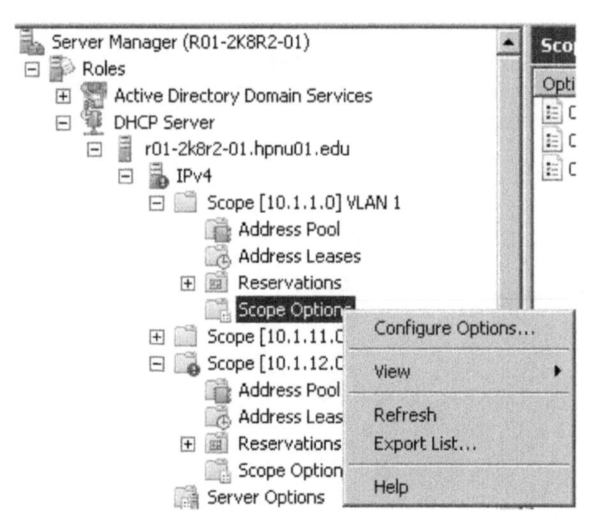

Figure 11-105: Scope Options > Configure Options

c. Scroll through the General options and select the check box for option 43, Vendor Specific Info.

d. Click under ASCII and copy the exact string that you see in the figure below:

Lab APs:Sunnyvale,10.1.11.13,password

(The characters separating elements are commas.)

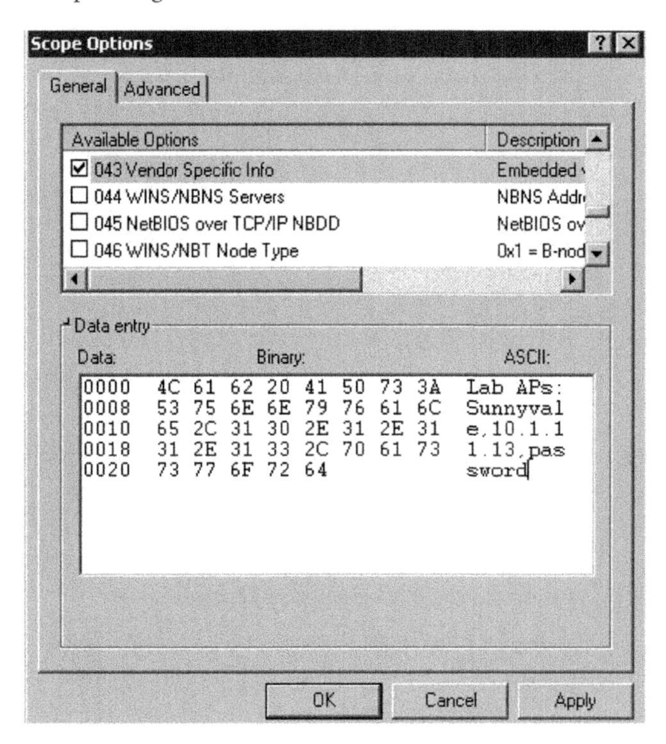

Figure 11-106: Copy the ASCII string

e. Scroll down and select the check box for option 60.

f. For String value, type **ArubaInstantAP**.

g. Click **OK**.

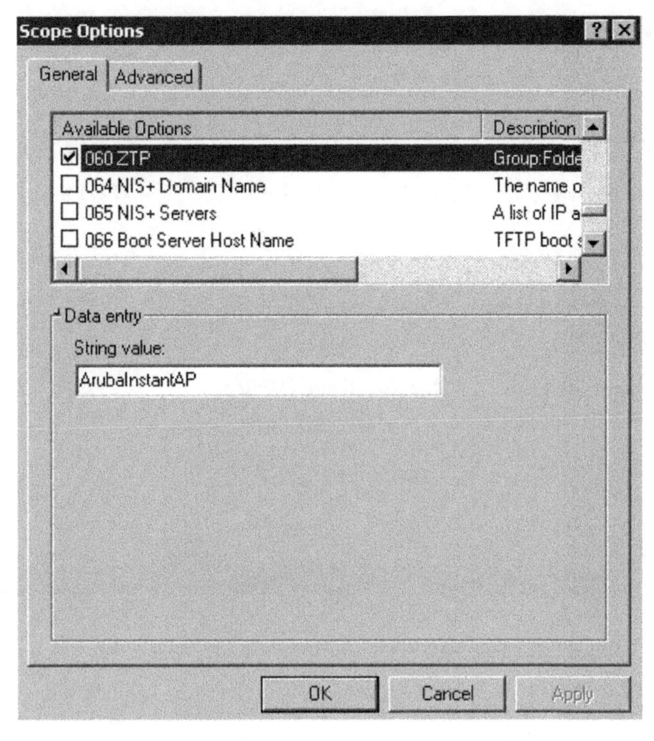

Figure 11-107: Add String value for Option 60

10. The VLAN 1 scope options should match the figure below.

Scope Options

Option Name	Vendor	Value	Class
003 Router	Standard	10.1.1.1	None
006 DNS Servers	Standard	10.1.11.11	None
015 DNS Domain Name	Standard	hpnu01.edu	None
043 Vendor Specific Info	Standard	4c 61 62 20 41 50 73 3a 53 75 6e 6e...	None
060 ZTP	Standard	ArubaInstantAP	None

Figure 11-108: Confirm VLAN 1 scope options

Core-1

You can now connect the IAPs to the network. They will receive the AirWave information with their IP addresses, and, after they form a cluster, the VC will contact AirWave.

11. Access the terminal session with Core-2.

12. Enable the interfaces that connect to the IAPs.

```
Core-1(config)# interface 2/a3,2/a4 enable
```

Windows server

13. Return to the Windows server desktop and the browser connected to the AMP.
14. Authorize the IAP.
    a. Wait a couple minutes. You should then see the VC in the APs/Devices > New page.
    b. Select the VC's check box.
    c. You can leave the Group and Folder as they are because the settings from the DHCP option 43 will take precedence.
    d. Set the Management Level to **Manage Read/Write**.
    e. Click **Add**.

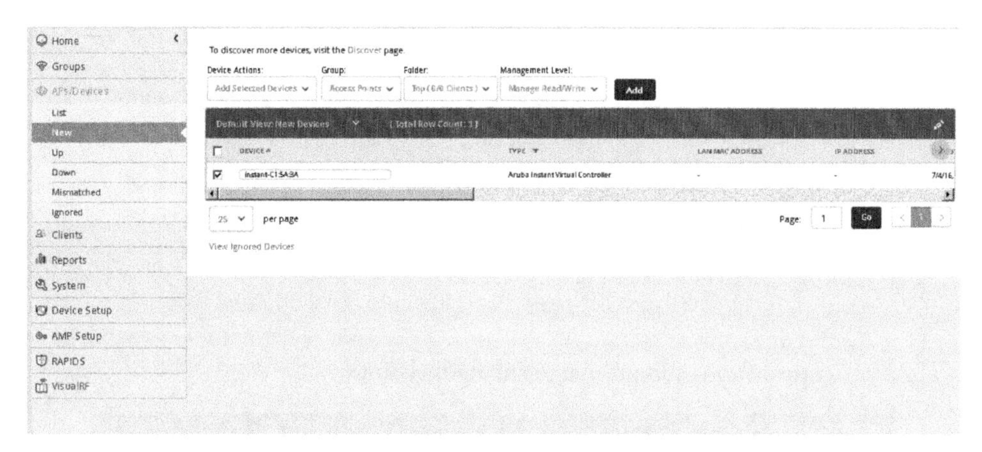

Figure 11-109: Authorize the IAP

    f. You can ignore the warning. You want the IAP to receive its configuration from the template. Click **Apply Changes Now**.

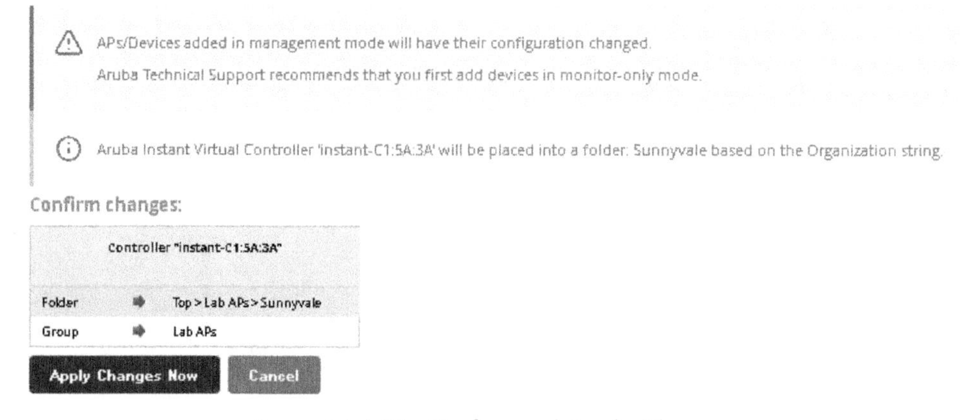

Figure 11-110: Confirm and Apply Changes

Wireless Windows PC

15. Wait a couple minutes and then verify that the Instant cluster has received its configuration and is broadcasting the ATPNk-Ss SSID. Access the wireless client and look for your SSID in the list.

Windows server

The template had some variables such as system name, so the Instant VC is using the default settings for those. You will now see how to define the variables when the Instant GUI is enabled.

Note

You are using the Instant GUI to manage the APs. If the Instant GUI were disabled, though, you could still define the values for variables. You would apply select the VC in the device list and a device action to for defining VC variables.

16. Return to the server desktop and the browser connected to AMP.
17. Click Groups and select Lab APs.
18. You should be taken to the Monitor page. Scroll down to the device list.
19. The Instant cluster received the configuration so smoothly that you might almost wonder if the IAPs actually reset their configuration. But you can see that they did because the IAPs no longer have the device names that you configured on them.

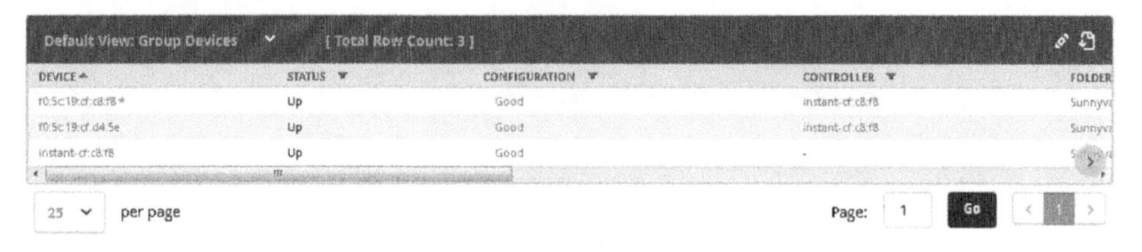

Figure 11-111: IAP device names after the configuration reset

20. Change the Instant cluster's system name.
    a. Click **Instant Config**.
    b. Expand **System** and click **General**.
    c. Click **Edit** next to Name.

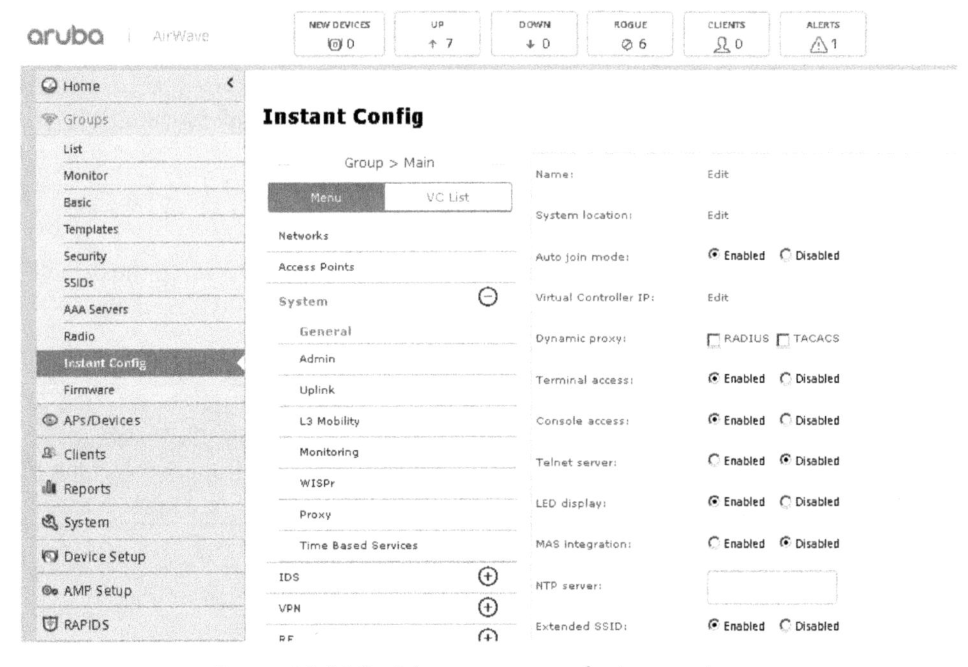

Figure 11-112: Edit system name for Instant cluster

21. Set an intuitive name such as Sunnyvale-IAPs.
22. Click **OK**.

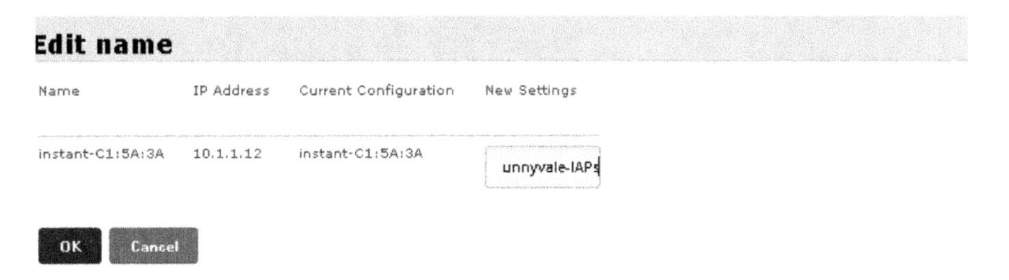

Figure 11-113: Click OK to confirm new name

23. Click **Apply** then click **Yes**

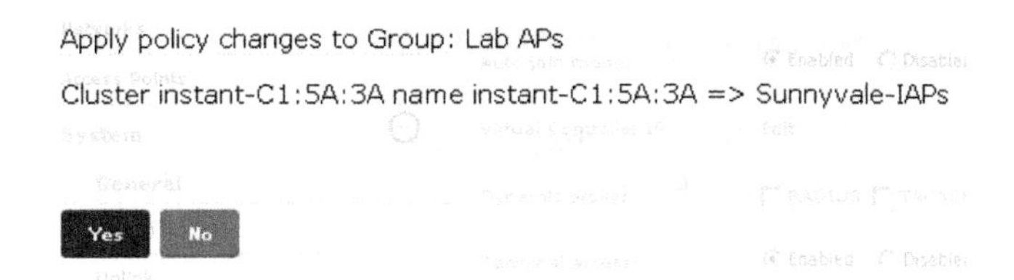

Figure 11-114: Apply policy changes

Task 3: Use ZTP to deploy Access-2

In this scenario, the Access-3800 group has the configuration template that you want to use on switches at all branches, and Access-2 is a switch at a new branch. You will set up ZTP. You will then erase Access-2's configuration, and it will use ZTP to become provisioned.

The ArubaOS switch and the IAPs are receiving DHCP settings on the same subnet and from the same scope, but each needs a different organization string in option 43 because the devices belong in different AirWave groups. In a scenario like this, you use the vendor class option to tie the particular ArubaOS switch type to the correct DHCP option for it.

Access-2

1. Access a terminal session with Access-2.
2. Find the string that this switch looks for in vendor-specific DHCP options.

```
Access-2(config)# show dhcp client vendor-specific

Vendor Class Id = HP J9575A 3800-24G-2SFP+ Switch

Processing of Vendor Specific Configuration is enabled
```

Windows server

3. Define the vendor class for your access layer switch.
   a. Return to the Windows server desktop and open the Server Manager.
   b. Expand **Roles** > **DHCP Server** > **server_name**.
   c. Right-click IPv4 and select Define Vendor Classes.

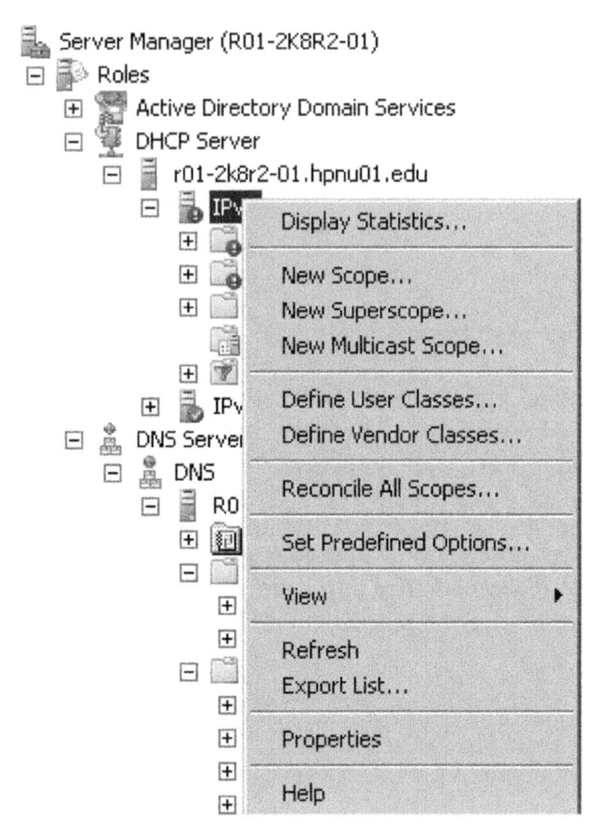

Figure 11-115: Define vendor class for access layer switch

d. Click **Add**.

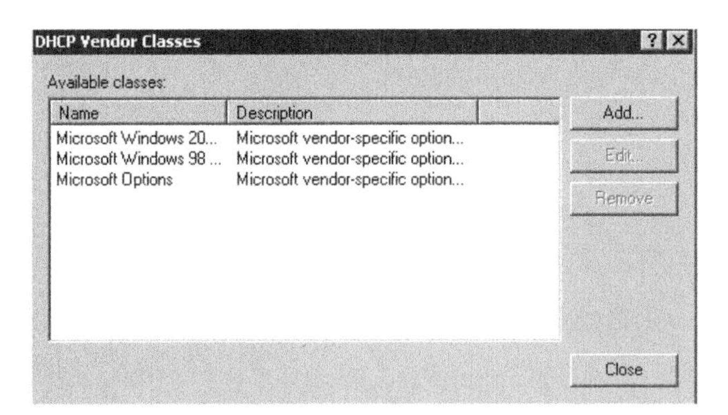

Figure 11-116: View new DHCP vendor class definitions

e. Define an intuitive Display name like **Aruba 3800**.

f. Under ASCII, exactly copy the vendor class ID that you found on Access-2.

g. Click **OK**.

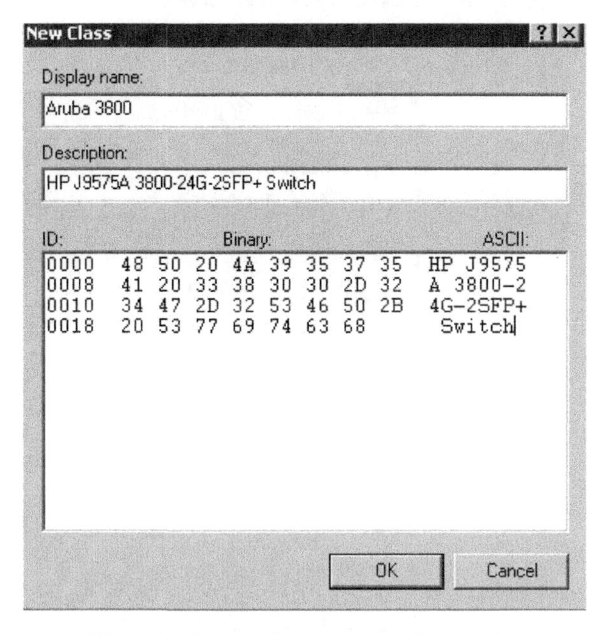

Figure 11-117: Add the Display name and copy the ASCII string

h. Verify the option and click **Close**.

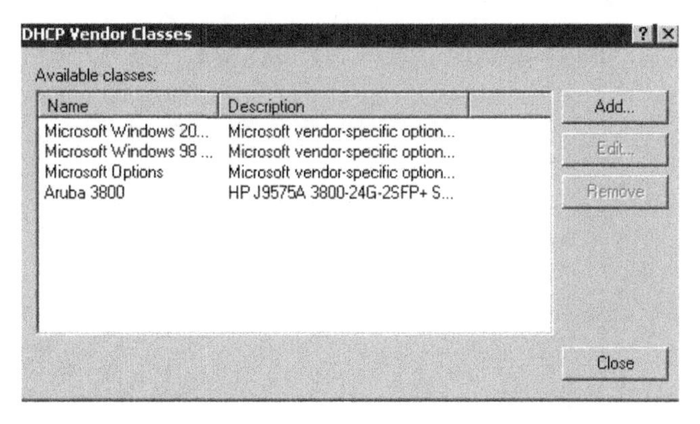

Figure 11-118: Verify new DHCP Vendor Class entry

4. Next create a DHCP option 146 that is tied to this vendor class.

   a. Right-click IPv4 and select Set Predefined Options.

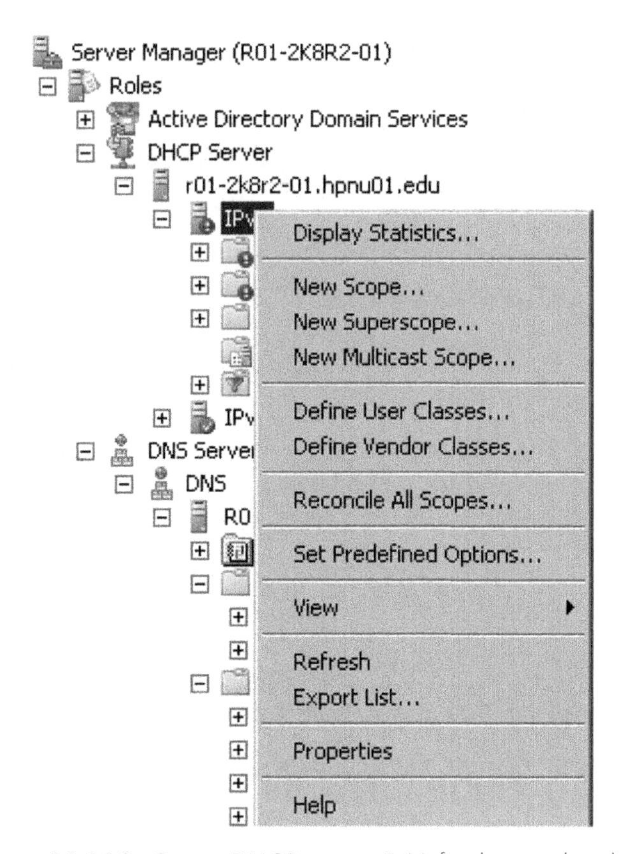

Figure 11-119: Create DHCP option 146 for this vendor class

b. For Option class, choose the class that you just defined.

c. Click **Add**.

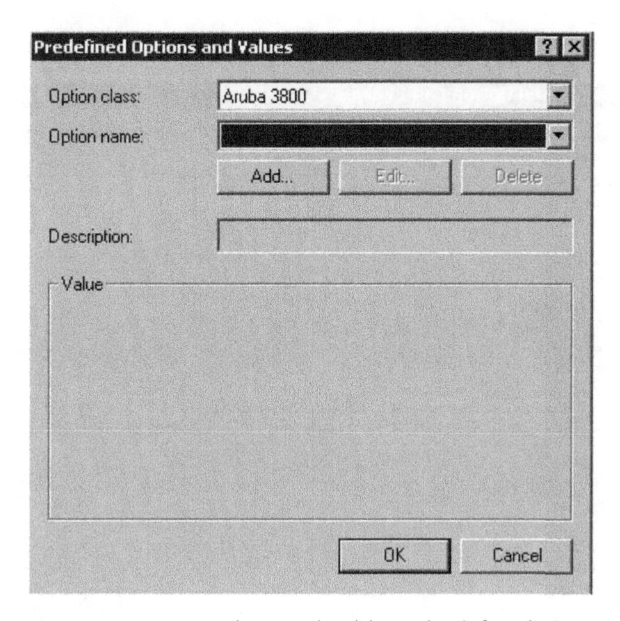

Figure 11-120: Select and add newly defined class

d. For Name, type an intuitive name.

e. For Data type, select **String**.

f. For Code, type **146**.

g. Click **OK**.

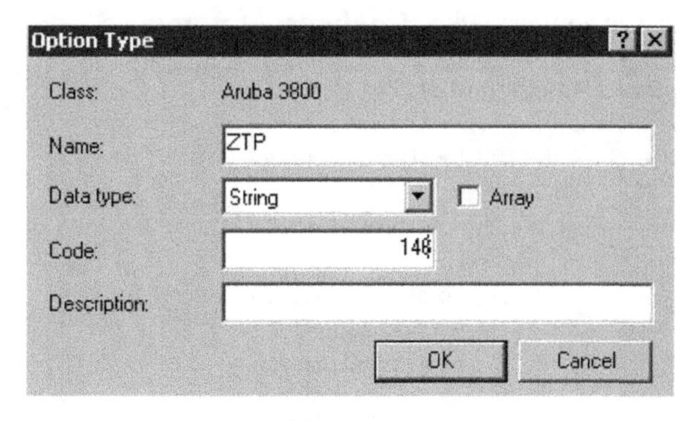

Figure 11-121: Add Option Type information

h. For the value string, copy the exact string shown below:

Access-3800:Lab switches:Sunnyvale,10.1.11.13,password

The first part of the string is the group. The next parts are the folder and subfolder for the device. (Colons separate the first elements. Commas separate Sunnyvale, the IP address, and the password.) Make sure to include the hyphen (-) between Access and 3800.

i. Click **OK**.

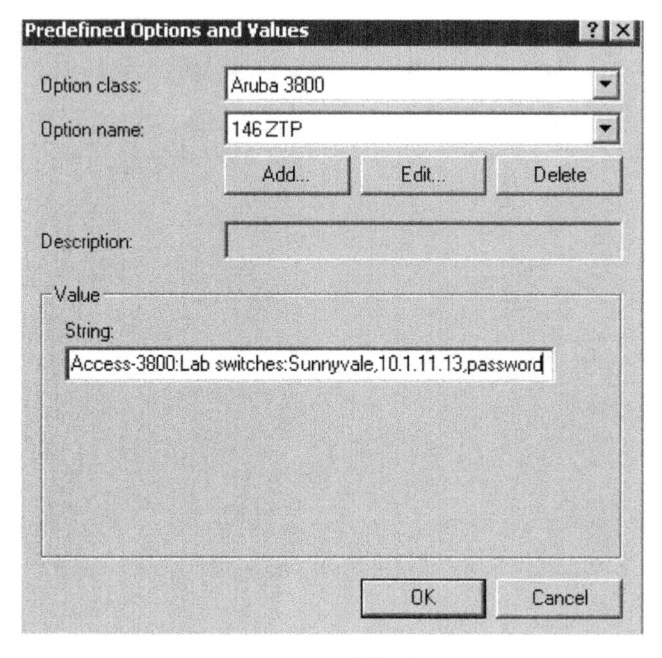

Figure 11-122: Add value string

5. Now add the option to the VLAN 1 scope.
   a. Expand Scope [10.1.1.0] VLAN 1.
   b. Right-click Scope Options and click **Configure Options**.

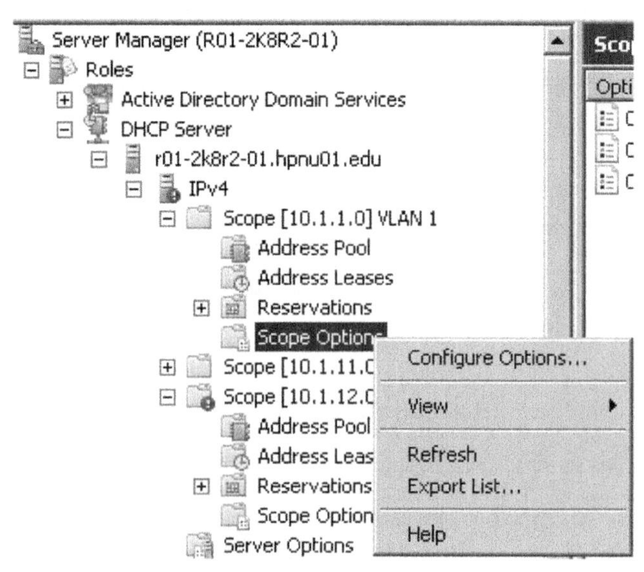

Figure 11-123: Access Scope Options > Configure Options

c. Click the **Advanced** tab.

d. Choose your vendor class.

e. Select your option's check box.

f. Click **OK**.

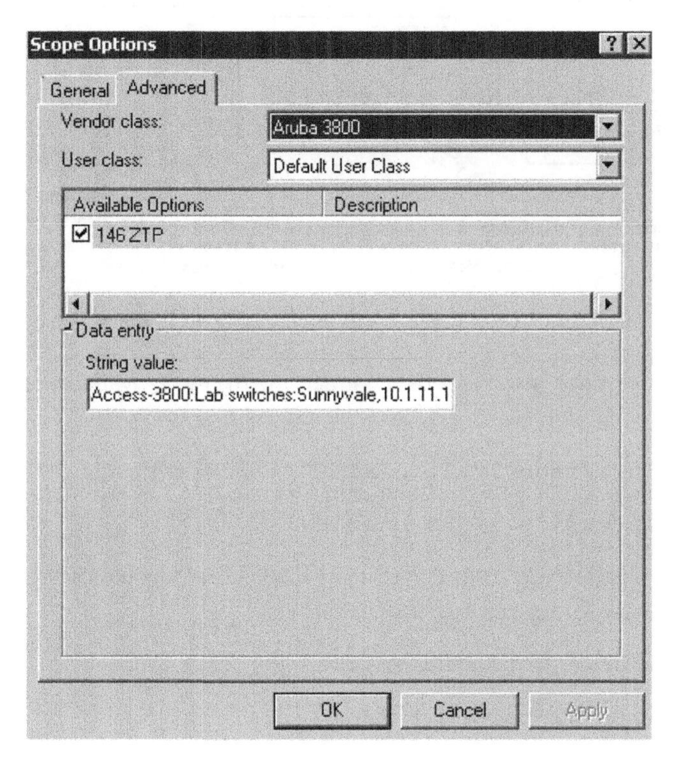

Figure 11-124: Choose Vendor class > select your option's check box

Important

Option 60 is also required with this method, but you've already added that option to this scope.

Core-1

In your test network environment, Core-1 is acting as the device that connects to the branch switch. (In the real world, this might be a WAN router, or it could be a core routing switch in an Ethernet-based metro WAN.) You will now set up Core-1 to connect to Access-2. It is very important that you enable just one interface at this point. Otherwise, in a test environment with permanent, redundant connections, Access-2 could create loops when it returns to its factory default settings.

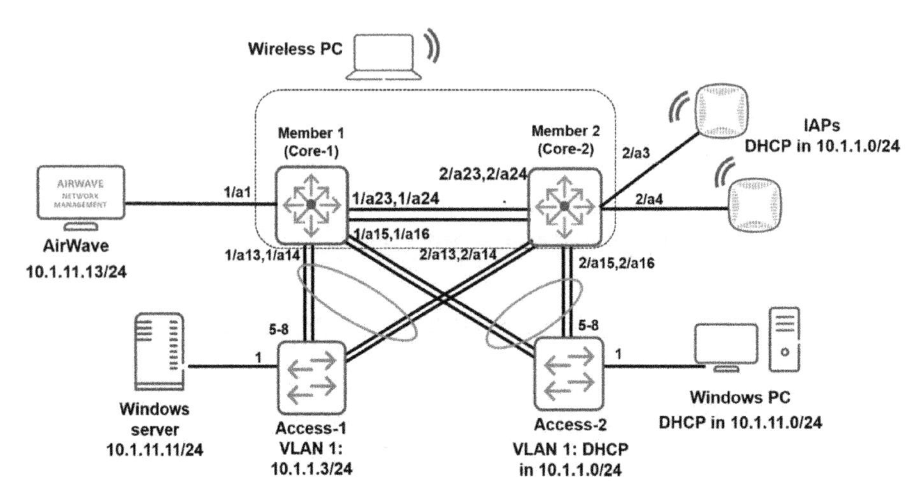

Figure 11-125: Set up Core-1 to connect to Access-2

6. Disable all but one of the interfaces that connect to Access-2. Make sure that one interface is enabled.

```
Core-1(config)# interface 1/a16,2/a15,2/a16 disable
Core-1(config)# interface 1/a15 enable
```

7. Make the link aggregation trk3 is untagged for VLAN 1 and tagged for VLAN 11, which are the settings that you want for branch switches in this task.

```
Core-1(config)# vlan 1 untagged trk3
Core-1(config)# vlan 11 tagged trk3
```

Access-2

8. You will now reset Access-2's configuration. It will receive a DHCP address and the AirWave management string in option 146.

   a. Access a terminal session with Access-2.

   b. Reset the configuration.

```
Access-2(config)# erase startup-config
The current configuration will be deleted, existing login passwords
removed, and the device rebooted.
Continue (y/n)? y
```

9. Wait for the switch to reboot. Wait another thirty seconds or a minute and then verify that the switch has received its AirWave settings.

```
HP-3800-24G-2SFPP# show amp

 Airwave Configuration details

  AMP Server IP              : 10.1.11.13

  AMP Server Group           : Access-3800

  AMP Server Folder          : Lab switches:Sunnyvale

  AMP Server Secret          : password

  AMP Server Config Status   : Configured
```

Important

If the switch does not receive AMP settings, check if it has a DHCP address (enter **show ip**). If it has a DHCP address, double-check the vendor class that you configured for typos. If the switch does not have a DHCP address, shut down all interfaces except 5. A loop could be occurring due to interfaces that connect to switches used by other classes.

Windows server

10. Authorize the switch in AirWave.

    a. Return to the Windows server browser connected to the AMP.

    b. Wait a couple minutes. Move to the APs/Devices > New page.

    c. Select the switch.

    d. Set its Management Level to **Manage Read/Write**.

    e. Click **Add**.

Figure 11-126: Select switch > Management Level

f. Apply the changes.

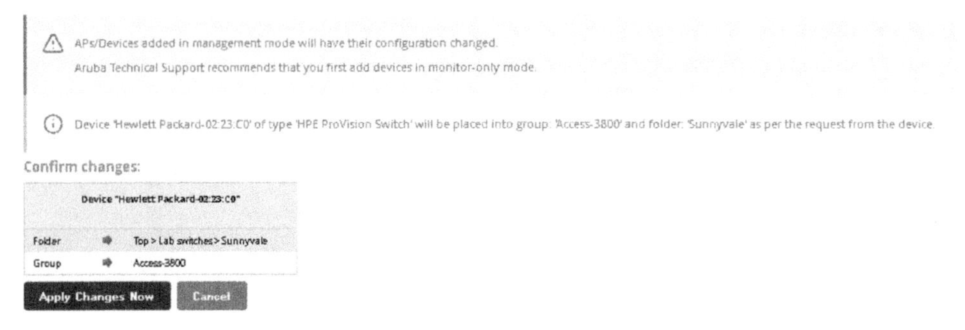

Figure 11-127: Confirm and apply changes

Access-2

11. Monitor the Access-2 CLI. The switch should reboot after a couple of minutes to apply the new configuration.

12. Wait until the switch reboots. Log into the CLI (use **manager** and **hpe** for the credentials or simply press **[Enter]**).

13. Make sure that the switch has applied the new settings. (Enter, for example, **show vlan 11** and see that the switch has this VLAN.)

Important

If the switch does not reboot, an error might be preventing AirWave from applying the configuration. You can look for clues in the AMP. Check the Group folder and make sure that a typo in the DHCP option didn't put the switch in the wrong folder. In this case, move the switch to the right folder now.

In the APs/Devices > List page, select Lab switches > Sunnyvale for the folder. Scroll to the Devices List and click the status under Configuration for this switch. You should see an audit of how the switch's configuration compares to the template. Click View Telnet/SSH Command log for a detailed view of how AirWave has attempted to push the template.

Windows server

14. Modify the switch's management communication settings and fill in some variables for it.

Note

Configuration templates allow you to configure communication settings in advance, and AirWave then automatically applies those settings after applying the template. However, this feature is in beta for ArubaOS switches in the version of AirWave used in this task. Therefore, you should change the settings as described below.

a. In the AMP, move to the APs/Devices > List page.

b. Select the Lab switches > Sunnyvale folder.

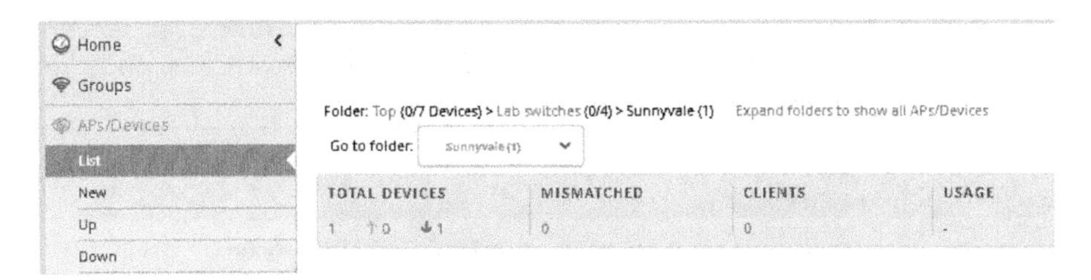

Figure 11-128: Access the Sunnyvale folder

c. Scroll down to the Devices List.

d. Click the switch name.

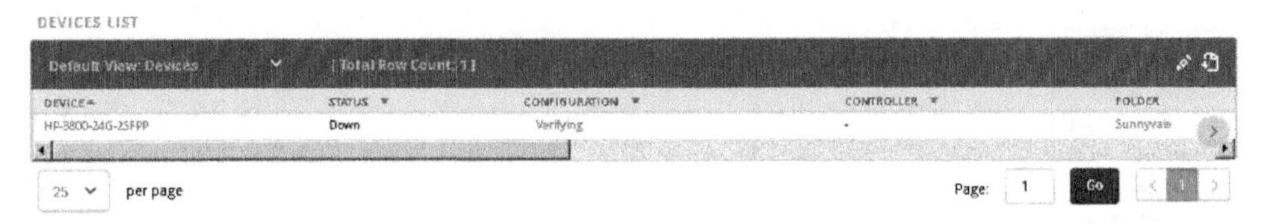

Figure 11-129: Select the switch from the Devices list

e. Click **Manage** under **APs/Devices** in the left navigation bar.

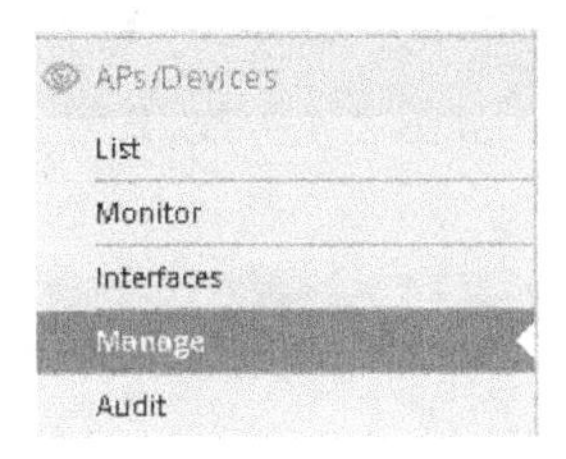

Figure 11-130: Access the Manage page

f. Scroll down to the Device Communication section.

g. For Community String and Confirm Community String, type **password**.

h. For Telnet/SSH Username, type **manager**.

i. For Telnet/SSH Password and Confirm Telnet/SSH password, type **hpe**.

Note

If you needed to change the credentials for multiple devices at once, you could use the process given in the "Discover and Monitor Devices in Aruba AirWave" section of this chapter (in the Device List, select the pencil icon, select multiple switches, and choose the Device Action for updating communication credentials.

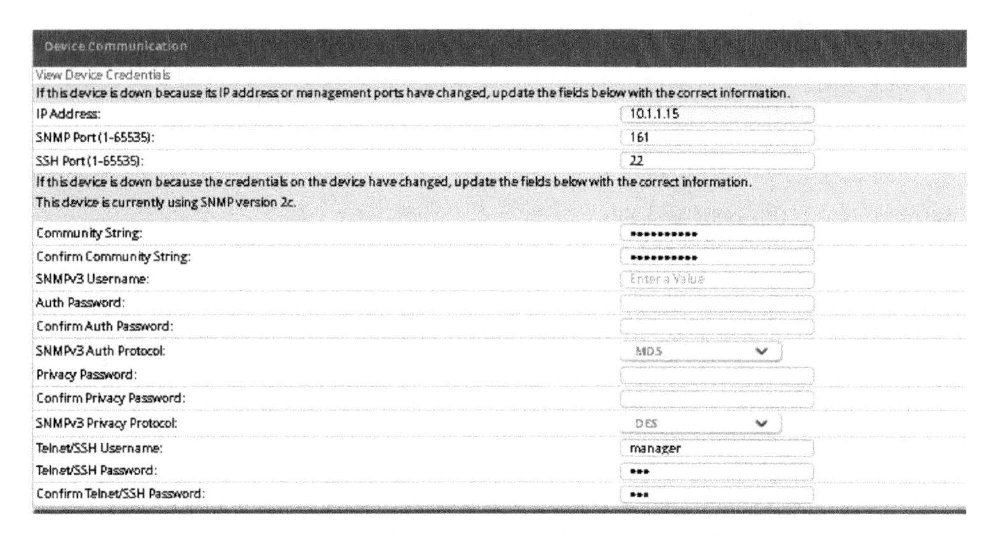

Figure 11-131: Add password and username information

j. Scroll to the Settings section. Change Name to **Access-2**.

k. Change VLAN 1 Untagged Ports, to **2-4,9-26,Trk1**.

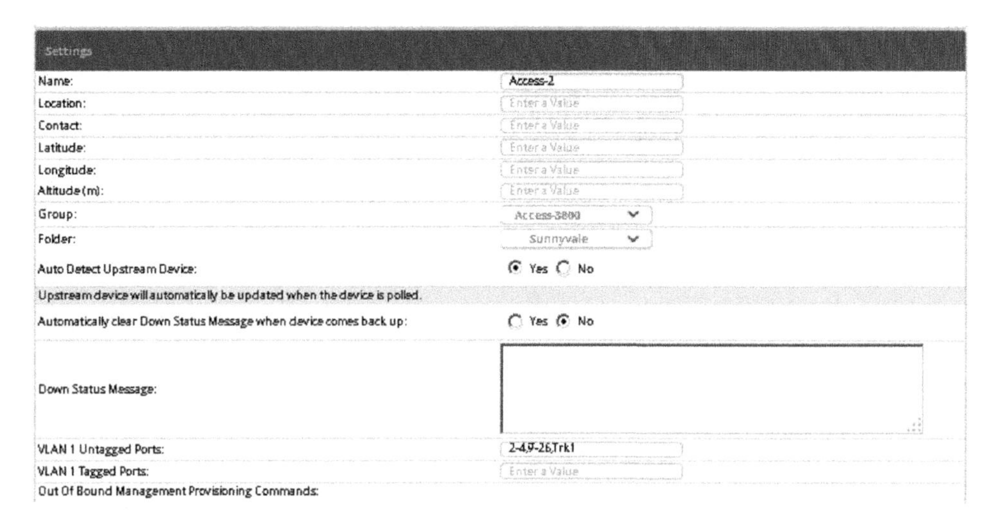

Figure 11-132: Change VLAN 1 Untagged Ports

l. Click **Save and Apply**.

m. Click **Apply Changes Now**.

15. AirWave's attempts to reach the device while you updated the settings might have timed out. If so, click Poll Now to allow AirWave to try again.

Figure 11-133: Click Poll Now if AirWave attempts timed out

16. AirWave might reboot the switch to apply the changes. (In the real world, you can schedule applying changes for a maintenance window.)
17. You can confirm that the switch is connected correctly and placing the client in VLAN 11 by accessing the client, renewing its IP address, and seeing that it receives an address in 10.1.11.0/24.

Important

You might need to configure the client to use DHCP for its IP address.

18. You can also access the Access-2 CLI and see that AirWave has changed the hostname.

Task 4: Remove passwords from switches

In your test environment, you might want to access the Core-1, Access-1, and Access-2 switches and enter this command:

```
Switch_name(config)# no aaa authentication local-user manager
```

Learning check

1. How did AirWave create the templates for the groups?

2. How did you use the template to control the configuration for the switches in the Access-3800 group? How might you apply this process to other tasks?

3. Why did you need to use a vendor class option to send the AirWave management settings to the ArubaOS switch?

Answers to Learning check

1. How did AirWave create the templates for the groups?

   It automatically created each template based on the configuration of the first device that you added to the group. It automatically adjusted the template to make it more generic so that all devices in the group can share the template. It did this by replacing some settings with variables.

   You might have noticed that AirWave adjusted the configuration in a few other ways to make it more suitable as a template. For example, AirWave added **include-credentials** to the ArubaOS configuration template so that it can check for the proper credentials.

2. How did you use the template to control the configuration for the switches in the Access-3800 group? How might you apply this process to other tasks?

   You simply adjusted the configuration much like you might adjust a configuration file. AirWave then pushed the change to the devices in the group. You might have listed many different ideas for using this capability. You can make any change that you can make in the CLI. You can also schedule applying the new template for a later time to coordinate changes and ensure they take place during a maintenance window.

3. Why did you need to use a vendor class option to send the AirWave management settings to the ArubaOS switch?

   Typically, you can use option 43 with the ArubaOS switches, just as on the Aruba IAPs. However, you needed to use a vendor class because the IAPs and the switches were sharing a subnet and DHCP scope. You needed to assign one string to the IAPs and a different one to the ArubaOS switches, and the vendor class let you do that.

Summary

This chapter introduced you to Aruba AirWave, a unified management solution for a company's complete wired and wireless access infrastructure. You learned how to discover and authorize IAPs and ArubaOS switches in AirWave. You also learned how to use template-based configuration management and how to implement ZTP.

– Discovering devices in AirWave
– Monitoring and managing devices in AirWave
– Using configuration templates
– Implementing ZTP

Figure 11-134: Summary

Learning check

1. You have several Instant clusters that are already configured. You want to begin monitoring them with AirWave. How do you enable AirWave to discover the clusters?
    a. Configure a scan set in AirWave.
    b. Set up CDP on the individual IAPs.
    c. Configure System > Admin settings in the clusters' Instant UI.
    d. Add the correct DHCP options 60 and 43 to the IAPs' DHCP server.

2. You are deciding how to group network infrastructure devices on AirWave. What is one common element for devices in the same group?
    a. management level
    b. communication credentials
    c. folder
    d. configuration templates

3. Which protocol does AirWave use to discover ArubaOS switches?
    a. Telnet
    b. SNMP
    c. SSH
    d. HTTPS

Answers to Learning check

1. You have several Instant clusters that are already configured. You want to begin monitoring them with AirWave. How do you enable AirWave to discover the clusters?
    a. Configure a scan set in AirWave.
    b. Set up CDP on the individual IAPs.
    c. Configure System > Admin settings in the clusters' Instant UI.
    d. Add the correct DHCP options 60 and 43 to the IAPs' DHCP server.

2. You are deciding how to group network infrastructure devices on AirWave. What is one common element for devices in the same group?
    a. management level
    b. communication credentials
    c. folder
    d. configuration templates

3. Which protocol does AirWave use to discover ArubaOS switches?
    a. Telnet
    b. SNMP
    c. SSH
    d. HTTPS

12 Practice Exam

INTRODUCTION

The Aruba Certified Switching Associate certification validates your knowledge of the features, benefits, and functions of Aruba networking components and technologies used in the Aruba Mobile-First architecture. The exam for achieving this certification tests your ability to implement and validate a basic secure wired and wireless network with the use of ArubaOS switches, CLI, various technologies, and Aruba Instant Access Points (IAPs). This exam also tests your ability to manage and monitor the network with AirWave.

The intent of this ebook is to set expectations about the context of the exam and to help candidates prepare for it. Recommended training to prepare for this exam can be found at the HPE Certification and Learning website (https://certification-learning.hpe.com). It is important to note that, although training is recommended for exam preparation, successful completion of the training alone does not guarantee that you will pass the exam. In addition to training, exam items are based on knowledge gained from on-the-job experience and application and on other supplemental reference material that may be specified in this guide.

Minimum qualifications

To achieve the Aruba Certified Switching Associate certification, you must pass the HPE6-A41, HPE2-Z39, or HPE2-Z40 exam. (The correct exam depends on factors such as your existing certifications.) Candidate should have a thorough understanding of Aruba switch and access point implementations in small to medium businesses (SMBs). To pass the exam, you should have at least six months experience deploying small-to-medium enterprise-level networks. You should also have an understanding of wired technologies used in edge and simple core environments and fundamental knowledge of wireless technologies.

Exams are based on an assumed level of industry-standard knowledge that may be gained from the training, hands-on experience, or other prerequisite events.

HPE6-A41 exam details

The following are details about the HPE6-A41 exam:

- Number of items: 60
- Exam time: 1 hour 55 minutes
- Passing score: 68%
- Item types: Multiple choice (single-response) and multiple choice (multiple response)
- Reference material: No online or hard copy reference material is allowed at the testing site.

HPE6-A41 testing objectives

This exam validates that you can successfully perform the following objectives. Each main objective is given a weighting, indicating the emphasis this item has in the exam.

- **40%** Identify, describe, and apply foundational networking architectures and technologies.
  - Describe the basics of Layer 2 Ethernet to include broadcast domains and ARP messages.
  - Interpret an IP routing table and explain default routes, static routing, and dynamic routing such as OSPF.
  - Identify the roles of TFTP, SFTP, FTP, Telnet, SNMPv2, and SNMPv3 in the management of Aruba network devices, and apply the appropriate security for these features.
  - Describe Layer 2 redundancy technologies such as STP, RSTP, MSTP and VSF, and recognize the benefits of each.
  - Describe and apply link aggregation.
  - Identify, describe, and explain VLANs.
  - Describe, identify, and explain wireless technologies.
- **8%** Identify, describe, and differentiate the functions and features of Aruba products and solutions.
  - Identify and explain how Aruba, a Hewlett Packard Enterprise company, delivers solutions that enable the digital workplace.
  - Identify basic features and management options for Aruba wired and wireless products.
  - Compare and contrast Aruba Networking solutions and features, and identify the appropriate product for an environment.
- **37%** Install, configure, set up, and validate Aruba networking solutions.
  - Configure basic features on ArubaOS switches to include initial settings and management access.

- Configure ArubaOS switches with Layer 2 technologies such as RSTP and MSPT, link aggregation, VLANs, LLDP, and device profiles.
- Configure basic IP routing with static routes or OSPF on ArubaOS switches.
- Manage the software and configuration files on ArubaOS switches, and manage ArubaOS switches and APs with Aruba AirWave.
- Validate the installed solution with the use of debug technology, logging, and show and display commands.

- **5%** Tune, optimize, and upgrade Aruba networking solutions.
  - Optimize Layer 2 and Layer3infrastructuresthrough broadcast domain reduction, VLANs, and VSF.
- **10%** manage, monitor, administer, and operate Aruba solutions.
  - Perform network management according to best practices.
  - Perform administrative tasks such as moves, adds, changes, deletions, and password resets for managed devices.

Test preparation questions and answers

Note

The following questions help you measure your understanding of the material presented in this book. Read all of the choices carefully, since there may be more than one correct answer. Choose all correct answers for each question.

Questions

1. Which characterizes out-of-band management?
   a. Uses the same Ethernet link that the management station uses to transmit traffic onto the network
   b. Is the only secure method to manage switches
   c. Limits the network administrator's ability to configure switch features and functions
   d. Separates management traffic from data traffic

2. Which interface identifier correctly specifies the second port on the second module of an Aruba 5400R zl2 switch?
   a. 2/2
   b. A2/2
   c. B2
   d. 2/0/2

3. If a network manager enters the **include-credentials** command on an Aruba switch, what happens?
   a. Login credentials for switch access are stored in internal flash and are part of the configuration file.
   b. The switch sends login credentials to a network RADIUS server and grants access only to network administrators that the server approves.
   c. The local manager and operator usernames and passwords are stored in internal flash but are not part of the configuration file.
   d. The switch enforces 802.1X authentication for specified ports using the manager credentials configured on the switch.

4. Refer to this command output:

```
Switch# show flash
Image                 Size (bytes) Date     Version
----------------- ------------ -------- --------------
Primary Image     :     15540244 03/23/16 KA.16.01.0006
Secondary Image   :     15890772 08/24/15 KA.15.15.0014

Boot ROM Version
----------------
Primary Boot ROM Version   : KA.15.09
Secondary Boot ROM Version : KA.15.09

Default Boot Image   : Primary
Default Boot ROM     : Primary

Switch# show config files
Configuration files:
 id | act pri sec | name
 ---+-------------+------------------------------------------------
  1 |      *      | config1
  2 |          *  | config2
  3 |   *         | config3
```

A network administrator changes several settings on this Aruba switch. The administrator then enters **write memory**. To which configuration or configurations does the switch save the new settings?

a. config1 only

b. config1 and config3

c. config3 only

d. config1, config2, and config3

5. Examine the network topology in Figure 12-1.

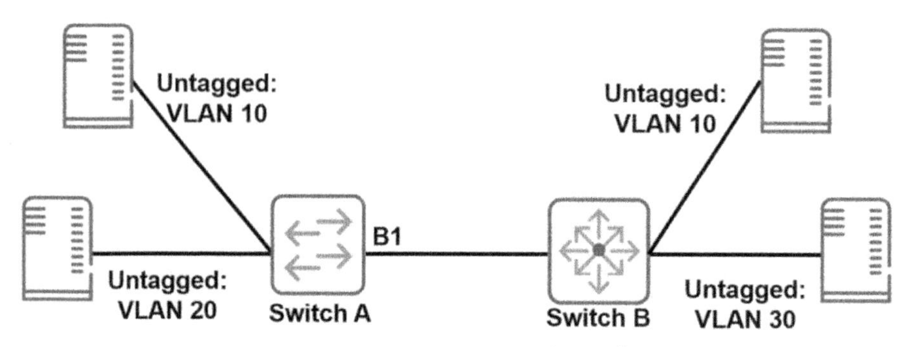

Figure 12-1: Question 5 network topology

Switch A is an Aruba switch without IP routing enabled, and Switch B is an Aruba switch with IP routing enabled. Based on this topology, what is the correct configuration for port B1 to ensure connectivity within VLAN 10 and inter-VLAN connectivity between all three VLANs?

a. vlan 10 tagged B1

   vlan 20 tagged B1

b. vlan 30 tagged B1

   vlan 10 tagged B1

c. vlan 20 tagged B1

   vlan 10 tagged B1

   vlan 30 tagged B1

d. vlan 10 untagged B1

   vlan 20 untagged B1

6. Examine the network topology in Figure 12-2.

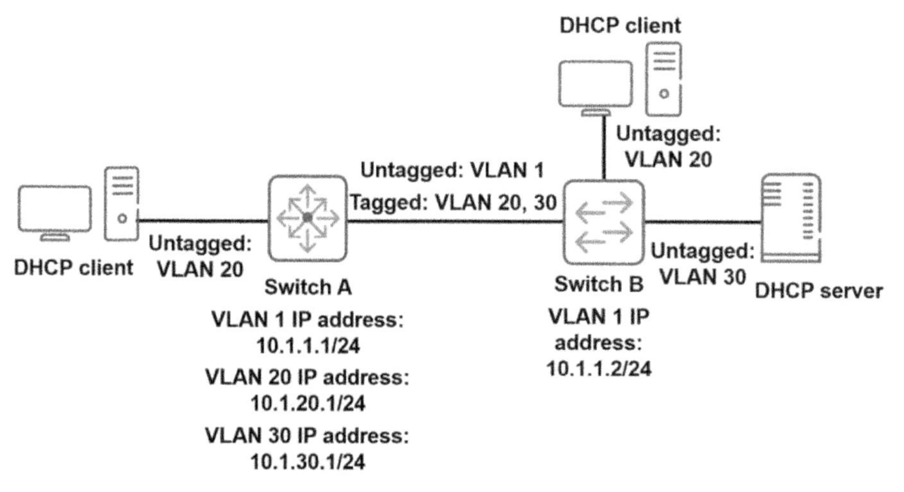

Figure 12-2: Question 6 network topology

Users in VLAN 20 need to acquire addressing information from the DHCP server in VLAN 30. Where should the network administrator configure the IP helper address?

a. on VLAN 30 on Switch A

b. on VLAN 20 on Switch A

c. on VLAN 30 on Switch B

d. on VLAN 20 on Switch B

7. Examine the network topology in Figure 12-3.

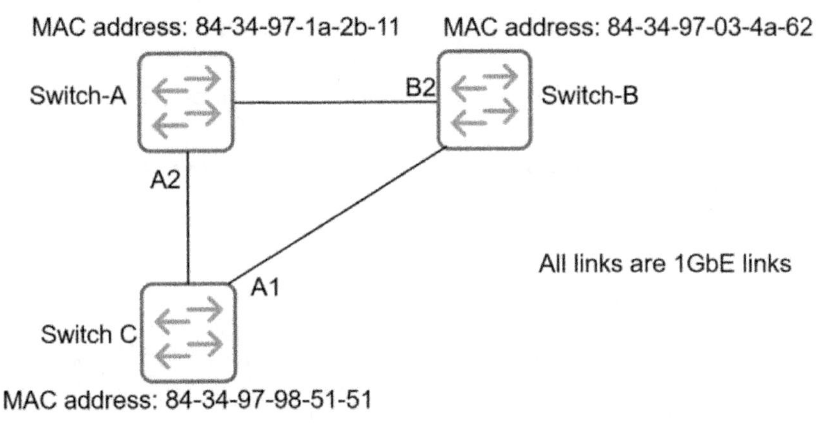

Figure 12-3: Question 7 network topology

Switch-A, Switch-B, and Switch-C are Aruba switches. All switches are at their factory default configurations with the exception of spanning tree being enabled on all three switches. Based on this information, which statements about this topology are correct? (Select two.)

a. Switch-A is the root switch.

b. Switch-B is the root switch.

c. Switch-C is the root switch.

d. Port A1 is a designated port.

e. Port A2 is a designated port.

f. Port B2 is a root port.

8. Examine the network topology in Figure 12-4.

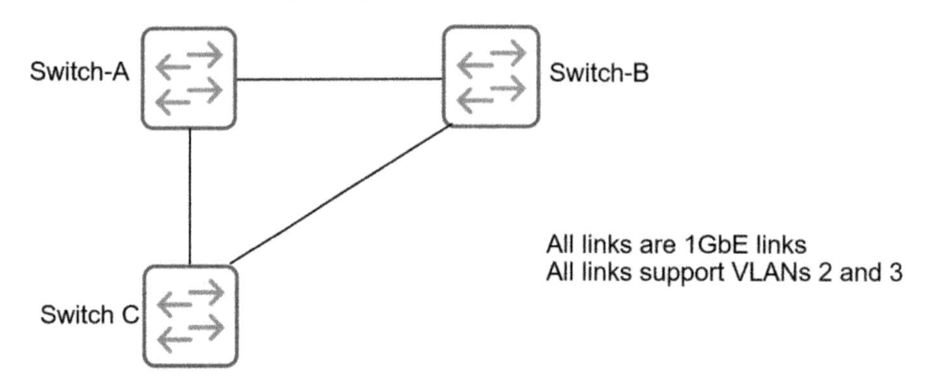

Figure 12-4: Question 8 network topology

These commands were executed on all three switches, which were at their factory default configuration:

```
spanning-tree
spanning-tree config-name exam
spanning-tree config-revision 1
spanning-tree instance 1 vlan 2
spanning-tree instance 2 vlan 3
```

These commands were then executed on Switch-A:

```
Switch-A(config)# spanning-tree priority 0
Switch-A(config)# spanning-tree instance 1 priority 0
Switch-A(config)# spanning-tree instance 2 priority 1
```

These commands were then executed on Switch-B:

```
Switch-B(config)# spanning-tree priority 1
Switch-B(config)# spanning-tree instance 1 priority 1
Switch-B(config)# spanning-tree instance 2 priority 0
```

This command was then executed on Switch-C:

```
Switch-C(config)# spanning-tree pending instance 2 vlan 3-4
```

Which statement is true?

a. A Layer 2 loop occurs exists.

b. Switch-C is the root switch for all three instances.

c. Switch-C carries traffic in all VLANs on the link to Switch-A.

d. Switch-C load balances traffic on both links on a per-VLAN basis.

9. An Aruba switch connects to a server on port 1 and port 2, which are configured as a link aggregation. Examine the output for this switch:

```
Switch# show trunk

   Load Balancing

   Port | Name    Type        | Group Type
   ---- + ----    --------- + ----- -----
   1    | Server 100/1000T | Trk1  LACP
   2    | Server 100/1000T | Trk1  LACP

Switch# show lacp

                         LACP

   PORT LACP    TRUNK PORT   LACP    LACP
   NUMB ENABLED GROUP STATUS PARTNER STATUS
   ---- ------- ----- ------ ------- -------
   1    Active  Trk1  Up     Yes     Success
   2    Active  Trk1  Up     Yes     Success

Switch# show interface display

                   Status and Counters - Port Counters

                                                          Flow B*
   Port    Total Bytes Total Frames Errors Rx Drops Rx Ctrl L*
   ------- ----------- ------------ --------- -------- ---- -*
   1-Trk1      375,648       40,516         0        0 off  0
   2-Trk1        3,434          450         0        0 off  0

   <-output omitted->
```

Based on this configuration and output, which statement is true concerning the link aggregation configuration and load-sharing operation?

a. The switch and server have failed to form an aggregated link.

b. The server is not using LACP, so the aggregated link has formed but is not load balancing.

c. The link aggregation is not carrying enough conversations for effective load balancing, and Layer 4 balancing might work better.

d. The link aggregation should use dynamic LACP, rather than static LACP, so that it can provide better load balancing.

10. A network administrator is setting up an LACP aggregated link between two Aruba switches. What must the administrator match so the switches set up the link successfully? (Select two.)

    a. Media type

    b. MSTP settings

    c. VLAN settings

    d. Duplex mode

    e. QoS settings

11. Examine the topology in Figure 12.5:

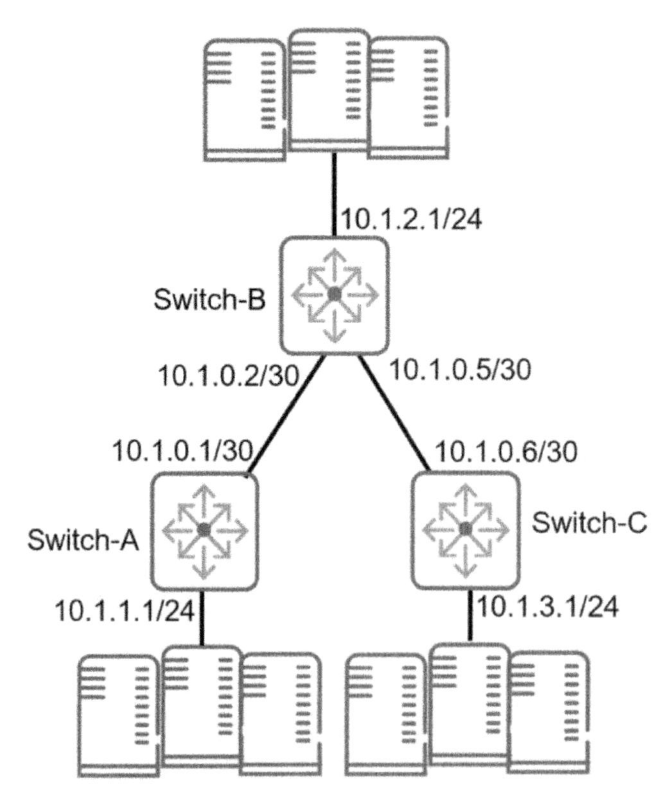

Figure 12-5: Question 14 network topology

Also examine this command output:

```
Switch-A# show ip ospf interface

OSPF Interface Status

IP Address   Status    Area ID    State  Auth-type  Cost   Pri  Passive
----------   ------    --------   -----  ---------  ----   ---  -------
10.1.1.1     enabled   backbone   WAIT   none       1      1    yes
10.1.0.1     enabled   backbone   BDR    none       1      1    no

Switch-A# show ip ospf neighbor

OSPF Neighbor Information

Router ID   Pri   IP Address   NbIfState   State   QLen   Events   Status
---------   ---   ----------   ---------   -----   ----   ------   ------
10.0.0.2    1     10.1.0.2     DR          FULL    0      7        None

Switch-B# show ip ospf interface

OSPF Interface Status

IP Address   Status    Area ID    State  Auth-type  Cost    Pri    Passive
----------   ------    --------   -----  ---------  -----   ---    -------
10.1.0.2     enabled   backbone   DR     none       1       1      no
10.1.0.5     enabled   backbone   DR     none       1       1      no

Switch-B# show ip ospf neighbor

OSPF Neighbor Information

Router ID   Pri   IP Address   NbIfState   State    QLen    Events   Status
---------   ---   ----------   ---------   -----    -----   ------   ------
10.0.0.1    1     10.1.0.1     BDR         FULL     0       7        None
10.0.0.3    1     10.1.0.6     BDR         FULL     0       7        None
```

Based on this information, for which networks will Switch-C receive an OSPF route?

a. 10.1.0.0/30 and 10.1.1.0/24

b. 10.1.0.0/30 and 10.1.2.0/24

c. 10.1.1.0/24 and 10.1.2.0/24

d. 10.1.0.0/30, 10.1.1.0/24, and 10.1.2.0/24

12. An Aruba switch knows two routes to 10.1.2.0/24 with the same administrative distance. How does the Aruba switch determine which routes to add to its IP routing table?

    a. It always adds both routes.

    b. It adds the route with the lower metric, and adds both routes, if the metric is same.

    c. It adds the route with the higher metric, and then the higher next hop IP address, if the metric is the same.

    d. It adds the route with the lower next hop IP address.

13. Examine the network topology in Figure 12-6.

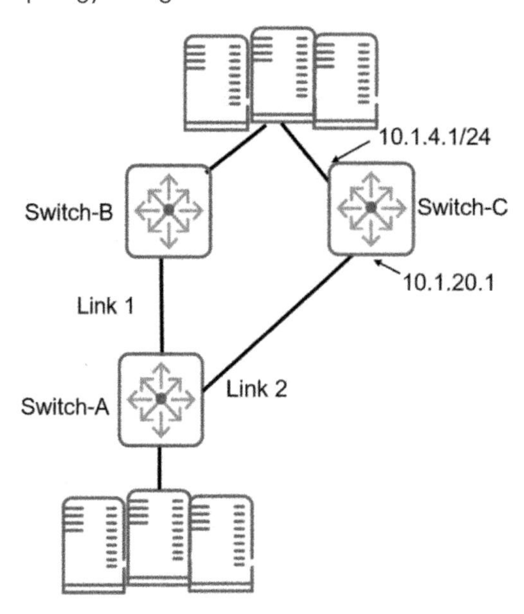

Figure 12-6: Question 16 network topology

Switch-A is currently routing traffic to 10.1.4.0/24 over Link 1 using a static route entered with the default administrative distance and metric. A network administrator wants to create a redundant IP route to 10.1.4.0/24 on Switch-A. The redundant route should use Link 2, and the switch should load balance traffic over both routes.

What is the correct setup?

a. An IP route through next hop 10.1.4.1 with distance 1 and metric 1

b. An IP route through next hop 10.1.4.1 with distance 60 and metric 1

c. An IP route through next hop 10.1.20.1 with distance 1 and metric 1

d. An IP route through next hop 10.1.20.1 with distance 60 and metric 1

14. Examine the command output on a VSF fabric:

```
VSF# show vsf lldp-mad parameters
MAD device IP            : 10.1.1.3
MAD readiness status     : Success
MAD device MAC           : 6c3be5-6208c0
Reachable via Vlan       : 101
Local LAG interface      : Trk2
MAD-probe portset        : 1/A1,1/A2,2/A1,2/A2,
LAG connectivity         : Full
```

Member 1 is currently commander, and member 2 has the standby role. The VSF link fails. What occurs next?

a. Both member 1 and member 2 take the commander role, but member 2 uses 10.1.1.3 as its IP address.

b. Both member 1 and member 2 take the standby role.

c. Member 2 sends a probe to 10.1.1.3.

d. Member 1 sends a keepalive to member 2's VLAN 101 IP address.

15. What advantages does VSF provide over multiple spanning tree protocol (MSTP)? (Select two.)

a. Faster failover

b. The ability to provide link redundancy

c. Device redundancy for Layer 3 functions

d. Support for multiple VLANs

e. Deployment at both access and core layers

16. Which are the responsibilities of only the commander in a VSF fabric? (Select two.)

a. Run IP routing protocols

b. Forward traffic

c. Build MAC addressing table

d. Support ports that learn MAC addresses

e. Build new Layer 2 headers for forwarding routed traffic

17. A network administrator connects a new Aruba Instant AP (IAPs) to a PoE switch in a network that has no other IAPs. What happens when the IAP boots up?
    a. It searches for a mobility controller to take control of it.
    b. It connects to Aruba ClearPass and downloads a configuration from ClearPass.
    c. It elects itself virtual controller (VC) and starts advertising the default SSID.
    d. It enters an error state and must be provisioned through its console port.

18. Match the wireless security setting with the appropriate description.
    a. WEP with 802.1X
    b. WPA2-PSK
    c. WPA2 with 802.1X
    d. MAC-Auth alone

        _______i. Often used by small companies that want a simple solution
        _______ii. Not recommended for secure authentication because it can be easily spoofed
        _______iii. Not supported with 802.11n or ac
        _______iv. Recommend for enterprises that want the most secure option

19. Which credentials does Aruba AirWave use to discover and configure Aruba switches?
    a. SNMP to discover and Telnet/SSH login to configure
    b. Telnet/SSH login to discover and SNMP to configure
    c. SNMP to discover and configure
    d. Telnet/SSH login to discover and configure

20. Aruba AirWave has discovered several Aruba Instant clusters. What is one step that administrators must complete to enable configuration of the clusters from the Instant GUI in AirWave?
    a. Generating certificates on AirWave and deploying the certificates to the Instant devices
    b. Ensuring that AirWave has the correct Telnet/SSH credentials to log into the Instant devices
    c. Ensuring that AirWave has the correct SNMP read-write credentials for the Instant devices
    d. Placing the Instant devices to Manage Read/Write mode

Answers

1. **D** is correct. Out-of-band management completely isolates management traffic from data traffic. Administrators can obtain out-of-band management through a switch's console port, using the serial cable that ships with the switch, or through the OOBM port, using an Ethernet cable.

   A is incorrect because out-of-band management, even when using the Ethernet OOBM port, uses a different connection from that used to transmit data traffic. **B** is incorrect because you can protect in-band management communications, using security mechanisms such as Secure Shell (SSH). **C** is incorrect because out-of-band management does not limit the network administrator's access to the switch features and functions.

   For more information, see Chapters 2 and 3.

2. **C** is correct. The Aruba modular switches identify modules with letters and starting number ports at 1, so the correct answer is B2.

   A, **B**, and **C** are incorrect because this is an invalid nomenclature for Aruba switches.

   For more information, see Chapter 2.

3. **A** is correct. On Aruba switches, login credentials are not stored in flash by default. If you want to store the credentials in the internal flash and view them in the config, you should enter the **include-credentials** command.

   B and **D** are incorrect. The **include-credentials** command does not require the switch to send login credentials to an external RADIUS server or affect the 802.1X authentication process. C is incorrect because after the **include-credentials** command is entered, the login credentials for manager and operator are part of the configuration file as well as being stored in flash.

   For more information, see Chapter 3.

4. **C** is correct. The **write memory** command saves configuration changes to the active configuration file. The command output indicates that config3 is the active file.

   A, **B**, and **D** are incorrect because the switch only saves the changes to the active file, not to any others.

   For more information, see Chapter 4.

5. **B** is correct. Because VLAN 10 extends between the two switches, it must be carried on the switch-to-switch link. Switch A supports devices in VLAN 20, and Switch B is the default router for this VLAN. Therefore, the switch-to-switch link must also support this traffic.

   A is incorrect because VLAN 30 is local to Switch B and, therefore, doesn't have to be included on the VLAN trunk. Switch B will route traffic to/from it as necessary for the other VLANs. **C** is incorrect because VLAN 20—not 30—needs to be included on the trunk. **D** is incorrect because a link cannot support two untagged VLANs.

   For more information, see Chapter 5.

6. **B** is correct. You should specify the **ip helper-address** command for the VLAN where the clients reside on a switch that has an IP address in that VLAN. The switch must also have IP connectivity to the DHCP server. Switch A meets both of those criteria.

   A and **C** are incorrect because VLAN 30 is the VLAN on which Switch A reaches the server, no the clients. **D** is incorrect because Switch B does not have an IP address on VLAN 20, so it cannot support DHCP relay on this VLAN.

   For more information, see Chapter 5.

7. **B** and **E** are correct. Switch-B has the lowest switch ID (lowest MAC address) and, therefore, is the root switch. (The default priority on all Aruba switches is 32,768.) The link with A2 is blocked because it doesn't offer the lowest cost path to the root. One side is an alternate port, and the other side is a designated port. A2 is the designated port since both Switch-A and Switch-B advertise the same root path cost, but Switch-A has a lower bridge priority (which is the tie breaker).

   A and **C** are incorrect because these switches have higher MAC addresses. **D** is incorrect because A1 is a root port. **F** is incorrect because all active ports on the root switch are designated ports.

   For more information, see Chapter 6.

8. **D** is correct. The first commands configured the switches with identical MSTP region settings. Then Switch-A and Switch-B were configured with priorities that resulted in Switch-A being elected root in instance 0 and 1 and Switch-B being elected root in instance 2. This means that Switch-C has a different active link in each instance. A change to Switch-C's VLAN-to-instance mappings would place Switch-C in a different region. However, the command was entered with the pending keyword, and the pending changes have not yet been applied.

   A is incorrect because MSTP is configured and operating correctly. **B** is incorrect because Switch-C remains in the same MSTP region as the other switches for now, and Switch-A and Switch-B are the root switches based on their changed priorities**.** **C** is incorrect for the same reason; after the changes are applied, Switch-C would use RSTP to communicate with other switches, eliminating load balancing. (Only the link the Switch-A would be active.) However, the pending changes have not yet been applied.

   For more information, see Chapter 6.

9. **B** is correct. Based on the **show lacp** command, both sides have successfully negotiated link aggregation. If only one link is carrying most traffic most of the time, the link probably carries just a few conversations, which happen to be assigned mostly to one link. Aruba switches use Layer 3 load balancing for IP traffic, by default. This server seems to most communicate with a couple IP addresses, and Layer 4 load balancing might work better.

   A and **B** are incorrect because they contradict the output of the **show lacp** command, which shows that the links have an LACP partner and have been successfully added to the aggregation. **D** is incorrect because dynamic LACP link aggregations handle load balancing in the same manner as static LACP link aggregations.

   For more information, see Chapter 7.

10. **A** and **D** are correct. When using LACP to establish an aggregated link, Aruba switches check media speed and duplex mode.

    B and **C** are incorrect. Aruba switches do not check MSTP and VLAN settings when an administrator adds interfaces to the aggregated link. Instead, any settings configured on the individual interface are removed, and administrators must configure them on the link aggregation (trk) interface instead. **E** is incorrect. Although QoS settings should be consistent, Aruba switches do not check these settings when establishing an LACP aggregated link.

    For more information, see Chapter 7.

11. **A** is correct. The command output shows that Switch-B has become fully adjacent neighbors with both Switch-A and Switch-C, which means that all three switches will have identical databases. Switch-C learns OSPF routes to all the networks on which Switch-A and Switch-B enable OSPF. Based on the **show ip ospf interface** command, these are 10.1.0.0/30 and 10.1.1.0/24. (Switch-C is directly connected to 10.1.0.4/30.)

    B, **C**, and **D** are incorrect. The **show ip ospf interface** command output for Switch-B indicates that Switch-B does not enable OSPF on 10.1.2.0/24. Therefore, Switch-B does not advertise this network in OSPF, and Switch-C does not learn a route to it.

    For more information, see Chapter 8.

12. **B** is correct. A routing switch uses metrics to distinguish between routes that are learned in the same way such as two routes learned through OSPF. These routes have the same administrative distance, so the switch selects the route that has the lower metric. If the metric is the same, the switch adds both routes.

    A is incorrect because the OSPF routes will have the same administrative distance. In this case, the metric breaks the tie. **C** is incorrect because the switch uses metrics to differentiate between two OSPF routes, and OSPF metrics are not based on the number of hops to the destination network. **D** is incorrect because the VLAN is not a factor in selecting routes.

    For more information, see Chapter 8.

13. **C** is correct. To use Link 2, the route must use 10.1.20.1, an IP address on a neighboring router interface, as the next hop. To ensure that the switch will load share across the redundant routes, the second route must use the same administrative distance and metric as the first route: 1.

    A and **B** are incorrect because they use the wrong next hop IP address. **D** is incorrect because the new route must have the same administrative distance and metric as the existing route or the switch will not load share traffic across the two routes. The default distance and metric are both 1.

    For more information, see Chapter 8.

14. **C** is correct. The failure of the port causes the VSF link to fail. The command output shows that LLDP-MAD is active, so member 2, the standby member, sends a probe to 10.1.1.3 to determine whether member 1 is still up.

    A is incorrect; this would be the behavior if LLDP-MAD or OOBM-MAD were not functional. **B** is incorrect; a VSF link failure does not cause both members to take the standby role. **D** is incorrect; member 2 does not have its own IP addresses, and LLDP-MAD does not work in this manner.

    For more information, see Chapter 9.

15. **A** and **C** are correct. VSF fabrics have a much faster failover than MSTP, which was designed for legacy networks and can take as long as one second to reconverge. A VSF fabric also provides device redundancy for Layer 3 functions, which MSTP does not offer.

    B, **D**, and **E** are incorrect because both VSF and MSTP provide link redundancy, support multiple VLANs. MSTP can be deployed at the access and core layers.

    For more information, see Chapter 9.

16. **A** and **C** are correct. In a VSF fabric, the commander's responsibilities include running the IP routing protocols and building the MAC addressing tables. The commander forwards the MAC addressing tables to the other member of the VSF fabric.

    B and **E** are incorrect because both members in a VSF fabric forward traffic and create new headers to forward routed traffic. **D** is incorrect because MAC addresses are learned on ports on both members. The standby member forwards this information to the commander, which creates the MAC addressing tables.

    For more information, see Chapter 9.

17. **C** is correct. If an IAP does not receive any beacons from a VC in an active cluster, the IAP elects itself as the VC. Because this is a new IAP, it advertises the default Instant SSID, which allows administrators to connect to the SSID and provision the IAP.

    A is incorrect. IAPs do not operate in controlled mode by default. B is incorrect; ClearPass does not provide this feature. (Aruba Activate and AirWave do provide a zero touch provisioning feature.) **D** is incorrect. An IAP does not enter an error state if it cannot join a cluster; it simply elects itself VC in a cluster of one.

    For more information, see Chapter 10.

18. Answers are listed below.

    b i. Often used by small companies that want a simple solution

    d ii. Not recommended for secure authentication because it can be easily spoofed

    a iii. Not supported with 802.11n or ac

    c iv. Recommend for enterprises that want the most secure option

    For more information, see Chapter 10.

19. **A** is correct. AirWave uses SNMP to discover Aruba switches and Telnet/SSH login to configure them.

    B, **C**, and **D** are incorrect. These options do not correctly identify the discovery or configuration mechanism.

    For more information, see Chapter 11.

20. **D** is correct. Configuration in the Instant GUI applies to Instant devices at the Manage Read/Write level.

    A is incorrect. AirWave and IAPs do use HTTPS to communicate, but administrators do not need to generate certificates on AirWave. **B** and **C** are incorrect. AirWave does not use Telnet, SSH, or SNMP to apply configurations to Instant APs.

    For more information, see Chapter 11.

Index

A

B

C

O

P

Q

R

W

Z